Marine and Coastal Biodiversity in the Tropical Island Pacific Region

Volume 1
Species Systematics and Information
Management Priorities

edited by

James E. Maragos, Melvin N. A. Peterson,
Lucius G. Eldredge, John E. Bardach,
and Helen F. Takeuchi

October 1995

Program on Environment
East-West Center
Honolulu, Hawaii

Ocean Policy Institute
Pacific Forum/CSIS
Honolulu, Hawaii

Pacific Science Association
c/o Bishop Museum
Honolulu, Hawaii

LIBRARY OF CONGRESS CATALOGING-IN-PUBLICATION DATA

Marine and coastal biodiversity in the tropical island Pacific region
/ edited by James E. Maragos . . . [et al.].
    p. cm.
    Proceedings of two workshops held at the East-West
Center, Honolulu, in November 1994.
    Includes bibliographical references.
    Contents: v. 1. Species systematics and information management
priorities
    ISBN 0-86638-175-9 (paperback)
    1. Marine organisms—Oceania—Congresses. 2. Species diversity—
Oceania—Congresses. 3. Biological diversity conservation—
Oceania—Congresses. I. Maragos, James E., 1944–
QH198.A1M37 1995
574.92'576—dc20   95-34642

About the volume editors: James Maragos, senior fellow, and Helen Takeuchi, senior editor, are with the East-West Center Program on Environment; Melvin Peterson was director, Ocean Policy Institute, of the Pacific Forum/Center for Strategic and International Studies, Washington, D.C.; Lucius Eldredge is executive secretary of the Pacific Science Association; and John Bardach is senior fellow emeritus at the East-West Center.

Cover photo: A pair of transparent *Periclimenes* shrimp on the stinging bubble coral *Plerogyra sinuosa* in Marovo Lagoon, New Georgia group, Solomon Islands. (Photo taken in 1994 by James E. Maragos.)

To secure copies, contact:
  Publication Sales Office
  East-West Center
  1777 East-West Road
  Honolulu, HI 96848 USA
  Phone: 808-944-7145
  Fax: 808-944-7376
  E-mail: ewcbooks@ewc.bitnet

Contents

Appendices

Figures and Tables

Figures

Tables

Preface

This book presents the edited papers and proceedings of the first of two workshops in November 1994 at the East-West Center on "Marine and Coastal Biodiversity in the Tropical Island Pacific Region." The theme for the first workshop was "Species Systematics and Information Management Priorities," while the theme for the second was "Population, Development, and Conservation Priorities." The second book of this series will present the proceedings of the second workshop. The first workshop attracted more than fifty participants and contributors and included twenty-four presentations and several group discussions. The first day at the workshop covered taxonomic status reviews for groups of important plants and animals occupying nearshore and coastal tropical ecosystems in the insular Pacific. The second day of the workshop covered presentations on various biodiversity database and information management systems potentially applicable to the tropical island Pacific. The third (and final) day of the first workshop concentrated on group discussions that led to the development of an outline for an action plan bridging species taxonomy/systematics and information management priorities for use in the conservation and sustainable use of nearshore marine and coastal resources in the region.

The idea for the first workshop originated at an earlier conference in Honolulu during the summer of 1992 that focused on the *Diversity of Oceanic Life: An Evaluative Review* (Peterson 1992), sponsored by the Ocean Policy Institute of the Pacific Forum/Center for Strategic and International Studies (CSIS). The primary purpose of the earlier conference was to "assess the quality of scientific knowledge on oceanic life and the level of effort needed to advance the field." Oceanic biodiversity remains relatively vast and unchartered, and the earlier group ultimately conceded that a realistic strategy is to think globally but to act locally. The tropical island Pacific was identified as a "local" region where further advances in the state of marine and coastal biodiversity knowledge could be reasonably pursued and eventually applied to broader global-level initiatives. The group agreed that species-level systematics/taxonomy is the fundamental basis for assessing biodiversity but that scientific funding and granting institutions were reducing financial support in these areas in favor of more high technology research.

Nevertheless great advances in computer technology and the new science of information management have begun to facilitate greater support and accomplishments in the fields of taxonomy and systematics. More importantly several different biodiversity database systems had been launched during the past few years, both within and outside the Pacific region. The timing seemed appropriate to bring together taxonomists, information managers, and other biodiversity and conservation scientists. The purpose was to assess the state of the knowledge and develop a framework specific to the insular tropical Pacific for biodiversity information management, focusing on marine and nearshore species and ecosystems. It was also the intention to incorporate appropriate existing database systems within the framework of the proposed insular tropical Pacific system.

As one workshop participant eloquently stated: "Biodiversity is the variability of living things as individuals, species, and other assemblages. It occurs as local variants in space and time. Taxonomy is the interface between the variants of living things, and information technology makes it possible to code diversity into a form amenable for information management." Such a system would render marine biodiversity information more useful, more organized, and more accessible to a greater science and nonscience audience, including decision makers.

Species-level biodiversity assessments continue to be the most important, if not the most fundamental, basis for biodiversity research and management. Although great advances are being made in the fields of genetic and ecosystem-level biodiversity assessments, neither of these will likely replace the Linnaean taxonomic system of binomial nomenclature. First introduced in 1758, it is still the only universal system for linking basic (species) biodiversity information among the diverse cultures and languages of an increasingly anthropocentric world. As a result, the central component of any biodiversity information management system must include species-level taxonomy and systematics, and such a system can be easily designed to interface with a variety of other biological, cultural, ecological, conservation, and development information and objectives.

The workshop and the proceedings of the workshop on "Marine and Coastal Biodiversity in the Tropical Island Pacific: Species Systematics and Information Management Priorities" were largely sponsored by a U.S. Environmental Protection Agency grant to the Ocean Policy Institute of the Pacific Forum/CSIS through the National Sea Grant Program. We gratefully acknowledge the support and patience of the representatives of the federal sponsors of this work, especially Jane Khanna and Jim Kelly of the Pacific Forum/ CSIS and Jane Ball and Jack Davidson of the University of Hawaii Sea Grant Program. Other funding support was contributed by the East-West Center (EWC), which covered some of the editorial and publication costs.

We also acknowledge the exceptional work of Daniel Bauer and Helen Takeuchi for editorial, graphics, layout, and other production support for the publication. We thank Mary Hayano of the Program on Environment (ENV), EWC, for word processing support. The University of Hawaii Marine Options Program also graciously provided student volunteers who helped organize and run the workshop sessions. We also thank EWC degree fellows Alyssa Miller and Leanne Fernandes for serving as recorders and Dr. John Bardach, EWC Senior Fellow Emeritus, for serving as rapporteur for one session and reviewing the draft papers.

We especially thank the participants who attended and volunteered many hours in preparing for their presentations and writing up their proceedings. We also thank several contributors who could not attend the workshop but who were still willing to contribute papers for the proceedings: Sam Gon III, Paul Holthus, Michelle Kelly-Borges, John Kuo, Maria Lourdes D. Palomares, and Gordon Nishida. (The list of all participants, including address and phone and facsimile numbers, is in Appendix D.) We sincerely thank all members of the workshop "team" for their accomplishments. We give special recognition to the efforts of June Kuramoto, Sally Yap, and student Shayne Hasegawa of the Program on Environment (ENV), EWC, for flawless and outstanding logistical management of the workshops and the travel and accommodation needs of overseas participants. Finally, we thank Fannie Lee Kai of ENV for decorating the workshops with flowers and bringing a spirit of Aloha.

James E. Maragos
Melvin N. A. Peterson
Honolulu, 1995

REFERENCE

Peterson, M. N. A., ed. 1992. *Diversity of oceanic life: An evaluative review.* The Center for Strategic and International Studies, Significant Issues Series, Vol. 14, No. 12. Washington, D.C.

IN MEMORY OF MELVIN N. A. PETERSON

As this book was going to press, co-editor Melvin N. A. Peterson suffered a massive heart attack while fishing offshore from San Diego and died peacefully at sea on 20 September 1995. Dr. Peterson served as director of the Ocean Policy Institute, Pacific Forum/CSIS in Honolulu, Hawaii. He was former chief scientist of the National Oceanic and Atmospheric Administration within the U.S. Department of Commerce.

After receiving his Ph.D. from Harvard University, Mel spent almost three decades on the faculty of the Scripps Institution of Oceanography where he also served as chief scientist, program director, and principal investigator of the Deep Sea Drilling Project, with its scientific drilling vessel *Glomar Challenger.* Mel was not only responsible for raising the funds for the workshop and the present volume, but he also edited the earlier related volume *Diversity of Oceanic Life: An Evaluative Review.*

It is fitting to dedicate this book to the memory of Mel Peterson, who devoted his entire life to the ocean. We will all remember his kindness, enthusiasm, and inspiration, and it was a great privilege for me to work with Mel during the past few years. Many of us will miss him dearly.

James E. Maragos
29 September 1995

Marine Biodiversity at Risk

Sylvia A. Earle

There seems to be a widely held belief that the sea, reservoir for 97 percent of earth's water, is so enormous that humankind can do little to change its nature. Lord Byron expressed this view two centuries ago when he said, poetically,

> Roll on thou deep and dark blue ocean, roll!
> Man marks the earth with ruin—
> His control stops with the shore.

That was before thousands of miles of drift nets strip-mined the sea; before mega-trawlers with maws large enough to consume a dozen 747 aircraft scraped the ocean floor, taking entire ecosystems in extracting the desired fish. In 1800, there were no nuclear wastes, no nerve gas–disposal sites, no oil spills, no plastic debris, no pesticides or herbicides, or runoff of chemical fertilizers from lawns, fields, and farms into the sea. From prehistory to the edge of the year 2000, great chunks have been removed from the Earth's treasury of genetic and ecosystem diversity because of habitat destruction, overkill, and the introduction of exotic chemicals and other pollutants.

In Byron's time, the sea was largely a wilderness, the distillation of four-and-a-half billion years—still largely intact. Since then, numerous species and entire complex ecosystems millions of years in the making have been significantly altered, from the obvious— populations of whales and other large mammals—to dozens of commercially valued fish species, all marine turtles, many sharks, and numerous small creatures including certain krill, crabs, and shrimp. Policies that subsidize and encourage the unsustainable taking of wild animals from the sea on a commercial basis have brought the fishing industry—and many populations of fish—to a state of near collapse. Despite efforts to curb the fishermen's enthusiasm for taking more tuna, cod, herring, pollock, halibut, and other species each year than populations can reproduce, many more have been taken than populations can tolerate without serious decline. And declining they are. Currently, 80 percent of the world's commercially taken marine species are regarded as seriously depleted. Worldwide, the living network of organisms that shape the basic ingredients of the ocean's "living soup" has been tugged, and the supporting ecosystems have been nudged in new directions.

Too little is known about the earth's living processes to know or predict the results of our tinkering, but disrupting the nature of the sea, both by changes in ocean chemistry and in the composition of the living elements, is not likely to be in our best interest. The ocean is, after all, the cornerstone of the planet's life-support system, the place that comprises more than 95 percent of the planet's living space, the biosphere. It is the realm that contains by far the greatest genetic diversity, considering the basic broad categories of life, and it is the primary source of much that is generally taken for granted—climate, weather, basic planetary chemistry. When I try to imagine earth without its mantle of salt water, I think of Mars. No ocean there. And no life. Without an ocean, earth would not be a congenial place for our species, nor, perhaps, for any of earth's other inhabitants.

Clearly, species have come and gone through time, and so have orders, classes, and even phyla. But there is no precedent for the rapid loss of biodiversity caused by humankind— superimposed on the changes brought about by forces over which our species has no control. For a biologist, it is satisfying to see the growing awareness of the significance of biodiversity and of healthy natural ecosystems to the survival and well-being of human-kind. As a marine biologist, it is puzzling to witness the disproportionate attention given to terrestrial creatures and terrestrial ecosystems.

Concern about destruction of well-known areas of high diversity on the land is certainly justified. While comprising only about 7 percent of the earth's land mass, rainforests are thought to provide home for more than half the species known. However, most of this incredible richness can be accounted for by creatures who share a common genetic program that insists they have six legs and otherwise fit the criteria established for insects. It may be that half the planet's species are insects and that most of them are beetles, but sweeping generalizations are premature until the oceans are at least as well known as terrestrial environments. This is especially true of microbial forms, to date virtually unexplored. It is estimated that less than one-tenth of 1 percent of the deep sea, below 50 meters or so, has been explored, at all. Most of it remains unknown, except with respect to gross anatomy, and much of that is based on educated guesswork.

New technology developed in the past few decades has provided unprecedented access to the sea and has given us a glimpse of the magnitude of the "unknowns" about most of the biosphere. Widespread use of self-contained underwater breathing apparatus—scuba—in the past several decades makes it possible for modern scientists to explore the ocean directly, instead of viewing only the scrambled remains of samples taken in nets and dredges. Underwater camera systems, remotely operated vehicles such as *Phantom* and the famous *Agro-Jason* system, have revolutionized the nature of underwater research and documentation. Recent discovery of hydrothermal vents in the deep sea and their diverse assemblages of a marine life awaited use of remotely deployed underwater vehicles and the manned submersible, *Alvin.* Subsequent investigations have been limited to brief visits by a privileged few using one of the small number of manned and robotic vehicles now in operation. Currently, only five submersibles exist that can transport observers to as much as half the ocean's depth, about 6,000 meters. None now exist to go to full-ocean depth, although once, in 1960, the now-retired bathyscaphe *Trieste* transported two men for a brief look at the nature of creatures living at the bottom of the Marianas Trench near the Philippines.

By using a submersible, *Deep Rover*, in 1986 I explored a steep ocean wall along the edge of Lee Stocking Island in the Bahamas and gained a new appreciation for the magnitude of biodiversity in the sea. A chunk of rock about 0.5 m² taken from 200 meters underwater along the wall appeared, superficially, to be as barren as a moonrock. In fact, I soon discovered upon inspection that within an area that I could embrace with my arms there was represented a fair slice through the history of life on earth, alive and well. The rock provided living space for three divisions of plants—red, green, and blue-green algae—and representatives of eleven phyla of animals—foraminifera, sponges, coral, nematodes, brachiopods, bryozoans, arthropods, mollusks, polychaetes, sipunculids, and echinoderms—brittle stars, a tiny urchin, and a number of highly specialized crinoids. Such broad-scale diversity is normal in the ocean, whether a sample is scraped from a piling, the deep seafloor, or a thriving forest of kelp.

Although coral reefs have few insects (i.e., sea-going water striders skate over the water's surface), they are rich in species, from numerous variations of red, green, brown, and blue-green algae and representatives from nearly all the other major divisions of plants known to numerous kinds of invertebrates. Most phyla of animals can be found on coral reefs and other marine habitats; only about half have ever made it to terrestrial areas.

Like rainforests, coral reefs are now in jeopardy, but the specific causes of global decline of reef systems are not as immediately obvious as the clear-cutting and burning of forests. While some reefs are being mined for building materials or smothered by obvious silt and debris from coastal development, and others are being dynamited by fishermen for extracting the resident fish, even in areas accorded protection such as Australia's Great Barrier Reef and reefs within the Florida Keys National Marine Sanctuary, serious degradation has occurred in the past thirty years. Causes may include pollution from agriculture and other sources, loss of critical fish species, and more subtle, overarching factors such as a possible global warming trend and increased ultraviolet radiation. Whatever the cause, or causes, if the decline continues at the same pace for another thirty years, there will be little living reef remaining in many parts of the world where coral has thrived for hundreds of millennia. The consequences of losing conspicuous living ecosystems such as tropical forests and coral reefs are not easy to predict, but no amount of clever human engineering, scientific brilliance, or financial commitment can restore such systems once they are gone.

Kelp forests, not coral, should perhaps be awarded the title "rainforests of the sea." They are even wetter than rainforests and have a crown of foliage nearly as high above the seafloor as the rainforest trees do from the leafy ground below. As in coral reefs and most other marine systems, nearly all the major divisions of plant and animal life are represented. Fish, sea hares, jellyfish, and the planktonic stages of numerous groups of animals flit, fly, or balloon among the blades of kelp. Mollusks, echinoderms, brachiopods, bryozoans, crustacea, and various other invertebrates creep, crawl, burrow, or simply attach to the substrates below. And there are mammals, of course.

Sea otters are among the residents of the central California coast so sought after for their warm, furry hides that they came perilously close to being eliminated many decades ago, as did the large cetacean species that once traveled along the western coast of North America in large migrating pods. It once was thought that large numbers of marine mammals could be removed from the sea on a sustained basis—that the sea is so vast, the number of seals, whales, walruses, and otters so great, that humankind could do little to reduce their kind, let alone eliminate them forever. Yet marine mammals proved to be vulnerable even to rather crude killing techniques used in the 1800s and early 1900s. Within a century, all the great whale populations were drastically reduced, and even with widespread protection in recent years, most have shown little evidence of recovery. Despite policies to manage the taking of seals, walruses, and other marine mammals, these, too, have sharply declined. E. O. Wilson remarked in *The Diversity of Life* (1992) that as human population spread, "Mankind soon disposed of the large, the slow and the tasty!" In North America, more than 70 percent of large mammal genera that existed at the time human hunter-gatherers crossed the Bering Strait from Siberia are extinct. A similar pattern is emerging in the sea, where marine mammals, large fish, and anything edible, no matter how critical to other species or ecosystems, are most vulnerable to human predation.

To encourage the recovery of species that are threatened with extinction, an obvious, good first step is to stop killing them. Beyond that, it makes sense to protect vital habitats. For whales, this is tricky because most species require huge tracts of healthy ocean where they can feed, socialize, and reproduce their kind. Humpbacks range over tens of thousands of miles, from polar feeding areas to tropical calving lagoons. Mindful of such requirements, recent measures have been taken to establish all the Indian Ocean and much of the ocean surrounding Antarctica as "whale sanctuary." The concept is also embodied in a growing network of more than 1,000 protected areas under national jurisdiction, worldwide, including the U.S. National Marine Sanctuary Program and a system of National Marine Estuarine Research Preserves. Protecting endangered species and ecosystems is part of the rationale for establishing such areas.

However, "protection" has many interpretations. Commercial fishing continues in Antarctica with little restriction on most species taken. Some guidelines are set for taking krill, *Euphausia superba*, but the limits established are based on guesswork including far-from-perfect understanding of what requirements are vital for krill to maintain their populations. Depleting krill in the southern oceans is sure to have profound implications not only for the krill and whales, birds, squid, fish, and numerous others who depend on one or more of the dozen or so stages young krill go through on their way to becoming a three-inch long adult, but also for the adults themselves.

There are thousands of species of crustacea in the sea, but only about eighty species in the genetically related group of creatures known as krill. While it appears that *Euphausia superba* has special ecological significance, each one of the eighty or so others has a magnified importance in terms of genetic information as compared to, say, a family of creatures that is represented by hundreds of variations. Among those most diverse of all creatures, the insects, each species lost is irreplaceable, and each vacancy jars some part of the ecosystem. However, there is likely to be more genetic resiliency ensuring the continuity of insectkind when there is a starting pool of perhaps a million species. The magnitude of basic genetic information lost is arguably greater with the demise of one member of a group represented by very few species. There are, for example, only four species of horseshoe crabs in an entire class of organisms. The genetic equivalent of losing one of them might be equated in some ways to eliminating a quarter of a million kinds of insects.

There are only about 350 kinds of sharks—creatures described by the National Marine Fisheries Service as "underutilized" only a decade ago. Fishermen were encouraged to take what they could, however they could, in part to make up for low catches of depleted species. But like whales, sharks tend to mature and reproduce slowly, and thus are especially vulnerable. Within a decade, it has become clear that sustained taking of these creatures on a large, commercial scale is unrealistic, and some limits are being set for some species in U.S. waters to protect what remains, but there are few protective measures elsewhere.

Another group of creatures with a distinguished lineage and a possible short future is the cephalopod mollusks—octopuses, squids, nautiluses—in all, only about 350 kinds are known. Some individual species are numerous enough to be sought by commercial fishermen, and as a consequence, once-thriving populations are being swiftly depleted. Despite their apparent fecundity, it is conceivable that some heavily fished species could go the way of passenger pigeons—and for the same reason that it takes a critical mass under just the right circumstances, not just a handful, to keep the population dynamics going. Squid, in particular, are vulnerable to human-catch techniques. In the Pacific, the extraction of enormous quantities of squid from the sea not only threatens their survival but also disrupts ocean ecosystems where large numbers of squid are vital to the survival of numerous other creatures including fish, marine mammals, albatrosses, and other seabirds—all with few choices about what they can consume.

This is true, too, of other kinds of sea creatures: arthropods, echinoderms, mollusks, ctenophores, coelenterates, and so on. The list is long and growing. Marine biodiversity is threatened by deliberate extraction of huge quantities of living sea creatures—currently more than 90 million tons a year—by humankind. Just as serious a threat is the long and growing list of dangerous substances that are dumped into the sea. Some substances such as plastic debris, lost or discarded fishing gear, cans, bottles, and other forms of trash are visible; most of them are a legacy of the past half century. When Thor Heyerdahl crossed the Pacific aboard the Kon Tiki in 1947, he was impressed by the pristine nature of the wild, open sea. By 1969 and 1970, during his famous *Ra* expeditions, Heyerdahl encountered conspicuous evidence of human garbage, debris, tar balls, oil, and other substances floating throughout the open sea. Since then, the quantity and variety of trash and toxic substances entering the sea have increased dramatically. Shorelines around the world are littered with trash, largely plastics. Pesticides, heavy metals, and other toxic substances dumped into the ocean have made their way into the tissues of fish, birds, marine mammals, and the humans who dine upon them. Some species are more vulnerable than others to the recent changes. Plastic bags engulfed by sea turtles who mistake them for jellyfish, one of their favored foods, die from starvation as their intestines become jammed with indigestibles. Death by entanglement, mostly with lost fishing gear, significantly impacts the fur seals and other marine mammals of the northwestern Pacific. It poses a serious threat to the Hawaiian monk seal, where every individual lost or gained tips the scales one way or the other toward survival or extinction.

Toxic materials more lethal than oil are being spilled into the ocean, but oil spills—especially big ones in pristine areas such as Prince William Sound—are hard to ignore. Say what you will about the impact of oil on marine life (some say it really is not so bad); 250,000 barrels of spilled oil were not exactly a prescription for health at Valdez, Alaska, in 1989 when the tanker *Exxon Valdez* went aground. After the spill, at the Reluctant Fisherman Inn in Cordova, a sign was posted: "Flags are at half mast because of the death of our environment." While overly pessimistic, for individual otters, birds, fish, and other creatures trapped in the oil, the message was no exaggeration, and for all, the spill brought into focus the difficulty of trying to put things back together once damage has occurred. It is difficult enough to restore familiar terrestrial ecosystems, whether the challenge is reestablishing a forest or returning a hawk to the wild after being in captivity for a while. It is virtually impossible to know what to do to restore a marine area such as Prince William Sound—or even a single depleted species of squid or shark, let alone minute microbeasts whose life histories are still a mystery.

Without pausing to understand the consequences to the living systems that support us, humankind has over the ages relentlessly diminished the genetic and ecological diversity to the point where full recovery is no longer possible. Each species lost diminishes the chance that we can "get it right," that is, to find an enduring place for ourselves within the living matrix that sustains us. If humankind proves its ability to be wise as well as clever, we will implement new and durable measures to maintain the living assets that we have too often taken for granted.

How can individuals stop the actions that are degrading the quality of life, closing the doors not only for future generations but also for those of us now alive? Whales are still being killed by wealthy nations as luxury food, despite worldwide outrage and awareness of their alternative values—economic, aesthetic, scientific, moral. Thousands of once and nevermore species are being rendered extinct each year, perhaps as many as four a minute through rainforest destruction and loss of other unique ecosystems. With a brazen indifference to world opinion and international agreements, some nations still dump/emit **highly toxic materials into the sea and sky, a deadly legacy** for the future, with immediate consequences for those here today. Despite evidence that ocean ecosystems are collapsing, that populations of fish, squid, and other species cannot sustain commercial taking, huge sets, trawlers, and factory ships are still being deployed—and more are being built, often subsidized by public taxes.

It is not possible to go back and redirect the course of history, but now—and not for long—there is a brief window of opportunity to restore and protect the remaining healthy ecosystems that support us. Far-reaching anthropogenic changes are sweeping earth's aquatic atmosphere—chemically, physically, and biologically. In the past few decades—

my lifetime—the sea has changed. Every day, the number and demands of humankind increase; the size of the planet does not.

Traditionally, the sea has been regarded as the common heritage for all humankind; now its care must be acknowledged as a common responsibility. To ensure a decent quality of life for the rest of our lives, as well as for all those who follow, we must develop policies that recognize the interdependence of life—and the need for nations to agree on mutually beneficial measures to protect and maintain the basic elements of life support—on a planetary scale. But it is difficult to be concerned about what we do not know. That is why it is absolutely vital to explore and come to understand the nature of the oceans— still largely unknown. Organizing complex information and making it readily and universally available—the subject of this meeting—is a vital part of knowing. With knowing comes caring, and with caring, an impetus toward the needed sea change of attitude, one that combines the wisdom of science and the sensitivity of art to create an enduring ethic. This is the time as never before, and perhaps never again, to establish policies—on a small personal scale as well as on broad public ones—to protect the ocean and the diversity of life that the sea sustains. How to achieve this most desirable goal is not clear, but the need to protect large marine areas from destructive actions seems to be fundamental.

Graeme Kelleher, for many years the Director of the Great Barrier Reef Marine Park Authority, suggested in a recent letter:

> It is my observation that administrative and bureaucratic problems often equal biophysical problems in importance when we try to maintain a healthy earth. . . . I believe that ultimately all of the world's seas should be zoned. In many places, marine protected areas are small islands of protection in an unregulated sea. Such small protected areas must be highly protected if they are to have any effect. On the other hand, large protected areas, covering complete marine ecosystems, can be zoned so that various human activities such as fishing can proceed within parts of them while other parts are totally protected from extractive and other exploitive activities. Such large protected areas can be seen as stepping stones towards the goal of integrated management of all the world's seas, with sustainable use being the overriding objective.

One thing is clear. We have an opportunity now as never before, and perhaps never again, to achieve for humankind, and for those with whom our species shares the planet, a prosperous and enduring future. If we fail, through inability to resolve thorny issues or by default born of indifference, greed, or lack of knowledge, our kind might well be a passing short-term phenomenon, a mere three- or four-million-year blip in the ancient and ongoing saga of life on earth.

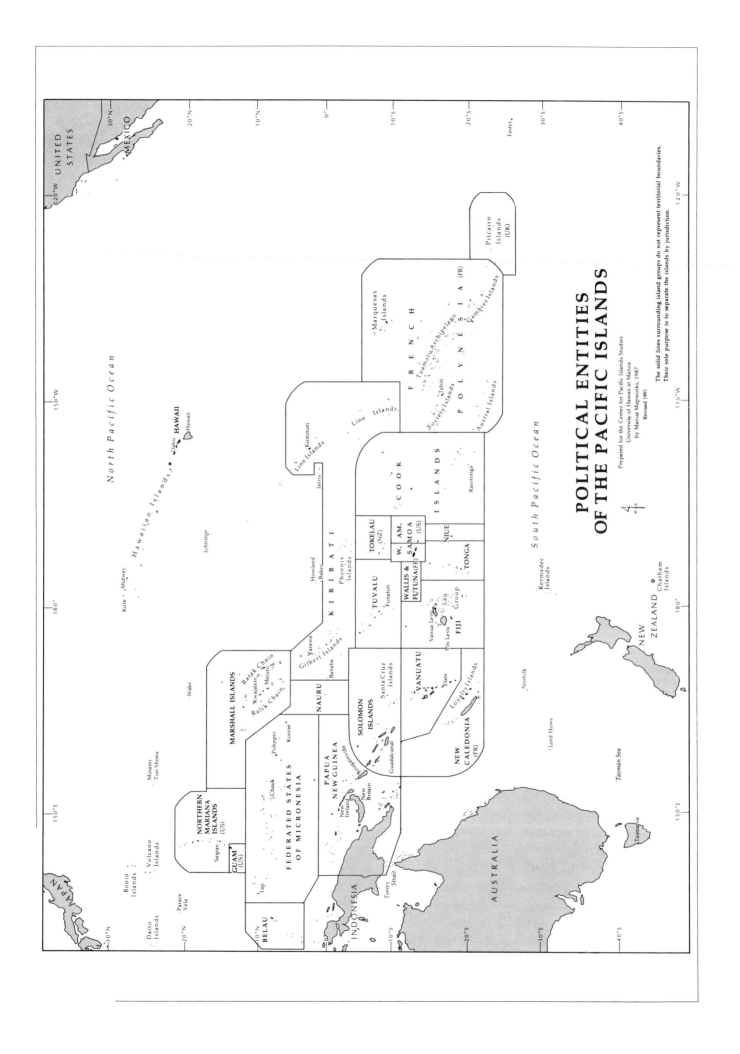

POLITICAL ENTITIES
OF THE PACIFIC ISLANDS

Prepared for the Center for Pacific Islands Studies
University of Hawaii at Manoa
by Manoa Mapworks, 1987
Revised 1991

The solid lines surrounding island groups do not represent territorial boundaries.
Their sole purpose is to separate the islands by jurisdiction.

Introduction

CHAPTER 1

Executive Summary

James E. Maragos

INTRODUCTION

The first of two workshops on Marine and Coastal Biodiversity in the Tropical Island Pacific Region was held at the East-West Center in Honolulu on 2–4 November 1994. The goal of the first workshop was to review the status of species systematics and database management studies and develop an action plan for marine and coastal biodiversity information management for the region. The workshop was divided into three parts. Day one was devoted to presentations on the status of species systematics by recognized specialists for key animal and plant taxa inhabiting nearshore marine and coastal habitats in Oceania (the tropical insular Pacific region). Day two was devoted to presentations by specialists on the status of biodiversity information management systems being developed for or applicable to Oceania. Day three was devoted to working group sessions that led to the formulation of an action plan for biodiversity information management that would assist in marine and coastal conservation and sustainable development in Oceania. The presenters, as well as additional participants, contributed to the discussions and development of the action plan. Over fifty participants attended the workshop, which included twenty-four separate presentations and several group discussions. See Appendix A for the flyer sent to participants prior to the workshop. Appendix B summarizes the results of the working group sessions, and Appendix D is a list of all participants. Appendix C is a glossary, and the remaining chapters cover the edited written presentations at the workshop.

The second workshop on Marine and Coastal Biodiversity in the Island Pacific was held during the following week, on 7–9 November 1994 at the East-West Center. That workshop focused on "Population, Development, and Conservation Priorities." Many of the presenters and participants at the second workshop arrived early to attend and contribute to the first workshop. The proceedings of the second workshop will be published in a separate volume.

EXCERPTS FROM THE KEYNOTE ADDRESS

Sylvia Earle delivered the keynote address for the workshop, calling attention to the unprecedented level of exploitation and pollution now affecting what was a century ago thought to be an enormous and boundless ocean. In Lord Byron's time, the sea was largely a wilderness, the distillation of over four billion years of evolution and change, but today fisheries exploitation has exceeded sustainability to the point of collapse for many fisheries stocks. Many of the ocean's species, especially the "large, slow, and tasty" have been depleted to the edge of extinction, including many marine mammals, sea turtles, sharks, squid, fish, and shellfish. Not until the middle of this century has technology provided the opportunity to begin serious exploitation and assessment of changes to the ocean and other aspects of the biosphere controlled by it. The biodiversity of the ocean has not yet been estimated to any acceptable degree of confidence. We do know that some marine ecosystems, such as coral reefs, support high levels of taxonomic diversity comparable to or exceeding that of the most complex of rainforests.

A good first step in the recovery of depleted species is to stop killing them, and for other species to set accountable harvesting guidelines. Other threats need to be controlled including pollution from oil spills, tar, lost fishing gear, cans, bottles, other trash, and plastics. It is not possible to go back and redirect the course of history, but there is a brief window of opportunity to restore and protect the remaining healthy ecosystems that support us. Every day the demand of humankind increase while the size of the planet does not. Traditionally the sea was regarded as the common heritage for all humankind but now its care must be acknowledged as a common responsibility. However, it is difficult to be concerned about what we do not know.

It is absolutely vital to explore and come to understand the nature of the ocean. Organizing complex information and making it readily and universally available—the subject of the workshop—is a vital part of knowing. With knowing comes caring and with caring an impetus toward the needed change in attitude and the establishment of policies to protect the ocean. Especially needed is the protection of large marine areas from destructive actions.

STATUS OF MARINE PLANT SYSTEMATICS

Algae

Isabella Abbott reviewed the systematics of marine algae in the tropical island Pacific (Chapter 2). Algal species diversity has been adequately assessed in only a few islands in the broader Pacific: Hawaii, Maui, and the Northwest Hawaiian Islands (Hawaii), Samoa, Enewetak (Marshalls), Nha Trang (Vietnam), Tahiti and Mo'orea (Societies), and Easter Island. It is not yet possible to assess the patterns of endemism, given the paucity of region-wide information. Only a few species of algae are reported at all sites. Many new algal species, genera, and families are still being described, but there are only a few trained scientists working on tropical algae. Many of the older Pacific studies are flawed because species names were used without reference to type material and voucher specimens. Within the adequately studied areas, species richness ranged from about sixty-seven species in Samoa to over 200 species each at Enewetak and the Northwest Hawaiian Islands. To the west of Oceania, the Philippines may have as many as 800 species. Additional investigations will be needed in southeast Polynesia and the most of Micronesia, Melanesia, and Indonesia before the distribution of marine algae in the Pacific is adequately understood. Abbott listed 3 divisions (red, green, and brown algae), 22 orders, 56 families, and 172 genera of marine algae occurring in the tropical Pacific.

Seagrasses

Robert Coles and John Kuo reviewed the systematics of seagrasses in the tropical island Pacific (Chapter 3). Seagrasses are true flowering plants that live in seawater. In the tropics they most commonly occur in depths at less than 10 meters but may be found down to a depth of 50 meters. Out of a world total of fifty-eight species, fourteen species belonging to eight genera and two families have been documented from Oceania. There is a pronounced attenuation of species when moving in a west-to-east direction across the Pacific. Only a few species have been reported in the eastern tropical Pacific. Endemism is rare in seagrasses, although one species—*Halophila hawaiiana*—appears limited to Hawaii. More studies on seagrass collections and taxonomy is needed in the Pacific, especially within most of Polynesia and Micronesia. A central organization such as the Bishop Museum is needed for storing and maintaining a seagrass reference collection and computer database. Formal standardized survey collection and storage protocols are needed for the region, patterned after Southeast Asian or Australian protocols. An assessment of important seagrass areas warranting protection is also needed.

Mangroves

Joanna C. Ellison reviewed the systematics and distribution of mangroves in the insular Pacific (Chapter 4). A total of thirty-four species and three hybrids belonging to sixteen genera and thirteen families have been reported from Oceania in the literature. There is a pronounced gradient of declining diversity from west to east across the tropical Pacific, reaching an eastern limit at American Samoa. To the east of Samoa, mangroves have been intentionally introduced in Hawaii and Tahiti. There are uncertainties in the distribution of some species, owing to sporadic collections and misidentifications. The insular Pacific region lags behind the Caribbean and Southeast Asia in developing a regional monitoring system and database of mangrove community structure and ecosystem health, which would allow full assessment and management of mangrove resources.

STATUS OF MARINE INVERTEBRATE SYSTEMATICS

Stony Corals

John E. N. Veron reviewed the taxonomy and biodiversity of Scleractinian corals in the tropical island Pacific (Chapter 5). He makes the distinction between *systematic order*, which is genetically based and conceptual, and *taxonomic order*, which is mostly morphologically based and operational. Genera and species are taxonomic units. Most coral species display plant-like morphological variations in response to local environments, which make species identification difficult and a task for specialists. Any consideration of coral biodiversity within a region as broad as the Pacific must take full account of *within-species* variations. About 20 percent of the coral species display little or no geographic variation. Other species show disjunct distributions in high latitudes. A third group of corals form geographic subspecies in isolated regions. A fourth group includes species with increasing variation with distance. Some species pairs or groups also appear to display hybrid patterns. An essential question is, How close does "species" taxonomy reflect the natural world and how much of it is a human construct? Collectively those issues have complicated identification and listing of species in field investigations and have led to numerous mistakes and incomplete studies. As a result, Veron concludes that species boundaries are arbitrary and that the systematic states of one species may be different from that of another. Methods used to date to determine taxonomic or systematic relationships include cladistics, colony transplants, physiological and behavioral studies, reproduction studies, and molecular methods.

Veron lists eighty-four genera and fifteen families of extant zooxanthellate corals of the order Scleractinia likely to be found in the tropical Pacific. The total number of species in the Indo-Pacific exceeds 700, and it is difficult at this stage to determine the approximate subtotal for the Oceania region of the Pacific. As with seagrasses and mangroves, coral species diversity attenuates in a west-to-east direction across the tropical Pacific.

Sponges

Michelle Kelly-Borges could not attend the workshop but agreed to prepare a sponge taxonomic review paper for presentation at the workshop (Chapter 6). More than one thousand sponge species are listed collectively from eighteen of the approximately twenty-five island groups in the tropical Pacific. Sponges are a significant component of all tropical, temperate, and polar marine benthic communities. Sponges are divided into three classes. A total of 6,000 to 7,000 *described* species of sponges occur in the world, and all but 150 species are marine. However, there are many undescribed species, and Australian island and continental faunas alone may total 5,000 *undescribed* species. It is also becoming clear that regional endemism of sponges is substantially higher than that of many other marine invertebrate phyla.

Most of the earliest publications on Indo-Pacific sponge fauna are extremely unreliable. With the exception of Hawaii, Micronesia, Papua New Guinea, and New Caledonia, most sponge faunas of the tropical island region of Oceania are poorly known. Ninety-eight species have been reported from Hawaii, 184 species from Papua New Guinea, 274 species from New Caledonia, and 277 species from the Federated States of Micronesia. The higher species totals in part is due to more intensive collection and taxonomic research, and many more species likely occur in each of these as well as in other island groups.

Geographic and habitat scale is extremely important in the consideration of sponge distribution, and the sponge fauna of Oceania can be divided into four broad groups: (1) regionally endemic sponges that are habitat- or locality-specific, (2) regionally endemic sponges that occur within a single atoll or island group, (3) species that are found in super regions (e.g., Micronesia alone or Solomon Islands alone), and (4) species that are found throughout the broad Indo-West Pacific region.

The minimum information required for management of sponge species data include species, site, and bibliographic information. Additional information could include protection status, commercial use or application, and specimen information. The database should allow the storage of hierarchical Linnaean taxonomy, coding of synonymous species names, incorporation of literature references, and personnel data (i.e., whether the person is a collector, taxonomist, donor, author). Photographic and site information is also useful. There is an urgent need to expand sponge biodiversity initiatives within Oceania.

Polychaetes

Julie H. Bailey-Brock reviewed the systematics and ecology of polychaete worms from the Pacific islands (Chapter 7). Most tropical islands in the central and western Pacific region support speciose polychaete assemblages that are mostly cryptic and endolithic or infaunal. Carbonate rocks, rubble, sands, muds, live corals, algal turf, and sponges all provide habitat for different assemblages of polychaetes. They also are found in tide pools of rocky intertidal areas, on terraced benches, reef flats, and subtidal reef slopes with extensive coral cover. Some polychaetes are also common in man-made harbors and natural embayments.

Bailey-Brock lists the characteristics of seventeen families of polychaetes commonly present on reefs and the number of genera at each of six well-studied Pacific island locations (Hawaii, Fiji, Tonga, Solomons, Cooks, and Enewetak in the Marshalls). Ideally generic and species-level diagnoses are preferable for taxonomic studies, but much can be done to characterize an area or habitat by grouping polychaete worm assemblages at the family level. A total of forty-three families and fifty-one genera of polychaetes have been reported for Hawaii, the best-studied area in Oceania. The number of genera from the five other studied areas varies from seven (Fiji) to thirty-five (Enewetak).

Mollusks

E. Alison Kay reviewed the systematics and database requirements for Pacific island marine mollusks (Chapter 8). The first approximation of species diversity among the islands of the tropical Pacific is approached by listing families, genera, and numbers of species known from Guam and the Northern Mariana Islands, Kwajalein, and Enewetak (Marshall Islands), Hawaii, Societies, Tuamotus, Pitcairn and Easter Islands. About 2,600 species belonging to 183 families are collectively associated with these islands. Species, genera, and families attenuate when moving in a west-to-east direction across the Pacific. The total fauna for Oceania may approach 6,000 species, and Kay reports approximately 1,000 species from each of the Hawaiian Islands and Enewetak in the Marshall Islands, and nearly 1,500 species from Kwajalein in the Marshall Islands. Kay believes that the minimum biodiversity database needs for marine mollusks should include the species list with author and date of description, habitat (infaunal, epifaunal), trophic habit, spatial habit (free living, sessile), date of collection, substrate, depth, and references (including figure number).

Crustaceans

Lucius G. Eldredge reviewed the status of crustacean systematics for the tropical and subtropical islands of the Pacific (Chapter 9). More than 2,000 species of crabs, alone, are estimated to occur within the region, and no estimates have been made for other crustacean groups due to limited distributional and taxonomic data. Over the past fourteen years, an average of 220 *new* genera and subgenera are described annually in the *Zoological Record.* Only a few of the 113 known families of decapod crustaceans have been monographed, and an additional 543 crustacean families are recognized among the other taxonomic groups. There are no complete crustacean studies in the area of the tropical Pacific, and comparisons among island groups within the region are impossible to make at this time. Specialists cannot even agree on whether the Crustacea should be classified as a phylum, subphylum, or superclass. There are six total classes of crustaceans. For database management purposes, the taxonomic hierarchy should be reduced to class, order, family, genus, and species. The database should also include an appropriate authority file so that intervening taxonomic levels can be searched efficiently. Aside from the lack of monographs and review articles, there is a severe lack of crustacean systematists, and the situation is expected to worsen with time.

Echinoderms

David L. Pawson reviewed the status of echinoderm systematics, ecology, and biogeography (Chapter 10). The Phylum Echinodermata comprises more than 7,000 species and is represented in all seas and at all depths. Six classes are recognized today. The tropical Indo-Pacific marine region supports a nearshore/inshore echinoderm fauna of approximately 1,300 species and about 452 species belonging to 187 genera and 65 families. The twenty-three recently described new tropical island Pacific species represent 18 percent of the new records and indicates that more intensive exploration virtually anywhere in the Pacific will result in numerous additions to the faunal list. Any new field exploration, however, should be coupled with the study of existing collections in major museums where many unidentified or unpublished collections are housed. The island Pacific echinoderm fauna remains poorly known, and additional studies must be encouraged. In light of the growing economic importance of echinoderms (e.g., bêche-de-mer), production of a computerized island database is a necessary next step.

STATUS OF MARINE FISH SYSTEMATICS

John E. Randall reviewed the zoogeography of the inshore Hawaiian fish fauna (Chapter 11). The number of species of reef and shore fishes now known for Hawaii is 557, disregarding additional species of fish that have been intentionally introduced during the past half century. The 557 species belong to 99 families. The number of species within the various families of fish in Hawaii is often out of proportion to the number of species of these families in the rest of the Indo-Pacific region in general. The paucity of species in some families may be related to Hawaii's geographic isolation from neighboring island groups and the associated limitations imposed on species with short planktonic larval stages.

Richard Pyle reviewed the zoogeography of reef and shore fishes in the remainder of the insular Pacific (Chapter 12). At least 3,392 species within 839 genera, and 166 families have been reported from Oceania. Previous estimates of the total number of Indo-Pacific reef and shore fish species may be too low. The most frequently used taxonomic levels in marine fish literature are family, genus, and species, and the taxonomy of the fishes at these levels is relatively stable and complete compared to higher taxonomic levels. A declining rate of new species descriptions in recent years is not the result of fewer discoveries because 13 percent of the known species remain undescribed. To utilize reef and shore fish species as indicators of biodiversity patterns will require that published and unpublished information be combined in a single database management system (see Eschmeyer, Chapter 19). A database of larger scope is needed, including information on biology, ecology, and distribution of reef and shore fish species. Because much misinformation persists in published literature, extensive documentation of information sources must be included with such data. The sophisticated database management system "FishBase" (see Froese and Palomares, Chapter 20) should be adopted as the standard repository of data pertaining to fishes and should receive full cooperation from ichthyologists worldwide. In addition, comprehensive species checklists of certain regions in Oceania are needed as is further investigation of the fish fauna inhabiting deep reefs.

STATUS OF OTHER TAXONOMIC GROUPS

It was not possible to present status reviews for all tropical Pacific plant and animal groups occurring in the nearshore marine and coastal environments of Oceania, owing to funding constraints and the unavailability of recognized experts. Reviews needed for other major taxonomic groups include phytoplankton and zooplankton, blue-green cyanobacteria, other stony corals, soft corals, jellyfish, anemones, tunicates, bryozoans, and a variety of invertebrate and worm phyla. The species of marine turtles, marine mammals, and seabirds present in Oceania will be covered in the proceedings of the second volume of this series.

STATUS OF THE MARINE ECOSYSTEM CLASSIFICATION FOR THE TROPICAL ISLAND PACIFIC

Paul F. Holthus and James E. Maragos gave a progress report on the development of a marine ecosystem classification for Oceania (Chapter 13). A comprehensive hierarchical classification system for coastal and marine ecosystems in Oceania was developed as a response to the need for systematic conservation-oriented inventory of ecosystem biodiversity information. The classification system was developed with input from a wide range of scientists and resource managers and can be used as a part of field surveys or as part of the review and evaluation of existing data and literature. A preliminary inventory of the ecosystems within the U.S.-affiliated islands in Oceania helped to refine and improve the classification system. The revised system is presented in its entirety, including a full glossary and diagrams of technical terms. The revised classification system should be useful during the inventory and classification of coastal and marine ecosystems in other tropical Indo-Pacific areas outside the U.S.-affiliated islands.

STATUS OF EXISTING OR PROPOSED BIODIVERSITY INFORMATION MANAGEMENT SYSTEMS POTENTIALLY APPLICABLE TO OCEANIA

Global Task Team on the Implications of Climate Change on Coral Reefs

Charles Birkeland discussed the mission and philosophy of the UNEP-IOC(UNESCO)-ASPEI-IUCN Global Task Team on the Implications of Climate Change on Coral Reefs and its relationship to information management needs (Chapter 14). The focus of the Task Team is on rapidly determining *changes* that might be currently occurring on coral reefs. This focus will require broad-scale time-series assessments at fixed or permanent reef locations. A key strategy is to utilize a paired comparisons design for monitoring coral reefs to differentiate changes attributed to "natural" causes from those attributed to anthropogenic causes. Resurveying of previous sites is especially high priority, and size distributional data for important reef organisms (e.g., corals, fish, mollusks) would be more informative in determining the status of reef conditions than relying on surface cover or community composition data. It is urgent to begin the fieldwork monitoring phase as soon as possible, but it is also critical that an efficient protocol for data storage and retrieval be designed before the monitoring program is put into effect; otherwise, the data may never be adequately analyzed or retrieved. It is also critical that species identifications are accurate and certified. If errors in identification without voucher specimens are accepted into a database, it is quite likely that the species could be falsely and permanently recorded as present at some place during an earlier survey, then later interpreted as becoming locally extinct when the place is later resurveyed. The trained personnel of the database facility must develop data storage and checking protocols and prepare training methods for data collection and entry to ensure that all data are verified upon submission, and, if necessary, report back to the scientist submitting the data whenever errors are found. The database facility must also prepare regular summaries and status reports in order for the monitoring data to be used and analyzed.

The Global Task Team recommends that the basic structure of data management be a central dedicated database, connected by electronic mail to a network of databases within each of the participating countries. The Australian Institute of Marine Science was suggested as a location for the central database.

The information management requirements and criteria of the Task Team will differ from those needed for other types of biodiversity inquiries and purposes. Because it may not be practical to achieve differing criteria simultaneously within a single database system, it is most important to code all data with indicators of the level of taxonomic certification in order to allow the data to be sorted and retrieved automatically according to the levels of certification needed for the questions at hand. For example, the levels of taxonomic certification could be categorized among the following codes: 1 = field observation, 2 = identified by a qualified taxonomist, 3 = voucher specimen number for a museum collection.

World Conservation Monitoring Centre

Mark D. Spalding presented an overview of the World Conservation Monitoring Centre's (WCMC) handling of global biodiversity data (Chapter 15). The WCMC compiles and disseminates substantial global information on biological resources and their conservation and sustainable use. These data include information on threatened and endemic species, protected areas, and habitats. They are stored, managed, and maintained on purpose-built databases using a variety of software packages where they can be accessed by a wide range of users. Three of the main databases include the Species Database, the Protected Areas Database, and the Biodiversity Map Library. Information in these databases is now available on-line over computer networks. None of these databases are specifically marine or coastal, but all contain considerable volumes of information on marine/coastal species sites and habitats.

One clear need that arises from this conference was for better management of taxonomic data for marine and coastal species in the region. The development of a database to serve these needs should make full use at both existing taxonomic database structures and the data already held in these and other databases.

A key product of the workshop should be the development of a more coordinated and efficient approach to biodiversity data management for the tropical island Pacific region. This should include networking and data exchange, thereby reducing costs and preventing repetition in developing data management systems. Thus, the "new" Oceania system might instead be one that facilitates data exchange and wider use of existing data by scientists and decision makers, and one that would also provide encouragement for gathering new data. Such a database could be a single system based in one organization but freely available to the wider community or a more loosely linked series of existing regional databases with a number of specialist nodes throughout the region.

THE UNITED STATES CORAL REEF INITIATIVE

Michael Crosby and James Maragos prepared a status report on the U.S. Coral Reef Initiative (USCRI), Chapter 16, supplementing the brief progress report on the USCRI presented by Maragos at the workshop. The long-term vision of the USCRI is to build a comprehensive national program for conservation and effective management of coral reef ecosystems, including mangroves and seagrass beds, combining new and existing activities and programs. This initiative is intended to reverse the current trend toward degradation of these valuable ecosystems, both domestically and internationally. The initiative initially split into two subprograms—the International Coral Reef Initiative and the U.S. Coral Reef Initiative. The USCRI will be based on partnerships developed among federal agencies, states, territories, and commonwealths of the United States, and other interested and affected parties including non-government organizations, industry, the scientific community, the public, other countries, and international and regional organizations. The International Coral Reef Initiative will forge comparable relationships for coral reefs in other (non-U.S.) areas. The USCRI will provide significant benefits to the United States by strengthening domestic capabilities and programs, strengthening existing coastal and marine ecosystem management and conservation, and integrating the private sector into efforts to provide solutions to environmental problems and expand markets for U.S. environmental technologies, management strategies, and techniques, and demonstrating U.S. commitment to sustainable development and biodiversity conservation in fragile tropical marine ecosystems.

The USCRI will foster the development of a national and global monitoring program and network involving scientists and managers. Initial contributions to this network will be (1) establishment of an International Coral Reef Research and Monitoring Office, (2) establishment of a U.S. Coral Reef Monitoring and Assessment Program, and (3) establishment of a Pacific Regional Coral Reef Assessment and Monitoring Program (PACICOMP) to complement the existing Caribbean Program (CARICORP). The monitoring program will be closely coordinated and under the auspices of the International Oceanographic Commission, UNESCO, IUCN, and others (refer also to Chapter 14 on the Global Task Team). The USCRI will also focus on coordinating strategic research on coral reefs.

In the insular tropical Pacific, a regional planning meeting was held in December 1994 to implement the coral reef initiative and to formalize community and local/state government involvement in the design and implementation of the program. This meeting later led to a more localized coral reef initiative in the State of Hawaii. The Hawaii Coral Reef Initiative is now in the planning stage for a comprehensive assessment of coral reefs in the State of Hawaii, which will eventually lead to a comprehensive monitoring program and other management activities for Hawaii's coral reefs. Higher education and research institutions in the U.S. Pacific region have also begun to network more closely in developing specific coral reef research, education, and conservation programs.

ReefBase

A status report on ReefBase was presented by John W. McManus and others (Chapter 17). ReefBase, which is currently being prepared, is an international database on the coral reefs of the world. The system will include maps designed specifically for it, as well as other available maps and imagery. The basic unit of information will be the reef by year. Most data are acquired from reports and publications, and the system evolves according to data trends and questions to be answered. ReefBase will include interactive linkages to other databases including CoralBase and FishBase (see Chapters 18 and 20, respectively). Major spin-offs of the project include improved reef terminology; determinations of reef health; estimates of reef contributions to food, income, and atmospheric processes; and standardization of reef survey and monitoring methods. The current phase will end with the release of a preliminary CD-ROM version by 1996, and future expanded phases are planned. Databases of this type must be adaptive to trends in available data as well as to an evolving set of questions for which the database should provide answers. Such databases should involve a centralization of data entry efforts and the occasional use of trial queries to ensure that data demands are being met efficiently.

Management of ecosystems in Oceania would be greatly facilitated by development of regional databases. A possible biodiversity database is discussed by Froese and Palomares (Chapter 20) and involves determining the distributions of a wide range of species based on museum records and other data sources. However, regional management would be further enhanced with the development of an environmental database for Oceania. Such a database would focus on ecosystems as units and permit inquiries on the ecological and management studies. Environmental databases tend to fall somewhere on a continuum between two types: (1) those intended for a specific sampling program and (2) those intended to accommodate existing and future data sources. The data fields for the first type are primarily determined before implementation of the database. The second type of databases must be adaptive, and the Oceania database would probably be primarily of this type. Regardless, there must be a continuous effort to ensure that the data being input can specifically answer the questions the database is designed to answer. It is also important that strict controls be made on the use of the database prior to planned "official" release dates. There may also be too much concern initially about what the database "looks like." The "face" of the database for input purposes should be designed specifically for those responsible for input of data. On the other hand, the most important face of the database will be that designed to access the data for user queries.

CoralBase

Kim F. Navin and John E. N. Veron presented the status of CoralBase, a taxonomic and biogeographic information system for Scleractinian corals (Chapter 18). CoralBase is a distributable Scleractinian coral information system of taxonomic and biogeographic data being developed at the Australian Institute of Marine Science. While concentrating on the Indo-West Pacific Region, it contains a database of approximately 700 coral species from 350 locations worldwide, a large image collection, and a 4,000 reference bibliography. Utilities to plot distributions, carry out common analyses, draw graphs, and compile reports are also included. Systematic data are available at the species, genus, and family levels. For each species the family, genus, species, and authority are recorded. Site data include site name, latitude, longitude, ReefBase reef codes, and for East Australia the Great Barrier Reef Marine Park Authority's gazetteer reef code. Sites are grouped into zones based on species composition, and zones are further grouped into regions based broadly on biogeographic division.

The "AIMS-Base" software used for CoralBase is generic enough to be used with any taxonomic group with little modification and has been written with that future purpose in mind. Because of its modular construction, it is not necessary to have all the data types to use the database. CoralBase aims to overcome many of the problems associated with printed publications by providing a distributable, updatable, and easy-to-use method for assessing the latest information on Scleractinian corals from all over the world. With cooperation from researchers worldwide, and frequent updates, CoralBase will become the standard reference on Scleractinian corals and will expand to cover all areas of the globe.

ROLE OF TAXONOMIC DATABASES

William N. Eschmeyer made a presentation on the role of taxonomic databases with special emphasis on fishes (Chapter 19). Managing information about animals and plants depends on accurate taxonomic databases. Understanding taxonomy and the role of taxonomists is important for all aspects of biological information management. Taxonomists do two important tasks: name organisms and make classifications. The system of hierarchical classification and a two-word system for naming species began with Linnaeus in 1758 and was later codified and adopted by all zoologists worldwide from 1843 to the present. Another major activity (or duty) of taxonomists is to make "synonymies" that summarize prior accumulated knowledge about each species. Unfortunately scientific names change for several reasons that make inventory especially difficult since information about a single species may be found under several scientific names. Thus, it is the duty for taxonomists to sort out and determine the preferred or correct names, identify the other names (junior synonyms) for the same species, and explain the reasons for name changes.

The building of accurate taxonomic databases is time consuming and requires excellent library facilities and considerable funding; there are not many shortcuts. The California Academy of Sciences (CAS) with National Science Foundation funding began building a database for fish genera. Eventually, a list of genera was allocated to a recent classification. The work was published in 1990 and consisted of 10,000 fish genera arranged alphabetically with their associated information and a bibliography of about 4,500 references. The genera database is continually updated, and in early 1994, it and associated appendices were placed on the Internet Gopher System of the CAS. It was also adopted as an authority standard by the *Zoological Record*. While the genera database was being prepared for publication, Eschmeyer's team began preparing the database for fish species. This database consists of about 55,000 records and 20,000 references and eventually will be available electronically as a CD-ROM. The approximate cost and time for building the fish database will total $490,000, which is the equivalent of about 14 person years.

Once an authoritative database is available, other databases can be associated with it. Databases are of tremendous value to taxonomists and will make taxonomists much more efficient. Accurate taxonomic databases will be essential for many aspects of studying and managing the information on the aquatic biota of the tropical island Pacific region. Building them is something that taxonomists and museums are ideally suited to undertake.

FishBase

Rainer Froese and Maria Lourdes D. Palomares made a presentation on FishBase and its role as part of an Oceania biodiversity information system (Chapter 20). FishBase is an electronic encyclopedia on fish being developed at the International Center for Living Aquatic Resources Management in cooperation with the FAO of the United Nations and funded by the Commission on European Community. To date, FishBase contains key information for about 12,000 fish species globally, including about 2,800 species of the Oceania region. FishBase is available on CD-ROM for computers with Windows 3.1 or later. FishBase contains key information on fish populations such as nomenclature, ecology, population dynamics, aquaculture, genetics, physiology, and occurrence of fishes. It was conceived as a "tool" to help fisheries researchers and managers to better understand and manage their natural resources. FishBase is designed to answer a variety of questions for the fishes of a given country and also offers a structure for national information such as occurrence, importance, use, or restrictions in a given country.

FishBase also makes use of published literature. Contributions are given proper citations and acknowledgments in FishBase; every datum is attached to the reference from which it was derived and every record bears a stamp for identifying who entered, contributed, modified, or validated the information. The quality of information in the database is monitored in-house through several loops of verification by research assistants. As of October 1994, the FishBase team has incorporated 12,000 species extracted from more than 7,000 references, representing half of the estimated 25,000 species of fish in the world. In addition, FishBase has become the repository of several outstanding collections of data including population dynamics, ecology, electrophoresis, common names (by country and language), the genera from Eschmeyer (see Chapter 19), metabolism, larval dynamics, introduced species, and 1,000 line drawings or color pictures of adults, larvae, eggs, or diseases.

Before a database system can be designed for Oceania, one has to be clear about the purpose(s) it should serve. The background information for the workshop objectives call for a "data management system to assist in a long-range program to conserve biodiversity" and to "summarize information on the systematics and taxonomy of nearshore marine and coastal species in Oceania." Froese and Palomares recommend that an Oceania database system contain the following tables (refer to Figure 20.2 of Chapter 20):

- a *specimen* table serving as the center of the database;

- the taxonomic group, genus, and species fields of the *specimen* table are linked to *taxonomic dictionaries*;

- the *specimen* table is linked to the *identification* tables through the *taxonomic group* and *catalogue code*;

- the *station code* links the *specimens* to the *stations* table;

- the *locality code* links the *stations* table to the *localities* table (e.g., gazetteer of aquatic localities);

- the *acquisitions code* links the *stations* table to the *acquisitions* table;

- the *country code* links the *localities* table to the *countries* table;

- the *ecosystem code* links the *localities* table to the *ecosystems* table;

- the *contributor code* in the *specimen* table establishes the link to the *contributors* table; and

- a *reference* table is linked to the *references* mentioned in the other tables containing the full citation.

This structure consisting of more than ten tables with many code fields will be largely hidden from the user who will just click on buttons to select options. The core information to set up the database system is the species name, date, locality, and source (name of the provider of the information). Such a system would be able to answer a number of species-related questions. A crucial point in the suggested system is the quality of the identifications, and it will be necessary to determine expert centers for each species group who would classify the probability of correctness. It would be best to start a database activity with an international project which, if successful, will be turned into a permanent activity of an appropriate international body.

Hawaii Biological Survey

Allen Allison, Scott E. Miller, and Gordon M. Nishida made a presentation on the Hawaii Biological Survey (HBS) as a model for the Pacific region (Chapter 21). The Bishop Museum in Honolulu houses the world's largest natural history collection from Hawaii with nearly four million specimens. Designated as the Hawaii Biological Survey by the State Legislature in 1992, the museum has implemented a six-point plan to fully develop its information resources to support a wide range of applications in Hawaii including research and resource management. For each major group of plants and animals, the comprehensive plan consists of:

- developing a computerized database of the literature;

- preparing a checklist based on the literature, collections, and consultations with experts;

- databasing museum collections with localities geocoded to facilitate GIS analysis and presentation;

- databasing information from other collections on organizations;

- filling gaps in information through additional fieldwork and research; and

- making this information widely available.

In actual practice, many of these activities take place simultaneously or out of the above sequence and some are undertaken with cooperating agencies and partners such as The Nature Conservancy of Hawaii, Natural Heritage Program, Center for Plant Conservation, University of Hawaii, state agencies, National Biological Service (NBS), and other federal agencies.

The museum is now extending the above approach to other parts of the Pacific, including development of a national biological survey for Papua New Guinea. The HBS team sees its role as providing a service to scientific and local communities as an international clearing center. The success of the NBS serves to illustrate the importance of museum collections to document biodiversity and provide information crucial for environmental planning and natural resources management. A biological survey provides the framework to bring together widely dispersed or inaccessible information to support a wide range of applications. This approach is crucial to the widely dispersed Oceania region, and many large collections from the Pacific are in museums around the world. Many of those museums are also now databasing their collections, and it is crucial to develop a mechanism to link this information with other biologically important data (e.g., marine surveys, individual research projects). Museums also need to take a more active role in identifying regional environmental priorities.

Hawaii Natural Heritage Program

Samuel Gon III and Roy Kam described the mission of the Hawaii Natural Heritage Program (HINHP) as a natural diversity database (Chapter 22). The HINHP compiles and maintains a set of databases and geographic information systems (GISs) on the location and status of rare and endangered species and ecosystems in Hawaii. This information is used to synthesize, interpret, and distribute accurate, current, and comprehensive information to a wide set of appropriate users for the purpose of making a positive impact on biodiversity protection. The HINHP tracks rare, threatened, and endangered terrestrial and wetland taxa of the Hawaiian Archipelago which are referred to as rare elements. Currently over 1,000 different native taxa and types of ecosystems are tracked, and over 10,000 rare element location records are mapped statewide. The basic database for rare taxon/ecosystem tracking is called the *Element Occurrence Database* with individual records referred to as *Element Occurrence Records* (EORs). Each EOR is a summary of all the available information for a single element at a single location or occurrence. The EOR is produced by combining information from museum collections, published and unpublished reports, communications from knowledgeable individuals, and field surveys. Manual maps of the HINHP use standard U.S. Geological Survey 7.5-minute topographic quadrangles as a base upon which numbered mapped symbols are affixed. The shapes of the symbols reflect the relative precision of the location of source information, and the colors of the symbols reflect the four basic types of rare elements (plants = yellow, vertebrates = orange, invertebrates = purple, and ecosystems/natural communities = green). The GIS for the HINHP uses a UNIX workstation running ArcINFO software and networked to a Pentium PC running ArcCAD, AutoCAD, and ArcVIEW. Map digitizing tables are used for full USGS quad scale input of data. Plotters and scanners are also utilized for additional data input and display.

A main strength of the HINHP mapping convention is that databases are arranged primarily by location, which allows rapid location-wise queries by landowners, managers, scientists, and other users. The main limitations of the system are the labor-intensive database updates every two years and quality control procedures associated with each update. Another limitation external to HINHP is that much of the information relies on third-party data that sometimes require confirmation. The GIS for the HINHP is compatible with those in use by the State of Hawaii Office of State Planning and the federal government (U.S. Fish and Wildlife Service, U.S. Department of Defense). The database system used (Advanced Revelation) allows for import/export of data with dBASE and other emerging database standards.

For establishing a new database and mapping system, several specific recommendations are warranted: (1) involve an advisory committee of pertinent individuals, agencies, and organizations to help guide development and set information priorities, (2) explore the scope of other related database and mapping efforts to strive for complementarity, (3) continually assess the true cost of database maintenance beyond start-up, and (4) continuously assess the users of the database to service both their short- and long-term needs.

MARINE BIOSYSTEMATIC/ BIODIVERSITY PRIORITIES: A CANADIAN PERSPECTIVE

Don E. McAllister presented his perspective on marine biodiversity and systematic priorities based on his experience in Canada (Chapter 23). Biological surveys, presentation of voucher specimens, and geographic distribution analysis using GIS can help to pinpoint potential protected areas. Sound GIS analyses depend on geographically adequate sampling, correctly identified specimens (preferably verifiable via voucher collections in museums), and ongoing support for systematic research and natural history. Useful GIS methods include analysis of all species or a subset in a taxonomic group, endemic species, and the variety of higher taxa in which lie many kinds of diversity not expressed by single species counts. Information from coral reef fishes was used to illustrate these points. All-taxa biodiversity inventories (ATBIs) are also advocated to measure the total diversity in one or more areas in the ocean. Equal-area global grid analysis provides a useful tool for comparing biodiversity within different grids, for discovering biodiversity hot spots, and for recognizing geographic trends.

Biosystematics serves societal roles in science, conservation, sustainable use, and new bio-industries. However, biosystematic research in some cases and knowledge of distribution are hampered by inadequate collections and poor geographic sampling. Research is also held back by the declining number of taxonomists and graduate students in taxonomy, decreasing positions for taxonomists, and declining support for collections. Behind these declines are swings in support for emerging fields in biology, but also involved is the failure of biosystematists to communicate the nature and societal roles of their research to a broader public.

Recommendations by McAllister include the following:

- promote, prepare, and publish national country studies on biodiversity as recommended by the Biodiversity Convention;

- become involved in Biodiversity Convention negotiations;

- popularize systematics with stories about your work;

- develop plans for global, national, and state/provincial equal-area grid-based biological inventories and develop plans for at least one marine, aquatic, and terrestrial all-taxa biodiversity inventory (ATBI) in every country;

- seek funding for taxonomic study of major taxa and poorly known groups;

- develop geographic, environmental, and conservation priorities for funding biosystematic research;

- creatively solve potential conflicts between conservation organizations and museum databases;

- urge creation of new posts for systematists in museum and universities using new priorities;

- support harmonization of the taxonomic codes for zoology, botany, and microbiology;

- promote use of cladistic, species, and taxonomic data in biodiversity and ecological inventories;

- support development of an on-line global computer catalogue of all species of plants, animals, and microorganisms;

- develop popular guides to marine organisms for education, recreation, and eco-tourism; and

- support and cooperate with non-government organizations involved in conservation and public awareness for biodiversity research.

SUMMARY OF THE WORKING GROUP SESSIONS

Appendix B summarizes the group notes from the working group sessions. Workshop contributors and participants met on the third day (4 November 1994) to develop an action plan for species systematics and information management priorities. The first plenary session identified the concerns, objectives, and goals to help focus the two working group discussions. Following the first plenary session, the participants divided into two working groups—one on species systematics and the other on information management. At the end of the day, the two working groups joined in a second plenary session to review the decisions reached by the working groups and make recommendations on an action plan.

The principal points raised during the first plenary session included the following:

- The central component of any biodiversity information management system is species-level taxonomic and systematic information. Core information for species-level biodiversity includes the quadruplet: species name, date, locality, and data source.

- Although the workshop had a science-driven agenda and limited budget, biodiversity research and conservation must involve the active participation of Pacific islanders and must include cultural information to preserve local taxonomy, place names, and conservation practices; local names for species, ecosystems, habitats, and processes need to be linked to global scientific equivalent terminology and names.

- Information on ecosystems should be an integral part of any marine biodiversity database system for the tropical insular Pacific.

- Any biodiversity database management system needs to have geographical units or boundaries based on natural features.

- Biodiversity database management systems should address both the immediate and long-range needs, which have cultural as well as natural resource dimensions.

- Provided that monitoring data are accessible to biodiversity database management systems, there is an opportunity to conduct analyses on changes in species and ecosystem status at specific sites over time.

- Database management systems need to include attribute data on species, ecosystems, environmental conditions, monitoring, distribution, etc., tied to mapped data.

- Collaboration is strongly needed among the scientific, cultural, and educational systems within the region and outside the region.

The principal conclusions reached during the working group session on species systematics are as follows:

- Species systematic data must include what, where, when, and why data were collected.

- Public support is needed for taxonomy and biodiversity conservation.

- Critical biodiversity information needs to be communicated to the public and policy makers in order to advance conservation.

- Taxonomists and systematists need to work closely with computer specialists in order to develop data storage and analytical capabilities for taxonomic data.

- There is an obvious need for compatible and user-friendly worldwide databases for biodiversity, including the ability to network with other databases.

- Biodiversity is the variability of living things as individuals, species, and other assemblages. It occurs as local variants in space and time. Taxonomy is the interface between the variants of living things, and information technology makes it possible to code diversity in a form amenable to information management.

- There is an obvious shortage of active taxonomists and related research.

- Taxonomists need to work with computer experts/information management specialists to design a taxonomic database management system.

- Coastal areas and species groups need priority attention due to the proximity of humans and their impacts to nearshore and coastal biodiversity; in addition, priority attention is needed on many invertebrate and microorganism groups and deep-sea organisms.

The consensus of the working group on information management is summarized below.

1. There are several important fields (taxonomy/biodiversity, biology, ecology, management, etc.) which are integral to applying sound decisions on biodiversity management, and these need to be accessible in a database management system for Oceania.

2. An information management system would at least include data at the taxonomy/biodiversity level.

3. The minimum information needed for an acceptable single record was species, locality (longitude, latitude, depth), date, and scientific source (and evidence of the correctness of the identification).

4. A relational database structure should be applied with major tables including *(a)* a taxonomic dictionary, *(b)* recording the occurrence of specimens in space and time, *(c)*

a gazetteer providing standards for locality or "place" names, and *(d)* a link to ecosystem information.

5. There are potentially millions of existing records that fulfill the minimum standards listed in point 3 above, which continues to be collected and which is continually in danger of being lost.

6. It would be of enormous value for biodiversity studies to have access to these data preferably in one coherent database. CD-ROM technology is the appropriate medium for storing and distributing such a database.

7. The implementation strategy should involve a relatively small project-based core group that would create the database and operate it.

8. Institutions participating in this exercise would form a network with some institutions taking on coordinating tasks for certain taxonomic groups or regions.

RECOMMENDATIONS ON THE FRAMEWORK FOR AN INFORMATION MANAGEMENT SYSTEM FOR MARINE AND COASTAL BIODIVERSITY FOR THE OCEANIA REGION

- A regional database management system should be developed for Oceania but with linkages to *external* database systems to promote interchange of information and avoid unneeded duplication of effort. The system should be linked to the following established or developing databases: *WCMC*—for maps, protected areas, rare species information; *ReefBase*—for maps and reef data and terminology; *FishBase*—for various information on fishes; *CoralBase*—taxonomic dictionary for corals; *Eschmeyer's database*—taxonomic dictionary for fish; and *Global Task Team database*—coral reef monitoring data.

- The core data for the individual records in the database should include the following minimum information: species name; source of data (name of collector, identifier, author/citation); location (latitude, longitude, and depth); date (of observations or collection). In addition, the ecosystem category based on the SPREP Ecosystem Classification would also be added if known.

- A cultural data field should also be added to the basic system in order to link local place names to gazetteers, local species names to the "universal" scientific name, and to provide additional information on reef status, conservation, ecology, exploitation levels, use, importance, and other threats.

- The specific purposes and uses of the Oceania database must be defined and periodically revised based on the needs of system users.

The figures in Appendix B show the details and interrelationships among the components of a proposed database system. Figure 1.1 of this chapter summarizes the principal conceptual components of a proposed Oceania database on nearshore marine and coastal biodiversity and relationships to external databases.

INPUTS

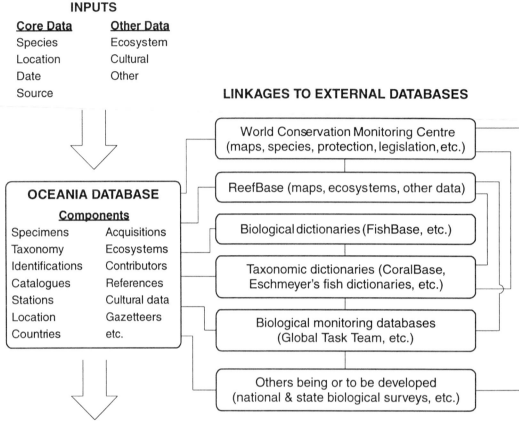

OUTPUTS & QUERIES

Species richness & distribution

Endemic & rare species distribution

Threatened & endangered species distribution

Ecosystem richness & distribution

Conservation accomplishments

Environmental legislation & regulations

Trends over time (size, abundance, etc.)

Reference lists

Local/cultural data (species names, ecosystem
    names, place names, etc.)

Other uses (educational, etc.)

Figure 1.1 Conceptual framework for a regional Oceania database for coastal and nearshore marine biodiversity

SECTION 1
Systematics and Classification Systems

CHAPTER 2

The State of Systematics of Marine Algae in Tropical Island Pacific

Isabella A. Abbott

ABSTRACT

*The systematics of marine algae in the tropical island Pacific are hampered by a lack of adequate collections and of trained persons to study them. Although many expeditions have visited islands along the major sailing routes (Hawai'i, Tahiti, Guam, parts of the Philippines), only a few algae were collected compared to the large number of vascular plants. In modern times, the algae of few tropical Pacific islands have been adequately investigated, but the number is increasing owing to greater interest and opportunistic surveys. Among ten far-flung locations in the Pacific (Northwest Hawaiian Islands, Maui and Hawai'i islands, Samoa, the Marshall Islands, Enewetak, Nha Trang [Vietnam], Tahiti, Moorea, and Easter Island), only two green algae (*Ventricaria ventricosa *and* Halimeda opuntia*), only one brown algal species (*Lobophora variegata*), and only two red species (*Centroceras clavulatum *and* Asparagopsis taxiformis*) occur at all sites (except for* H. opuntia, *which is not present on Maui and Hawai'i islands). As detailed knowledge of various species is accumulated, it may soon be possible to predict the algae most common in the warm Pacific.*

INTRODUCTION

Many systematists find it difficult to believe that taxonomy of marine algae is still in the alpha-taxonomic mode. New species, genera, and families are still being described even while other genera are being monographed and the number of names are being reduced. In Hawai'i, a new genus and fourteen new species among the red algae have been described in the last five years (1989–94). In the same period, forty-five species have been reduced to nine valid species. Descriptions of two new genera and eight new species of red algae have been submitted for publication in the last few months of 1994.

Because of the relatively small number of phycologists in the tropics or who work on tropical algae, most of the taxonomic algal problems are slow to solve, even though most of them have been obvious for some time. The bulk of green, brown, and red algae are poorly known with respect to their morphological and anatomical features, their life history phases, and their responses to ecological parameters. The application of the International Code of Botanical Nomenclature is still new to many who publish on marine algae.

Taxonomy in the algae includes more than a morphological description of the plants in hand. It should now include culture studies since the discovery in the last three decades that some green, brown, and red algae have different ways of getting through a life history and that the different phases are often morphologically dissimilar. For example, an erect, branched plant may alternate with one that is a small ball-like plant, or an erect, much-branched plant may alternate with a crust-like phase. In some cases the alternate phase already has a name, and under that name was placed in an order different from the erect plant. Here, nomenclatural rules require that the oldest name has priority in use. For example, the well-known *limu kohu* in Hawai'i (known for its high price of about US$36/kg wet weight as well as for its pungent odor and taste) is collected as fuzzy-topped plants about 10 centimeters high, whereas its alternate phase consists of nearly microscopic filaments that grow in a different subtidal habitat.

Another aspect, which has proven to be effective in grouping red algae phylogenetically, involves transmission electron microscopy whereby the configuration of cell membranes can show conclusive evidence of relationships. These arrangements are thought to be conservative and reflective of the very ancient ages of these plants.

Our attempts to examine phylogenetic relationships molecularly are only beginning. Most algae have a large number of strange chemicals, some of them having large molecules of polysaccharides that must be removed before extraction of nuclear material can be successful. There are few tried-and-true techniques to follow or to test.

Compared to phycology in the previous fifty years, many more trained scientists are publishing on the systematics of marine algae in the Pacific than previously. But most of them do not study *tropical* marine algae. For example, the scholarly and comprehensive studies on Southern Australian taxa by Womersley—the green algae in 1984, the brown algae in 1987, and the first part of the red algae in 1994—deal mainly with temperate and subantarctic species, having very little overlap with tropical taxa, even though parameters are drawn for many generic limits among the tropical groups. It is only through such critical contributions that those working in the tropics will come to know that names of species that have been applied over the world such as *Ulva lactuca* and *Gracilaria verrucosa* cannot be applied to tropical material without careful study. To my knowledge, these two species are not in the tropical Pacific, although there are many reports of their presence.

Many of the older studies from the Pacific islands are flawed because species names were used without reference to type material or were compared to material that had been previously identified by someone familiar with the group. Worse than this for conscientious phycologists is that voucher specimens were not kept or have been lost. Thus, in any given region of the Pacific through which exploring expeditions have gone for over 200 years, only very few herbaria have representatives of specimens purportedly collected on those expeditions. Anyone working on the marine algae of Hawai'i and Samoa, for example, must carry names of <u>reported</u> species that cannot be verified because the whereabouts of specimens are unknown.

ISLANDS OR ISLAND GROUPS PREVIOUSLY REPORTED

Before 1985, the only places in the warm Pacific that could be reasonably compared with respect to its published algal flora, and because voucher material could be studied to make such comparisons, were Samoa (Setchell 1924), the Marshall Islands (Taylor 1950; Dawson 1956), Enewetak (Dawson 1957), Nha Trang, Vietnam (Dawson 1954), Tahiti (Setchell 1926), Moorea (Payri and Meinesz 1985), and Easter Island (Børgesen 1924; Abbott and Santelices 1985). It should be noted that Hawai'i is not one of these locations, nor are the Philippine Islands, because only special groups have been published and not floras, which could be compared. More information has accumulated, however, since 1985, and includes studies by Tsuda (1987) on Enewetak, and Abbott (1989) on the Northwest Hawaiian Islands. In a flurry of activity prior to publication of a marine flora of the Hawaiian Islands, the following have been recently published: Hodgson and Abbott 1992; R. E. Norris and Abbott 1992; Abbott and Norris 1993; R. E. Norris 1993, 1994a, 1994b; J. N. Norris, Abbott, and Agegian 1995; and Meneses 1995.

MOST COMMON SPECIES

A tabulation of the five most commonly occurring species out of a list of fifteen from seven Pacific locations previously treated by Abbott and Santelices (1985) shows only one green alga (*Ventricaria ventricosa*, formerly *Valonia ventricosa*), only one brown species (*Lobophora variegata*), and only two red species (*Centroceras clavulatum* and *Asparagopsis taxiformis*) present in all seven locations. Reports from Enewetak (Tsuda 1987), Northwest Hawaiian Islands (Abbott 1989), Maui (Hawai'i; Hodgson and Abbott 1992), and Hawai'i Island (Smith 1992) support the previous "most common" categories. *Ventricaria ventricosa* occurs at all nine sites. *Halimeda opuntia* is found at all locations except Maui and Hawai'i islands, although it is also present in the Northwest Hawaiian Islands; additionally, *Hincksia breviarticulata* (formerly *Ectocarpus breviarticulata*) occurs at all sites except Moorea. Table 2.1 shows presence (+) or absence (0) of the fifteen most common species. Because I have examined most of the specimens listed here (except from Moorea), I am confident of the identification of these taxa except for *Lithothamnion samoense*; however, others more familiar with crustose coralline algae identified these and their judgment can be accepted. It may soon be possible to predict the most common algae in the warm Pacific, although they are expected to remain relatively few. The classification followed, and taxa expected to be found in the Pacific islands are given in the Appendix to this paper.

The nine locations treated here are within the heart of the sunny Pacific, and observation suggests that islands in the hot tropics present very few shady places that most algae would prefer as habitats. A recent study (Abbott 1989) of the marine algae of the leeward and oldest Hawaiian Islands, a chain of islands stretching about 1,200 miles northwest of the main islands, showed that in areas where there is very little exposed intertidal space, if there are surfaces furnished by eroded limestone, or macroalgae at depths up to 70

Table 2.1 Commonest species of Pacific tropical marine algae

Division and species	Easter Island	Samoa	Marshalls	Enewetak	Vietnam	Tahiti	Moorea	HI	NWHI
CHLOROPHYTA									
Enteromorpha clathrata	+	+	+	+	+	?	+	+	?
Cladophora socialis	+	0	+	+	0	+	0	+	+
Cladophoropsis sundanensis	+	0	+	+	0	+	0	+	0
Ventricaria ventricosa	+	+	+	+	+	+	+	+	+
Halimeda opuntia	+	+	+	+	+	+	+	0	+
PHAEOPHYTA									
Hincksia breviarticulata	+	+	+	+	+	+	0	+	+
Colpomenia sinuosa	+	0	0	0	+	+	+	+	+
Hydroclathrus clathratus	+	0	0	+	+	+	+	+	+
Lobophora variegatus	+	+	+	+	+	+	+	+	+
Sphacelaria novae-hollandiae	+	0	0	+	+	0	0	+	+
RHODOPHYTA									
Asparagopsis taxiformis or *Falkenbergia*	+	+	+	+	+	+	+	+	+
Amphiroa beauvoisii	+	+	0	0	+	+	+	+	+
Centroceras clavulatum	+	+	+	+	+	+	+	+	+
Taenioma perpusillum	+	0	0	+	+	0	0	+	+
Lithothamnion samoense	+	+	0	0	+	+	0	0	0

After Abbott and Santelices (1985).

+ = presence of species; 0 = absence of species

meters, a large number of small epiphytic algae occur. Similarly, with depths up to 50 meters, larger algae, averaging 15 centimeters long, can be found. In other Pacific islands where there are reef flats fully exposed to sunlight, small numbers of species might be expected, but over the reef crest, shaded by spurs and grooves and depth, larger numbers of species occur (Doty, Gilbert, and Abbott 1976) and make up a group of new species or of new records that are surprising for their long-distance distribution. Nonetheless, for most tropical islands, species of short stature or cryptic habitats can be expected to make up the bulk of a flora.

Table 2.2 compares the total number of genera and species in eight Pacific locations, differing only in substituting the Northwest Hawaiian Islands for HI in Table 2.1, which represented the Maui and Hawaii island collections of Hodgson and Abbott (1992) and

Table 2.2 Biodiversity of Pacific marine algae: Selected Pacific areas compared to Easter Island

	Total genera	Genera in common	Total species	Species in common
Easter Island	85	–	144	–
Samoa (Setchell 1924)	43	31	67	13
Marshalls (Dawson 1956)	75	42	141	30
Enewetak (Tsuda 1987)	114	40	222	24
NWHI (Abbott 1989)	84	38	205	16
Nha Trang (Dawson 1954)	98	53	132	18
Tahiti (Setchell 1926)	68	46	123	33
Moorea (Payri and Meinesz 1985)	69	37	101	16

After Abbott and Santelices (1985).

Smith (1992), respectively. Table 2.2 also shows that Samoa, with 43 genera and 67 species, is woefully undercollected and understudied, while Tahiti with 68 genera and 123 species requires more study. Even tiny Howland and Baker islands in the central Pacific near 0°12 to 48' N, and 176° W, barely visible above the horizon, have been reported to have 19 genera and over 20 species (Tsuda and Trono 1968) in habitat not comparable to the high volcanic islands of Samoa. In other words, high volcanic islands can be expected to have more species than low coral atolls (Smith 1992), and such is, in fact, the case.

ARE THERE ENDEMIC TROPICAL ALGAE?

When new genera and species are described, they are usually isolated and labeled by some as "endemic." Many examples of marine algae could be thought to be endemic because they have never been found beyond the boundaries of the topotype locality. But many more can be cited as being isolated when first described and are now known to have a wide distributional pattern; indeed, so wide as to be a little unbelievable, calling for first-hand examination of the specimens. One such species is *Gibsmithia hawaiiensis*, first described from Hawai'i (Doty 1963), and thirty-two years later known from southern Japan, the Philippines, New Caledonia, Tahiti, New Guinea, and at seven locations in the Great Barrier Reef (Kraft 1986). A corollary is the intensity of collections. No less spectacular is the distribution of *Dudresnaya hawaiiensis* R. K. S. Lee (Kane'ohe Bay, Hawai'i, type locality), now known from many places in the Hawaiian Islands, but also from Lord Howe Island, Norfolk Island, and several places in the Great Barrier Reef (Robins and Kraft 1985) and from South Africa (R. E. Norris 1992). I have described two genera, *Peleophycus* (Abbott 1984) and *Reticulocaulis* (Abbott 1985), which have not yet been found beyond the Hawaiian Islands. The closest relative to *Peleophycus* is *Gloeophycus*, known in Korea and northern Japan. Since both species are "spring annuals," they are found for only about six weeks. These are not expected to occur in the very warm, exposed-to-sunlight habitats that abound in the central Pacific.

On the other hand, five species of *Sargassum* (Phaeophyta) are thought to be endemic to the Hawaiian Islands (Magruder 1988), although one of them, *S. echinocarpum*, has been reported from a wide variety of locations from the Red Sea to Indonesia in older literature. These outlying specimens, however, have never been examined by a student of the genus; the present consensus among these specialists is that these outliers will probably be identified as other species or are already known under other names.

In Hawai'i, famous for geographic isolation that provides numerous microhabitats in which speciation among flowering plants took place at a high rate, the surrounding marine waters seem not to have provided the same set of circumstances that spur speciation and endemism among marine plants. One of the exceptions to this statement is furnished by the six Hawaiian species of *Sargassum* (Phaeophyta), which appear to be endemic. Their distribution within the Hawaiian archipelago is roughly northwest to southeast, involving the southeastern Northwest Hawaiian Islands and the main Hawaiian Islands. Though ocean currents can theoretically limit distribution between the southern Pacific and Hawaii, here we may expect the seasonal shifts of small gyres within the southern Northwest Hawaiian Islands in the region of the French Frigate shoals southeast to Necker and Nihoa to possibly affect the distribution through drift specimens in both directions. It is only a short distance between Nihoa and Kaua'i and Ni'ihau in the main Hawaiian Islands. Molecular evidence is being sought in our Botany Department to probe the relationships of the Hawaiian *Sargassum* species among themselves and among other species in the warm Pacific.

The total marine flora of the Hawaiian Islands at this time cannot be characterized as *tropical*; the large numbers of species of those taxa that are known from closer to the equator are absent or are in very small numbers if present (especially the green algae *Caulerpa* and *Halimeda*). Of the locations from which several comprehensive collections have been made and with competent phycologists studying them, I would select Enewetak, an outlier northwest of the Marshall Islands and about 10° south of the Hawaiian Islands, as typically <u>tropical</u> for the Pacific Basin. Tsuda (1987) tallied 222 species from Enewetak, and Abbott (1989) 205 from the Northwest Hawaiian Islands. The floras differed in a preponderance of green algae from the south, and a larger number of red algae in the northern collections. A study of the several intertidal and subtidal collections of Lynn Hodgson at Ant Atoll near Pohnpei in November 1994 revealed that the green algae *Caulerpa* and *Halimeda* dominated all other algae.

PERIPHERAL PACIFIC, OR THE SOURCES OF SPECIES

Outside the central Pacific, important contributions have been made in recent years by Kraft and his students from Australia's Great Barrier Reef and the shores of western Australia along the Indian Ocean, as well as from northern New South Wales and Lord Howe Island, the latter representing a tropical flora in relatively cool waters. From the Great Barrier Reef region, several taxa, not well known but of great interest because of their extremely wide distributional range, were described or recorded: *Predaea* (Kraft 1984), *Acrosymphyton* (Millar and Kraft 1984), *Dudresnaya* (Robins and Kraft 1985), *Gibsmithia* (Kraft 1986). These four taxa contain species that are also known in Hawai'i. Other contributions are those of Huisman (1986, 1987) and Huisman and Borowitzka (1990) where a family of red algae (Galaxauraceae), which contains genera found in nearly every tropical location, was monographed for Australia. The results, however, affected the integrity of most of the species known and distributed throughout the Indian Ocean, the warm Atlantic, and the Caribbean, as well as the warm Pacific.

From Lord Howe Island, brown algae, including such well-known tropical genera as *Padina, Stypopodium, Lobophora, Dictyota*, and *Dictyopteris*, were reported (Allender and Kraft 1983), and the common genus of green algae, *Codium*, was also described (Jones and Kraft 1984) with five species, one of which was new. A catalog that lists marine algal species on Lord Howe Island, together with those from New South Wales, shows a rich flora containing only a small number of "typical" tropical taxa (Millar and Kraft 1993).

Similarly, only a small number of tropical species is recorded (Huisman and Walker 1990) from Rottnest Island, Western Australia, among the more numerous cold-water species. Included is a species of *Dotyophycus* (Kraft 1988), a genus first described from Hawai'i (Abbott 1976).

Taiwan, an area with many tropical species, but with a temperate intrusion along the northwestern coast, has been reported by Chou and Chiang (1981), Yang and Chiang (1982), Chiang and Chen (1982), and Chiang and Chen (1983). Lewis and Norris (1987) gave an historical account and annotated list of the benthic marine algae.

The Philippine Islands have perhaps the largest number of habitats of any island group in the Pacific, from basalt to coral and principally coralline algal ridges, with both an island and continental flora to be expected. Relatively small areas have been exploited for algae. I speculate that for the tropics the largest species lists and many taxa unknown in other tropical areas will be found there. In the catalog of Philippine algae (Silva, Meñez, and Moe 1987), 911 specific and infraspecific names are recorded, of which I suggest that closer to 800 might be correct, since the records were made on unevaluated opinions, and critical studies remain to be done. This remark should not in any way detract from this very useful compilation.

A boon to taxonomic studies in the tropical Pacific has been five workshops, sponsored by the California Sea Grant College Program with the cooperation of Sea Grant programs from Washington, Oregon, and Hawai'i. Specialists in various seaweed genera of economic importance have met in Guam, Qingdao (China), Sapporo (Japan), La Jolla (California), and Honolulu (Hawai'i) and exchanged specimens and literature. Four volumes (Abbott 1988, 1992, 1994; Abbott and J. N. Norris 1985), which include the results of these workshops, have been published, with one in press. Agar-producing genera such as *Gelidium, Gelidiella, Pterocladia,* and *Gracilaria* have been studied critically from a variety of aspects—so have carrageenan-producing genera such as *Eucheuma* (and its recently recognized segregates *Kappaphycus* and *Betaphycus*), which are included in these publications. Perhaps the most far-reaching contributions will be on the systematics of the numerous species of *Sargassum*, which has attracted the largest number of phycologists. Initially, many persons were reluctant to tackle this formidable genus (more than 400 described species) on their own. However, as a group, great headway was made and is continuing to be made on what appears to be a stable classification. This could be the single, most useful contribution in the study of tropical marine algae in the 1990s. Although the findings probably will not add to life history or reproductive knowledge, the numerous species will be properly classified after nearly two centuries of chaos.

Because of the geographic isolation of the Hawaiian Islands and, more importantly, their marine isolation due to major oceanic currents, the marine flora is proving to have a significantly different composition from the average or "typical" tropical Pacific island group. Earlier in this paper, Enewetak was proposed as having a "typical" tropical marine flora. The main reason for this choice is that several competent phycologists have collected algae there, and each has contributed to the comprehensive report by Tsuda (1987). Tsuda stated, "There are more species of marine benthic algae known from this atoll than are known from any other Indo-Pacific atoll." This statement still stands. The species lists in Tsuda's paper were compiled from data in Taylor (1950), Dawson (1957), Gilmartin (1960), and several minor papers. A secondary reason is that in species composition, more green algae were reported (89 species versus 48 from the Northwest Hawaiian Islands), compared to 109 red algal species from Enewetak versus 124 species from the

Northwest Hawaiian Islands. In contrast, earlier studies by Trono (1968, 1969) from the Caroline Islands reported 90 species of green algae versus 50 of the red.

In the many papers by van den Hoek (e.g., 1975, 1984) the phytogeography of marine algae has been examined and segregated for most of the world latitudinally, longitudinally, and by temperature ranges, but the warm waters of the central and south Pacific have been omitted. Although affinities of many Hawaiian marine animals, such as fishes and marine mollusks have been thought to be "Indo-Pacific," it is not yet possible to make this assertion for the marine algae. In terms of ocean currents, apparently one of the sources of propagules or possible long-distance dispersal is from Baja California, which is the path of the California Current as it turns westward. However, a comparison of species from that area of the eastern Pacific with that of the central Pacific shows almost no species in common. Despite formidable obstacles posed by the Equatorial and its Counter Current, a northward distribution across these Currents from the South Pacific, and perhaps south-eastern Pacific, or at some specific geologic time, would appear to be potentially possible. The Southern Oscillation events following the El Niño years may have affected the distribution of marine algae in a broad band, but the pertinent island flora has not been analyzed, and it appears that the Hawaiian Islands lie outside of that broad region.

THE STATUS OF KNOWLEDGE OF THE ALGAE AND WHAT REMAINS TO BE DONE

A survey of what is known about marine algae from tropical island Pacific from published and unpublished sources shows that the main Hawaiian Islands, the Marquesas, Australs, Gambier, Tuamotu, the Society Islands, and Samoa, as well as most of Melanesia, Indonesia, and Micronesia, need attention before we can say that we understand the marine algae of the Pacific islands. At this point, for example, there are more adequate data published for the Northwest Hawaiian Islands than for the main Hawaiian Islands, and that the marine algae of Guam, although incompletely studied, are the best known for the general area of Micronesia. A marine flora for the Hawaiian Islands is expected to be published in the near future. It will show a flora that would be accurately described as subtropical rather than tropical, and therefore will show a greater richness of species than the tropical islands to the south. Some genera not usually thought to be tropical such as *Peleophycus* (closest relative from Korea), *Sporochnus* (closest relative in Baja California), and *Cryptopleura* (closest relative in California) are found in Hawai'i. More than 400 species will be reported in the Hawaiian flora.

Databases are available for Hawaiian marine algae at the B. P. Bishop Museum, Honolulu, and the Department of Botany's laboratory, University of Hawaii. Besides the nomenclatural history, geographic distribution is recorded for the Hawaiian Islands, as well as the Pacific Basin in detail, and more generally in the tropics. An extended tropical algal base will be added as time permits.

CONCLUSIONS

This review of what is known about the systematics of marine algae in the tropical Pacific shows that the collection of algae has been poorly conducted in many areas. For example, among some of the older but still available reports, the low numbers of species and genera given indicate fragmented collections. Nonetheless, extensive archipelagos such as the Tuamotus in the Pacific have no marine floras reported from them. Many areas (like New Caledonia), which historically have important collections, should be re-examined for their rich species diversity. When the collections from the Dutch "Snellius" expedition have been studied, and other collections that have been made in the warm Pacific examined, we may have a better picture of not only the diversity of marine algae, but their biogeography and affinities.

APPENDIX

Classification: Hierarchies

In the classification of the marine greens, browns, and reds that follows, only verifiable generic names (tied to specimens in the author's herbarium, or the herbarium of the Bishop Museum, Honolulu) as occurring in the Pacific island tropics are used. Because of the subtropical nature of some of the taxa known from Hawai'i, these taxa are noted with asterisks; they are not expected to occur in most of the Pacific islands.

Note: The category <u>Division</u> is equivalent to <u>Phylum</u>.

CLASSIFICATION OF GREEN ALGAE

Division Chlorophyta
Class Chlorophyceae (only marine macrophytes are treated here)
Orders and families with genera that may be expected to be found widely distributed in
 the warm Pacific and adjacent seas

Order Ulvales
Family Percursariaceae: *Percursaria**
Family Monostomataceae: *Monostroma* (or *Ulvaria*)
Family Ulvaceae: *Enteromorpha, Ulva*

Order Cladophorales
Family Anadyomenaceae: *Anadyomene, Microdictyon*
Family Boodleaceae: *Boodlea, Struvea*
Family Valoniaceae: *Valonia, Ventricaria*
Family Siphonocladaceae: *Dictyosphaeria, Siphonocladus*
Family Cladophoraceae: *Chaetomorpha, Cladophora, Rhizoclonium, Rhipidophyllon*

Order Bryopsidales
Family Bryopsidaceae: *Bryopsis, Trichosolen*
Family Codiaceae: *Codium*
Family Derbesiaceae: *Derbesia (Halicystis)*
Family Caulerpaceae: *Caulerpa, Caulerpella*
Family Halimedaceae: *Halimeda*
Family Udoteaceae: *Avrainvillea, Boodleopsis, Chlorodesmis, Pseudochlorodesmis,*
 Rhipidosiphon

Order Dasycladales
Family Dasycladaceae: *Bornetella, Halicoryne, Neomeris*
Family Acetabulariaceae: *Acetabularia*

CLASSIFICATION OF BROWN ALGAE

Division Phaeophyta
Class Phaeophyceae
Orders and families with genera that may be expected to be found widely distributed in
the warm Pacific and adjacent seas

Order Ectocarpales
Family Ectocarpaceae: *Ectocarpus, Feldmannia, Hincksia*
Family Ralfsiaceae: *Mesospora*

Order Chordariales
Family Chordariaceae: *Nemacystus**

Order Sporochnales
Family Sporochnaceae: *Sporochnus**

Order Scytosiphonales
Family Scytosiphonaceae: *Chnoospora, Colpomenia, Endarachne, Hydroclathrus, Rosenvingea*

Order Sphacelariales
Family Sphacelariaceae: *Sphacelaria, Halopteris,* * Discosporangium**

Order Dictyotales
Family Dictyotaceae: *Dictyopteris, Dictyota, Lobophora, Padina, Spatoglossum, Stypopodium*

Order Fucales
Family Sargassaceae: *Sargassum, Turbinaria*

CLASSIFICATION OF MARINE RED ALGAE

Division Rhodophyta
Classes, orders, and families with genera that may be expected to be widely distributed in
the warm Pacific and adjacent seas
Class Rhodophyceae
Subclass Florideophycidae

Order Nemaliales
Family Acrochaetiaceae: *Acrochaetium, Liagorophila*
Family Nemaliaceae: *Nemalion, Trichogloea, Trichogloeopsis*
Family Liagoraceae: *Helminthocladia, Helminthora,* * Liagora*
Family Dermonemataceae: *Dermonema, Dotyophycus, Liagoropsis, Yamadaella*
Family Galaxauraceae: *Actinotrichia, Galaxaura, Scinaia,* * Pseudogloiophloea,* * Tricleocarpa*

Order Gelidiales
Family Gelidiaceae: *Gelidium, Pterocladia*
Family Gelidiellaceae: *Gelidiella*

Order Corallinales
Family Corallinaceae: *Amphiroa, Corallina, Haliptilon, Jania, Lithophyllum, Meso-phyllum, Porolithon*

Order Cryptonemiales
Family Dumontiaceae: *Acrosymphyton, Dudresnaya, Gibsmithia*
Family Rhizophyllidaceae: *Portieria*
Family Peyssonneliaceae: *Peyssonnelia*
Family Halymeniaceae: *Carpopeltis,* * *Cryptonemia, Grateloupia, Halymenia, Prionitis* *
Family Kallymeniaceae: *Kallymenia* *

Order Gigartinales
Family Polyidaceae: *Stenopeltis*
Family Nemastomataceae: *Predaea, Titanophora*
Family Plocamiaceae: *Plocamium*
Family Solieriaceae: *Betaphycus, Eucheuma, Kappaphycus, Meristotheca, Solieria*
Family Caulacanthaceae: *Caulacanthus, Catenella*
Family Hypneaceae: *Hypnea, Hypneocolax*

Order Gracilariales
Family Gracilariaceae: *Gracilaria, Gracilariopsis*

Order Ahnfeltiales
Family Ahnfeltiaceae: *Ahnfeltiopsis*
Family Gigartinaceae: *Chondrocanthus*

Order Rhodymeniales
Family Rhodymeniaceae: *Botryocladia, Chrysymenia, Coelothrix, Erythrocolon, Gelidi-opsis, Gloiocladia, Halichrysis, Rhodymenia* *
Family Champiaceae: *Champia*
Family Lomentariaceae: *Lomentaria*

Order Ceramiales
Family Ceramiaceae: *Acrothamnion, Aglaothamnion, Anotrichium, Antithamnion, Antithamnionella, Balliella, Centroceras, Ceramium, Corallophila, Crouania, Diplo-thamnion, Griffithsia, Gymnothamnion, Haloplegma, Mesothamnion, Monosporus, Pleonosporium, Ptilocladia, Ptilothamnion, Spyridia, Tiffaniella.*
Family Delesseriaceae: *Branchioglossum,* * *Caloglossa, Hypoglossum, Martensia, Myrio-gramme, Taenioma, Vanvoorstia*
Family Dasyaceae: *Dasya, Dictyurus, Heterosiphonia*
Family Rhodomelaceae: *Acanthophora, Bostrychia, Chondria, Dipterosiphonia, Exo-phyllum, Herposiphonia, Laurencia, Leveillea, Lophocladia,* * *Lophosiphonia, Osmun-daria, Polysiphonia, Rhodolachne, Spirocladia, Symphiocladia, Tayloriella, Toly-piocladia, Womersleyella*

Subclass Bangiophycidae:

Order Bangiales

Family Bangiaceae: *Bangia*,* *Porphyra*
Family Eyrthropeltidaceae: *Erythrotrichia*

Order Porphyridiales

Family Goniotrichaceae: *Chroodactylon, Goniotrichum*

REFERENCES

Abbott, I. A. 1976. *Dotyophycus pacificum* gen. et sp. nov., with a discussion of some families of Nemaliales (Rhodophyta). *Phycologia* 15:125–32.

———. 1984. *Peleophycus multiprocarpium* gen. et sp. nov. (Gloiosiphoniaceae, Rhodophyta). *Pac. Sci.* 38:324–32.

———. 1985. Vegetative and reproductive morphology in *Reticulocaulis* gen. nov. and *Naccaria hawaiiana* sp. nov. (Rhodophyta, Naccariaceae). *J. Phycol.* 21:554–61.

———, ed. 1988. *Taxonomy of economic seaweeds with reference to some Pacific and Caribbean species.* Vol. 2. California Sea Grant College Program, Report No. T-CSGCP-018, University of California, La Jolla, California.

———. 1989. Marine algae of the Northwest Hawaiian Islands. *Pac. Sci.* 43:223–33.

———, ed. 1992. *Taxonomy of economic seaweeds with reference to some Pacific and Western Atlantic species.* Vol. 3. California Sea Grant College Program, Report No. T-CSGCP-023, University of California, La Jolla, California.

———, ed. 1994. *Taxonomy of economic seaweeds with reference to some Pacific species.* Vol. 4. California Sea Grant College Program, Report No. T-CSGCP-031, University of California, La Jolla, California.

Abbott, I. A., and J. N. Norris, eds. 1985. *Taxonomy of economic seaweeds with reference to some Pacific and Caribbean species.* California Sea Grant College Program, Report No. T-CSGCP-011, University of California, La Jolla, California.

Abbott, I. A., and R. E. Norris. 1993. New species of Ceramiaceae (Rhodophyta) from the Hawaiian Islands. *Phycologia* 32:451–61.

Abbott, I. A., and B. Santelices. 1985. The marine algae of Easter Island (Eastern Polynesia). Vol. 5, *Proc. Fifth Intern. Coral Reef Congress, Tahiti,* ed. B. Delesalle, R. Galzin, and B. Salvat, 71–75. Antenne Museum EPHE, Moorea.

Allender, B. M., and G. T. Kraft. 1983. The marine algae of Lord Howe Island (New South Wales): The Dictyotales and Culteriales (Phaeophyta). *Brunonia* 6:73–130.

Børgesen, F. 1924. Marine algae of Easter Island. In *Natural history of Juan Fernandez and Easter Island,* ed. C. Skottsberg 2:247–309. Göteborg, Sweden.

Chiang, Y. M., and C. Chen. 1982. The genus *Liagora* of Taiwan. *Acta Oceanogr. Taiwanica,* no. 13:181–96 (in Chinese with English summary).

———. 1983. Studies on *Dotyophycus yamadae* (Ohmi et Itono) Abbott et Yoshizaki (Nemalionales, Rhodophycophyta) from southern Taiwan. *Jpn J. Phycol.* (Sorui) 31:10–15.

Chou, H. N., and Y. M. Chiang. 1981. The *Sargassum* of Taiwan. *Acta Oceanogr. Taiwanica,* no. 12:132–49.

Dawson, E. Y. 1954. Marine plants in the vicinity of Nha Trang, Viet Nam. *Pac. Sci.* 8:373–470.

———. 1956. Some marine algae of the southern Marshall Islands. *Pac. Sci.* 10:25–65.

———. 1957. An annotated list of marine algae from Eniwetok Atoll, Marshall islands. *Pac. Sci.* 11:92–132.

Doty, M. S. 1963. *Gibsmithia hawaiiensis* gen.n. et sp.n. *Pac. Sci.* 17:458–65.

Doty, M. S., W. J. Gilbert, and I. A. Abbott. 1976. Hawaiian marine algae from seaward of the algal ridge. *Phycologia* 139:345–57.

Gilmartin, M. 1960. Ecological distribution of the deep water algae of Eniwetok Atoll. *Ecology* 41:210–21.

Hodgson, L. M., and I. A. Abbott. 1992. Near-shore benthic marine algae of Cape Kina'u, Maui. *Bot. Mar.* 35:535–40.

Hoek, C. van den. 1975. Phytogeographic provinces along the coasts of the northwestern Atlantic Ocean. *Phycologia* 14:317–30.

———. 1984. World-wide latitudinal and longitudinal seaweed distribution patterns and their possible causes as illustrated by the distribution of Rhodophytan genera. *Helgol. Meeresunters.* 41:261–72.

Huisman, J. M. 1986. The *Scinaia* assemblage (Galaxauraceae, Rhodophyta): A re-appraisal. *Phycologia* 25:403–18.

———. 1987. The red algal genus *Scinaia* (Galaxauraceae, Nemaliales) from Australia. *Phycologia* 25:271–96.

Huisman, J. M., and M. A. Borowitzka. 1990. A revision of the Australian species of *Galaxaura* (Rhodophyta, Galaxauraceae), with a description of *Tricleocarpa* gen. nov. *Phycologia* 29:150–72.

Huisman, J. M., and D. I. Walker. 1990. A catalogue of the marine plants of Rottnest island, Western Australia, with notes on their distribution and biogeography. *Kingia* 1:349–459.

Jones, R., and G. T. Kraft. 1984. The genus *Codium* (Codiales, Chlorophyta) at Lord Howe Island (N.S.W.). *Brunonia* 7:253–76.

Kraft, G. T. 1984. The red algal genus *Predaea* (Nemastomataceae, Gigartinales) in Australia. *Phycologia* 23:3–20.

———. 1986. The genus *Gibsmithia* (Dumontiaceae, Rhodophyta) in Australia. *Phycologia* 25:423–47.

———. 1988. *Dotyophycus abbottiae* (Nemaliales), a new red algal species from Western Australia. *Phycologia* 27:131–41.

Lewis, J. E., and J. N. Norris. 1987. A history and annotated account of the benthic marine algae of Taiwan. *Smithsonian Contributions to the Marine Sciences*, no. 29, Washington, D.C.

Magruder, W. H. 1988. *Sargassum* (Phaeophyta, Fucales, Sargassaceae) in the Hawaiian Islands. Vol. 2, *Taxonomy of economic seaweeds with reference to some Pacific and Caribbean species*, ed. I. A. Abbott, 65–87. California Sea Grant College Program, La Jolla, California.

Meneses, I. 1995. Notes on *Ceramium* (Rhodophyta: Ceramiales) from the Hawaiian Islands. *Pac. Sci.* 49:165–74.

Millar, A. J. K., and G. T. Kraft. 1984. The red algal genus *Acrosymphyton* (Dumontiaceae, Cryptonemiales) in Australia. *Phycologia* 23:135–45.

———. 1993. Catalogue of marine and freshwater red algae (Rhodophyta) of New South Wales, including Lord Howe Island, south-western Pacific. *Aust. Syst. Bot.* 6:1–90.

Norris, J. N., I. A. Abbott, and C. R. Agegian. 1995. *Callidictyon abyssorum* gen. et sp. nov., a new deep-water net-forming red alga from Hawai'i. *Pac. Sci.* 49:192–201.

Norris, R. E. 1992. Six marine macroalgal genera new to South Africa. *S. Afr. J. Bot.* 58:2–12.

———. 1993. Taxonomic studies on Ceramieae (Ceramiales, Rhodophyta) with predominantly basipetal growth of corticating filaments. *Bot. Mar.* 36:389–98.

———. 1994a. Notes on some Hawaiian Ceramiaceae (Rhodophyceae), including two new species. *Jpn J. Phycol.* 42:149–55.

———. 1994b. Some cumophytic Rhodomelaceae (Rhodophyta) occurring in Hawaiian surf. *Phycologia* 33:434–43.

Norris, R. E., and I. A. Abbott. 1992. New taxa of Ceramieae (Rhodophyta) from Hawaii. *Pac. Sci.* 46:453–65.

Payri, C. E., and A. Meinesz. 1985. Algae. Vol. 1, French Polynesian coral reefs, *Proc. 5th Intern. Coral Reef Congress Tahiti*, ed. B. Delesalle, R. Galzin, and B. Salvat, 498–518. Antenne Museum EPHE, Moorea.

Robins, P. A., and G. T. Kraft. 1985. Morphology of the type and Australian species of *Dudresnaya* (Dumontiaceae, Rhodophyta). *Phycologia* 24: 1–34.

Setchell, W. A. 1924. American Samoa: Part 1. Vegetation of Tutuila Island; Part 3, Vegetation of Rose Atoll. *Carnegie Inst. Washington Pap.* 20: 1–88, 225–72.

———. 1926. Tahitian algae and Tahitian spermatophytes collected by W. A. Setchell, C. B. Setchell, and H. E. Parks. *Univ. Calif. Publ. Bot.* 12:61–240.

Silva, P. C., E. L. Meñez, and R. L. Moe. 1987. Catalog of the benthic marine algae of the Philippines. *Smithsonian Contributions to the Marine Sciences*, no. 27, Washington, D.C.

Smith, C. M. 1992. Diversity in intertidal habitats: An assessment of the marine algae of select high islands in the Hawaiian archipelago. *Pac. Sci.* 46:466–79.

Taylor, W. R. 1950. *Plants of Bikini, and other northern Marshall Islands*. Ann Arbor: Univ. Mich. Press.

Trono, G. C., Jr. 1968. The marine benthic algae of the Caroline Islands, I. *Micronesica* 4:137–207.

———. 1969. The marine benthic algae of the Caroline Islands, II. Phaeophyta and Rhodophyta. *Micronesica* 5:25–119.

Tsuda, R. T. 1987. Marine benthic algae of Enewetak Atoll. Vol. 2, Biogeography and systematics, *Natural History of Eniwetak Atoll*, 1–9. Washington, D.C.: U.S. Dept. of Energy.

Tsuda, R. T., and G. Trono, Jr. 1968. Marine benthic algae from Howland Island and Baker Island, Central Pacific. *Pac. Sci.* 22:194–96.

Womersley, H. B. S. 1984. *The marine benthic flora of Southern Australia. Chlorophyta*. Part 1. South Australia: D. J. Woolman, Government Printer.

———. 1987. *The marine benthic flora of Southern Australia. Phaeophyta*. Part 2. Adelaide: South Australian Government Printing Division.

———. 1994. *The marine benthic flora of Southern Australia. Rhodophyta*. Part 3a. Australian Biological Resources Study, Canberra.

Yang, H. N., and Y. M. Chiang. 1982. Taxonomical study on the *Gracilaria* of Taiwan. *J. Fisheries Soc. Taiwan* 9:55–71 (in Chinese, with English summary).

CHAPTER 3

Seagrasses

Robert Coles and John Kuo

ABSTRACT

Seagrasses are flowering plants that can live immersed in seawater. They are most common in tropical waters in depths of less than 10 meters below mean sea level but may be found down to at least 50 meters below mean sea level. Fifty-eight species of seagrass are recorded for the world, in twelve genera, four families, and two orders. Fourteen species, including eight genera and two families, are represented in the island Pacific region. These species, and their distribution based on records of collections, are described with information on morphology, anatomy, flowering, and fruiting. A key for sterile material is included. The need is emphasized for further studies of seagrasses in the island Pacific region to estimate the total areas of seagrasses, their distribution, and taxonomy, fisheries values, and the extent of areas under threat.

INTRODUCTION

The importance of seagrass meadows as structural components of coastal ecosystems has been recognized during the past twenty years, resulting in more research interest on the biology and ecology of seagrasses. These marine angiosperms are important for stabilizing coastal sediments; for providing food and shelter for diverse organisms; as a nursery ground for many shrimp and fish of commercial importance; and for nutrient trapping and recycling (Larkum, McComb, and Shepherd 1989; Edgar and Kirkman 1989).

Seagrasses are unique among flowering plants in that they can live entirely immersed in seawater. Several species are found at depths of down to 50 meters (den Hartog 1977), but tropical species are most common in less than 10 meters below mean sea level (MSL) (Lee Long, Mellors, and Coles 1993). Adapting to a marine environment imposes major constraints on morphology and structure. The restriction to seawater may have also influenced their geographic distribution and speciation.

A major catalyst for the recent surge in seagrass research was the publication in 1970 of *Sea-grasses of the World* by C. den Hartog. Forty-nine species from throughout the world were described, providing a solid base for taxonomic work. This publication remains the major source of information on seagrass species distribution in the Island Tropical Pacific region. Since 1970 additional species described have brought the total to fifty-eight species in twelve genera, four families, and two orders (Kuo and McComb 1989). It is almost certain that new species remain to be described from the Pacific Ocean countries as collections from this region are relatively few.

Destruction or loss of seagrasses has been reported from most parts of the world, often from natural causes such as "wasting disease" (den Hartog 1987) or high energy storms (Patriquin 1975; Poiner, Walker, and Coles 1989). More commonly, destruction has resulted from human activities, for example, as a consequence of eutrophication (Bulthuis 1983; Orth and Moore 1983; Cambridge and McComb 1984) or land reclamation and land-use changes (Kemp et al. 1983). Anthropogenic impacts on seagrass meadows continue to destroy or degrade coastal ecosystems and decrease their yield of natural resources (Walker 1989).

It is important to document seagrass species diversity and distribution and to identify areas requiring conservation measures before significant areas and species are lost. Determining the extent of seagrass areas and the ecosystem values of seagrasses in the island Pacific region is the initial information required for coastal zone managers. This information is essential for planning and making development decisions that will minimize future impacts on coastal seagrass habitat.

This paper summarizes the existing published knowledge of seagrass species and distribution in the island Pacific region. It is extended with our views and cautions as to the reliability of the taxonomic information and highlights the requirements for data management and collection.

METHODS

We have based most of this paper on existing scientific literature available through electronic searches and our personal libraries. Where possible, this included papers from the Tropical Island Pacific region as defined for this workshop, although this information is sparse for most topics. Information gaps were filled by consulting literature from surrounding regions and from our own research and experience in the tropical Pacific. The confirmation of occurrence of specific seagrass species in the island Pacific region where possible was based on John Kuo's examination of seagrass material at various herbaria. These included the British Natural History Museum, London; Bishop Museum, Department of Botany, Honolulu; Botanisk Museum and Herbarium, Copenhagen, Denmark; Guam University, Department of Botany; Herbarium of the Royal Botanical Gardens, Kew; Kyoto University, Botany Department Herbarium; Rijkksherbarium, Leiden, The Netherlands; Makino Herbarium, Tokyo Metropolitan University; Melbourne University, Botany Department; National Science Herbarium, Tokyo; New York Botanical Gardens; Paris Natural History Museum; National Herbarium of New South Wales; University of California, Berkeley Herbarium; Smithsonian Institution, Washington, D.C.; and the University of Tokyo Herbarium, Hongo. In compiling records of seagrass species for the region, we have been conservative and have only used validated records from herbarium collections or from our collection and information.

The key for identifying sterile material in the Appendix to this paper is modified for the Tropical Island Pacific region and is based on previously published keys for tropical seagrass species. These include the work of Miki (1934), den Hartog (1970), Lanyon (1986), and Kuo and McComb (1989). Because of this approach most records from non-refereed or unpublished sources have been omitted. Future publication and verification will almost certainly expand the number of species recorded and their distribution in the Tropical Island Pacific region.

RESULTS

Systematics and Taxonomy

Diversification and Speciation   Fourteen species of seagrass are identified from the Tropical Island Pacific region (see Table 3.1), representing approximately 20 percent of the world's identified species. These seagrasses are a diverse group, including eight genera in three families. They range from the large strap-like species standing 1-meter tall or more, such as the *Enhalus acoroides*, to the small pioneering leaf-like species that may be less than 1-centimeter tall, such as the *Halophila minor*. This diversity would be expected if we accept that seagrasses as a group have developed from several separate invasions of the sea from freshwater relatives (Larkum and den Hartog 1989). Seagrasses probably experienced an early rapid phase of speciation leading to high diversity, but since the Eocene have remained largely unchanged. Larkum and den Hartog (1989) have suggested that the evolution of full hydrophily would have freed seagrasses from the nearshore environment. This may have resulted in a subsequent slower rate of speciation in the relatively stable conditions of coastal waters. Speciation may also have been slowed if, as seems likely, living immersed in seawater reduces the effectiveness and frequency of sexual reproduction. The end of the Cretaceous Period saw the greatest mass extinction of plant and animal species documented, and it is also possible that many proto-seagrasses became extinct at this time.

In modern seagrasses, high rates of variation occur in *Halophila* with at least eleven species and in *Zostera* also with eleven recognized species worldwide. Other genera have fewer species, although this may result from the lack of taxonomic work. Species and races may yet be identified, particularly from little-studied areas such as the island Pacific region.

Several existing keys (in addition to the key in the Appendix) provide for identification. An easy-to-use key with species descriptions that include most south Pacific species is in Lanyon (1986).

Classification of Tropical Island Pacific Seagrasses   Identification of seagrass species is the first step in the process of preservation of biodiversity and in the conservation of significant habitats for this group. Seagrass identification is historically dependent on structural studies and comparative morphology and anatomy. Past work has concentrated

Table 3.1 Seagrasses recorded from the Pacific Islands

	Ea	Th	Hd	Hh	Hm	Ho	Hu	Hp	Cs	Cr	Si	Tc	Zc	Zj
RYUKYU IS														
Osumi														
Tanagashima					x	x								x
Amami														
Amamioshima					x	x	x	x	x	x	x			x
Kikai									x					
Yoron		x			x	x		x			x			x
Okinawa														
Kumo	x	x			x	x		x		x	x			
Okinawa		x	x		x	x	x	x	x	x	x			x
Miyako														
Miyako		x			x	x	x	x	x	x	x			
Yaeyama														
Ishigaki	x	x			x	x	x	x	x	x	x			x
Yonaguni		x			x	x		x		x	x			
Tatokomi		x			x	x			x					
Iriomote	x	x			x	x	x	x	x	x	x			x
MICRONESIA														
Marianas														
Guam	x				x	x	x							
Saipan	x				x	x		x						
Caroline Is														
Palau	x	x			x	x	x	x	x	x	x	x		
Yap	x	x				x			x	x	x			
Truk	x	x				x					x	x		
Woleai		x												
Ifaluk	x	x							x					
Pohnpei	x	x							x	x				
Kussio		x												
Esulipik		x												
Nomwin		x												
Satawan		x												
Mokil		x												
Marshall Is														
Ujolang		x												
Jaluit		x												
Kiribati														
Onotos		x												
POLYNESIA														
Hawaiian Is														
Oahu				x										
Kauai				x										
Molokai				x										
Samoa						x					x			
Tonga						x					x			
Society Is			x											
MELANESIA														
Papua New Guinea	x	x			x	x	x	x	x	x	x	x		
Vanuatu	x	x	x		x	x	x	x	x	x	x	x		

Table 3.1. *(continued)*

	Ea	Th	Hd	Hh	Hm	Ho	Hu	Hp	Cs	Cr	Si	Tc	Zc	Zj
New Caledonia	x	x	x		x	x	x	x	x	x	x			
Fiji						x	x	x			x			
OTHERS														
Lord Howe						x							x	
Norfolk						x								

Sources: den Hartog (1970); Tsuda and Kamura (1990); Mukai (1993).

Note: This list may be incomplete due to lack of comprehensive surveys of seagrasses in the island Pacific region. Records reflect species that have been cited in published refereed sources or have been seen by one of the authors.

Abbreviations: Ea, *Enhalus acoroides;* Th, *Thalassia hemprichii;* Hd, *Halophila decipiens;* Hh, *Halophila hawaiiana;* Hm, *Halophila minor;* Ho, *Halophila ovalis;* Hu, *Halodule uninervis;* Hp, *Halodule pinifolia;* Cs, *Cymodocea serrulata;* Cr, *Cymodocea rotundata;* Si, *Syringodium isoetifolium;* Tc, *Thalassodendron ciliatum;* Zc, *Zostera capricorni;* Zj, *Zostera japonica.*

on temperate species resulting in a dearth of information for tropical species that include most of those from the Tropical Island Pacific region. We have based our following classification and identification for the species found in the region on the work of Kuo and McComb (1989).

DIVISION MAGNOLIOPHYTA Cronquist, Takhtajan & Zimmermann "Angiosperms"

CLASS LILIOPSIDA Cronquist, Takhtajan & Zimmermann "Monocotyledons"

ORDER HYDROCHARITALES Lindley

Perennial or annual herbs. Emergent, floating, or submerged aquatics in fresh to marine water. Leaves eligulate. Flowers epigynous, inflorescence cymose with bracts often fused to form a spathe. Pollination anemophilous or hydrophilous. Carpels 2 or more with laminar or parietal placentation. The order consists of a single family.

FAMILY HYDROCHARITACEAE Jussieu

Perennial or annual herbs, either freshwater or marine; partially or totally submerged; dioecious, monoecious, or hermaphrodite. Leaves spirally or distichously arranged. Squamules present in each leaf axil. Stomata present or absent. Inflorescence pedunculate or sessile, cymose. Flowers solitary to several and enclosed by a spathe. Flowers are actinomorphic; male flowers with anthers sessile or on slender filaments. Pollen grains globose to ellipsoid; usually in chains. Female flowers sessile or on long pedicles with a long hypanthium. Ovary inferior, compound, unilocular; ovules numerous. Fruits fleshy, usually indehiscent; seeds several to numerous. About 100 species in fifteen genera with three of these genera considered seagrasses. All three occur in the Tropical Island Pacific region.

Genus *Enhalus* L. C. Rich.

Perennial, marine, dioecious. Rhizome with 2 unbranching cork-like roots and 1 shoot with 3–4 leaves at each rhizome node. Older rhizome covered with numerous long, persistent, black, stiff fibers. Strap leaves eliguate; leaves arranged distichously; longitudinal veins numerous and stomata absent. Inflorescence with bracts. Male flower pedicellate; sepals 3, petals 3, stamens 3, anthers subsessile, bilocular, the open anthers float on the water surface; pollen grains large. Female flowers with a long peduncle, sepals 3, styles 6; each forked from the base. Ovary rostrate, carpels 6, ovules several. Fruits ovoid, acuminate, with several to numerous angular seeds. Flowering often occurs in monthly cycles apparently in response to tidal fluctuation. This is a monotypic genus, widely distributed in the tropical Indian and west Pacific oceans including Okinawa Islands, Micronesia, Papua New Guinea, and New Caledonia.

Enhalus acoroides (L. *f*) Royle

> *Stratiotes acoroides* L.f. 1781
> *Enhalus koenigi* Rich. 1812
> *Vallisneria sphaerocarpa* Blanco 1837
> *Enhalus marinus* Griff. 1851

Genus *Halophila* Du Petit Thouars

Perennial or annual, marine or estuarine, monoecious or dioecious. Rhizome with 1 unbranching root, 1 lateral shoot, and with 2 scales at each node. Leaves in pairs, in pseudo-whorls or distichously arranged, sessile or petiolate. Blade surfaces naked or hairy, margins entire or serrate, cross veins present, stomata absent. Inflorescence with bracts. Male flower pedicellate; tapals 3, stamens 3; pollen grains ellipsoid in long chains. Female flowers sessile, hypanthium with 3 reduced tapals, styles 3–6. Ovary inferior, ovules numerous, fruits ellipsoid to globose, unilocular, with several to numerous globose seeds. There are more than 15 species in this genus, which is widely distributed in tropical to warm temperate waters. Six species occur in the Tropical Island Pacific region.

This genus is one of the most taxonomically complex groups among seagrasses. Den Hartog (1970) classified nine species in this genus with four subspecies in *H. ovalis*, which he considered as a "collective species." Since 1970, three new species and a new subspecies have been described. Phillips and Meñez (1988) recognized 11 species without subspecies in *Halophila*. We treat *H. ovalis*, *H. hawaiiana*, and *H. minor* as different species. We anticipate more new species, including some from the Pacific Ocean, will be added when the treatment of the genus is completed in the near future.

Halophila decipiens Ostenfeld
    Halophila decipiens var. *pubescens* den Hartog 1957

Monoecious, petioles shorter than blades, blade margins serrate, surfaces usually hairy. Blade oblong-elliptic, 10–25 mm long, 3–6 mm wide, L:W ratio 1.5–2; with 6–9 pairs of unforked cross veins. This species is widely distributed in the tropical Indian Ocean and the Pacific Ocean, as well as along the Caribbean and Brazilian coast.

Halophila hawaiiana Doty & Stone
    Halophila ovalis subspecies *hawaiiana* (Doty & Stone) den Hartog 1970

Dioecious, leaf blades narrowly obovate or spatulate, 10–30 mm long, 2–7 mm wide, L:W ratio 3–10; with 7–10 pairs of ascending rarely forked cross veins. Restricted to Hawaiian Islands.

Halophila minor (Zollinger) Den Hartog

> *Lemnopsis minor* Zoll. 1854
>
> *Halophila lemnopsis* Miq. 1855

Dioecious, blades obovate, 5–15 mm long, 3–5 mm wide, L:W ratio 1.8–2; with 9–12 pairs of forked cross veins. The species is widely distributed in the west Pacific Ocean, but not in the Hawaiian Islands.

Halophila ovalis (Br.) D. J. Hook.

> *Caulinia ovalis* R. Br. 1810
>
> *Lemnopsis major* Zoll. 1854
>
> *Halophila major* (Zoll.) Miq. 1855
>
> *Halophila euphlebia* Makino 1912

Dioecious, blades usually ovate to elliptic, 12–25 mm long, 6–13 mm wide, L:W ratio 1.5–3; with 12–22 pairs of usually forked cross veins. This species is widely distributed in the Indian and west Pacific oceans but is not found in the Hawaiian Islands.

Genus *Thalassia* Banks ex König

Perennial, marine, dioecious. Rhizome with a scale at every node, 1 unbranching root and 1 erect shoot with several leaves at regular intervals between nodes. Strap-like, distichous, eliguate leaves, 9–17 longitudinal veins, stomata absent. Squamules 2 per node. Inflorescence pedunculate, 1 or 2 per plant. Male flower pedicellate; tapals 3; stamens 3–12, sessile; anthers oblong, loculi 4, pollen grains spherical, arranged in chains. Female flowers with 3 tepals, styles 6–8, each split into 2 filiform stigmata. Fruits dome-shaped with 1–3 pyriform seeds.

This genus contains two closely related species in tropical waters. No distinct vegetative characteristics can be used to separate these species. *Thalassia hemprichii* is widely distributed in the tropical Indian and western Pacific oceans including Okinawa Islands, Micronesia, Papua New Guinea, and New Caledonia. *Thalassia testudinum* Banks ex König occurs along the coast of the Caribbean Sea and the Gulf of Mexico.

Thalassia hemprichii (Ehrenb.) Aschers.

> *Schizotheca hemprichii* Ehrenb. (1871)

Fruit splits into 8–20 irregular valves, beak 1–2 mm long, seeds 3–9.

Thalassia hemprichii is common in reef- or island-associated sites. It is not usually found at sites above MSL. This species generally grows in coral sand and often in association with *Cymodocea*, *Halophila*, and *Halodule* species. It is widely distributed in the Indian Ocean and the western Pacific region up to New Caledonia and also in the Melanesian and Micronesian coral-island groups.

ORDER POTAMOGETONALES Tomlinson

Perennial, rarely annual, submerged aquatic herbs, sometimes with floating leaves and emergent inflorescences. Leaves usually ligulate, squamules present. Monoecious or dioecious. Flowers rarely with tapal floral envelopes. Carpels several or more, uniovulate. Pollination anemophilous or hydrophilous. Fruit usually achenous. There are six families around the world; three of these families are seagrasses and two occur in the Pacific Ocean.

FAMILY CYMODOCEACEAE Taylor

Perennial, dioecious, marine herbs; rhizome creeping and erect stems with open or closed circular scars. Leaves distichous, ligulate, blade linear or nearly so. Inflorescence in a cyme or raceme of cymes or paired axillary flowers. Bracts present, perianth absent. Male flower with 2 stamens or representing 2 reduced flowers each of 1 stamen; anther 2-celled, longitudinally dehiscent; pollen thread-like, hydrophilous. Female flower with 2 carpels or representing 2 flowers each of 1 carpel, free, ovule 1, style 1, stigmas 2–3, filiforme. Fruiting carpels indehiscent, 1 seeded, viviparous in *Amphibolis* and *Thalassodendron*. The family consists of two sections, five genera and about sixteen species, mainly on tropical to warm temperate coasts. Six species, in three genera, and one section occur in the Pacific region.

Genus *Cymodocea* König

Herbaceous rhizome with several roots and a short, erect stem at each rhizome node. Leaves normally 2–7 per erect stem; blade linear, 7–17 longitudinal veins, apex usually rounded with teeth. Squamules 10 or more in two groups at each node. Inflorescence on terminating lateral shoots, naked. Male flowers stalked, anthers 2; fused at the same level. Female flower with 2 carpels, free, each with one ovule extended to a short style dividing to 2 long stigmas. Fruit with a hard pericarp.

A genus of four species with a wide but disjunct distribution in tropical and warm temperate seas of the Old World. Two species occur in the Pacific Ocean.

***Cymodocea rotundata* Ehrenb. & Hempr. ex Aschers.**
   Phucagrostis rotundata Ehrenb. & Hempr. 1868
   Cymodocea acaulis Peter 1928

Erect stem with close circular scars, strap-like leaves linear to curved flat, 9–15 longitudinal veins, leaf apex obtuse with very faint serrations. This species is also widely distributed along the coasts of the Indian Ocean and Malesia (Indonesia, Borneo, and New Guinea) into the western Pacific up to New Caledonia.

***Cymodocea serrulata* (R. Br.) Aschers. & Magnus**
   Caulina serrulata R. Br. 1810
   Phucagrostis serrulata Kuntze 1891
   Cymodocea asiatica Makino 1912

Erect stem with open circular scars, strap-like leaves linear to falcate, 13–17 longitudinal veins, leaf apex obtuse with distinct serrate to dentate. The species has a similar distribution to *C. rotundata* in the Indian Ocean and the western Pacific Ocean.

Genus *Halodule* Endl.

Herbaceous rhizome with 1 or more roots and a short, erect stem at each rhizome node. Leaves normally 1–4 per erect stem; blade linear, 3 longitudinal veins, apex usually dentate, rounded or emarginate. Squamules 2. Inflorescences solitary, naked on terminating lateral shoots. Male flowers stalked, anthers 2; fused at different levels. Female flower with carpels 2, free, each with a long style. Fruit with a hard pericarp.

Taxonomy of the genus *Halodule* is confusing. C. den Hartog (1970) recognized six morphologically very similar species. Four of these are found in the Caribbean Sea. Phillips and Meñez (1988) recognized only three species. Two species, *H. uninervis* and *H. pinifolia*, have been recorded in the Pacific Ocean, with the literature often reporting two morphological forms (narrow and broad-leaved forms) in *H. uninervis*. The relationship between the narrow-leafed *H. uninervis* and *H. pinifolia* has yet to be clearly defined.

Halodule pinifolia (Miki) den Hartog
Diplanthera pinifolia Miki 1932

Leaf width 0.25–1.2 mm, with leaf length up to 20 cm. It always has a leaf apex more or less rounded, somewhat irregularly serrated. The species also is widely distributed in the western Pacific through Malesia to New Caledonia.

Halodule uninervis (Forsk.) Aschers.
Zostera uninervis Forsk. 1775
Halodule australis Miq. 1855
Halodule tridentata F. v. M. 1882
Cymodocea australis Miq. 1885
Diplanthera uninervis Aschers. 1897

One to 6 roots and a short, erect stem at each node. Leaf sheath 1–3.5 cm long. Leaf blade 6–15 cm long and 0.25–3.5 mm wide. Leaf apex with 2 linear lateral teeth and an obtuse median tooth in which the mid-longitudinal vein ends.

Halodule uninervis is widely distributed along the coasts of the Indian Ocean and the western Pacific Ocean. This species occurs from MSL to 30 m deep, but more commonly from the intertidal zone to a depth of 5–6 m on fine mud to shell grit. *Halodule* and *Halophila* species frequently occur together in shallow coastal sites and form a preferred food source for dugongs.

Genus *Syringodium* Kützing

Herbaceous rhizome with one or more roots and a short, erect stem at each node. Leaves normally 2–3 per erect stem; blade distinctly terete; sheath persistent after the blade has been shed. Inflorescence has a rather loose cyme and is enclosed in bracts containing two anthers or two carpels. Mature fruit with a hard pericarp. There are two species: *Syringodium filiforme* is restricted to the Caribbean Sea; the other species, *S. isoetifolium*, is widely distributed in the tropical Indian Ocean and western Pacific Ocean.

Syringodium isoetifolium (Aschers.) Dandy

    Cymodocea aequoreq var. Wight ex Kunth 1841
    Cymodocea isoetifolia Aschers. 1868
    Phuucagrostis isoetifolia Kuntze 1891
    Phycoschoenus isoetifolia Nakai 1943

Leaf blade 10–30 mm long, 1–2 mm in diameter; 7–10 peripheral veins. Male flowers with anthers ovate, 4 mm long. Female flower sessile, ovary ellipsoid, stigmata 4–8 mm long. Fruit obliquely ellipsoid, 3–4 mm long.

This species occurs from low intertidal down to approximately 10 m. It may form pure meadows or occur as sparse plants among *Thalassia, Cymodocea, Halodule,* and *Enhalus,* and *Halophila* species.

Genus *Thalassodendron* den Hartog

Robust rhizomes bearing 1 or 2 erect unbranched or little-branched stems at every fourth node. Leaves occur in a terminal cluster, blade linear, numerous veins; apex truncate with teeth. Blade and sheath are shed together leaving an annular scar. Inflorescences contain two flowers protected by bracts on terminating short lateral shoots. Each male flower has two anthers. Each female flower also has two carpels, one of which is aborted and the inner bract modified into a succulent false fruit to protect the developing fruit. Reproduction is viviparous. The genus consists of two species, one of which is distributed widely in the tropical Indian Ocean and also extends into the western Pacific Ocean.

Thalassodendron ciliatum (Forsk.) den Hartog

    Zostera ciliata Forsk. 1775
    Thalassia ciliata König 1805
    Cymodocea ciliata Ehrenb. ex Aschers. 1867
    Phycagrostis ciliata Ehrenb. & Hempr. 1868
    Amphibolis ciliata Moldeke 1940

Robust rhizome up to 0.5 cm thick. Erect stem, unbranched or little branched at every fourth rhizome node. Roots 1–5 and much branched. Leaf blade linear, falcate, 10–15 cm long and 6–13 mm wide. Leaf margins entire, apex rounded and denticulate.

FAMILY ZOSTERACEAE Dummortier

Monoecious or dioecious marine to estuarine herbs with creeping perennial rhizomes. Roots unbranched and leaves with a distinct blade and sheath. The leaf blade is flat with 3–7 parallel longitudinal veins. Flowers with perianth absent are protected by a modified spathe. Flowers are arranged in 2 longitudinal rows, or male and female flowers arranged alternately in monoecious species. Male flowers have a stamen and 2 anthers; pollen is filiforme and pollination is hydrophilous. The female flower has a unilobular ovary with one ovule, a short style, and two long stigmatic branches. Each fruit has only one seed.

This family is not exclusively marine, and some species may grow in estuary and inlet conditions. The majority of species in this group are temperate. It has three distinct genera, all have a spathe-like reproductive arrangement unique to this family. Plants in the genus *Phyllospadix* are dioecious and grow on rocky substrates, with distribution restricted to the temperate northern Pacific region. The other two genera, *Heterozostera* and *Zostera*, are monoecious, grow on sand to mud substrates, and occur in mainly temperate regions in the world.

Genus *Zostera* L.

Monoecious, perennial; a monopodial-branched herb. Rhizome internode with two lateral vascular bundles. Roots in two groups of one to several. The erect stem is short and bears 2–6 distichous leaves with a linear blade and a variable leaf apex.

The genus consists of two subgenera: *Zostera* s.s. Ascherson and *Zosterella* (Ascherson) Ostenfeld. The former subgenus contains four species and occurs only in the temperate Northern Hemisphere, whereas the latter genus has six described species and is found in both hemispheres. One species extends the distribution of this genus into the Ryukyu Islands in the western Pacific Ocean.

Zostera japonica Aschers. & Graebn.

> *Zostera nana* Mertens ex Roth. 1868

Rhizome 0.5–1.5 mm in diameter with 2 roots at each node. Leaf blade 3–30 cm long and 0.75–1.5 mm wide; veins 3; apex obtuse, slightly emarginate. Seeds covered with smooth testa. This species occurs from Sakhalin Island to Vietnam, including the Ryukyu Islands.

Zostera capricorni Aschers.

Rhizome 0.75–2 mm in diameter, with two groups of roots at each node. Leaf blade 7–50 cm long and 2.5 mm wide; five longitudinal veins, tip truncate, slightly denticulate. Seeds with striate testa. This species mainly occurs in eastern Australia (Queensland and New South Wales) and extends to Lord Howe Island.

Morphology and Anatomy

Seagrasses are marine monocotyledons possessing roots, rhizomes, and leaves that are interconnected with a vascular system. Some genera (e.g., *Thalassodendron, Amphibolis, Cymodocea, Halodule,* and *Syringodium*) have distinct erect stems. Except for *Halophila*, which usually has petiolate leaves, and *Syringodium*, which has terete leaves, all other seagrasses have linear strap-shaped leaves. Only *Halophila decipiens* has hairs on blade surfaces. Several species have marginal dentition, in particular at the apex, and this is often used for species identification. There are usually three to more than twenty parallel longitudinal veins in the strap-like leaves. Both erect stems and leaves, particularly the older portions, support numerous epiphytes. Too many epiphytes can reduce seagrass photosynthetic rates.

Vegetative organs of all seagrasses consist of epidermis, parenchyma, fiber, aerenchyma, and vascular tissues. The detailed arrangement of these tissues, particularly in a leaf, may vary with genera and species.

Roots and rhizomes act as an anchor for seagrasses. The morphology of roots may vary with species, and this is probably related to the ecological habitat of the species. The presence of branching and root hairs also differs between seagrass species. Besides serving as an anchor and for storage, the rhizomes are responsible for vegetative expansion of seagrasses. The rhizosphere of both roots and rhizomes provides an excellent substrate for micro-organisms, which may aid plant nutrition.

The shape of the epidermal cells in seagrass leaves are easily visible and sometimes offer a useful taxonomic character. Stomata are absent in seagrasses, and the cuticle is extremely thin. Epidermal cells always have thickened walls and contain numerous chloroplasts for photosynthetic functions. The prominent air lacunae contain both oxygen and carbon dioxide and, in view of the absence of stomata, are considered important in photosynthesis. A vascular bundle in seagrasses consists of both xylem and phloem tissues. The number and size of xylem elements are much reduced compared to those in terrestrial plants. Xylem walls in seagrasses also have little lignification and secondary wall thickening. The structure of the phloem tissue in seagrass is similar to that of terrestrial plants. The thick-walled fiber cells probably provide tensile strength while retaining a high degree of flexibility.

Flowering and Fruiting

The timing and sequence of events for seagrass floral induction vary between species and are presumably adapted to environmental conditions. Temperature and light effects, probably day length, are known to be involved in floral induction and seed germination in some *Halophila, Cymodocea, Thalassia,* and *Syringodium* (Kuo et al. 1991; Kuo, Lee Long, and Coles 1993; McMillan 1976, 1979, 1980). Flowering may be irregular, and in species such as *Cymodocea serrulata* may be a very rare event (Kirkman 1975). In contrast some *Halophila* species are annuals and regrow from seed each year with seed production as high as 70,000 seeds m^{-2} year^{-1} (Kuo, Lee Long, and Coles 1993). For many species, however, asexual reproduction via rhizome growth may be the primary method of reproduction.

Seagrasses rely completely on water for pollen dispersal. This presents special problems as terrestrial angiosperm reproductive processes incorporate highly sensitive recognition mechanisms that are disrupted by seawater (McConchie and Knox 1989). Seagrasses have adapted with structures such as waterproof pollen and stigmas or by synchronizing flowering with tidal cycles so pollen can be carried on the water surface (*Enhalus acoroides*). The flowers of the different seagrass families have little in common, reflecting their diverse origin (Tomlinson 1982). Compared to angiosperms, all seagrass flowers have an unusual association of floral parts and all lack a petaloid whorl. Leaf-like structures may serve to direct pollen-bearing currents past the stigma in a similar manner to wind-pollinated species (McConchie and Knox 1989).

Eleven of the fourteen seagrass species in the Tropical Island Pacific region are dioecious, the remainder monoecious or hermaphrodite. One Pacific species, *Thalassodendron ciliatum*, is viviparous—the seed germinates and commences development while still attached to the mother plant. Details of reproductive development are species-specific and cannot be addressed within the scope of this chapter. Reproductive development of seagrasses is reviewed in more detail in den Hartog (1970) and McConchie and Knox (1989).

Tropical Island Pacific Region Species

Seagrass species can be found on reef platforms, intertidal coastal regions, and in inter-reef waters down to at least 50 m. Recent research (Derbyshire 1994) demonstrates, however, that associated fauna is greatly reduced at 10 m or more below MSL compared with samples taken at 3 m. Reef platform seagrasses were found to be the most highly productive for penaeid fisheries in the Torres Strait, supporting this evidence (Mellors 1990). Reef platform species found in the Pacific include *Thalassia hemprichii*, *Thalassodendron ciliatum*, and *Halophila* species. Other species such as *Halodule uninervis* and *Cymodocea serrulata* are common in intertidal and reef lagoon habitats. Table 3.1 lists the species recorded from the island Pacific region. As one of us (Kuo) has been revising the taxonomy of the genus *Halophila*, records for this group are in greater detail. The taxonomy of other genera need revision, and we suspect that additional records, and possibly new species, exist for this region. There have also been no comprehensive surveys of seagrass distribution within the island Pacific region. Most information is from taxonomic collections or from impact assessment studies. Few collections have been made from water deeper than 15 m below MSL.

The majority of Pacific Ocean seagrasses are predominantly tropical species. Most of these species are also widely distributed in the Indian Ocean. The number of seagrass species decreases from the eastern land masses to the western Pacific islands and drops markedly eastward of Palau and Vanuatu (Table 3.1) (Coles and Lee Long 1995).

One accepted (*Halophila hawaiiana*), and possibly more, endemic species of *Halophila* occurs in the island Pacific region; which is evidence of continuing speciation. One common Australian and Southeast Asian species, *Halophila spinulosa*, has so far not been found in the island Pacific region, and its absence is difficult to explain.

At least two theories (vicariance and center of origin) have been used to explain the present distribution on an ocean-wide scale of island Pacific region seagrasses (Coles and Lee Long 1995). Our view, however, is that further taxonomic and species distribution work is required in the Pacific before sufficient, reliable, basic information is available on which a satisfactory model can be built.

Role of Pacific Seagrasses in Increasing Productivity and Biodiversity

Seagrass meadows make significant contributions to coastal productivity (Orth, Heck, and Van Montfrans 1984), and their primary productivity ranks among the highest recorded for marine ecosystems. As a rich source of food and shelter, seagrass meadows support an abundance of fauna.

Higher vertebrates (turtles, birds, and marine mammals) are significant consumers of living seagrass tissue (Lanyon, Limpus, and Marsh 1989). McRoy and Helfferich (1980) have assembled a list of 154 species including fish and invertebrates that consume living seagrass tissue. Most likely, one of the important roles of seagrasses is in increasing the complexity of habitat and substrate surface area. This would allow for the growth of trophically important epiphytic algae and micro flora and fauna associated with detrital food chains.

Although the amount of published research on tropical seagrass systems is increasing, published literature records for Pacific seagrasses are sparse. Undoubtedly, there are records from reports and from programs of non-governmental organizations that are not accessible from library search procedures. It would be a valuable exercise to compile and publish data of this form and to use them to identify areas in the Tropical Island Pacific region where the seagrass knowledge base urgently requires improvement.

DISCUSSION

The records of Tropical Island Pacific seagrass collections and research are spread throughout the world and are mostly the result of individual collections and short-term research. The value of seagrass to fisheries productivity and its potential for loss due to coastal development have prodded funding agencies to support seagrass research in tropical regions (Coles et al. 1987). However, the island Pacific region was excluded, and the total impact of seagrasses lost to coastal development and eutrophication has not been estimated. Taxonomic records to date indicate about nine or ten species in the western Pacific (within the area defined for this chapter), reducing to one or two in the central and eastern Pacific. However, the total area of seagrass meadows has never been estimated, and no general information exists on the overall ecological importance of seagrass habitat in the region. The importance to fisheries productivity; the total areas of seagrass that may be threatened by coastal development; coastal population increase; and potential sea-level rises have yet to be studied, let alone quantified.

Part of this assessment/data-collecting process is in having available baseline taxonomic collections, published research, and adequate species descriptions. Protocols are required for lodging of preserved or dried specimens, particularly fruit and flower records. Collections should be kept with a regional herbarium, along with locality depth and substrate information and details of the collector and collection date. A record in an electronic database is of little use if it cannot be validated. The ASEAN-Australian program is a useful model for such a seagrass-collecting protocol (English, Wilkinson, and Baker 1994).

Patterns of seagrass abundance and species are of direct importance for managing coastal habitats and in maintaining the integrity of coastal zone systems and plant and animal communities. Differences in seagrass species, distribution, and abundance are important to assess the value of seagrasses to dugong and turtle populations and for fisheries stocks. Physical factors such as water quality (salinity, etc.) and seagrass habitat structure (Heck and Orth 1980; Gore et al. 1981; Bell and Westoby 1987), as well as biological factors such as shelter from predation (Bell and Pollard 1989), competition, food availability, and sediment type, all contribute to determining the value of a seagrass bed in increasing coastal fisheries productivity in the tropics.

Only limited information is available for tropical seagrass species on seasonal and year-to-year changes in species and biomass (e.g., Mellors, Marsh, and Coles 1993). For Tropical Island Pacific species, records of seasonal changes such as of seagrass fruiting and flowering and the availability of seed bank resources to provide for recovery from loss of seagrass are almost nonexistent. Urgent steps need to be taken to ensure this information is collected.

Mathews (1993) has outlined the problems of expense and simple human frailty in preserving data in both developed and developing countries. We support his suggestion of maintaining regional databases and ensuring all research is published in a form where it is accessible. Accessing data on seagrasses and their value to regional communities will always be difficult for an area as diverse as the Pacific, but setting protocols now will ensure information exists on which to base management decisions to prevent future loss of these habitats.

On world terms seagrasses rank collectively as one of the major ecosystems. Their conservation significance is greatly increased by the complexity of the communities they support. Seagrass beds are wetland communities under the definition of the Ramsar Convention on Wetlands of International Importance. The IUCN *World Conservation Strategy* (IUCN 1980) draws attention to seagrass meadows and comments that as a priority governments should ensure the principal management goal for coastal wetlands and shallows critical for fisheries, which is the maintenance of habitats on which the fisheries depend. In the Pacific the first step in this process is the collection of more rigorous taxonomic and distribution information.

CONCLUSIONS AND RECOMMENDATIONS

Seagrasses are identified as a significant coastal habitat type in the Tropical Island Pacific region. Information on the extent of this habitat and its species composition is patchy for the region. To ensure biodiversity is maintained, more precise information is required and the foundation for this is adequate taxonomic information. The major requirements are as follows:

- Seek support for funding further seagrass collection for distribution and taxonomic studies in the Pacific region.

- Designate a central organization such as the Bishop Museum in Hawaii to store and maintain a Pacific region seagrass reference collection and computer database.

- Establish a formal standardized collection, data storage, and survey protocols for the region, based on the ASEAN-Australia handbook standards developed by the Australian Institute of Marine Science (English, Wilkinson, and Baker 1994).

- Identify seagrass areas in the Tropical Island Pacific region that are important on a regional scale and/or that require protection. Identifying the most appropriate ways to provide that protection requires an immediate research initiative.

Acknowledgments—We acknowledge the assistance of Warren Lee Long and Len McKenzie in preparing this manuscript. Support was provided by the University of Western Australia, the Queensland Department of Primary Industries, the Co-operative Research Centre for the Ecologically Sustainable Development of the Great Barrier Reef, and the Pacific Science Association.

APPENDIX

Key for Sterile Material of Pacific Islands Seagrasses

1. Leaves petiolate or eligulate strap-shaped   **(Hydrocharitaceae) 2**

   Leaves not petiolate but ligulate and linear to
   strap-shaped   **(Cymodoceaceae, Zosteraceae) 7**

2. Leaves petiolate   ***Halophila* 3**

   Leaves strap-shaped, not petiolate   **(*Enhalus, Thalassia*) 6**

3. Leaf blade shorter than petiole; blade margin serrulate, blade
   surfaces usually hairy   ***H. decipiens***

   Leaf blade normally longer than petiole, blade margin entire, blade surface naked   **4**

4. Leaf blade 0.5–1.5 cm long and less than 5 mm wide; cross veins
   4–10 pairs   ***H. minor***

   Leaf blade 1–7 cm long; 5–20 mm wide; cross veins 6–30 pairs   **5**

5. Leaf blade ovate to elliptic, blade L:W ratio ca. 1–3; cross veins
   10–30 pairs usually branched   ***H. ovalis***

   Leaf blade elongated, narrow toward the base, blade L:W ratio ca. 3–6;
   cross veins 6–9 pairs, unbranched   ***H. hawaiiana***

6. Rhizome more than 1 cm in diameter, without scales; but
   covered with long black bristles (fiber strands); roots cord-like   ***Enhalus acoroides***

   Rhizome less than 0.5 mm in diameter, covered with scales,
   but no fibrous bristles; root normal   ***Thalassia hemprichii***

7. Rhizome bearing short, erect stems; leaf sheath finally falls
   and leaves a clean scar   **(Cymodoceaceae) 8**

   Rhizome without erect stems; leaf sheath persistent, remains
   as fibrous strands covering rhizomes   **(Zosteraceae) *Zostera* 13**

8. Leaf blade terete   ***Syringodium isoetifolium***

   Leaf blade flat or somewhat concavo-convex, not terete   **9**

9. Plants with elongated erect stem bearing terminally
   clustered leaves; rhizome stiff, woody; root stiff   ***Thalassodendron ciliatum***

Plants with a short or no erect stem, bearing linear leaf blades; rhizome herbaceous; root fleshy   **10**

10. Leaf blade 2 mm wide or less with three vascular bundles   *Halodule* **11**

Leaf blade more than 3 mm wide with nine or more vascular bundles   *Cymodocea* **12**

11. Leaf apex tridentate, with median tooth blunt and well-developed lateral teeth   *H. uninervis*

Leaf apex more or less rounded, lateral teeth weak   *H. pinifolia*

12. Leaf scars closed, rhizome internodes without cortical fiber bundles   *C. rotundata*

Leaf scars open, rhizome internodes with cortical fiber bundles   *C. serrulata*

13. Leaf blade apex truncate   *Z. capricorni*

Leaf blade apex slightly emarginate   *Z. japonica*

Glossary for Terms in Key

cortical   of the tissue between the epidermis and the rhizome stele (or vascular core)

emarginate   notched or indented at its tip

internodes   part of the rhizome lying between two adjacent nodes

ligulate   a tongue-like structure at the junction of leaf blade and sheath

petiolate   on a stalk

serrulate   leaf margins finely toothed with forward-pointing teeth

terete   round in cross-section

truncate   squared off at the apex

REFERENCES

Bell, J. D., and D. A. Pollard. 1989. Ecology of fish assemblages and fisheries associated with seagrasses. In *Biology of seagrasses: A treatise on the biology of seagrasses with special reference to the Australian region,* ed. A. W. D. Larkum, A. J. McComb, and S. A. Shepherd, 565–609. Elsevier: Amsterdam.

Bell, J. D., and M. Westoby. 1987. Effects of an epiphytic algae on abundances of fish and decapods associated with the seagrass *Zostera capricorni. Australian Journal of Ecology* 12:333–37.

Bulthuis, D. A. 1983. Effects of *in situ* light reduction on density and growth of the seagrass *Heterozostera tasmanica* in Western Port, Victoria, Australia. *Journal of Experimental Marine Biology and Ecology* 67:91–103.

Cambridge, M. L., and A. J. McComb. 1984. The loss of seagrasses in Cockburn Sound, Western Australia. 1. The time, course, and magnitude of seagrass decline in relation to industrial development. *Aquatic Botany* 20:229–43.

Coles, R. G., and W. J. Lee Long. 1995. Marine/coastal biodiversity in the Tropical Island Pacific region. 2. Population development and conservation priorities: Seagrasses. Report to the Joint Workshop for the Pacific Science Association Scientific Taskforce on Biodiversity and Conservation, November 1994.

Coles, R. G., W. J. Lee Long, B. A. Squire, L. C. Squire, and J. M. Bibby. 1987. Distribution of seagrass beds and associated juvenile commercial penaeid prawns in north-eastern Queensland waters. *Australian Journal of Marine and Freshwater Research* 38:103–19.

den Hartog, C. 1970. *The seagrasses of the world.* Amsterdam: North-Holland Publishing.

———. 1977. Structure, function and classification in seagrass communities. In *Seagrass ecosystems: A scientific perspective,* ed. C. P. McRoy and C. Helfferich, 89–121. New York: Marcel Dekker.

———. 1987. "Wasting disease" and other dynamic phenomena in *Zostera* beds. *Aquatic Botany* 27:3–14.

Derbyshire, K. J. 1994. Small prawn habitat and recruitment study. Report to the Queensland Fish Management Authority, Queensland, Australia.

Edgar, G., and H. Kirkman. 1989. Recovery and restoration of seagrass habitat of significance to commercial fisheries. Victorian Institute of Marine Science Working Paper No. 19, Melbourne.

English, S., C. Wilkinson, and V. Baker. 1994. *Survey manual for tropical marine resources.* ASEAN-Australia marine science project: Living coastal resources. Australian Institute of Marine Science, Townsville, Australia.

Gore, R. H., E. E. Gallaher, L. E. Scotto, and K. A. Wilson. 1981. Studies on the decapod Crustacea from the Indian River region of Florida. *Estuarine Coastal and Shelf Science* 12:485–508.

Heck, K. J., Jr., and R. J. Orth. 1980. Structural components of eelgrass (*Zostera marina*) meadows in the lower Chesapeake Bay—decapod Crustacea. *Estuaries* 3:289–95.

IUCN. 1980. *World conservation strategy.* International Union for the Conservation of Nature and Natural Resources, Gland, Switzerland.

Kemp, W. M., W. R. Boynton, R. R. Twilley, J. C. Stevenson, and J. C. Means. 1983. The decline of submerged vascular plants in upper Chesapeake Bay: Summary of results concerning possible causes. *Marine Technical Science Journal* 17:78–89.

Kirkman, H. 1975. Male floral structure in the marine angiosperm *Cymodocea serrulata. Bot. J. Linn. Soc.* 70:267–68.

Kuo, J., R. G. Coles, W. J. Lee Long, and J. Mellors. 1991. Fruits and seeds of *Thalassia hemprichii* from Queensland, Australia. *Aquatic Botany* 40:165–73.

Kuo, J., W. J. Lee Long, and R. G. Coles. 1993. Occurrence and fruit and seed biology of *Halophila tricostata* Greenway. *Australian Journal of Marine and Freshwater Research* 44:43–58.

Kuo, J., and A. J. McComb. 1989. Seagrass taxonomy, structure and development. In *Biology of seagrasses: A treatise on the biology of seagrasses with special reference to the Australian region,* ed. A. W. D. Larkum, A. J. McComb, and S. A. Shepherd, 6–67. Elsevier: Amsterdam.

Lanyon, J. 1986. Seagrasses of the Great Barrier Reef. Great Barrier Reef Marine Park Authority Special Publication Series No. 3, Townsville.

Lanyon J., C. J. Limpus, and H. Marsh. 1989. Dugongs and turtles; grazers in the seagrass system. In *Biology of seagrasses: A treatise on the biology of seagrasses with special reference to the Australian region,* ed. A. W. D. Larkum, A. J. McComb, and S. A. Shepherd, 610–27. Elsevier: Amsterdam.

Larkum, A. W. D., and C. den Hartog. 1989. Evolution and biogeography of seagrasses. In *Biology of seagrasses: A treatise on the biology of seagrasses with special reference to the Australian region,* ed. A. W. D. Larkum, A. J. McComb, and S. A. Shepherd, 112–45. Elsevier: Amsterdam.

Larkum, A. W. D., A. J. McComb, and S. A. Shepherd, eds. 1989. *Biology of seagrasses: A treatise on the biology of seagrasses with special reference to the Australian region.* Elsevier: Amsterdam.

Lee Long, W. J., J. E. Mellors, and R. G. Coles. 1993. Seagrasses between Cape York and Hervey Bay, Queensland, Australia. *Australian Journal of Marine and Freshwater Research* 44:19–31.

Mathews, C. P. 1993. On preservation of data. *NAGA: The ICLARM Quarterly,* April–July, 39–41.

McConchie, C. A., and R. B. Knox. 1989. Pollination and reproductive biology of seagrasses. In *Biology of seagrasses: A treatise on the biology of seagrasses with special reference to the Australian region,* ed. A. W. D. Larkum, A. J. McComb, and S. A. Shepherd, 74–101. Elsevier: Amsterdam.

McMillan, C. 1976. Experimental studies on flowering and reproduction in seagrasses. *Aquatic Botany* 2:87–92.

———. 1979. Flowering under controlled conditions by *Cymodocea rotundata* from the Palau Islands, Micronesia. *Aquatic Botany* 6:397–401.

———. 1980. Reproductive physiology in the seagrass *Syringodium filiforme* from the Gulf of Mexico and the Caribbean. *American Journal of Botany* 67:104–10.

McRoy, C. P., and C. Helfferich. 1980. Applied aspects of seagrasses. In *Handbook of seagrass biology—An ecosystem approach,* ed. R. C. Phillips and C. P. McRoy, 297–342. New York: Garland Publications.

Mellors, J. E., ed. 1990. Torres Strait prawn project—A review of research 1985–1987. Queensland Department of Primary Industries Information Series No. QI90018, Australia.

Mellors, J. E., H. Marsh, and R. G. Coles. 1993. Intra-annual changes in seagrass standing crop, Green Island, northern Queensland. *Australian Journal of Marine and Freshwater Research* 44:19–32.

Miki, S. 1934. On the sea-grasses in Japan III: General consideration on the Japanese seagrasses. *Botanical Magazine* (Tokyo) 48:171–219.

Mukai, H. 1993. Biogeography of the tropical seagrasses in the western Pacific. *Australian J. Marine and Freshwater Research* 44(1): 1–17.

Orth, R. J., K. L. Heck, and J. Van Montfrans. 1984. Faunal communities in seagrass beds: A review of plant structure and prey characteristics on predatory prey relationships. *Estuaries* 7:339–50.

Orth, R. J., and K. A. Moore. 1983. Chesapeake Bay: An unprecedented decline in submerged aquatic vegetation. *Science* 222:51–53.

Patriquin, D. G. 1975. "Migration" of blowouts in seagrass beds at Barbados and Carriacou, West Indies, and its ecological and geological implications. *Aquatic Botany* 1:163–89.

Phillips, R. C., and E. G. Meñez. 1988. Seagrasses. Smithsonian Contributions to Marine Science No. 34. Washington, D.C.: Smithsonian Institution Press.

Poiner, I. R., D. I. Walker, and R. G. Coles. 1989. Regional studies—Seagrasses of tropical Australia. In *Biology of seagrasses: A treatise on the biology of seagrasses with special reference to the Australian region,* ed. A. W. D. Larkum, A. J. McComb, and S. A. Shepherd, 279–96. Elsevier: Amsterdam.

Tomlinson, P. B. 1982. Helobiae (Alismatidae). In Vol. 7, *Anatomy of Monocotyledons,* ed. C. H. Metcalfe. Oxford: Clarendon Press.

Tsuda, R. T., and S. Kamura. 1990. Comparative review on the floristics, phytogeography, seasonal aspects and assemblage patterns of the seagrass flora in Micronesia and the Ryukyu Islands. *Galaxea* 9:77–93.

Walker, D. I. 1989. Methods for monitoring seagrass habitat. Victorian Institute of Marine Science Working Paper No. 18, Melbourne.

CHAPTER 4

Systematics and Distributions of Pacific Island Mangroves

Joanna C. Ellison

ABSTRACT

A review of regional floras and mangrove taxonomic literature indicates that in the Pacific islands thirty-four species of mangroves occur, and three hybrids, from sixteen genera and thirteen families. These are of the Indo-Malayan assemblage (with one exception), and decline in diversity from west to east across the Pacific, reaching a limit at American Samoa. To the east of Samoa there are no mangroves at present, though they are introduced in Hawaii and Tahiti. The taxonomy of the recognized species and their synonyms are given, with recorded distributions. There are uncertainties in the distribution of some species, owing to sporadic collections and misidentifications.

The Pacific island region lags behind the Caribbean and Southeast Asia in developing a regional monitoring system and database of mangrove community structure and ecosystem health, which would allow full assessment and management of mangrove resources.

INTRODUCTION

Mangroves are a taxonomically diverse group of perennial angiosperms that grow in the intertidal zone of sheltered shores in the tropics. True mangrove species occur exclusively in this saline wetland environment (Tomlinson 1986), with adaptations such as aerial roots, halophytic strategies, and vivipary. These distinguish mangrove trees as a specialized minority within their families. Also included are minor mangrove components, which are not restricted to the mangrove habitat but also occur in other habitats such as dry littoral forest.

The Pacific island region is largely composed of the Pacific island countries and territories of the South Pacific Commission and the South Pacific Regional Environment Programme (SPREP), with the addition of Hawaii and the subtropical Pacific islands of Japan (Marcus, Bonins, Ryukyus), Chile (Easter and Sala y Gómez), Australia (Norfolk and Lord Howe), and New Zealand (Kermadecs).

A review of the taxonomy and distribution of mangroves in the Pacific island region has not been compiled before. Fosberg (1975) achieved this for Micronesia, which is incorporated here. The regional review by Woodroffe (1987) only considered mangrove genera, owing to the confusion arisen as a result of changes in nomenclature and the perpetuation of errors of misidentification and mislabeling of mangrove collections in the region. This confusion is clarified here, from a review of published floras in the region and the use of mangrove taxonomic literature. In the Pacific islands thirty-four species of mangroves occur, and three hybrids, from sixteen genera and thirteen families. The taxonomy of the recognized species and their synonyms are given here.

TAXONOMY

The genera of true and minor mangrove species that occur in the Pacific islands are shown in Table 4.1. The total number of genera and species existing in each family is given to indicate that mangroves are a specialized minority within their family and that each has independently developed its specialized adaptations to this environment.

MANGROVE SPECIES AND SYNONYMS

Recognized names and synonyms of Pacific island mangroves are given in this section, in the order used in Table 4.1. Names used by Pacific island botanists are itemized and mistakes clarified.

Plumbaginaceae

Aegialitis annulata R. Brown 1810 Prodromus Flora Novae Hollondae: 426. Identified by Percival and Womersley (1975) and Frodin and Leach (1982) in Papua New Guinea.

Table 4.1 Classification of mangrove genera occurring in the Pacific islands

Subclass	Order	Family	No. in family Genera/(sp.)	Mangrove genus	Mangrove sp./(hybrids) Total	Pacific islands
Phylum Angiosperms—Division MAGNOLIOPHYTA						
Caryopyllidae	Plumbaginales	Plumbaginaceae	10 (560)	*Aegialitis*	2	1
Dilleniidae	Malvales	Sterculiaceae	70 (1200)	Heritiera	3	1
		Bombacaceae	31 (225)	*Camptostemon*	2	1
	Primulales	Myrsinaceae	1 (2)	*Aegiceras*	2	1
Rosidae	Myrtales	Sonneratiaceae	2 (6)	*Sonneratia*	6 (3)	3 (1)
		Myrtaceae	147 (3000)	*Osbornia*	1	1
		Combretaceae	20 (500)	*Lumnitzera*	2 (1)	2
	Rhizophorales	Rhizophoraceae	16 (120)	*Rhizophora*	6 (3)	4 (2)
				Bruguiera	6	6
				Ceriops	3	2
				Kandelia	1	1
	Euphorbiales	Euphorbiaceae	300 (7000)	Excoecaria	2	1
	Sapindales	Meliaceae	50 (6000)	Xylocarpus	3	3
Asteridae	Lamiales	Avicenniaceae	1 (8)	*Avicennia*	8	4
	Rubiales	Rubiaceae	500 (7000)	*Scyphiphora*	1	1
Class LILIOPSIDA (Monocots)						
Arecidae	Arecales	Palmae (Arecaceae)	210 (2800)	*Nypa*	1	1

Adapted from Duke 1992; classification, following Conquist 1981.

Note: Underlined genera are true mangroves; others are associate mangroves.

Sterculiaceae

Heritiera littoralis Dryander, in Aiton 1789 Hort. Kew. 3:546. A complete synonymy and citation are given in Kostermans (1959). Type specimen was collected in Tonga by Newson on Cook's third voyage. Identified by Hemsley (1894) in Tonga; Guillaumin (1948) in New Caledonia; Yuncker (1959) in Tonga; Whitmore (1966) in the Solomon Islands; Stone (1970) in Guam; Parham (1972) in Fiji; Percival and Womersley (1975) in Papua New Guinea; Marshall and Medway (1976) in Vanuatu; Stemmerman (1981) in the Federated States of Micronesia; Smith (1981) in Fiji; Frodin and Leach (1982) in Papua New Guinea; Tomlinson (1986) in New Caledonia; Ellison (1990) in Tonga; Devoe (1992) in Kosrae.

Synonym:
H. minor Lam.

Bombacaceae

Camptostemon schultzii Masters 1872, *Hook. Icon. Pl.* 12:18. Identified by Percival and Womersley (1975), Frodin and Leach (1982), and Tomlinson (1986) in Papua New Guinea. It is only known from the southern coast (Percival and Womersley 1975).

Myrsinaceae

Aegiceras corniculatum (L.) Blanco, 1837 in Fl. Filip.: 79. Identified by Whitmore (1966) in the Solomon Islands; Percival and Womersley (1975) and Frodin and Leach (1982) in Papua New Guinea.

Synonyms:
A. majus Gaertn.
A. fragrans Konig

Sonneratiaceae

Sonneratia alba J. E. Smith, 1819 in Rees Cycl. 33(2). A complete synonymy and citation are given in Duke and Jackes (1987). Identified by Guillaumin (1948) in New Caledonia; Whitmore (1966) in the Solomon Islands; Percival and Womersley (1975) in Papua New Guinea; Marshall and Medway (1976) in Vanuatu; Stemmerman (1981) in the Federated States of Micronesia; Frodin and Leach (1982) in Papua New Guinea; Tomlinson (1986) in New Caledonia, Vanuatu, and the Solomon Islands; Fosberg and Sachet (1987) in Kiribati; Duke and Jackes (1987) in Papua New Guinea, New Caledonia, Vanuatu, and the Solomon Islands; Devoe (1992) in Kosrae.

Sonneratia caseolaris (L.) Engler, in Engler and Pratl, *Nachtr.* 261, 1897. A complete synonymy and citation are given in Duke and Jackes (1987). Identified by Whitmore (1966) in the Solomon Islands; Percival and Womersley (1975) in Papua New Guinea; Marshall and Medway (1976) in Vanuatu; Frodin and Leach (1982) in Papua New Guinea; Tomlinson (1986) in the Solomon Islands and Vanuatu; Duke and Jackes (1987) in Papua New Guinea, Vanuatu, and the Solomon Islands.

Synonyms:

S. acida L.

S. lanceolata Bl.

Sonneratia x gulngai N.C. Duke, in Austrabaileya 2 (1984) 103. A putative hybrid between *S. alba* and *S. caseolaris*. A complete synonymy and citation are given in Duke and Jackes (1987), who identified the hybrid in northern Papua New Guinea and the Solomon Islands.

Sonneratia ovata Backer 1929, *J. Asiat. Soc. Bengal* 2:56. A complete synonymy and citation are given in Duke and Jackes (1987). Identified by Percival and Womersley (1975), Frodin and Leach (1982), Tomlinson (1986), and Duke and Jackes (1987) on the southern coast of Papua New Guinea.

Sonneratia lanceolata Blume, *Mus. Bot. Lugd.-Bat.* 1 (1851) 337. A complete synonymy and citation are given in Duke and Jackes (1987). Identified by Duke and Jackes (1987) on the southern coast of Papua New Guinea.

Myrtaceae

Osbornia octodonta F. Muell., in *Fragmenta* 3, 30, 1862. Identified by Whitmore (1966) in the Solomon Islands; Frodin and Leach (1982) in Papua New Guinea.

Combretaceae

Lumnitzera littorea (Jack) Voigt., in *Hort. Suburb. Calcutt.* 39, 1845. Identified by Guillaumin (1948) in New Caledonia; Yuncker (1959) in Tonga; Whitmore (1966) in the Solomon Islands; Stone (1970) in Guam; Parham (1972) in Fiji; Percival and Womersley (1975) in Papua New Guinea; Richmond and Ackermann (1975) in Fiji; Marshall and Medway (1976) in Vanuatu; Stemmerman (1981) in the Federated States of Micronesia; Frodin and Leach (1982) in Papua New Guinea; Woodroffe and Moss (1984) in Tuvalu; Smith (1985) in Fiji; Fosberg and Sachet (1987) in Kiribati; Ellison (1990) in Tonga; Devoe (1992) in Pohnpei; Whistler (1992) in Tonga, Fiji, and Guam.

Synonyms:

L. coccinea (Gaud.) Wight and Arnold. Name used by Hemsley (1894) in Tonga.

Pyrrhanthus littoreus Jack

Lumnitzera racemosa Willd. in Neue Schr. Ges. Naturf. Fr. Berl. 4 (1803), 187. Guillaumin (1948) in New Caledonia records *L. racemosa*. Identified by Percival and Womersley (1975) and Frodin and Leach (1982) in Papua New Guinea.

The status of *L. racemosa* east of New Guinea is uncertain. Exell (1954) describes both *L. littorea* and *L. racemosa* as occurring in Polynesia, though the distribution map given does not extend east of New Guinea. Chapman (1975) shows a distribution map in the Pacific islands, where *L. littorea* extends east as far as New Guinea, while *L. racemosa* extends east to the Solomon Islands and Federated States of Micronesia. As shown earlier, most Pacific island botanists identify *L. littorea*, not *L. racemosa*. Tomlinson et al. (1978) do not give details on distributions, except that it is similar for the two species. However, there are limited records of *L. racemosa* by Pacific island botanists. It is possible, therefore, that the species occurs in the Pacific islands east of New Guinea but has been overlooked owing to rarity and similarity to *L. littorea*.

Rhizophoraceae

Rhizophora apiculata Blume, in En. Pl. Jav. 1:91, 1827. Identified by Whitmore (1966) in the Solomon Islands; Stone (1970) in Guam; Percival and Womersley (1975) in Papua New Guinea; Marshall and Medway (1976) in Vanuatu; Tomlinson (1978) in New Caledonia, Vanuatu, and the Solomon Islands; Frodin and Leach (1982) in Papua New Guinea; Devoe (1992) in Kosrae.

Synonyms:
R. conjugata Arn. (non L.) Name used by Guillaumin (1948) in New Caledonia.
R. candelaria DC.

Rhizophora stylosa Griff., in Nat. Pl. As. 4:665, 1854. Identified by Whitmore (1966) in the Solomon Islands; Marshall and Medway (1976) in Vanuatu; Stone (1970) in Guam; Parham (1972) in Fiji; Richmond and Ackermann (1975) in Fiji; Percival and Womersley (1975) in Papua New Guinea; Tomlinson (1978) in Fiji, New Caledonia, Vanuatu, and the Solomon Islands; Taylor (1979) in French Polynesia; Smith (1981) in Fiji; Frodin and Leach (1982) in Papua New Guinea; Woodroffe and Moss (1984) in Tuvalu; Ellison (1990) in Tonga; Whistler (1992) in Guam, Marianas, Caroline Is., Fiji, and Tonga; Thaman (pers. com.) in Nauru.

Synonym:
R. mucronata Lamk. var. *stylosa* (Griff.) Schimp. Name used by Fosberg (1975).

Schimper (1891) reduced *R. stylosa* to a variety of *R. mucronata*. This was used by Fosberg (1975), identifying *R. mucronata* var. *stylosa* in Micronesia distinguished by length of style from *R. mucronata* var. *mucronata*. Stemmerman (1981) does not separate *R. mucronata* and *R. stylosa* in the Federated States of Micronesia. Subsequently, it has become more common to use the separation of the two species by Ding Hou (1960), and verified in Queensland by quantitative comparison of *R. apiculata*, *R. stylosa*, *R. mucronata*, and *R. lamarckii* (Duke and Bunt 1979).

Hemsley (1894) misidentifies *R. stylosa* in Tonga as *R. mucronata*, as does Seeman (1865) in Fiji, Guillaumin (1948) in New Caledonia, Yuncker (1959) in Tonga, and Parham (1972) in Fiji. Parham (1972) in Fiji separates *R. stylosa* and *R. mucronata*, though only *R. stylosa* occurs in Fiji (Smith 1981).

The range of *R. stylosa* in the N.W. Pacific is not certain (Fosberg 1975; Tomlinson 1986). It is identified in Guam by Stone (1970) but not in Pohnpei, Yap, or Kosrae (Devoe 1992).

Rhizophora x lamarckii Montrouzier, in Mem. Acad. Sci. Lyon 10:201, 1860. Putative hybrid between *R. apiculata* and *R. stylosa*. Identified by Tomlinson (1978) in New Caledonia and the Solomon Islands; Frodin and Leach (1982) in Papua New Guinea.

Synonym:
R. conjugata var. *lamarkii* (Montr.) Guill.

Rhizophora mucronata Lamk., in Encycl. 6, 189, 1804. Identified by Whitmore (1966) in the

Solomon Islands; Stone (1970) in Guam; Marshall and Medway (1976) in Vanuatu; Tomlinson (1978) in the Solomon Islands and Vanuatu; Stemmerman (1981) in the Federated States of Micronesia; Frodin and Leach (1982) in Papua New Guinea; Devoe (1992) in Kosrae.

R. mucronata was intoduced to Hawaii in 1922 from the Philippines but did not establish. It was last recorded in 1928 (Wester 1981).

Synonyms:
R. mucronata Lamk. var. *typica* Schimp.
R. mucronata var. *mucronata*. Name used by Fosberg (1975).

Rhizophora mangle L., in Sp. Pl. ed. 1:443, 1753. Identified by Hemsley (1894) in Tonga; Guillaumin (1948) in New Caledonia; Yuncker (1959) in Tonga; Ding Hou (1960); Parham (1972) in Fiji; Richmond and Ackermann (1975) in Fiji; Whistler (1976, 1984) in Samoa; Stemmerman (1981) in Samoa; Ellison (1990, 1991) in Tonga; Whistler (1992) in Tonga, Samoa, Fiji, and New Caledonia.

Synonyms:
R. samoensis (Hochr.) Salvoza. Name used by Tomlinson (1978) in New Caledonia, Fiji, Samoa, and Tonga; Smith (1981) in Fiji.
R. mangle var. *samoensis* Hochr. Name used by Christophersen (1935) in American Samoa.

Hochreutiner (1925, 447) proposed that the S.W. Pacific *R. mangle* as described by Jouan (1865, 149), Schimper (1891, 92), Hemsley (1894, 176), and Guppy (1906, 441) be distinguished as a variety, calling it *R. mangle* var. *samoensis* Hochr., based on the geographical separation. Salvoza (1936, 201) distinguished the populations of *R. mangle* on the west and east coasts of America, reclassifying the west coast plants as *R. samoensis* (Hochr.) Salv., and including the S.W. Pacific occurrence with these.

Tomlinson (1978, 158–59) found Fijian and American populations of *R. mangle* to be morphologically distinct, calling the former *R. samoensis*. Later he described *R. samoensis* as a S.W. Pacific geographical outlier of *R. mangle*, distinguished with difficulty (Tomlinson 1986, 333–34). Smith (1981, 605) and Blasco (1984, 85–86) followed Tomlinson (1978) in distinguishing *R. mangle* of tropical America and West Africa from *R. samoensis* of the S.W. Pacific.

Ding Hou (1960, 631) tested by measurement Salvoza's separation of Pacific and Atlantic *R. mangle* based on length of style and found sufficient overlap to make distinction impossible. Graham (1964, 289) described *R. mangle* from the Pacific coast of America and islands of the S.W. Pacific as having slightly smaller flowers with shorter styles than the Atlantic *R. mangle*, but found this also a questionable basis on which to separate the species. Breteler (1977, 227) stated that the small differences mentioned by Salvoza to distinguish *R. mangle* and *R. samoensis* do not justify its segregation, which is the approach adopted in this chapter.

Setchell (1924) misidentifies *R. mangle* as *R. mucronata* in American Samoa, as does

Lloyd and Aiken (1934) in Western Samoa, who also mistakenly illustrate the description with a *Bruguiera* flower.

R. mangle from Florida was introduced in Hawaii in 1902 (Wester 1981) and has become well established. The species is also recorded as introduced on Enewetak (St. John 1960).

Rhizophora x selala (Salvoza) Tomlinson, in Arnold Arbor. 59:159, 1978. Putative hybrid between *R. stylosa* and *R. mangle*. A complete synonymy and citation are given in Tomlinson (1978). Identified by Tomlinson (1978) in New Caledonia and Fiji; Smith (1981) in Fiji.

Synonyms:

R. mucronata Lamk. var. *selala* Salv.

R. selala Tomlinson & Womersley

Bruguiera gymnorrhiza (L.) Lamk., in Encycl. Meth. Bot. 4:696, 1797–98. A complete synonymy and citation are given in Ding Hou (1958). Identified by Whitmore (1966) in the Solomon Islands; Stone (1970) in Guam; Percival and Womersley (1975) in Papua New Guinea; Smith (1981) in Fiji; Frodin and Leach (1982) in Papua New Guinea; Tomlinson (1986) in Ryukyu Islands to Samoa; Ellison (1990) in Tonga; Whistler (1992) in Tonga, Fiji, Samoa, and Guam.

Spelled *B. gymnorhiza* by Marshall and Medway (1976) in Vanuatu, Richmond and Ackermann (1975) in Fiji, and Chapman (1975) in a regional review; Whistler (1976) in American Samoa; Stemmerman (1981) in the Federated States of Micronesia; Wester (1981) in Hawaii; Thaman (1992) in Nauru.

Walsh (1967) mistakenly uses the name *B. sexangula* in Hawaii for *B. gymnorrhiza*, which was introduced from the Philippines in 1922 (Wester 1981). The species has become well established.

Synonyms:

B. gymnorhiza Savigny. Name used by Fosberg (1960) in Micronesia; Parham (1972) in Fiji; Devoe (1992) in Kosrae.

B. rheedii Blume. Name used by Seeman (1865) in Fiji; Hemsley (1894) in Tonga; Setchell (1924) in American Samoa; Lloyd and Aiken (1934) in Western Samoa; Christophersen (1935) in Samoa; Taylor (1950) in the Northern Marshall Islands; Yuncker (1959) in Tonga.

Synonyms:

B. conjugata (L.) Merr.

Rhizophora gymnorrhiza L.

Rhizophora conjugata L.

Bruguiera parviflora Wight & Arn., in Trans. Med. Phys. Soc. Cal. 8:10, 1936. A complete synonymy and citation are given in Ding Hou (1958). Identified by Whitmore (1966) in the Solomon Islands; Percival and Womersley (1975) in Papua New Guinea; Marshall and Medway (1976) in Vanuatu; Frodin and Leach (1982) in Papua New Guinea; Tomlinson (1986) in the Solomon Islands and Vanuatu.

B. parviflora was introduced to Hawaii in 1922 from the Philippines but did not establish. It was last recorded in 1948 (Wester 1981).

Bruguiera sexangula (Lour.) Poir., in Lamk. Encycl. 4:262, 1816. A complete synonymy and citation are given in Ding Hou (1958). Identified by Whitmore (1966) in the Solomon Islands; Percival and Womersley (1975) and Frodin and Leach (1982) in Papua New Guinea.

Synonyms:

B. eriopetala Wight & Arnold. Name used by Guillaumin (1948) in New Caledonia.
B. malabarica F. -Vill. (non Arn.)

Bruguiera cylindrica (L.) Blume 1827, *En. Pl. Jav.* 1:93. Identified by Percival and Womersley (1975) and Frodin and Leach (1982) in Papua New Guinea.

Synonyms:

Rhizophora caryophylloides Burm.f.
B. malabarica Arnold.

Bruguiera exaristata Ding Hou 1957, *Nova Guinea n.s.* 8:166. Identified by Percival and Womersley (1975) and Frodin and Leach (1982) in Papua New Guinea.

Bruguiera hainesii C. G. Rogers 1919; *Kew Bull.* 225. Identified by Percival and Womersley (1975) and Frodin and Leach (1982) in Papua New Guinea.

Synonym:

B. eriopetala Wight of Watson 1928 (*Malay. For. Rec.* 6:109).

Ceriops tagal (Perr.) C. B. Rob., Robinson in Philipp. J. Sci. Bot. 3:306, 1908. A complete synonymy and citation are given in Ding Hou (1958). Identified by Whitmore (1966) in the Solomon Islands; Percival and Womersley (1975) in Papua New Guinea; Marshall and Medway (1976) in Vanuatu; Stemmerman (1981) in Palau; Frodin and Leach (1982) in Papua New Guinea; Tomlinson (1986) in New Caledonia, the Solomon Islands, and Vanuatu.

Synonyms:

C. candolleana Arn.
C. timoriensis Domin. Name used by Guillaumin (1948) in New Caledonia.
C. boiviniana Tulasne
C. tagal was introduced to Hawaii in 1922 from the Philippines but did not establish and was not recorded after that year (Wester 1981).

Ceriops decandra (Griff.) Ding Hou 1958, *Fl. Males.* Series I, Vol. 5:471. Identified by Percival and Womersley (1975) and Frodin and Leach (1982) in Papua New Guinea.

Synonym:

C. roxburghiana Arn.

Kandelia candel (L.) Druce, in Rep. Bot. Exch. Club Brit. Isles 1913(3), 420, 1914. A complete synonymy and citation are given in Ding Hou (1958). Identified by Hosokawa, Tagawa, and Chapman (1977) in the Ryukyus; Tomlinson (1986) in the Ryukyu Islands.

Synonym:

K. rheedii Wight & Arnold

Euphorbiaceae

Excoecaria agallocha L., in Syst. Nat. Ed., 10:1288 (1759). A complete synonymy and citation are given in Webster (1975). Identified by Hemsley (1894) in Tonga; Yuncker (1943) in Niue; Guillaumin (1948) in New Caledonia; Yuncker (1959) in Tonga; Whitmore (1966) in the Solomon Islands; Stone (1970) in Guam; Parham (1972) in Fiji; Percival and Womersley (1975) in Papua New Guinea; Marshall and Medway (1976) in Vanuatu; Frodin and Leach (1982) in Papua New Guinea; Smith (1981) in Fiji; Stemmerman (1981) in the Federated States of Micronesia and Guam; Tomlinson (1986) from Ryukyu Islands to Samoa and Niue; McPherson and Tirel (1987) in New Caledonia; Whistler (1992) in the Marianas, Caroline Is., Guam, Tonga, Fiji, and Niue.

Meliaceae

Xylocarpus granatum Koenig, in Naturforscher 20:2, 1784. A complete synonymy and citation are given in Pennington and Styles (1975). Identified by Yuncker (1959) in Tonga; Whitmore (1966) in the Solomon Islands; Parham (1972) in Fiji; Percival and Womersley (1975) in Papua New Guinea; Richmond and Ackermann (1975) in Fiji; Marshall and Medway (1976) in Vanuatu; Stemmerman (1981) in the Federated States of Micronesia; Frodin and Leach (1982) in Papua New Guinea; Smith (1985) in Fiji; Mabberley (1988) in New Caledonia; Devoe (1992) in Kosrae; Whistler (1992) in the Caroline Is., Fiji, and Tonga.

Stone (1970) in Guam includes *X. granatum* as a synonym of *X. moluccensis*.

Synonyms:
Carapa obovata Blume. Name used by Guillaumin (1948) in New Caledonia.
X. obovatus (Blume) Juss.
X. benadirensis Mattei
Carapa granatum Keonig
C. indica Adr. Juss.
Granatum obovatum (Blume) Kuntze
Carapa moluccensis auct. non Lam.

Xylocarpus moluccensis (Lamarck) Roemer, in Syn. Hesper. Fasc. 1:124, 1846. Identified by Setchell (1924) in American Samoa; Christophersen (1935) in Samoa; Yuncker (1959) in Tonga; Stone (1970) in Guam; Parham (1972) in Fiji; Percival and Womersley (1975) in Papua New Guinea; Richmond and Ackermann (1975) in Fiji; Whistler (1976) in American Samoa; Stemmerman (1981) in Palau; Smith (1985) in Fiji; Frodin and Leach (1982) in Papua New Guinea; Whistler (1984) in Samoa; Ellison (1990) in Tonga; Whistler (1992) in Guam, Marianas, Fiji, Tonga, and Samoa. Not identified by Lloyd and Aiken (1934) in Samoa.

Synonym:
Carapa moluccensis Lamark. Name used by Hemsley (1894) in Tonga; Burkill (1901) in Vava'u, Tonga.

Xylocarpus mekongensis Pierre 1897, *Fl. For. Cochinch.* Identified by Tomlinson (1986) in Papua New Guinea.

Synonym:

X. australasicus Ridley. Name used by Percival and Womersley (1975) and Frodin and Leach (1982) in Papua New Guinea.

Avicenniaceae

Avicennia alba Blume, in Bijdr. Flor. Ned. Ind. 14, 821 (1826). A complete synonymy and citation are given in Duke (1991). Identified by Whitmore (1966) in the Solomon Islands; Stone (1970) in Guam; Percival and Womersley (1975) and Frodin and Leach (1982) in Papua New Guinea; Duke (1991) in the Solomon Islands and northern coast of Papua New Guinea.

Synonym:

A. marina (Forsk.) Vierh. var. *alba*

Avicennia marina (Forsk.) Vierh. in Denkschr. Akad. Wiss. Wein Math. -Nat. 71, 435 (1907). A complete synonymy and citation are given in Duke (1991). Identified by Whitmore (1966) in the Solomon Islands; Percival and Womersley (1975) in Papua New Guinea; Stemmerman (1981) in Guam and Palau. The species is geographically separated into two varieties in the Pacific islands (Duke 1991).

Avicennia marina var. *eucalyptifolia* (Zipp.) N.C. Duke. Duke (1991) identified the variety in Vanuatu and the Solomon Islands. Earlier described by Marshall and Medway in Vanuatu as *Avicennia marina* (Forsk.) Vierh. var. *australasica* (Forst. f.) Moldenke.

Synonyms:

Avicennia eucalyptifolia Zipp. ex. Moldenke. Identified by Percival and Womersley (1975) and Frodin and Leach (1982) in Papua New Guinea.

Avicennia marina var. *australasica* (Walp.) Moldenke. Identified by Duke (1991) in New Caledonia, where it was misidentified as *A. officinalis* by Guillaumin (1948) and Thollot (1987). In New Caledonia, Tomlinson (1986) described it as *Avicennia marina* var. *resinifera* (Forst.) Bakhuizen.

Avicennia officinalis L. Identified by Percival and Womersley (1975) and Frodin and Leach (1982) in Papua New Guinea. Limited to the south coast of Papua New Guinea (Duke 1991). Whitmore (1966) identifies *Avicennia officinalis* L. in the Solomon Islands, though other records show that the species does not occur east of New Guinea (Duke 1991). The eastern limit is therefore uncertain.

Avicennia rumphiana Hallier f. Frodin and Leach (1982) identify *Avicennia rumphiana* Hallier f. in southern Papua New Guinea, and Duke (1991) extends the known distribution to include the north coast as far as Lae and New Britain. Percival and Womersley (1975) do not record the species in Papua New Guinea, and Tomlinson (1986) limits the eastern extent to Sarawak.

Rubiaceae

Scyphiphora hydrophyllacea Gaertn.f., in Fruct. Sem. Pl. s, 91 (1805). Originally spelled *hydrophylacea* by Gaertner. Identified by Guillaumin (1948) in New Caledonia; Whitmore (1966) in the Solomon Islands; Stemmerman (1981) on Yap and Palau; Frodin and Leach (1982) in Papua New Guinea; Tomlinson (1986) in New Caledonia, the Solomon Islands, and Palau.

Synonym:
Ixora manila Blanco

Palmae (Arecaceae)

Nypa fruticans (Thunb.) Wurmb., in Verh. Batav. Genootsch, 1781. A complete synonymy and citation are given in Moore (1973). Identified by Whitmore (1966) in the Solomon Islands; Stone (1970) in Guam; Percival and Womersley (1975) in Papua New Guinea; Stemmerman (1981) in the Federated States of Micronesia; Tomlinson (1986) in the Ryukyu Islands and Solomon Islands; Devoe (1992) in Kosrae.

Synonym:
Nipa fruticans Thunberg

Stemmerman (1981) noted that *N. fruticans* is introduced on Guam but native elsewhere in the Caroline Islands.

DISCUSSION

Distribution of mangrove species in the Pacific islands from the preceding review are summarized in Table 4.2, incorporating the distribution reviews of Fosberg (1975) and Hosokawa, Tagawa, and Chapman (1977). Except for *Rhizophora mangle*, mangroves of the Pacific islands are of the Indo-Malayan assemblage, and decline in diversity from west to east across the Pacific, reaching a limit at American Samoa. To the east of Samoa there are no mangroves at present, though they are introduced in Hawaii and Tahiti.

Mangrove species are morphologically similar, and misidentifications have occurred. Many herbarium collections are sporadic or incomplete and poorly catalogued and maintained. Information needs to be compiled on the existing herbarium collections held in the Pacific and elsewhere, and the descriptions updated and published. There is considerable potential for genetic study of population separations in the mangroves of the Pacific islands, and as some mangrove areas are threatened by human pressure, collections of all species from all island groups should be taken to enable future biogeographic study.

A regional mangrove monitoring system is needed within the Pacific islands that would allow assessment of levels of sustainable use of mangrove resources and ecosystem health relative to the pressures of human impacts, climate change, sea-level rise, and extreme events. Such a system has already been established in Southeast Asia (English, Wilkinson, and Baker 1994) and the Caribbean (CARICOMP 1992). This would incorporate mangrove species and associates present in each island nation, and their relative abundance over time, as well as physical site parameters. A database would be necessary to ensure comparability and compatibility among participating countries and to allow interpretation of regional trends in the data. The database design for mangrove monitoring in Asian countries (English, Wilkinson, and Baker 1994) is suitable for the Pacific islands, including the relevant species and country codes.

Note: This is contribution no. 725 from the Australian Institute of Marine Science.

Table 4.2 Distribution of mangrove species in the Pacific islands

Mangrove species	Papua New Guinea	Solomon Islands	Vanuatu	New Caledonia	Ryukyu Islands	Palau	Northern Mariana Islands	Guam	Federated States of Micronesia
Aegialitis annulata	x								
Heritiera littoralis	x	x	x	x	x	x		x	x
Camptostemon schultzii	x								
Aegiceras corniculatum	x	x							
Sonneratia alba	x	x	x	x	x	x			x
S. caseolaris	x	x	x						
S. x gulngai	x	x							
S. ovata	x								
S. lanceolata	x								
Osbornia octodonta	x	x							
Lumnitzera littorea	x	x	x	x	x	x		x	x
L. racemosa	x								
Rhizophora apiculata	x	x	x	x		x		x	x
R. stylosa	x	x	x	x			x	x	x
R. x lamarkii	x	x		x					
R. mucronata	x	x	x		x	x		x	x
R. mangle				x					
R. x selala				x					
Bruguiera gymnorrhiza	x	x	x	x	x	x	x	x	x
B. parviflora	x	x	x						
B. sexangula	x	x		x					
B. cylindrica	x								
B. exaristata	x								
B. hainesii	x								
Ceriops tagal	x	x	x	x		x			x
C. decandra	x								
Kandelia candel	x				x				
Excoecaria agallocha	x	x	x	x	x	x	x	x	x
Xylocarpus granatum	x	x	x	x		x			x
X. moluccensis	x	x	x	x		x		x	x
X. mekongensis	x								
Avicennia alba	x	x						x	
A. marina	x	x	x	x	x	x		x	x
A. officinalis	x								
A. rumphiana	x								
Scyphiphora hydrophyllacea	x	x		x		x			x
Nypa fruticans	x	x			x	x		x	x

Marshall Islands	Nauru	Kiribati	Tuvalu	Fiji	Tonga	Western Samoa	American Samoa	Niue	French Polynesia	Hawaii
				x	x					
x		x								
x		x	x	x	x					
	x		x	x	x				x	
x		x								
x				x	x	x	x			x
				x						
x	x	x		x	x	x	x			x
				x	x			x		
				x	x					
				x	x	x	x			

REFERENCES

Blasco, F. 1984. Taxonomic considerations of the mangrove species. In *The mangrove ecosystem: Research methods*, ed. S. C. Snedaker and J. G. Snedaker, 81–90. Paris: UNESCO.

Breteler, F. J. 1977. America's Pacific species of *Rhizophora*. *Acta Botanica Neerlandica* 26:225–30.

Burkill, I. H. 1901. The flora of Vava'u, one of the Tonga islands. *Botanical Journal of the Linnean Society* 35:20–65.

CARICOMP. 1992. Manual of methods for mapping and monitoring of physical and biological parameters in the coastal zone of the Caribbean. Caribbean Coastal Marine Productivity Program, University of Florida.

Chapman, V. J. 1975. Mangrove biogeography. In Vol. 1, *Proceedings of the International Symposium on Biology and Management of Mangroves*, ed. G. E. Walsh, S. C. Snedaker, and H. J. Teas, 3–21. Gainesville: University of Florida.

Christophersen, E. 1935. Flowering plants of Samoa. *Bernice P. Bishop Museum Bulletin* 128:1–221.

Cronquist, A. 1981. *An integrated system of classification of flowering plants.* New York: Columbia University Press.

Devoe, N. N. 1992. Country report mangrove forests in the Federated States of Micronesia. In *Proceedings Seminar and Workshop on Integrated Research on Mangrove Ecosystems in Pacific Islands Region and Reported Survey on Mangrove Ecosystems in the Pacific Islands*, ed. T. Nakamura. Japan International Association for Mangroves Workshop, 15–17 January, University of the South Pacific.

Ding Hou. 1958. Rhizophoraceae. *Flora Malesiana*, Series 1, 5:429–93.

———. 1960. A review of the genus *Rhizophora*. *Blumea* 10:625–34.

Duke, N. C. 1991. A systematic revision of the mangrove genus *Avicennia* (Avicenniaceae) in Australasia. *Australian Systematic Botany* 4:299–324.

———. 1992. Mangrove floristics and biogeography. In *Tropical mangrove ecosystems*, ed. A. I. Robertson and D. M. Alongi, 63–100. Washington, D.C.: American Geophysical Union.

Duke, N. C., and J. S. Bunt. 1979. The genus *Rhizophora* (Rhizophoraceae) in north-eastern Australia. *Australian Journal of Botany* 27:657–78.

Duke, N. C., and B. R. Jackes. 1987. A systematic revision of the mangrove genus *Sonneratia* (Sonneratiaceae) in Australasia. *Blumea* 32:277–302.

Ellison, J. C. 1990. Vegetation and floristics of the Tongatapu outliers. *Atoll Research Bulletin* 332:1–36.

———. 1991. The Pacific palaeogeography of *Rhizophora mangle* L. (Rhizophoraceae). *Botanical Journal of the Linnean Society* 105:271–84.

English, S., C. Wilkinson, and V. Baker. 1994. *Survey manual for tropical marine resources.* Townsville: Australian Institute of Marine Science.

Exell, A. W. 1954. Combretaceae. In *Flora Malesiana*, ed. C. G. G. J. Van Steensis, 1(4):533–89.

Fosberg, F. R. 1960. The vegetation of Micronesia. *Bulletin of the American Museum of Natural History* 119:1–75.

———. 1975. Phytogeography of Micronesian mangroves. In Vol. 1, *Proceedings of the International Symposium on Biology and Management of Mangroves*, ed. G. E. Walsh, S. C. Snedaker, and H. J. Teas, 23–42. Gainesville: University of Florida.

Fosberg, F. R., and M. H. Sachet. 1987. Flora of the Gilbert Islands, Kiribati, checklist. *Atoll Research Bulletin* 295.

Frodin, D. G., and G. J. Leach. 1982. Mangroves of the Port Moresby region. University of Papua New Guinea Department of Biology Occasional Papers 3. Revised ed. PNG.

Graham, S. 1964. The genera of the Rhizophoraceae and Combretaceae in the southeastern United States. *Journal of the Arnold Arboretum*, 45:285–301.

Guillaumin, A. 1948. *Flore Analytique et Synoptique de la Nouvelle Calédonie Phanerograms.* Paris, Office de la Recherche Scientifique Coloniale.

Guppy, H. B. 1906. *Observations of a naturalist in the Pacific between 1896 and 1899. Vol. 2, Plant dispersal.* New York: Macmillan.

Hemsley, W. B. 1894. The flora of the Tonga or Friendly Islands. *Botanical Journal of the Linnean Society* 30:158–217.

Hochreutiner, B. P. G. 1925. Plantae Hochreutineranae. Etude systematique et biologique des collections faites par l'auteur du monde pendant les annees 1903 a 1905. Fascicule II. *Condollea* 2:313–517.

Hosokawa, T., H. Tagawa, and V. J. Chapman. 1977. Mangals of Micronesia, Taiwan, Japan, the Philippines, and Oceania. In *Wet coastal ecosystems,* ed. V. J. Chapman, 271–91. Amsterdam: Elsevier Scientific.

Jouan, M. H. 1865. Recherches sur l'origine et la Provenance de certains vegetaux phanerogames observes dans les iles du Grand-Ocean. *Memoires de la Societe Imperiale des Sciences Naturelles de Cherbourg* 11:81–178.

Kostermans, A. J. G. J. 1959. Monograph of the genus *Heritiera aiton* (Stercul.) *Reinwardtia* 4:465–83.

Lloyd, C. G., and W. H. Aiken. 1934. Flora of Samoa. *Bulletin of the Lloyd Library and Museum* 33:1–113.

Mabberley, D. J. 1988. Meliaceae. In *Flore de la Nouvelle Caledonie et dependances,* ed. P. Morat and H. S. Mackee, 15:17–89. Paris: Museum National D'Histoire Naturelle.

Marshall, A. G., and L. Medway. 1976. A mangrove community in the New Hebrides, south-west Pacific. *Botanical Journal of the Linnean Society* 8:319–36.

McPherson, G., and C. Tirel. 1987. Euphorbiaceae I. In *Flore de la Nouvelle Caledonie et dependances,* ed. P. Morat and H. S. Mackee, 14:17–89. Paris: Museum National D'Histoire Naturelle.

Moore, H. E. 1973. The major groups of palms and their distribution. *Gentes Herbarium* 11:27–141.

Parham, J. W. 1972. *Plants of the Fiji islands.* Suva: Government Printer.

Pennington, T. D., and B. T. Styles. 1975. A generic monograph of the Meliaceae. *Blumea* 22:419–540.

Percival, M., and J. S. Womersley. 1975. Floristics and ecology of the mangrove vegetation of Papua New Guinea. Lae, *Papua New Guinea National Herbarium Botany Bulletin* 8:1–94.

Richmond, T. D., and J. M. Ackermann. 1975. Flora and fauna of mangrove formations in Viti Levu and Vanua Levu, Fiji. In *Proceedings of the International Symposium on Biology and Management of Mangroves,* ed. G. E. Walsh, S. C. Snedaker, and H. J. Teas, 147–52. Gainesville: University of Florida.

Salvoza, F. M. 1936. *Rhizophora. University of the Philippines Natural and Applied Science Bulletin* 5:179–273.

Schimper, A. F. W. 1891. *Die Indomalayische Strandflora.* Jena: Fischer.

Seeman, B. C. 1865. *Flora Vitiensis.* London.

Setchell, W. A. 1924. *American Samoa. Part I. Vegetation of Tutuila Island.* Carnegie Institute of Washington.

Smith, A. C. 1981. *Flora Vitiensis Nova.* Vol. 2. Hawaii, Pacific Tropical Botanical Garden.

———. 1985. *Flora Vitiensis Nova.* Vol. 3. Hawaii, Pacific Tropical Botanical Garden.

St. John, H. 1960. Flora of Enewetak Atoll. *Pacific Science* 14:313–36.

Stemmerman, L. 1981. *A guide to Pacific wetland plants.* Honolulu: U.S. Army Corps of Engineers.

Stone, B. C. 1970. The flora of Guam. *Micronesica* 6:1–629.

Taylor, F. J. 1979. *Rhizophora* in the Society Islands. *Pacific Science* 33:173–76.

Taylor, W. R. 1950. *Plants of Bikini and other northern Marshall Islands.* Ann Arbor: University of Michigan Press.

Thaman, R. R. 1992. Vegetation of Nauru and the Gilbert Islands: Case studies of poverty, degradation, disturbance and displacement. *Pacific Science* 46:128–58.

Thollot, P. 1987. *Importance de la mangrove pour l'ichthofaune du lagon de Nouvelle-Caledonie.* Diplome d'etude affronfondie en Oceanologie. Centre d'Oceanologie de Marseille. Noumea, New Caledonia: ORSTOM.

Tomlinson, P. B. 1978. *Rhizophora* in Australasia: Some clarification of taxonomy and distribution. *Journal of the Arnold Arboretum* 59:156–69.

———. 1986. *The botany of mangroves.* Cambridge: Cambridge University Press.

Tomlinson, P. B., J. S. Bunt, R. B. Primack, and N. C. Duke. 1978. *Lumnitzera rosea* (Combretaceae): Its status and floral morphology. *Journal of the Arnold Arboretum* 59:342–51.

Walsh, G. E. 1967. An ecological study of a Hawaiian mangrove swamp. In *Estuaries,* ed. G. H. Lauff, 420–31. Washington, D.C.: American Association for the Advancement of Science.

Webster, G. L. 1975. Conspectus of a new classification of the Euphorbiaceae. *Taxon* 24:593–601.

Wester, L. 1981. Introduction and spread of mangroves in the Hawaiian Islands. *Association of Pacific Coast Geographers Yearbook* 43:125–37.

Whistler, W. A. 1976. *Inventory and mapping of wetland vegetation in the territory of American Samoa.* Report for the U.S. Army Corps of Engineers.

———. 1984. Annotated list of Samoan plant names. *Economic Botany* 38:464–89.

———. 1992. *Flowers of the Pacific island seashore.* Hawaii: Isle Botanica.

Whitmore, T. C. 1966. *Guide to the forests of the British Solomon Islands.* Oxford: Oxford University Press.

Woodroffe, C. D. 1987. Pacific island mangroves: Distributions and environmental settings. *Pacific Science* 41:166–85.

Woodroffe, C. D., and T. J. Moss. 1984. Litter fall beneath *Rhizophora stylosa* Griff., Vaitapu, Tuvalu, South Pacific. *Aquatic Botany* 18:249–55.

Yuncker, T. G. 1943. The flora of Niue Island. *Bernice P. Bishop Museum Bulletin* 178:1–126.

———. 1959. Plants of Tonga. *Bernice P. Bishop Museum Bulletin* 220:1–283.

CHAPTER 5

Corals of the Tropical Island Pacific Region: Biodiversity

J. E. N. Veron

Coral biodiversity is usually portrayed (if not defined) in terms of generic and species "richness."[1] Genera and species are taxonomic, not systematic, units; thus, it is important at the outset to distinguish between systematics, taxonomy, and identification. Systematic order is genetically based and conceptual, and represents the true relationship between taxa. Taxonomic order is mostly morphologically based and operational. Identification is the application of morphological taxonomy to operational needs.

The primary object of taxonomy is to define the morphological limits of species, to separate species, to name them, and to describe them in a form meaningful to other taxonomists. The primary object of "identification" is to apply the results of taxonomy to other studies (i.e., to make them useful to nontaxonomists). With corals, identification depends on the appearance of the living organism, and this appearance varies greatly with physical environment. As a consequence, reliable identification requires considerable field knowledge and experience. Many keys have been written for coral identification, but except in limited circumstances, most are too simplistic to be effective.

STATUS OF TAXONOMIC KNOWLEDGE

It is now well understood by coral biologists that most species display plant-like morphological variations in response to local environments. This makes species identification a task for specialists and gives rise to much of the confusion found in the taxonomic literature. The nature of morphological variation is best appreciated by dissecting it into its principal components: most species do not show wide variation in all these components, but most of their variability are restricted to one or two. Veron (1993, 1995) gives relevant notes and references, a summary of which follows.

- *Corallite variation within a colony.* Almost all colonial corals show variation in skeletal structure between different corallites of the same colony. These variations are usually related to the particular location of the corallite within the colony and are due to factors such as space availability, age, predation, or microenvironment.

1. This chapter is about zooxanthellate corals that have photosynthetic endosymbiotic algae. In contrast, azooxanthellate corals are a physiologically very different group, which are uncommon on reefs and usually inhabit deep (euphotic) water.

- *Morphological variation within different parts of the same colony.* A small proportion of coral species show major growth-form modification with age or some other factor, so that one part of the colony becomes quite unlike another part. The majority of nominal species descriptions reflect only a small part of the true variation of species.

- *Variation between colonies within the same biotope.* Different colonies of the same species growing in proximity under uniform environmental conditions may show major variations in growth form, polyp, or corallite structure.

- *Contiguous variation between physical environmental zones.* Almost all colonial coral species exhibit major morphological change over a series of contiguous biotopes in response to physical environmental gradients. Over a wide depth range, such as between a reef flat and a lower slope, morphological variation in most species is enormous, so much so that a colony from a reef flat and another of the same species from a nearby lower slope may have less in common morphologically than colonies of different species from the same reef flat.

- *Variation within regions.* Most species with distributions extending to high latitudes undergo skeletal and other changes that can be correlated with decreasing temperature or genetic isolation, or both.

- *Nonskeletal variation.* As any field guide shows, the appearance of the living coral is of great importance in species recognition. Soft tissue morphologies are also used as taxonomic characters, especially with large-polyped genera that have polyps extended by day (notably *Goniopora, Alveopora,* and caryophylliids). So far, soft tissues in only six species of the caryophylliid genus *Euphyllia* are essential to species identification.

The Biogeographic Scale

Any consideration of biodiversity within an area the size of the Pacific must take full account of *within*-species variation on biogeographic (interregional) scales. With corals, as with plants, a species that is readily identified in one region may be much less readily identified in another (i.e., taxonomic distinctions between species become less reliable with distance). Reference collections and identification guides relevant to one region may be of limited value for another. Geographic variation is normally described as being interspecific *or* intraspecific. However, in practice, the two go together and are not mutually exclusive or fundamentally dissimilar: presence/absence (interspecific) differences between the corals of two regions are *indicative* of intraspecific differences.

Most taxonomic studies of Indo-Pacific corals are based on original studies within a single region. A few are based on original studies over a whole province. No original studies span the whole Pacific. The relevance of taxonomic decisions in one region to those of other regions is an important issue, especially because single taxonomic descriptions are of geographically limited use. The following categories of species variations form a continuum; they are not mutually exclusive, and many species can be placed into two or more categories.

- *Species that have little or no geographic variation.* Approximately 20 percent of the total species complement of (for example) the Great Barrier Reef apply equally to the southeast Pacific.

- *Species that have disjunct distributions in high latitudes.* High-latitude occurrences of species are found along continental coastlines or in regions isolated from the tropics. In the former case, poleward surface circulation ensures rapid one-way genetic connectivity from low to high latitudes. In the latter case, genetic isolation may continue into evolutionarily significant time.

- *Species that form geographic subspecies in isolated regions.* Depending on level of detail, the majority of species in any given region have unique points of distinction from the same species in other regions. Occasionally the same variant of a species found in one region appears in another very distant one, suggesting long-distance founder dispersion.

- *Species that have increasing variation with distance.* By its very nature, taxonomy is binary: two corals are either given the same name, or they are given different names. This gives false information if corals from two regions are separated by a continuum. Thus, one member of a geographic "species complex" may have one name on one side of the complex and another name on the other side. Where it occurs in intermediate regions, it might be given the one name or the other but it cannot be given *both* names, although it is, in effect, both "species." Conversely, the "complex" may be given one name, but the characters of the species in one region may be very unlike its characters in another.

- *Hybrid patterns.* When is a species not a species? Commonly, species pairs or groups appear to have differing levels of taxonomic separation in different regions. The answer to this question is a complex matter of biogeographic taxonomy and concept (Veron 1995).

SYSTEMATIC AFFINITIES   Over the past two decades, there has been an explosion of faunal, ecological, and physiological research on corals, funded in response to the needs of conservation and management. Almost all of this research is based on "species"; almost all requires taxonomic support in some form or other. Coral taxonomy is thus "operational," and it provides support for many hundreds of scientific publications every year. The essential question is how closely does this taxonomy reflect the natural world, and how much is it a human construct? The issues are ones of broad concept as well as of specific detail. They concern concepts of the nature of species, issues of identification, human-created nomenclatorial problems, mistakes and incomplete studies, unresolved taxonomic problems (especially over biogeographic scales), and actual and potential conflicts between taxonomic methods.

Veron (1995) concludes that species boundaries are arbitrary and that the systematic status of one species may be different from that of another. Different taxonomic methods will resolve different morphological and geographic boundaries, creating potential conflicts of interpretation.

The methods that have been used to date to determine taxonomic or systematic relationships among species are:

- *Cladistics.* Cladistics based on composites of morphological, environmental, and ecological data may point to faults or insufficiencies in data and data collecting, and impose rigor in both analysis and description. However, cladistics used as an endpoint will never be self-supporting with corals because of the overwhelming prevalence of homoplasy and loss of information derived from studies of co-occurring colonies. Cladistic parsimony analyses of molecular data are another matter and have the potential to test the validity of other taxonomic methods within definable boundaries.

- *Colony transplants.* Results of transplants, which demonstrate a lack of correlation between morphological variation and physical environment, may indicate faulty original taxonomy or identification, or both.

- *Physiological and behavioral studies.* Physiological and behavioral differences between populations, races, and geographic subspecies and species do not provide an independent test of taxonomy, but they do add to knowledge of variation.

- *Reproduction.* Studies of reproduction have begun to present an array of interesting conceptual problems and practical issues. Of greatest interest are barriers to cross-fertilization between co-occurring colonies of the same species or groups of "sibling species" and artificial hybridization between species.

- *Molecular methods.* Molecular techniques present a powerful array of tests of morphological taxonomy and, in the long term, may displace morphology as the primary basis for separating species. Hillis (1987) provides a general review of the subject, concluding that disagreements among morphological and molecular systematists over "species" definitions usually represent a disagreement of concept without due reference to biological realities. This is likely to increasingly be the case with corals.

The main source of conflict will come from the arbitrary nature of species boundaries, and hence debate as to what are species and what are subspecific taxa. The relevant points are that there may be no "correct" answers because species are arbitrary, that different information sources and purposes target different taxonomic levels or units, and that taxonomy must accommodate information from multiple sources.

FAMILIES AND GENERA

The fifteen families and approximately eighty-four genera of extant zooxanthellate corals of the order Scleractinia likely to be found in the tropical Pacific, excluding those confined to the Indian Ocean/Red Sea, Atlantic, Caribbean, and/or the Gulf of Mexico, are listed in the Appendix to this paper.

DATABASES

Veron (1993) is a global compilation of taxonomic and distribution data, which include species-level data for the Central Indo-Pacific and generic-level data for the world. This database, which is a compilation and syntheses of the taxonomic and biogeographic literature, is being extended and incorporated into the electronic database CoralBase (see Navin and Veron, this volume). The biogeographic and evolutionary issues that arise are discussed in Veron (1995).

REFERENCES

Hillis, D. M. 1987. Molecular versus morphological approaches to systematics. *Annu. Rev. Ecol. Syst.* 18:23–42.

Veron, J. E. N. 1993. *A biogeographic database of hermatypic corals: Species of the Central Indo-Pacific and genera of the world.* Australian Institute of Marine Science Monograph Series No. 9, Queensland.

———. 1995. *Corals in space and time: The biogeography and evolution of the* Scleractinia. Sydney: University of New South Wales Press. In press.

APPENDIX

Zooxanthellate Corals Likely to be Found in the Tropical Pacific

Family and genus	Extant species (no.)	Present distribution	General abundance
Astrocoeniidae *Stylocoeniella*	At least 3	Red Sea to central Pacific	Uncommon, cryptic
Pocilloporidae *Pocillopora*	Approx. 10	Red Sea and western Indian Ocean to far eastern Pacific	Very common, very conspicuous
Stylophora	Approx. 5	Red Sea and western Indian Ocean to southern Pacific	Very common, very conspicuous
Seriatopora	Approx. 5	Red Sea and western Indian Ocean to southern Pacific	Very common, conspicuous
Madracis	Zooxanthellate: Approx. 4	Western Indian Ocean to eastern Pacific and western to eastern Atlantic	Uncommon, mostly cryptic in the Indo-Pacific, much more common in the western Atlantic
Palauastrea	1	Central Indo-Pacific	Uncommon
Acroporidae *Montipora*	At least 80	Red Sea and western Indian Ocean to southern Pacific	Extremely common, some species inconspicuous
Anacropora	Approx. 6	Western Indian Ocean to western Pacific	Uncommon, mostly non-reefal
Acropora	At least 150	Cosmopolitan	Extremely common, very conspicuous, usually dominant in Indo-Pacific reefs
Astreopora	Approx. 15	Red Sea and western Indian Ocean to southern Pacific	Generally common, conspicuous
Poritidae *Porites*	Approx. 80	Cosmopolitan	Extremely common, conspicuous at generic level
Stylaraea	1	Red Sea and western Indian Ocean to western Pacific	Rare, occurs only in shallow, wave-washed biotopes

Family and genus	Extant species (no.)	Present distribution	General abundance
Goniopora	Approx. 30	Red Sea and western Indian Ocean to southern Pacific	Generally common, very conspicuous
Alveopora	Approx. 15	Red Sea and western Indian Ocean to southern Pacific	Sometimes common, very conspicuous
Siderastreidae			
Siderastrea	Approx. 6	Red Sea and western Indian Ocean to Philippines; far eastern Pacific, Caribbean, and Gulf of Mexico	Uncommon in the Indian Ocean, rare in the Pacific
Pseudosiderastrea	1	Western Indian Ocean to western Pacific	Uncommon, cryptic
Psammocora	Approx. 15	Red Sea and western Indian Ocean to far eastern Pacific	Generally common, sometimes cryptic
Coscinaraea	Approx. 12	Red Sea and western Indian Ocean to southern Pacific	Generally common, conspicuous
Agariciidae			
Pavona	Approx. 22	Red Sea and western Indian Ocean to far eastern Pacific	Very common, conspicuous
Leptoseris	Approx. 14	Red Sea and western Indian Ocean to far eastern Pacific and Caribbean and Gulf of Mexico	Sometimes common, mostly conspicuous
Gardineroseris	At least 2	Red Sea and western Indian Ocean to far eastern Pacific	Generally uncommon, sometimes cryptic
Coeloseris	1	Central Indian Ocean to central Pacific	Generally uncommon, sometimes cryptic
Pachyseris	Approx. 4	Red Sea and western Indian Ocean to western Pacific	Very common, very conspicuous
Fungiidae			
Cycloseris	Approx. 16	Red Sea and western Indian Ocean to far eastern Pacific	Generally uncommon, non-reefal
Diaseris	At least 3	Red Sea and western Indian Ocean to far eastern Pacific	Generally uncommon, non-reefal
Fungia	Approx. 33	Red Sea and western Indian Ocean to southern Pacific	Very common, very conspicuous
Ctenactis	3	Red Sea to western Pacific	Generally common, very conspicuous
Herpolitha	2	Red Sea and western Indian Ocean to western Pacific	Generally common, very conspicuous
Polyphyllia	3	Western Indian Ocean to southern Pacific	Sometimes common, very conspicuous
Sandalolitha	2	Central Indian Ocean to southern Pacific	Sometimes common, very conspicuous
Halomitra	2	Western Indian Ocean to southern Pacific	Generally uncommon, very conspicuous
Zoopilus	1	Central-west Pacific	Uncommon, very conspicuous
Lithophyllon	Approx. 4	Eastern Indian Ocean to western Pacific	Generally uncommon, conspicuous
Podabacia	2	Red Sea and western Indian Ocean to southern Pacific	Sometimes common, conspicuous
Cantharellus	3	Red Sea to western Pacific	Rare, cryptic
Oculinidae			
Galaxea	Approx. 5	Red Sea and western Indian Ocean to southern Pacific	Very common, very conspicuous
Acrhelia	1	Eastern Indian Ocean to southern Pacific	Generally uncommon, conspicuous
Simplastrea	1	Indonesia only	Rare

Family and genus	Extant species (no.)	Present distribution	General abundance
Parasimplastrea	1	Oman only	Rare
Pectiniidae			
Echinophyllia	Approx. 8	Red Sea and western Indian Ocean to southern Pacific	Very common, conspicuous
Oxypora	At least 3	Western Indian Ocean to southern Pacific	Generally common, conspicuous
Mycedium	At least 2	Red Sea and western Indian Ocean to southern Pacific	Generally common, conspicuous
Physophyllia	1	Western Pacific	Rarely common, conspicuous
Pectinia	Approx. 7	Western Indian Ocean to southern Pacific	Generally common, very conspicuous
Mussidae			
Blastomussa	3	Red Sea and western Indian Ocean to western Pacific	Generally uncommon, sometimes inconspicuous
Cynarina	1	Red Sea and western Indian Ocean to western Pacific	Generally uncommon, very conspicuous
Indophyllia	1	Indonesia only	Rare, inconspicuous
Scolymia	Approx. 4	Red Sea and western Indian Ocean to southern Pacific; Caribbean and Gulf of Mexico	Generally uncommon, conspicuous
Australomussa	1	Central Indo-Pacific	Sometimes common, very conspicuous
Acanthastrea	Approx. 6	Red Sea and western Indian Ocean to southern Pacific	Generally uncommon, *Favites*-like
Lobophyllia	Approx. 9	Red Sea and western Indian Ocean to southern Pacific	Very common, very conspicuous
Symphyllia	Approx. 6	Red Sea and western Indian Ocean to southern Pacific	Generally common, very conspicuous
Merulinidae			
Hydnophora	Approx. 7	Red Sea and western Indian Ocean to southern Pacific	Generally common, very conspicuous
Paraclavarina	1	Central Indo-Pacific	Generally uncommon, conspicuous
Merulina	3	Red Sea and western Indian Ocean to southern Pacific	Sometimes common, conspicuous
Boninastrea	1	Western Pacific	Rare
Scapophyllia	1	Eastern Indian Ocean to southern Pacific	Generally uncommon, conspicuous
Faviidae			
Caulastrea	Approx. 4	Red Sea and western Indian Ocean to southern Pacific	Generally common, very conspicuous
Favia	At least thirty	Cosmopolitan	Extremely common, conspicuous
Barabattoia	Approx. 3	Eastern Indian Ocean to southern Pacific	Sometimes common, readily confused with *Favia*
Favites	Approx. 15	Red Sea and western Indian Ocean to southern Pacific	Very common, conspicuous
Goniastrea	Approx. 12	Red Sea and western Indian Ocean to southern Pacific	Very common, generally conspicuous
Platygyra	Approx. 12	Red Sea and western Indian Ocean to southern Pacific	Extremely common, conspicuous but may be confused with *Goniastrea*
Australogyra	1	Central Indo-Pacific	Generally uncommon, very conspicuous

Family and genus	Extant species (no.)	Present distribution	General abundance
Leptoria	2	Red Sea and western Indian Ocean to southern Pacific	Sometimes common, conspicuous
Oulophyllia	Approx. 3	Red Sea and western Indian Ocean to western Pacific	Sometimes common, conspicuous
Montastrea	Approx. 13	Cosmopolitan	Generally common, conspicuous
Oulastrea	1	Central Indo-Pacific	Uncommon, occurs in non-reef biotopes
Plesiastrea	At least 2	Red Sea and western Indian Ocean to far eastern Pacific	Sometimes common
Diploastrea	1	Red Sea and western Indian Ocean to southern Pacific	Generally common, very conspicuous
Leptastrea	Approx. 8	Red Sea and western Indian Ocean to southern Pacific	Generally common, conspicuous
Cyphastrea	Approx. 9	Red Sea and western Indian Ocean to southern Pacific	Very common, conspicuous
Echinopora	Approx. 7	Red Sea and western Indian Ocean to southern Pacific	Very common, conspicuous
Moseleya	1	Central Indo-Pacific	Generally uncommon, very conspicuous
Trachyphylliidae *Trachyphyllia*	1	Red Sea and western Indian Ocean to southern Pacific	Uncommon, very conspicuous
Caryophylliidae *Euphyllia*	9	Red Sea and western Indian Ocean to southern Pacific	Generally common, very conspicuous
Catalaphyllia	Probably 1	Central Indo-Pacific	Generally uncommon, very conspicuous
Plerogyra	3	Red Sea and western Indian Ocean to southern Pacific	Generally uncommon, very conspicuous
Physogyra	3	Red Sea and western Indian Ocean to southern Pacific	Sometimes common, very conspicuous
Gyrosmilia	1	Western Indian Ocean	Uncommon
Heterocyathus	Probably 1 zoo-xanthellate species	Western Indian Ocean to central Pacific	Restricted to inter-reef habitats
Dendrophylliidae *Turbinaria*	Approx. 15	Red Sea and western Indian Ocean to southern Pacific	Very common, very conspicuous
Duncanopsammia	1	Central Indo-Pacific	Uncommon, very conspicuous
Heteropsammia	At least 2	Red Sea and western Indian Ocean to western Pacific	Uncommon, occurs on inter-reef sand flats

Note: List excludes those confined to the Indian Ocean/Red Sea, Atlantic, Caribbean, and/or the Gulf of Mexico.

CHAPTER 6

The Sponges of the Tropical Island Region of Oceania: A Taxonomic Status Review

Michelle Kelly-Borges and Clare Valentine

ABSTRACT

The taxonomic status of the sponges of the tropical island region of Oceania is reviewed and summarized to assist in the development of information management systems for biodiversity conservation initiatives within the region. More than 1,000 sponge species are listed for the Federated States of Micronesia, Tonga, Fiji, Samoa, Hawaii, the Solomon Islands, Vanuatu (New Hebrides), Kiribati and Tuvalu, French Polynesia, the Kermadec Islands of New Zealand, Norfolk and Lord Howe Islands of Australia, the Ryukyu Islands of Japan, Easter Island and Sala-y-Gómez of Chile, New Caledonia, and Papua New Guinea.

INTRODUCTION

Sponges are a significant component of all tropical, temperate, and polar marine benthic communities, having a large biomass above and within the substratum. They are a source of nutrition for fish, turtles, and molluscs and provide refuge for a diversity of micro- and macro-organisms. Far less is known about sponges and their basic bioecology than for most other sessile marine invertebrates; yet, as the oldest metazoans, studies on them have contributed a great deal to our understanding of life on earth. For example, the discovery of complex immune and cell aggregation systems in sponges has led to a better understanding of the cellular processes of higher organisms, and, in particular, the nature of their cell surfaces as active agents in cell and tissue recognition (Bergquist 1994; Kelly-Borges 1995).

The sponges are divided into three classes. Those in the largest class, the Demospongiae, possess a skeletal network of siliceous spicules, frequently supplemented or entirely replaced by spongin fibers. The Hexactinellida possess a unique skeletal structure of six-rayed siliceous spicules and a syncytial, rather than cellular organization, constructed of extensive regions of multinucleate cytoplasm (Reiswig and Mackie 1983). Fossil and recent sclerosponges, previously thought to be a fourth class of sponges, possess siliceous skeletons and calcareous basal skeletons. Recent sclerosponges are thought to represent relict forms of an ancestral sponge fauna, which includes the ancestry of several extant demosponge groups (Reitner 1991; Reitner and Engeser 1987).

The higher classification of the sponges has been based historically upon spicule shape, skeletal structure and organization, and gross morphology (Lévi 1956). Since the 1950s a substantial literature has been amassed, providing comparative data from ultrastructural histology, chemotaxonomy (Hooper et al. 1992; Kelly-Borges et al. 1994), and reproductive and benthic ecology of sponges (Fromont 1994; Fromont and Bergquist 1994; Kelly-Borges and Bergquist 1988). More recently, morphological (van Soest and Hooper 1993; Bergquist and Kelly-Borges 1991), biochemical (Kerr and Kelly-Borges 1994), and molecular data (Kelly-Borges, Bergquist, and Bergquist 1991; Kelly-Borges and Pomponi 1994; Rodrigo et al. 1994) have been used in phylogenetic analyses to enhance our understanding of the relationships between sponges.

Despite this recent progress, sponge classification is still relatively unstable. Within the Class Demospongiae alone, four out of the thirteen previous orders are currently undergoing revision. The "living fossil" sclerosponges and lithistids, and the axinellids, are now acknowledged to be polyphyletic (Reitner 1991; Kelly-Borges and Pomponi 1994; Hooper and Lévi 1993b); yet, it remains difficult to determine the affinities of their component species groups with other demosponge orders, and expert opinion is clearly divided (see Hooper and Wiedenmayer 1994).

Sponge Biodiversity

There are 6,000 to 7,000 described species of sponges (Minelli 1993; van Soest 1994), 150 of which are from freshwater. This is, however, an underestimate of the numbers of species that are known to exist; Hooper and Lévi (1994) estimate that there are about 5,000 undescribed species in the Australian island and continental faunas, alone. Very few regional sponge faunas are well known, except for those along the northeastern European coastline, in the Mediterranean Sea, the Caribbean Sea, and the north-west shelf of Australia. Only in rare cases are total regional faunas or taxonomic components of faunas thoroughly documented (e.g., Hooper and Lévi 1993a, b; Hooper 1991; Kelly-Borges and Vacelet in press; Bergquist and Kelly-Borges 1991; Kelly-Borges and Bergquist 1994; Hajdu and Desqueyroux-Faundez 1994; Bergquist, Ayling, and Wilkinson 1988).

Underestimates in worldwide sponge diversity are primarily due to difficulties in sponge classification and to a lack of comprehensive locality-specific collections. The most serious deterrent to obtaining accurate estimates of biodiversity is that many sponge genera lack an adequate range of taxonomic characters for species differentiation, and morphological and skeletal characters are plastic. Because of this, large numbers of established genera and "species" contain several entities (see Hooper 1994; Bergquist and Kelly-Borges 1991).

With the exception of a few studies, most older collections contain small numbers of specimens from diverse locations, rather than comprehensive collections from single habitats or regions. Most early publications are taxonomically unreliable and supply inadequate, unillustrated descriptions. There has been a tendency in the past for workers to use these few poorly described species names for sponges from diverse worldwide locations, giving the impression that large numbers of sponges are cosmopolitan (e.g., *Tethya aurantium* and *Aaptos aaptos*, but see Bergquist and Kelly-Borges 1991 and Kelly-

Borges and Bergquist 1994). It is becoming increasingly clear that regional endemism of sponges is substantially higher than for many other marine invertebrate phyla (Wilson and Allen 1987; van Soest 1989, 1990), and the size of local endemic faunas can be relatively small (remote islands or atolls, isolated patch reefs, bays, marine lakes) (Hooper 1994; Bergquist and Kelly-Borges 1991; Kelly-Borges, Bell, and Sharron 1994). What were once previously considered to be valid sponge "species" are increasingly being found to comprise heterogeneous populations, with biochemical and genetic diversity not necessarily manifested at the morphological level (Solé-Cava et al. 1991; Boury-Esnault, Solé-Cava, and Thorpe 1992; Hooper et al. 1992; Kerr and Kelly-Borges 1994; Kelly-Borges et al. 1994).

The Sponge Fauna of the Tropical Island Region of Oceania

Our knowledge of the general Indo-Pacific sponge fauna is derived from the early works of Baar (1903), Bowerbank (1877), Burton (1934), Hentschel (1912), Kirkpatrick (1900), Lendenfeld (1885a, b), Lindgren (1897), Ridley (1884a, b), Thiele (1898, 1899, 1900, 1903), Wilson (1925), and more recently the works of Bergquist (1965, 1979), Bergquist, Morton, and Tizard (1971), de Laubenfels (1935, 1949, 1950a, b, 1951, 1954, 1955, 1957), Desquey-roux-Faundez (1981, 1984, 1987, 1990), Fromont (1991, 1993), Kelly (1986), Kelly-Borges and Bergquist (1988), Hooper and Lévi (1993a, b, 1994), Hooper and Wiedenmayer (1994), Lévi (1961, 1967, 1993), Thomas (1982, 1991), and van Soest (1989, 1990). However, most of the earliest of these publications are extremely unreliable taxonomically, contain inadequate descriptions, and synonymies that require extensive revision. De Laubenfels (1954) expressed the difficulty of applying the older names in the literature to modern sponges:

> More than one hundred species are treated as new (for Micronesia). This is quite to be expected in view of the previously unstudied nature of the territory and the tendency of evolution to provide unique species in insular locations. Yet there is a problem here, because many of the students of sponges in the nineteenth century named species with utterly inadequate descriptions and no illustrations or poor ones. An example is found in Kieschnick's descriptions of East Indian species. Many of his descriptions are less than thirty words in total length, and some in less than twenty! They are completely devoid of measurements and provide no illustrations at all. If there are any types, their location is unknown. Because of the wars, it may well be that no specimens remain. Some names, thus unrecognizably given, may have been based on the species which also occur in my collections. Yet how can we ever know which, if any, are thus affected?

With the exception of Micronesia, Hawaii, Papua New Guinea, and New Caledonia, most regional faunas of the tropical island region of Oceania are poorly known. Prior to the 1950s, sponge species records are fragmentary, appearing in scientific reports of such major expeditions as that of the H. M. S. *Alert* (Ridley 1884a, b), the H. M. S. *Challenger* (Ridley and Dendy 1886, 1887), and the *Albatross* (Agassiz 1906; Lendenfeld 1910a, b, 1915; Wilson 1925), and in accounts of small numbers of species from isolated atolls and islands. The most valuable contribution to our knowledge of the sponge fauna of the tropical island region of Oceania was made by de Laubenfels (1954) in his monograph on the sponges of Micronesia. This was preceded and followed by numerous publications (de Laubenfels, 1935, 1950a, b, 1951, 1955) focusing on Micronesia and Hawaii. Bergquist

(1965, 1979) followed with valuable contributions on Belau (Palau) and Hawaiian sponges. Unfortunately her revisionary work on the sponges of Belau (Bergquist 1965) considered only a small number of the several hundred species that were described as new in de Laubenfels (1954). Although the 1954 monograph now requires extensive revision, the redeeming feature of de Laubenfels' work is that his species descriptions were based on the field characteristics of the living sponge. The clarity and accuracy of his descriptions have facilitated re-identification of these species in the locations from which they were described. The only other recent general systematic reviews available for the Oceania region cover the Solomons (Bergquist, Morton, and Tizard 1971), and Motupore Island in southern Papua New Guinea (Kelly-Borges and Bergquist 1988).

Today, sponge research in Oceania is pursued as part of general marine invertebrate programs at the University of Guam Marine Laboratory (chemical ecology), the University of Hawaii (sponge natural products chemistry), and the Center of Tropical and Subtropical Aquaculture, Hawaii, and College of Micronesia in Pohnpei (commercial bath-sponge culture). Current revisionary systematic studies (taxonomic monography), essential to our understanding and documentation of sponge biodiversity, are usually much broader in scope than the Oceania fauna, and so valid species records for the tropical island region of Oceania appear in a fragmented fashion in general taxonomic revisions (e.g., Kelly-Borges and Vacelet in press; Bergquist and Kelly-Borges 1991; Hooper and Bergquist 1992). While this later approach is crucial for the revision of diagnostic characters in an attempt to stabilize the systematics of the sponges, the accumulation of species information vital to biodiversity and conservation studies is slow.

However, some headway is being made in the general documentation of sponge biodiversity in the tropical island region of Oceania. A volume describing the sponges of New Caledonia is currently being prepared in conjunction with ORSTOM, New Caledonia (Contributors: P. R. Bergquist, C. Battershill, P. J. Fromont, J. N. A. Hooper, M. Kelly-Borges, C. Lévi, and J. Vacelet). The Coral Reef Research Foundation (CRRF), a non-profit organization based in Chuuk Atoll (Truk Atoll), Federated States of Micronesia, and Koror, Belau, is engaged in a long-term collection program in Micronesia and the western Pacific. One aspect of the foundation's work is to compile complete inventories of invertebrate diversity for selected Micronesian localities, but short-term expeditionary work has resulted in the recording of many new taxa and geographic records of sponges for Papua New Guinea, the Philippines, and other areas of the western Pacific (see Kelly-Borges and Vacelet in press; Colin and Arneson in press; Kelly-Borges, Bell, and Sharron 1994).

There is an urgent need to expand sponge biodiversity initiatives within the tropical island region of Oceania. It is a priority to review and define the diagnostic characters for sponge species and genera through comprehensive site-specific collections. Where genera contain numerous closely related species, greater emphasis needs to be placed upon photographic documentation of sponges in life, as details of important field characteristics such as color, habitat, and gross morphology are often lost upon preservation. This is particularly important for species in the genera *Niphates, Amphimedon, Haliclona, Callyspongia, Petrosia, Xestospongia, Oceanapia, Spheciospongia* (more commonly known as *Spirastrella*), *Axinyssa, Agelas, Clathria, Ircinia, Hyrtios,* and *Dysidea,* which are prevalent throughout Oceania. The primary aim of this review is to compile published and unpublished records of sponge species in the tropical island region of Oceania to assist in the development of information management systems for biodiversity conservation initiatives within the region. The resultant baseline taxonomic information will highlight problem areas for new revisionary systematic initiatives.

METHODS

More than 1,000 taxonomic, chemical, and ecological references to sponges from Oceania were compiled from the Zoological Record (Zoological Record: Porifera and Archaeocyatha, Vols. 1–130, Publishers: Zoological Society London and Biosis). The registers of the Porifera collection of the Natural History Museum of London (ca. 90,000 specimens) were searched for records of material from Oceania. Identifications of material in personal collections and several unpublished reports have also been included. Species records are listed by location within Tables 6.1–6.11 and are arranged alphabetically under their appropriate taxonomic order within these tables. The reference for each species record is given as a superscript numeral after the record, corresponding to the alphabetically arranged and numbered references in the bibliography. General references are also arranged alphabetically within the reference list but are not numbered.

For practicality this review is largely historical, and species names have been taken directly from the literature. No attempt has been made to revise them, except where they are well-known synonyms such as in the examples of *Chalina* (= *Haliclona*) and *Euspongia* (= *Spongia*). It is well known that sponge type material has been poorly described and mis-diagnosed in the past (Hooper and Wiedenmayer 1994), and sponge genera are currently undergoing intense revision. There is no value in altering any of the taxonomic assignments in this review without full histological examination of the material involved.

As the classification of the Porifera is relatively fluid, obviously polyphyletic groups such as the lithistid sponges and sclerosponges have been listed in the tables under these separate headings, rather than in their "new" systematic location, for ease of location. Although many calcareous and hexactinellid sponges are deep-water organisms, they have been included in this review for completeness. A classification of the demosponges reflecting major recent changes and systematic uncertainties is given in Table 6.12.

In recent years major initiatives have been established by pharmaceutical companies to discover biologically active compounds in marine invertebrates. This quest for new drugs focuses particularly on sponges, as they contain the widest diversity of biologically active compounds of any marine phylum (Garson 1994). Sponge secondary metabolite discoveries gradually began to dominate the sponge literature for Oceania and the general Indo-Pacific from the mid-1980s. There has been a tremendous surge of interest in the sponges of this region as it is a source of probably the greatest sponge diversity (van Soest 1989, 1990; Briggs 1974, 1987). Sponges recorded in the chemical literature are frequently identified only to the genus level reflecting the lack of recent revisionary systematic studies and new species descriptions, but have been included here as an aid to literature search for chemical structures in natural products research.

RESULTS

Easter Island and Sala-y-Gómez of Chile

Desqueyroux-Faundez (1990) provided the first full species account of sponges from Easter Island, but there are no records known for Sala-y-Gómez other than from a study of Hexactinellida by Tabachnik (1990) (Table 6.1). The general Chilean sponge fauna has been reviewed by Desqueyroux-Faundez (1994) but does not list specific information on Easter Island sponges. Prior to this, the ecological work of Di Salvo, Randall, and Cea (1988) identified nine sponges to genus and one to species. In a comprehensive bibliographic revision of the marine invertebrate fauna of the coastal zones of the Juan Fernandez Archipelago, Castilla and Rozbaczylo (1987) did not record any sponge species from Easter Island and Sala-y-Gómez, but provided general sponge references for the Archipelago (see Ridley and Dendy 1886, 1887; Breitfuss 1898; Thiele 1905, and Desqueyroux 1972). Agassiz (1906) visited Easter Island in the *Albatross* expedition of 1904 and remarked on the paucity of the sponge fauna of the region. A single sponge was described by Lendenfeld (1910b) from Easter Island.

French Polynesia

There is no general account of the sponge fauna of French Polynesia, but several records of sponges occur in scientific reports of the voyage of the H. M. S. *Challenger* (Ridley and Dendy 1887; Schulze 1887), and other papers on clionids (Topsent 1932, 1934), and "sclerosponges" (Vacelet 1977) (Table 6.2). Seurat (1934) and Salvat and Renaud-Mornant (1969) provide general ecological discussions of the region.

Hawaii

Prior to the 1950s, Hawaiian sponges were known only from remote deep-water collections of the H. M. S. *Challenger* (Schulze 1887; Ridley and Dendy 1887) and *Albatross* (Edmondson 1935; Baar 1903; Preiwisch 1903; Lendenfeld 1910a) expeditions (Table 6.3). The coastal fauna received considerably more attention after the 1950s by de Laubenfels (1950a, b, 1951, 1957) and Bergquist (1967, 1979). Bergquist (1979) revised de Laubenfel's species names and provided the most recent checklist of the sponge fauna to date, with

Table 6.1 Published records of sponges from Easter Island and Sala-y-Gómez of Chile

Class/Order	Species
DEMOSPONGIAE	
Astrophorida[73]	*Asteropus simplex, A. ketostea, Geodia amphistrongyla*[144]
Hadromerida[73]	*Cliona vastica, Pseudosuberites sulcatus, P. vakai, Spirastrella cuncatrix, Tethya deformis*
Poecilosclerida[73]	*Mycale paschalis, Tedania tepitootehenuaensis*
Haplosclerida[73]	*Cribrochalina dura, Callyspongia fusifera, Haliclona agglutinata, H. nitens, H. translucida, Reniera rapanui*
Dictyoceratida[73]	*Spongia virgultosa, Phyllospongia papyracea*
Verongida[73]	*Psammaplysilla purpurea*
HEXACTINELLIDA[211]	
	Caulodiscus polyspicula, Caulopacus abyssalis, C. vaniens juvenilis, Eurete lamellina, Farrea microclavula, F. occa polyclavula, Hexactinella divergens, Hyalonema campanula longispicula, Lophocalyx moscalevia, Malacosaccus leteropinlana, Pheronema megaglobosum, P. nasckaniensis, Platella polybasalia, Poliopogon amadon pacifica, P. maitai, Pseudoplectella dentatum, Regardrella peni, Schulzeviella gigas spinosum, Semperella alba, Tretopleura styloformis, Walteria leucarti longina

Table 6.2 Published records of sponges from French Polynesia

Class/Order	Species
DEMOSPONGIAE	
Astrophorida	*Pilochrota pachydermata*[206]
Hadromerida[221]	*Cliona carpenteri, C. vasifera, C. viridis, C. viridis caribbae, C. vermifera, C. celata, C. topsenti, C. ecaudis, Cliothoosa hancocki, Thoosa amphiasterina, Tethya seychellensis*[19]
Halichondrida	*Halichondria solida*[182]
Poecilosclerida	*Acarnus ternatus*[182], *Echinodictyum aspersum*[182], *Fausubera debrumi*[190]
Dictyoceratida	*Carteriospongia foliascens*[17],[141], *C. lamellosa*[17], *C. otahitica*[179], *Lufferiella variabilis*[140],[141],[169]
Dendroceratida	*Dysidea* sp[190]
CALCAREA[225]	
	Lelapiella incrustans sphaerulifera, Lepidoleucon inflatum, Murrayona phanolepis, Plectoninia radiata
HEXACTINELLIDA[197]	
	Dictyocalyx gracilis

Table 6.3 Published records of sponges from Hawaii

Class/Order	Species
DEMOSPONGIAE	
Homosclerophorida	*Oscarella tenuis*[15], *Plakina monolopha*[62],[15], *Plakortis simplex*[62],[15]
Astrophorida	*Asteropus kaena*[65],[15], *Erylus rotundus*[143],[15],[65], *E. sollasi*[143], *E. caliculatus*[143], *E. proximus*[62], *Geodia gibberella*[15],[62], *Rhabdastrella pleopora*[65], *Stelletta debilis*[15],[62], *Zaplethes digonoxea*[15],[60]
"Lithistida"	*Leidermatium* sp[15]
Hadromerida	*Anthosigmella valentis*[15],[65], *Cliona vastifica*[15],[60], *Chondrosia chucalla*[62], *Chondrosia* sp[176], *Diplastrella spiniglobata*[13],[15], *Kotimea tethya*[15], *Prosuberites oleteira*[15],[65], *Spheciospongia vagabunda*[13], *Spirastrella coccinea*[15],[13], *S. keaukaha*[62], *Terpios zeteki*[15],[62],[60], *T. granulosa*[13],[15],[108],[191], *Tethya diploderma*[62],[60],[15], *Timea xena*[15]
Halichondrida	*Axechina lissa*[15], *Axinella solenoides*[15],[65], *Ciocalypta pencillus*[13], *Eurypon nigra*[15],[13], *E. distincta*[15],[65], *Hymeniacidon chloris*[15],[60], *Homaxinella anamesa*[15],[65], *Halichondria coerulea*[13],[15], *H. dura*[15],[62], *H. melanadocia*[13],[15], *Phycopsis aculeata*[15],[62], *Raphisia myxa*[15],[62], *Ulosa rhoda*[65]
Poecilosclerida	*Axociella kilauea*[15],[62], *Clathria procera*[13],[15], *Damiriana hawaiiana*[15],[44],[45],[62],[60], *Esperiopsis anomala*[182], *Iotrochota protea*[15],[60],[62], *Kaneohea poni*[60], *Lissodendoryx calypta*[15],[65], *Mycale cecilia*[60],[15], *M. contarenii*[62],[15], *M. maunakea*[62], *Myxilla rosacea*[15],[62],[65], *Naniupi ula*[15],[60], *Microciona haematodes*[15],[65], *M. maunaloa*[15],[62],[65], *Strongylacidon* sp.[42], *Tedania ignis*[15],[60],[62], *T. macrodactyla*[13],[15], *Xytopsiphum kaneohe*[15],[60], *X. meganese*[15],[62], *Xytopsues zukerani*[65], *Zygomycale parishii*[15],[60]
Haplosclerida	*Adocia gellindra*[15], *Callyspongia diffusa*[15],[62],[80],[160],[202], *Haliclona flabellodigitata*[15],[65], *H. aquaeducta*[15],[62], *H. permollis*[15],[62], *Toxadocia violacea*[15],[60],[62]
Petrosida	*Pellina eusiphonia*[15], *P. sitiens*[65], *Petrosia puna*[15],[62]
Dictyoceratida	*Coscinoderma denticulatum*[169], *Hippospongia densa*[7], *Lufferiella* sp[109], *Spongia irregularis dura*[7], *S. irregularis tenuis*[7], *S. irregularis lutea*[7], *S. irregularis mollior*[7], *S. denticulata*[141], *S. oceania*[15],[62], *Stelospongia lordii*[7]
Dendroceratida	*Aplysilla rosea*[13],[15], *A. sulphurea*[13],[15], *A. violacea*[13],[15], *Dendrilla cactus*[15], *Dysidea herbacea*[13],[15], *D. avara*[60], *Dysidea* sp[109], *Pleraplysilla hyalina*[15],[60]
Verongida	*Psammaplysilla purpurea*[13],[15],[110]
CALCAREA	
	Clathrina sp[15], *Leucosolenia vesicula*[15],[65], *L. eleanor*[15], *Leucetta solida*[15],[60],[62], *Leuconia kaiana*[62], *Sycandra parvula*[171], *S. staurifera*[171], *S. coronata*[137]
HEXACTINELLIDA	
	Euplectella sp[8], *Stylocalyx elegans*[197]

sixty coastal species identified from Hawaii. Since this time, sponge species names have appeared sporadically within the chemical literature (see Erdman and Scheuer 1976; Carney, Scheuer, and Kelly-Borges 1993b; Jurek et al. 1993).

Kermadec Islands of New Zealand

Kirk (1911) described six of "apparently eleven or twelve species" represented in a collection of sponges from the Kermadec Islands off the northeast coast of New Zealand (Table 6.4). All were undescribed species, and he expressed surprise at this because the *Challenger* collections contained sponges from the Kermadec Islands and "Von Lendenfeld had described others." A search through Lendenfeld's publications failed to turn up those species names due to a lack of precise locality data. Published species names were re-listed in Barraclough-Fell (1950). Lévi (1964) provides the most recent account of extremely deep-water sponges from the Kermadec Trench.

Norfolk Island and Lord Howe Island of Australia

There are no published records of sponges from any of the Australian Territorial Islands (J. N. A. Hooper, pers. com.). The only sponge collections known from Lord Howe Island and Norfolk Island were made by N. Coleman between 1960 and 1970. These specimens were deposited in the Australian Museum but were recently disposed of due to a lack of locality data. The genus *Ianthella* has been recorded from Lord Howe Island (Bergquist and Kelly-Borges in press).

Fiji, Samoa, and Tonga

In 1965, a small collection of shallow-water sponges from Viti Levu in the Fiji Islands was made by Dr. T. Wolff during an expedition on the U.S. research vessel *Te Vega*. (Table 6.5). Previously only four sponges had been reported by Bowerbank (1873) and Lendenfeld (1889). A number of species have been added to this list in natural products publications, some of which are undescribed species. There is very little published information on the sponges of Samoa; two species were described in Selenka (1867), and de Laubenfels (1943) describes several sponges from the region. A single species was described from the *Challenger* collection by Schulze (1887). There is a single species recorded from Tonga by Lendenfeld (1906). One of the authors' (MKB) personal collections from Vava'u, Tonga, included several species common throughout the Indo-Pacific (*Hyrtios erecta*, *Stylotella aurantium*, *Dysidea avara*), in addition to species of *Latrunculia*, and a higher proportion

Table 6.4 Published records of sponges from the Kermadec Islands of New Zealand

Class/Order	Species
DEMOSPONGIAE	
Hadromerida	*Tethya lyncurium australis*[120], *T. deformis*[14],[120]
Poecilosclerida	*Asbestopluma wolffi*[148], *A. hadalis*[148], *A. brunni*[148], *A. biserialis*[148], *Chondrocladia asigmata*[148], *Clathria intermedia*[18],[120], *Myxilla frondosa*[182], *Raspailia rubrum*[120]
Haplosclerida	*Callyspongia fistulosa*[22],[120], *Callyspongia oliveri*[22],[120], *Haliclona reversa*[22],[120]
HEXACTINELLIDA[197]	
	Aulochone cylindrica, *Chonelasma lamela*, *C. hamatum*, *Euryplegma auriculara*, *Farrea occa*, *Polipogon gigas*, *Walteria flemmingii*

Table 6.5 Published records of sponges from Fiji[†], Samoa[§], and Tonga[‡]

Class/Order	Species
DEMOSPONGIAE	
Homosclerophorida	*Plakortis* sp[†50]
Astrophorida	*Ancorina acervus*[†212], *A. densa*[†27,179], *Pachymatisma contorta*[†144], *Myriastra purpurea*[†212], *Rhabdastrella dura*[†212], *Jaspis johnstoni*[†237], *jaspidae*[†1], *Geodia nux*[§144,198], *Halina plicata*[†212]
Spirophorida	*Paratetilla bacca*[†50], *Cinachyra australiensis*[†212], *C. alba-bidens*[‡142]
"Lithistida"	*Neosiphonia superstes*[†206], *Pleroma turbinatum*[†206]
Hadromerida	*Chondrilla australiensis*[†212], *Diacarnus spinipoculum*[†116], *Placospongia carinata*[†212], *Spheciospongia vagabunda*[†212], *Spirastrella solida*[†212], *Tethya diploderma*[†212], *Timea stellata*[†212], *Terpios hoshinoto*[§189]
Halichondrida	*Cymbastela* sp[†112]
Poecilosclerida	*Chondrocladia clavata*[†18,181,182], *Microciona eurypa*[†212]
Haplosclerida	*Callyspongia* sp[35], *Haliclona* sp[35], *Niphates* sp[†172]
Petrosida	*Xestospongia caycedoi*[†162], *Xestospongia* sp[†‡106]
Dictyoceratida	*Aplysinopsis* sp[†49], *Carteriospongia contorta*[†17], *C. lamellosa*[†17], *Carteriospongia* sp[†196], *Fascaplysinopsis reticulata*[†107], *Fascaplysinopsis* sp[†186], *Ircinia* sp[†160], *Spongia irregularis silicata*[†139], *S. discus*[†139,141], *S. irregularis lutea*[†139,141], *S. mycofijiensis*[†111,112], *thorectid*[87]
Dendroceratida	*Dysidea fragilis*[†163], *Dysidea* sp[†196]
Verongida	*Ianthella concentrica*[†103,140,141], *Psammaplysilla purpurea*[†48], *Psammaplysilla* sp[†‡173]
CALCAREA	
	Ascetta primordialis[†138], *Leucetta* sp[†5], *Leucandra bomba*[†137]
HEXACTINELLIDA	
	Taegeria pulchra[†197]

of dictyoceratid sponges than in the more western Pacific locations. These observations indicate a temperate-water influence on the sponge fauna. See also Wicksten (1981).

Papua New Guinea

It was not until 1988 that the first general account of sponges from Papua New Guinea became available (Kelly-Borges and Bergquist 1988), although Thomas (1982) earlier provided brief descriptions of some haploslerids (Table 6.6). Kelly (1986) identified 100 species, most of which were new, from Motupore Island within the southern Papua New Guinea Barrier Reef system. Prior to this recent literature, one or two species were recorded but these were either deep-water records (Lendenfeld 1906) or from the western part of the island, now known as Iryan Jaya (Bowerbank 1877). There are no recent papers that deal exclusively with Papua New Guinea sponges. Most records are parts of revision of monophyletic genera (Bergquist and Kelly-Borges 1991; Kelly-Borges and Vacelet in press; Sará 1992), ecological considerations (Bakus and Ormesby 1994), or natural products chemistry (see Uchio 1992; Iwagawa et al. 1992). There are, however, several comprehensive unpublished collections available from the northern and southern Papua New Guinea coastlines, made by one of the authors (MKB), CRRF, and the Australian Institute of Marine Science (McCauley et al. 1992). Unpublished species names are also listed in Table 6.6.

New Caledonia

The fauna of New Caledonia has been comparatively well studied with more than 200 species recently recorded for the region (Lévi 1967, 1979, 1983, 1993; Lévi and Lévi 1979, 1982, 1983, 1984, 1988; Desqueyroux-Faundez 1984, 1987; Hooper and Lévi 1993a, b;

Table 6.6 Published records of sponges from Papua New Guinea

Class/Order	Species
DEMOSPONGIAE	
Homosclerophorida	*Plakinastrella ceylonica**, *P. schulzei**, *Plakortis mammilaris*[161], *P. lita*[161],*
Astrophorida	*Ancorina acervus*[118],[113],*, *Geodia globostellifera*[113],*, *Jaspis johnstoni*[31],[30], *Melophlus sarasinorum*[161],[113],*, *Myriastra clavosa*[142],[113],[161],*, *Neamphius huxleyi*[68], *Plakinastrella schulzei**, *Rhabdastrella pleopora**, *Stelletta bouginvillea*[142]
"Lithistida"	*Leiodermatium pfeifferae**, *"Plakinalopha" mirabilis**, *Theonella cylindrica**, *T. invaginata**, *T. swinhoei**
Spirophorida	*Cinachyrella porosa**, *Cinachyra australiensis*[113], *C. schulzei**, *C. alba-obtusa*[142], *Paratetilla bacca*[113]
Hadromerida	*Chondrilla australiensis**, *Chondrosia corticata**, *Cliona jullieni**, *C. coralliophila**, *Spheciospongia vagabunda*[193],[161],[113],[114],*, *S. trincomalensis*[113], *S. tubulodigitata**, *S. gallensis*[113],*, *S. fungiodes*[113], *S. globularis**, *Terpios zeteki*[118],[113], *T. aploos**, *Tethya coccinea*[19],[113], *T. seychellensis*[19],[193],[113], *T. robusta*[19],[193],[113], *T. boeroi*[193], *T. densa*[193], *T. diploderma*[193], *T. microstella*[193], *T. pulchra*[193], *T. viridis*[193]
Halichondrida	*Axinyssa terpnis**, *A. fenestratus**, *A. aplysinoides**, *A. topsenti**, *Amorphinopsis foetida**, *Acanthella inflexa*[161], *A. costata**, *Ciocalypta sacciformis*[118],[113],*, *Cymbastela coralliophila**, *Cymbastela* sp[67], *Collocalypta* sp*, *Didiscus aceratus**, *Epipolasis stalagmites**, *Higginsia mixta**, *H. scabra**, *H. strigilata,** , *Katiba milnei**, *Phakellia flabellata*[161], *P. cavernosa*[113],*, *P. dendyi*[161], *P. carduus**, *P. stipitata**, *Pararaphoxya tenuiramosa**, *Reniochalina stalagmites**, *Stylotella aurantium*[118],[161],*, *S. aldis*[113], *Topsentia cavernosa**
Poecilosclerida	*Agelas mauritiana*[161], *A. robusta**, *Aegogropila digitata*[118],[113],*, *Acarnus ternatus*[113], *A. wolfgangi**, *Biemna fortis*[118],[113],* , *B. fistulosa**, *B. mnioeis**, *Coelocarteria singaporense*[118],[113],[161],*, *Clathria (Thalysias) reinwardti*[118],[113],[161],*, *C. vulpina**, *C. araiosa**, *Ectyodoryx frondosa**, *Echinodictyum asperum*[161], *E. antrodes**, *Iotrochota baculifera*[11],[113],[161], *I. purpurea*[220], *I. ditrochota**, *Liosina granularis*[118],[161],*, *L. paradoxa*[161],*, *Microciona eurypa*[118],[113], *Mycale armata**, *M. strongylophora**, *Neofolitispa ungiculata*[118],[113],[161],*, *Ophlitospongia meyeri*[29], *"Prianos" osiris**, *Psammopemma densum subfibrosa**, *Raspailia bifurcata**, *Sideroderma navicelligerum*[140],[181], *Zyzzya grata**
Haplosclerida	*Adocia pigmentifera*[161], *Aka mucosa*[118],*, *A. coralliphagum*[161], *Callyspongia rosa*[118], *C. fibrosa*[220], *C. confoederata*[161],[113],*, *C. aerizuza*[161],*, *C. schulzei*[161], *C. flammea*[201], *C. diffusa**, *C. muricina**, *Cribrochalina olemda**, *Euplacella ridleyi*[113], *E. annulata**, *E. elongata**, *Haliclona koremella*[118],*, *H. acoroides*[118],[113],[161], *H. cribricutis*[220], *H. viridis*[113],*, *H. scyphonoides*[113], *H. coerulescens**, *H. nigra**, *Haliclona* sp[9], *Gelliodes fibulata*[118],[220],[161],[113],*, *Niphates cavernosa*[118], *N. obtusa**, *N. chinensis**, *N. (Pachychalina) melior**, *Sigmadocia symbiotica*[118],[113],*, *S. amboinensis*[118],[113],*, *Siphonochalina extensa*[140], *S. fascigera**, *Toxochalina foliodes*[28],[179]
Petrosida	*Oceanapia saggitaria*[118],[113], *O. fistulosa*[82],[113], *O. amboinensis**, *O. media**, *Petrosia seriata*[220], *P. sphaeroida*[220], *P. constignata*[161], *P. pigmentosa*[161], *P. ficiformis**, *P. capsa**, *P. hebes**, *P. ficiformis**, *Strongylophora durissima*[220], *S. strongylata**, *S. septata*[220],*, *Xestospongia testudinaria*[220],[161],[113],*, *X. nigricans*[220], *X. pacifica*[118], *Xestospongia exigua*[161],[113],*
Dictyoceratida	*Carteriospongia foliascens*[17],[140],[219],[141], *C. otahitica*[169], *C. flabellifera*[29],[17],[118],[113],*, *Coscinoderma mathewsi**, *Fasciospongia cavernosa*[219], *F. chondrodes*[161], *Fascaplysinopsis reticulata**, *Hyrtios erecta*[219],[161],[113],*, *Hyatella intestinalis*[161], *Halispongia stellifera*[29], *Ircinia variabilis*[140], *Phyllospongia papyracea**, *Spongia irregularis*[113], *S. officinalis ceylonensis*[219], *Sarcotragus arbuscula**, *Smenospongia lamellata**
Dendroceratida	*Dysidea fragilis*[219],[161],[113], *D. herbacea*[219],[113], *D. avara**, *Dysidea* sp[184]
Verongida	*Aplysinella strongylata**, *A. pedunculata**, *A. tyroeis**, *Ianthella flabelliformis*[219],[113], *I. basta*[200],[113],[161],*, *Psammaplysilla purpurea*[118],[219],[161],[113]
"Sclerosponges"	*Astrosclera willeyana**
CALCAREA	
	Leucetta philippinensis[161], *Leuconia palaoensis**, *Pericharax heteroraphus*[161],[113]

* = Coral Reef Research Foundation identifications

Hooper and Bergquist 1992) (Table 6.7). A volume describing the common sponges of New Caledonia is currently in preparation (Contributors: P. R. Bergquist, C. Battershill, P. J. Fromont, J. N. A. Hooper, M. Kelly-Borges, C. Lévi, and J. Vacelet). Additional references include Rützler (1970) on freshwater sponges, and La Barre, Laurent, Sammarco, Williams, and Coll (1988) on sponge ichthyotoxicity.

Table 6.7 Published records of sponges from New Caledonia

Class/Order	Species
DEMOSPONGIAE	
Astrophorida[155]	*Ancorina acervus*[149], *Asteropus simplex*[149], *Characella flexibilis*[152], *Chelotropella neocaledonica*[152],[155], *Erylus fibrillosus*, *E. burtoni*, *Erylus* sp.[52], *Geodia vaubani*, *Monosyringia patriciae*[152], *Penares schulzi*, *P. micraster*[152], *P. palmatoclada*[152], *Pachataxa enigmata*, *Psammastra oxygigas*[152], *Poecillastra laminaris*, *P. stipitata*[152], *Sphinctrella orthotriaena*[152],[155], *Stelletta centroradiata*, *S. radicifera*, *S. hyperoxea*, *S. vaceleti*, *S. phialimorpha*[152], *S. toxiastra*[152], *Tylaspis topsenti*[156], *Thenea microspirastra*
"Lithistida"[155]	*Anaderma rancureli*, *Aulaxinia clavata*[157], *Corallistes fulvodesmus*, *C. multituberculatus*, *C. undulatus*, *C. microstylifer*, *Callipelta punctata*, *Costifer wilsoni*[152], *Discodermia proliferans*, *Herengeria auriculata*[157], *Iouea moreti*[157], *Jereicopsis graphidophora*[53],[155], *Lepidosphaera hindei*[153], *Macandrewia spinifoliata*, *Neosiphonia superstes*[155],[157],[166], *Neopelta plinthosellina*[157], *Pleroma menoui*[85],[155], *P. turbinatum*, *Reidispongia coerulea*[157], *Aciculites oxytylota*, *A. papillata*, *Microscleroderma herdmani*, *M. stonae*,*Scleritoderma camusi*
Spirophorida[155]	*Craniella neocaledonica*, *Tetilla falcipara*[152]
Hadromerida	*Atergia acanthoxa*[152], *Chondrilla australiensis*[149], *Cliona orientalis*[229], *Halicometes hooperi*[152], *Latrunculia brevis*[152], *L. crenulata*[152], *Podospongia similis*[152], *Polymastia tropicalis*[149], *Rhizaxinella dichotoma*[152], *Spheciospongia vagabunda*[149], *Spirastrella inconstans*[229], *Suberites pisiformis*[152], *Sphaerotylus exospinosus*[152], *Spinularia australis*[152], *Tethya japonica*[149], *T. sollasi*[19], *T. levii*[192], *T. novaecaledoniae*[192], *Tylexocladus hispidis*[152], *Trichostemma sarsi*[152], *Trachycladus stylifer*[152], *Trachostylea lamellata*[152]
Halichondrida[97]	*Axinella lifouensis*[152],[97], *A. carteri*, *Cymbastela cantherella*[95],[151], *C. concentrica*,*Higginsia anfractuosa*, *H. tanekea*, *H. massilis*, *Myrmekioderma granulata*, *Phakellia columnata*, *P. pulcherima*, *P. stipitata*, *P. plumosa*[156], *Ptilocaulis epakros*, *P. fusiformis*[149], *P. papillatus*, *Pseudaxinella debitusae*, *Parahigginsia phakellioides*, *Prostylissa radiata*[156], *Reniochalina plumosa*, *R.condylia*,*Raphoxya systrema*, *Stylissa flabelliformis*, *Spongosorites bubaroides*[152], *Trachycladus digitatus*
Poecilosclerida[96]	*Aulospongus clathrioides*[96],[149], *Antho novizelanica*, *Acarnus caledoniensis*, *Agelas mauritiana*[97],[149], *A. novaecaledoniae*[97],[156], *Agelas dendromorpha*[152], *Asbestopluma bilamellata*[152], *A. biserialis*[152], *Artemisina elegantula*[152], *Biemna granulosigmata*[152], *Crella spinulata*, *Clathria australiensis*[96],[149], *C. rugosa*, *C. litos*, *C. vulpina*, *C. flabellifera*, *C. cornoelia*, *C. araiosa*, *C. hirsuta*, *C. kylista*, *C. bulbosa*, *C. menoui*, *Clathria anthoides*[152], *C. macroisochela*[152], *Ceratopsion clavata*, *C. palmata*, *Coelosphaera bullata*[152], *C. chondroida*[152], *C. pedicellata*[152], *Coelodischela massa*[152], *Cladhoriza schistochela*[152], *C. similis*[152], *Chondrocladia concrescens*[152], *C. scolionema*[152], *C. pulvinata*[152], *Desmacella toxophora*[152], *Echinochalina intermedia*, *E. mollis*[86], *E. laboutei*, *E. bargibanti*, *Echinostylinos gorgonopsis*[152], *Esperiopsis macrosigma novaezelandiae*[18], *Esperiopsis challengeri*[152], *E. diasolenia*[152], *E. flava*[152], *E. inodes*[152], *E. magnifolia*[152], *Heterocornulum virguliferum*[156], *Hymedesmia brachyrhabda*[156], 152, *H, spiniarcuata*[156], *Hamacantha acerata*[152], *H. atoxa*[152], *H. forcipulata*[152], *Hoplikathara exoclavata*[152], *Iotrochota baculifera*[149], *Lissodendoryx bifacialis*[96],[152],[156], *L. stylophora*[96],[156], *L. catenata*[152], *L .tubiformis*[152], *Lithoplocamia dolichosclera*[96],[156], *Mycale myriasclera*[97],[156], *M. incurvata*[152], *Neofibularia hartmani*, *Plocamione pachysclera*[152],[96], *Phlyctaenopora bocagei*[152], *Phorbas erectus*[152], *Raspailia wilkinsoni*, *Stelodoryx chlorophylla*[66],[152], *S. phyllomorpha*[152], *Tedaniopsis turbinata*[152], *Yvesia acanthosclera*[96],[156]
Haplosclerida[71]	*Adocia* sp[92], *Amphimedon conica*, *A. viridis*, *Arenosclera heroni*, *A parca*, *A. rosacea*, *Cladacroce incurvata*[54], *Callyspongia carens*, *C. diffusa*[149], *C. communis*, *C. confoederata*, *C. fallax*, *C. fibrosa*, *C. laxa*, *C. peroni*, *C. subarmigera*, *C. tenerrima*, *C. villosa*, *C. aerizusa*[71],[83], *C. flammea*, *C. fruticosa*, *C. hispidoconulosa*, *C. parva*, *C. polymorpha*, *C. pseudoreticulata*[71],[83], *C. rigida*, *C. spinimarginata*, *Gelliodes fibulata*[71],[149], *G. carnosa*, *G. incrustans*, *G. fragilis*, *Gellius flagellifer*[152], *G. pedunculatus*[152], *G. cymiformis*[83],[149],[229], *Haliclona nodosa*[152], *Niphates amorpha*, *N. hispida*, *Rhizoniera strongylata*[156], *Toxochalina murata*, *T. robusta*, *T. fenestrata*, *T pseudofibrosa*, *T. staminea*
Petrosida[72]	*Inflatella perlucida*, *Foliolina vera*[152], *Oceanapia bartschii*, *O. fistulosa*[83],[72], *O. papula*, *O. tenuis*, *Oceanpia* sp[133], *Pellina triangulata*, *Petrosia ficiformis*, *P. hebes*, *P. capsa*, *P. granifera*, *Xestospongia coralloides*, *X. subtriangularis*, *X. testudinaria*, *Xestospongia* sp[175]
Dictyoceratida	*Fasciospongia* sp[119], *Lendenfeldia dendyi*[229]
"Sclerosponges"	*Acanthochaetetes wellsi*[228], *Astrosclera willeyana*[97],[228],[232], *Merlia deficiens*[228], *M. normani*[96], *Stromatospongia micronesica*[228], *Vaceletia crypta*[226],[227],[228]
CALCAREA[228]	*Anamixilla torresi*[26], *Grantessa syconiformis*[26], *Leucetta chagosensis*[26], *L. microraphis*[26], *Lelapiella incrustans*, *Lepidoleucon inflatum*, *Murrayona phanolepis*, *Minchinella kirkpatricki*, *Paramurrayona corticata*, *Plecronina neocaledoniense*, *P. microstyla*, *P. tetractinosa*, *P. lepidophora*, *P. hindei*, *P. minima*, *P. vasseuri*, *P. tecta*, *Tulearinia stylifera*
HEXACTINELLIDA[154]	*Aulochone clathroclada*, *Eurete farreopsis*, *Hyalonema* sp, *Pheronema semiglobosum*, *P. conicum*, *Semperella schulzei*, *Regadrella okinoseana*, *Pleurochorium cornatum*

Solomon Islands and Vanuatu (New Hebrides)

Bergquist, Morton, and Tizard (1971) described a small collection of common sponges collected during the 1965 Royal Society of London Expedition, emphasizing habitat descriptions and ecology in addition to provision of species records (Table 6.8). There is no general account of the Vanuatu sponge fauna, although several species were described from the *Challenger* expedition (Ridley and Dendy 1887; Lendenfeld 1889; Sollas 1888; Polejaeff 1884). Kirkpatrick (1908) described two sclerosponges from this region.

Federated States of Micronesia, Guam, Northern Mariana Islands, Belau (Palau), and the Marshall Islands

Knowledge of the sponge fauna of the Federated States of Micronesia is based upon the comprehensive but largely unrevised work of de Laubenfels (1954) (Table 6.9). Bergquist (1965) revised several of de Laubenfel's species and added new records. The area covered by de Laubenfel's important field-based study extends from 130° to 180° east longitude and from the equator to 20° north latitude consisting of the four largest island archipelagos—Guam, and Saipan in the Northern Mariana Islands, Belau (Palau) and Yap of the Western Caroline Islands, Pohnpei and Chuuk of the Eastern Caroline Islands, and Likiep, Majuro, Ebon, Ailinglaplap, Bikini, and Enewetak Atolls of the Marshall Islands. Intensive study was made at widely scattered points, and collections made by others (T. E. Bullock and J. P. E. Morrison) provided records for Bikini and Enewetak Atolls. The Coral Reef Research Foundation based in Chuuk and Belau are currently carrying out comprehensive habitat-based collections for selected Micronesian localities. Table 6.9 lists all known published and unpublished species (CRRF) from Micronesia. See also Tendal, Barthel, and Tabachnik (1993).

Table 6.8 Published records of sponges from the Solomon Islands and Vanuatu

Class/Order	Species
DEMOSPONGIAE	
Homosclerophorida	*Plakortis quasiamphiaster*[75]
Astrophorida[21]	*Melophlus sarasinorum, Ancorina acervus, Amphius huxleyi*[206]*, Jaspis coriacea, Stelletta durissima, S. tethyoides*[140]*, Thrombus challengeri*[206]
Spirophorida[21]	*Cinachyra australiensis*
Hadromerida[21]	*Placospongia melobesiodes, Spheciospongia vagabunda, Spirastrella coccinea, Tethya robusta*[19],[21]*, T. seychellensis*
Halichondrida[21]	*Halichondria solida rugosa*[182]*, Halichondria* sp, *Hymeniacidon aldis, Liosina paradoxa, Myrmekioderma granulata*
Poecilosclerida[21]	*Agelas mauritiana, Monanchora ungiculata, Clathria (Rhaphidophlus) reinwardti*
Haplosclerida[21]	*Callyspongia diffusa, Dactylia infundibuliformis, Gelliodes callista, Rhizochalina pedunculata*[182]
Petrosida	*Xestospongia exigua*[21]*, Xestospongia* sp[174]
Dictyoceratida	*Carteriospongia foliascens*[21]*, Cacospongia spinifera*[169]*, C. intermedia*[169]*, Hippospongia metachromia*[21]*, H. mauritiana*[169]*, Hyrtios erecta*[21]*, Hyatella intestinalis*[21]*, Hyatella sinuosa*[141]*, Ircinia foetida*[141]*, Lufferiella variabilis*[140],[141],[169]*, Lendenfeldia frondosa*[3]*, Phyllospongia papyracea*[21]*, Stelospongia cavernosa mediterranea*[141]
Dendroceratida[21]	*Aplysilla violacea, Dysidea herbacea, D. chlorea, Dysidea* sp[4]
Verongida[21]	*Psammaplysilla purpurea*
CALCAREA[122]	
	Minchinella lamellosa (Vanuatu)

Table 6.9 Published species records of sponges from the Federated States of Micronesia, Guam, the Northern Mariana Islands, Belau (Palau), and the Marshall Islands

* = Coral Reef Research Foundation identifications, X denotes records from published data, Saipan = Saipan, Tinian and Rota Islands (Northern Mariana Islands), Lik = Likiep Atoll, Maj = Majuro Atoll, Ebo = Ebon Atoll, All = Ailinglaplap Atoll, Ene = Enewetak Atoll, Bik = Bikini Atoll, Chuuk = Chuuk State. Several of the de Laubenfels[63] species have been excluded from this table due to their uncertain taxonomic status.

Class/Order/Species	Marianas		Marshalls						W. Carolines		E. Carolines	
	Saipan	Guam	Lik	Maj	Ebo	All	Ene	Bik	Belau	Yap	Pohnpei	Chuuk
DEMOSPONGIAE												
Homosclerophorida												
Plakortis lita[55],[63],[56],*											X	X*
P. simplex[63]					X							
*Plakinastrella ceylonica**												*
Plakina sp[187]											X	
Astrophorida												
Ancorina acervus[63],[12],*	*		X				X	X	X		X	
Dorypleres splendens[63],*	*								*		X	*
Jaspis coriacea[12]									X			
J. tuberculata[63]			X					X				
J. stellifera[63],*	*						X					X
Melophlus sarasinorum[12],[63],[131],[134],*		*							X		?X	X*
Penares nux[63],*									X			*
*Rhabdastrella pleopora**												*
Stelletta durissima[12]									X			
"Lithistida"												
*Aciculites papillata**	*											
Spirophorida												
*Acanthotetilla enigmatica**												*
Cinachyra australiensis[63],[12],*							X		X		X	X*
Cinachyra porosa[63],*	X								X			*
Craniella abracadabra[63],*											X	
Paratetilla bacca[12],*	*								X			*
P. lipotriaena[63]									X		X	
Tetilla microxea[12]									X			
Hadromerida												
Atergia purpurea[63]				X								
Aaptos unispicularis[63],*							X					*
A. chromis[63]												X
A. ana[63]												X
Cliona lobata[63]			X		X							X
C. schmidti[63]									X			
C. euryphylla[63]											X	
C. vastifica[63]				X								
Chondrilla australiensis[63]			X	X			X					
C. nucula[63]					X							
C. acanthastra[63]									X			
C. euastra[63],[59]										X		
C. grandistellata[63]			X									
Chondrosia chucalla[63]							X	X				
*C. corticata**	*											
Diacarnus spinipoculum[116],*												*
Prosuberites andrewsi[63]				X								
Placospongia melobesiodes[63],[12]									X		X	
*P. carinata**	*											
Ridleia pelia[63],*									X			

Table 6.9 *(continued)*

Class/Order/Species	Marianas		Marshalls						W. Carolines		E. Carolines	
	Saipan	Guam	Lik	Maj	Ebo	All	Ene	Bik	Belau	Yap	Pohnpei	Chuuk
Spheciospongia aurivilli[12],*	*								X			
*S. tubulodigitata**												*
*S. inconstans**												*
*S. globularis**												*
S. vagabunda[12,63],*	*		X		X		X		X		X	X
*S. purpurea**	*											
Spirastrella potamophera[63]			X	X	X	X			X		X	
S. decumbens[63]					X	X						X
Terpios fugax[63],*					X						X	*
Terpios sp[168]		X										
Terpios sp[39]		X										
T. aploos[63]		X			X							
T. hoshinoto[189]		X									X	
Timea granulata[12]									X			
Tethya seychellensis[63]			X									
T. viridis[63]					X							
T. coccinea[63],*	*				X		X					
*T. robusta**	*											
Halichondrida												
*Axinella proliferans**												*
*A. carteri**												*
Auletta bia[63],*											X	*
Axinyssa aplysinoides[90],*		*		*					X			*
A. pitys[63,12],*	*								X			*
A. terpnis[63,12],*	*								X		X	*
A. xutha[63],*	*											X
*Cymbastela marshae**												*
*C. vespertina**												*
*C. cantherella**	*											*
Ciocalypta sacciformis[63]									X			
*Dragmatella hispida**									*			*
Dictyonella dasyphyllia[63]									X			
Densa mollis[63],*	*											X
Homaxinella trachys[63,234]					X	X						
H. phrix[63]												X
Halichondria adelpha[63]					X		X					
Hymeniacidon maza[63]												X
H. dystacta[63]					X		X					
Halichondria sp[209]											X	
*Higginsia anfractuosa**	*											
H. mixta[12],*	*								X			
Katiba milnei[63]					X							*
*Liosina paradoxa**												*
Myrmekioderma granulata[63,12],*	*								X		X	X*
Pararaphoxya tenuiramosa[63]											X	
Phakellia costata[167]											X	
*P. stipitata**												*
*P. cavernosa**												*
*Pseudaxinella australis**	*										X	*
Quepanetsal madidus[63]					X							
*Rhabderemia sorokinae**	*											*

Table 6.9 *(continued)*

Class/Order/Species	Marianas		Marshalls						W. Carolines		E. Carolines	
	Saipan	Guam	Lik	Maj	Ebo	All	Ene	Bik	Belau	Yap	Pohnpei	Chuuk
Rhaphisia hispida[63]											X	
Stylotella aurantium[63,167],*	*	X							X		X	X
S. aldis[63],*	*					X[63]						
*S. maza**												*
*S. agminata**												*
Spongosorites porites[63,59]										X		
*Stylissa flabelliformis**												*
*Topsentia cavernosa**												*
Haplosclerida												
Adocia turquosia[63],*	*			X				X	X		X	X
Adocia sp[195]												X
A. viola[63],*		X									X	*
A. neens[63]			X	X				X			X	
Aka mucosa[12],*	*											*
A. cf *corallophagum**												*
Amphimedon sp[194]		X										
Arenosclera psammophera[63],*						X						*
*A. heroni**												*
Cribrochalina olemda[63,12],*									X			X*
Callyspongia subarmigera[12]									X			
C. ridleyi[12]									X			
C. fistularis[63]				X	X		X					
C. fibrosa[51]				?	?		?					
C. diffusa[63],*		X							X		X	*
*C. brucei**												*
*C. parva**	*											*
*C. aerizusa**	*											*
Gellius gracilis[63,12],*	*								X			*
Gelliodes callista[63],*									X			*
Haliclona velinea[63,12]					X				*			
Haliclona sp[81]									X			
H. koremella[63,12],*	*								X			*
H. monilata[63],*												X*
H. lingulata[63]		X										
H. streble[63]	X											
H. korema[63]								X	X			
H. coerulescens[63],*												X*
H. viridis[63],*	*					X			X		X	X
H. permollis[63]											X	
H. decidua[63]												X
H. parietilis[63]			X									
H. nigra[63]				X								
H. massilis[63]				X								
H. rotographura[63],*			X									*
*H. acoroides**	*											
*H. pigmentifera**												*
"*H. pellasarca*"*	*											
Kallypilidion poseidon[63,12]									X			X
*Niphates cavernosa**	*											
*N. spinosella**	*											
Nara nemitifera[63]				X	X							*

Table 6.9 *(continued)*

Class/Order/Species	Marianas		Marshalls						W. Carolines		E. Carolines	
	Saipan	Guam	Lik	Maj	Ebo	All	Ene	Bik	Belau	Yap	Pohnpei	Chuuk
Reniera implexa[63],*					X							*
Reniera sp[89]									X			
R. chrysa[63],*					X	X						*
Sigmadocia emphasis[63]									X			
*S. amboinensis**	*											
*S. symbiotica**	*											
Toxidocia tyroeis[63]									X			
T. violacea[12]									X			
Petrosida												
*Oceanapia fistulosa**												*
Oceanapia saggittaria[63],*									X			X*
*Petrosia pandora**												*
Pellina carbonaria[63],[12]									X			
P. eusiphonia[63],*	*				X							
P. pinella[63]					X							
P. carbonilla[63]					X							
P. pulvilla[63]		X										
*P. triangulata**												*
Xestospongia exigua[63],[12],[59],[185]						X		X	X	X	X	X
Xestospongia sp[203]		X										
Poecilosclerida[63]												
Axocielita linda			X			X			X			X
*Acamas caledoniensis**	*											
Agelas mauritiana			X	X			X	X				
*Amphinomia sulphurea**	*											
*Anaata lajorei**						X			*			*
Axociella arteria											X	
*Biemna fortis**	*								*		*	*
B. mnioeis											X	
Coelocartaria singaporense[63],[12],*	*								X			
Clathria cervicornis[12],*									X		X	X*
C. fasciculata[12],*									X			X*
C. frondifera		X										
C. cratita											X	*
C. abietina								X				X
*C. vulpina**	*											*
*C. reinwardti**												*
*Crella spinulata**	*								X			
*C. calypta**												*
*Chondropsis ceratosus**											X	X*
Carmia stegoderma											X	
Desmacella lampra[63],[12],*	*								X			X*
Damiria sp[207],[208]									X			
*Damiriana hawaiiana**												*
*Echinoclathria waldoschmitti**												*
*E. intermedia**												*
*Echinodictyum antrodes**		X	X					X	*		X	*
*E. asperum**												*
Fausubera debrumi			X						X			
Iotrochota baculifera[12],*	*								X			*
*Iotrochota pella**				X								*

Table 6.9 *(continued)*

Class/Order/Species	Marianas		Marshalls						W. Carolines		E. Carolines	
	Saipan	Guam	Lik	Maj	Ebo	All	Ene	Bik	Belau	Yap	Pohnpei	Chuuk
I. ditrochota											X	
I. hiatti												X
I. mystile			X									X
Iotrochostyla iota												X
Lissodendoryx oxytes*	*						X		X			
L. calypta							X					
Microciona eurypa[63,12,]*	*								X			*
M. plinthina*						X			X			*
M. micronesica				X					X			
M. placenta												X
Mycale lissochela[12]									X			
M. cavernosa[12,]*									X			
M. armata					X				X		X	*
M. cecile*	*											*
M. parishii*												*
Monanchora ungiculata[63,12]									X		X	X
Neofibularia irata*												*
Oxymycale stecarmia											X	
O. strongylophora						X						
Oxycarmia confundata											X	
Ophlitospongia mima				X					X			
Protolithospongia ada											X	
P. aga									X			
Prianos phlox	*		X	X		X						
P. melanos					X							
P. osiris												X
Stylotrichophora rubra					X							
Tedania oligostyla					X							
T. ignis									X			
Ulosa spongia*					X							
Xytopsiphum meganese*	*											
Dictyoceratida												
Aulena concertina[63]						X						
Coscinoderma mathewsi[63,139,141,115]											X	X
Dactylospongia elegans[12,]*									X			*
Fasciospongia chondrodes[63,12,]*	*								X			
Fenestraspongia sp[43]									X			
Hippospongia metachromia[63,]*	*								X			X
Hyrtios erecta[63,12,183,]*		X			X	X	X		X		X	X*
Ircinia ramosa[63,12,]*									X			X*
I. strobilina[63]									X		X	
I. halmiformis[63]							X	X				
Lendenfeldia dendyi[63,12]	*								X			
Lufferiella variabilis[78,170,]*		X							X			*
L. geometrica*												*
Leiosella levis*												*
L. silicata*												*
Phyllospongia papyracea[17]									X			
Strepsichordaia lendenfeldi*	*											
Sarcotragus arbuscula*												*

Table 6.9 (continued)

Class/Order/Species	Marianas		Marshalls						W. Carolines		E. Carolines	
	Saipan	Guam	Lik	Maj	Ebo	All	Ene	Bik	Belau	Yap	Pohnpei	Chuuk
Spongia matamata[63,12],*				X		X		X	X		X	X*
S. zimocca[63],*									X		X	*
Verongida												
Aplysinella tyroeis[63],*	*										X	*
*A. strongylata**	*											*
Ianthella basta		*										
Psammaplysilla purpurea[63,12,41]			X	X	X	X	X		X		X	X
*P. verongiformis**	*								*			
Dendroceratida												
Aplysilla sulfurea[63]				X							X	X
A. polyraphis[63]											X	
*A. rosea**		*										
Chelonaplysilla sp[24,25]									X		X	
Dysidea herbacea[63,224,12],*						X			X			X*
Dysidea sp[78]		X										
D. chlorea[63,12],*					X				X			*
D. granulosa[12],*									X			*
D. arenaria[12],*									X			*
D. fragilis[63],*	*					X						*
D. avara[63],*	*	X				X					X	*
D. rhax[63],*			X	X	X						X	*
D. crawshayi[63]			X	X	X						X	
Dendrilla nigra[63],*	*		X						X			
Dendrilla sp[23]									X			
Euryspongia lobata[12],*	*								X			*
E. phlogera[63]						X						
Halisarca melana[63]									X			
"Slerosponges"												
Astrosclera willeyana[11,164,178],*	*	X					*					*
*Acanthochaetetes wellsi**	*						X					*
Stromatospongia micronesica[88],*	*	X					X					
CALCAREA												
*Dendya prolifera**												*
Leucetta microraphis[47,132]									X		X	
Leucetta primigenia[63],*											X	X*
Leucetta avacado[63],*		X							X			X*
Leuconia tropica[63],*									X			*
Leuconia palaoensis[63,210],*									X			*
*Leucandra pacifica**												*
Murrayona phanolepis	*						X					
*Pericharax heteroraphus**												*
Scypha plumosa[63]									X			

Ryukyu, Bonin, and Marcus Islands of Southern Japan

Records of sponges from the Ryukyu Islands of southern Japan (Hoshino 1985a, b, 1987; Mori 1977) appear predominantly within the chemical literature (Table 6.10). Bruce (1985, 1988) recorded crustacean-associated sponges. There are no recorded sponge species from Bonin and Marcus Islands.

Onotoa (Gilbert Island Group, Kiribati) and Funafuti Atoll (Ellice Island Group, Tuvalu)

In July and August of 1951, twenty sponges were collected by P. E. Cloud and A. H. Banner and later identified by de Laubenfels (1955) (Table 6.11). Kirkpatrick (1900) described twenty-one sponges from the second expedition of the *Boring* to Funafuti Atoll provided

Table 6.10 Published records of sponges from the Ryukyu Islands of southern Japan

Class/Order	Species
DEMOSPONGIAE	
Homosclerophorida	*Plakortis* sp[129]
Astrophorida	*Jaspis* sp[126], *Theonella swinhoei*[128], *Theonella* sp[124], *Theonella* sp[125]
Hadromerida	*Terpios hoshinoto*[189]
Petrosida	*Xestospongia* sp[123]
Poecilosclerida	*Agelas nakamurai*[100], *A. nemechinata*[100]
Dictyoceratida	*Hyrtios erecta*[130], *Hyrtios* sp[77], *Lufferiella* sp[127]
Dendroceratida	*Dysidea minna*[101], *Dysidea* sp[104]
"Sclerosponges"	*Tabulospongia japonica*[165]

Table 6.11 Published records of sponges from Onotoa† (Gilbert Island Group, Kiribati) and Funafuti Atoll (Ellice Island Group, Tuvalu)

Class/Order	Species
DEMOSPONGIAE	
Homosclerophorida[121]	*Corticium candelabrum, Plakinalopha spinosa, Plakinastrella clathrata*
Astrophorida	*Erylus monticularis*[143], *E. nobilis*[121], *Jaspis stellifera*[164], *Myriastra debilis*[†64]*Onotoa amphiastra*[†64]
Hadromerida	*Chondrilla mixta*[121], *Chondrosia chucalla*[†64], *Cliona mucronata*[121], *C. lobata*[†64], *C. schmidti*[121], *Dyscliona davidi*[121], *Latrunculia clavigera*[121], *Polymastia dendyi*[233], *Spirastrella papillosa*[233], *S. potamophora*[†64]*Tethya diploderma*[†64], *Tethytimea stellagrandis*[†64]
Halichondrida	*Acanthella stipitata*[233], *A. pulcherrima*[233], *Ciocalypta incrustans*[233], *Halichondria solida rugosa*[233], *Spongosorites porites*[†64], *Stylotella aurantium*[†64]
Poecilosclerida	*Agelas gracilis*[121,233], *A. mauritiana*[†64], *Chondropsis ceratosus*[121], *Clathria pellicula*[233], *Echinodictyum asperum*[233], *Mycale armata*[†64], *Psammopemma purpureum*[121], *Tedania levis*[121]
Haplosclerida	*Callyspongia glomerata*[233], *C. fistularis*[†64], *Gellius aculeatus*[233], *Niphates fibrosa*[121], *Reniera australis*[233]
Petrosida	*Pellina carbonaria*[†64]
Dictyoceratida	*Carteriospongia sweeti*[121], *Hippospongia dura*[233], *Hyrtios erecta*[†64], *Lufferiella variabilis*[121], *L. geometrica*[121], *Spongia irregularis silicata*[233], *Spongia irregularis pertusa*[141], *S. officinalis matamata*[†64], *S. matamata canaliculata*[†64], *Stelospongus cavernosus*[121]
Dendroceratida	*Dysidea fragilis irregularis*[233], *D. chlorea*[†64]
"Sclerosponges"	*Astrosclera willeyana*[121,159]
CALCAREA	
	Clathrina depressa[121]
HEXACTINELLIDA	
	Plectroninia hindei[121]

to him for description by Professor Judd. These later sponges were mostly small and encrusting, being attached to coral that was dredged from 30–145 fathoms. Whitelegge (1897) described sixteen species from the reef flat, bringing the total to thirty-six described sponges for the region. First accounts of sclerosponges in the Pacific are given by Lister (1900) and Willey (1902).

DISCUSSION

There are over 1,000 published and unpublished species records for the tropical island region of Oceania. The sponge faunas of Easter Island (forty species), French Polynesia (twenty-six species), the Kermadec Islands of New Zealand (twenty species), the Ryukyu Islands of southern Japan (fifteen species), the Solomon Islands and Vanuatu (forty-eight species), Kiribati and Tuvalu (fifty-six species), and Fiji, Tonga, and Samoa (fifty-two species) are poorly represented in the published literature, and phyletic diversity of the collections is low. The sponge faunas of Hawaii (ninety-eight species), Papua New Guinea (184 species), New Caledonia (274 species), and the Federated States of Micronesia (277 species) are better represented, reflecting the greater effort that has been made in collections and taxonomic research in these locations, both historically and recently. However, even within these better-known faunas, the collections have been restricted to several widely separated locations and so are not geographically representative. Collections also frequently reflect research bias; the Sala-y-Gómez and New Caledonia species lists are dominated by records of hexactinellid sponges, and by tetractinomorph and poecilosclerid sponges, respectively.

The sponge fauna of the broad Indo-West Pacific region, with a center of diversity in the East Indies, is considered to be the most diverse in the world (Briggs 1987; van Soest 1989, 1990). However, as the sponge faunas of this broad region become better known, the notion of an East Indies center of diversity becomes less tenable. Hooper (1994) has found that the northwestern shelf of Australia to be equally as diverse. There are several regions within western Oceania also extremely rich in terms of species number and phyletic diversity. These include Lemetol Bay within the Chuuk Lagoon (de Laubenfels 1954; CRRF, unpublished data), Bootless Inlet within the southern Papua New Guinea Barrier Reef system (Kelly 1986; Kelly-Borges and Bergquist 1988; CRRF, unpublished data), and the marine lakes system in Belau (Kelly-Borges, Bell, and Sharron 1994). Nutrient-rich mangrove-lined inlets with pinnacle-like patch-reefs adjacent to mountainous coastlines such as within Bootless Inlet and Lemetol Bay are exceptionally diverse; more than 250 sponge species have been collected from Bootless Bay (MKB; CRRF, unpublished data), Papua New Guinea, and over 500 species are known for Chuuk Atoll (CRRF, unpublished data). A recent survey of six of the eighty known marine lakes in Belau resulted in more than 140 sponge species, each lake fauna being significantly different from the next, with the exception of the presence of several "cosmopolitan" Indo-Pacific species such as *Hyrtios erecta, Monanchora ungiculata, Stylotella aurantium,* and *Paratetilla bacca.* Several lakes held taxa that represent new genus records for Oceania (e.g., *Tetrapocillon*). (See Table 6.12 for classification of the recent demosponges.)

Table 6.12 Classification of the recent demosponges

CLASS DEMOSPONGIAE	
ORDER HOMOSCLEROPHORIDA	
FAMILY PLAKINIDAE (syn. Oscarellidae)	*Plakortis, Corticium, Plakina, Plakinastrella, Placinolopha, Diactinolopha, Astroplakina, Dercitopsis, Oscarella*
ORDER ASTROPHORIDA (syn. Choristida)	
FAMILY ANCORINIDAE (syn. Stellettidae)	*Rhabdastrella, Stelletta, Ancorina, Asteropus, Penares*
FAMILY GEODIIDAE	*Geodia, Erylus, Pachymatisma, Caminus, Sidinops, Isops*
FAMILY PACHASTRELLIDAE	*Pachastrella, poecillastra, dercitus, sphinctrella, triptolemus*
FAMILY COPPATIIDAE (syn. Jaspidae)	*Jaspis, Stellettinopsis*
FAMILY CALTHROPELLIDAE	*Calthropella, Pachastrissa, Chelotropella, Pachataxa*
FAMILY THENEIDAE	*Thenea, Papyrula*
FAMILY THROMBIDAE	*Thrombus*
"LITHISTID" FAMILY THEONELLIDAE	*Theonella, Discodermia, Racodiscula, Jereopsis*
"LITHISTID" FAMILY CORALLISTIDAE	*Corallistes, Macandrewia, Homophymia, Iouea, Herengeria,*
"LITHISTID" FAMILY PLEROMIDAE	*Pleroma*
"LITHISTID" FAMILY DORYDERMIIDAE	*Anaderma*
"LITHISTID" FAMILY PHYMATELLIDAE	*Neosiphonia, Reidispongia, Aulaxinia*
"LITHISTID" FAMILY PHYMARAPHINIDAE	*Kaliapsis*
"LITHISTID" FAMILY NEOPELTIDAE	*Neopelta*
ORDER SPIROPHORIDA	
FAMILY TETILLIDAE	*Cinachyra, Cinachyrella, Paratetilla, Craniella, Acanthotetilla*
"LITHISTID" FAMILY SCLERITODERMIDAE	*Scleritoderma, Amphibleptula, Microscleroderma, Aciculites*
"LITHISTID" FAMILY SIPHONIDIIDAE	*Gastrophanella, Siphonidium, Leiodermatium*
ORDER HADROMERIDA	
FAMILY TETHYIDAE	*Tethya, Burtonitethya, Tethycometes, Xenospongia*
FAMILY POLYMASTIIDAE	*Polymastia, Radiella, Tentorium, Weberella, Atergia, Proteleia*
FAMILY SUBERITIDAE	*Terpios, Suberites, Aaptos, Prosuberites, Pseudosuberites*
FAMILY TIMEIDAE	*Timea, Diplastrella*
FAMILY PLACOSPONGIIDAE	*Placospongia*
FAMILY STYLOCORDYLIDAE	*Stylocordyla, Tethycordyla, Oxycordyla*
FAMILY SPIRASTRELLIDAE	*Spirastrella, Anthosigmella*
FAMILY CLIONIDAE	*Cliona, Thoosa, Cliothoosa, Spheciospongia, Alectona*
FAMILY CHONDRILLIDAE (syn. Chondrosiidae)	*Chondrilla, Chondrosia*
"SCLEROSPONGE" FAMILY ACANTHOCHAETIDAE	*Acanthochaetetes*
?FAMILY HEMIASTERELLIDAE	*Hemiasterella*
?FAMILY LATRUNCULIIDAE	*Diacarnus, Latrunculia, Sigmosceptrella, Negombata*
?FAMILY TRACHYCLADIDAE	*Trachycladus*
ORDER HALICHONDRIDA	
FAMILY HALICHONDRIIDAE	*Topsentia, Amorphinopsis, Halichondria, Hymeniacidon, Stylotella, Ciocalypta, Spongosorites, Axinyssa, Epipolasis*
"LITHISTID" FAMILY PETROMICIDAE	*Petromica, Monanthus*
FAMILY DESMOXYIDAE	*Desmoxya, Higginsia, Myrmekioderma, Didiscus, Acanthoclada, Halicnemia, Heteroxya*
?FAMILY AXINELLIDAE	*Axinella, Phakellia, Acanthella, Auletta, Bubaris, Pararaphoxya, Homaxinella, Cymbastela, Ptilocaulis, Teichaxinella, Pseudaxinella, Axinectya, Axinosia, Dragmacidon, Reniochalina, Rhaphoxya*
?FAMILY DICTYONELLIDAE	*Dactylella, Dictyonella, Liosina, Scopalina, Tethyspira*
ORDER AGELASIDA	
FAMILY AGELASIDAE	*Agelas, Ectyonopsis*
"SCLEROSPONGE" FAMILY CERATOPORELLIDAE	*Ceratoporella, Stromatospongia, Hispidoptera, Goreauiella*
"SCLEROSPONGE" FAMILY ASTROSCLERIDAE	*Astrosclera*
ORDER POECILOSCLERIDA	
FAMILY IOPHONIDAE (syn. Cornulidae)	*Acarnus, Coelocarteria, Zyzzya, Damiria, Iophon*
FAMILY MICROCIONIDAE	*Antho, Clathria, Echinochalina, Echinoclathria, Pandoras*
FAMILY RHABDAREMIIDAE	*Rhabderemia*
FAMILY RASPAILIIDAE	*Raspailia, Endectyon, Ceratopsion, Hymeraphia, Thrincophora, Ectyoplasia, Echinodictyon, Amphinomia*

Table 6.12 *(continued)*

FAMILY ANCHINOIDAE (syn. Phorbasidae)	*Phorbas, Hamigera*
FAMILY COELOSPHAERIDAE	*Anomodoryx, Chaetodoryx, Coelocarteria, Coelosphaera, Histodermella, Amphiastrella, Inflatella, Manawa, Forcepia, Inflatella, Histodermella, Lissodendoryx*
FAMILY CRAMBIDAE	*Crambe, Discorhabdella, Monanchora, Psammochela*
FAMILY CRELLIDAE	*Crella, Naniupi, Barbozia, Yvesia*
FAMILY HYMEDESMIIDAE	*Acanthocora, Spanioplon, Stylopus, Hymedesmia*
FAMILY MYXILLIDAE	*Amphiastrella, Amphilectus, Desmacidon, Desmapsamma, Myxilla, Ectyomyxilla, Sigmarotula, Iotrochota, Stelodoryx*
FAMILY PHORIOSPONGIIDAE	*Batzella, Chondropsis, Strongylacidon, Psammoclemma*
FAMILY TEDANIIDAE	*Tedania, Tedaniopsis, Hemitedania, Kirkpatrickia*
FAMILY CLADHORIZIDAE	*Asbestopluma, Chondrocladia, Cladhoriza*
FAMILY GUITARRIDAE	*Guitarra, Tetrapocillon*
FAMILY DESMACELLIDAE (syn. Biemnidae)	*Desmacella, Biemna, Microtylostylifer, Kerasemmna, Neofibularia, Sigmaxinella*
FAMILY HAMACANTHIDAE	*Hamacantha*
FAMILY MYCALIDAE	*Mycale, Esperiopsis, Paraesperella, Sceptrella*
"SCLEROSPONGE" FAMILY MERLIIDAE	*Merlia, ?Chaetetes*
"LITHISTID" FAMILY DESMANTHIDAE	*Desmanthus*
?"LITHISTID" ORDER SPHAEROCLADINIDA	
FAMILY VETULINIDA	*Vetulina*
ORDER HAPLOSCLERIDA	
FAMILY CHALINIDAE (syn. Haliclonidae)	*Haliclona, Reniera, Gellius, Rhizoniera, Cladacroce*
FAMILY ADOCIIDAE	*Adocia, Sigmadocia, Orina, Toxadocia, Kallypilidion*
FAMILY NIPHATIDAE	*Amphimedon, Niphates, Gelliodes, Cribrochalina, Aka, Hemigellius, Haliclonissa, Siphonochalina, Microxina*
FAMILY CALLYSPONGIIDAE	*Callyspongia, Euplacella, Dactylia, Toxochalina*
"SCLEROSPONGE" FAMILY CALCIFIBROSPONGIIDAE	*Calcifibrospongia*
FAMILY SPONGILLIDAE (Freshwater sponges)	
FAMILY METANIIDAE (Freshwater sponges)	
FAMILY POTAMOLEPIDAE (Freshwater sponges)	
FAMILY LUBOMIRSKIIDAE (Freshwater sponges)	
ORDER PETROSIDA (syn. Nepheliospongida)	
FAMILY PETROSIIDAE	*Xestospongia, Petrosia, Strongylophora*
FAMILY PHLOEODICTYIDAE (syn. Oceanapiidae)	*Oceanapia, Pellina, Pachypellina, Biminia, Calyx, Vagocia, Foliolina, ?Prianos*
ORDER DICTYOCERATIDA	
FAMILY SPONGIIDAE	*Spongia, Hippospongia, Phyllospongia, Strepsichordaia, Coscinoderma, Hyatella, Leiosella, Carteriospongia*
FAMILY THORECTIDAE	*Hyrtios, Thorectandra, Cacospongia, Lufferiella, Thorecta, Fasciospongia, Fenestraspongia, Aplysinopsis,*
FAMILY IRCINIIDAE	*Ircinia, Sarcotragus*
ORDER DENDROCERATIDA	
FAMILY APLYSILLIDAE	*Aplysilla, Pleraplysilla, Chelonaplysilla, Dendrilla*
FAMILY HALISARCIDAE	*Halisarca*
FAMILY DYSIDEIDAE	*Dysidea, Euryspongia, Spongionella*
FAMILY DICTYODENDRILLIDAE	*Dictyodendrilla, Igernella*
ORDER VERONGIDA	
FAMILY APLYSINIDAE	*Aplysina, Verongula*
FAMILY APLYSINELLIDAE	*Pseudoceratina, Aplysinella, Psammaplysilla,*
FAMILY IANTHELLIDAE	*Ianthella, Anomoianthella, Bajalus*
"SCLEROSPONGE" ORDER VERTICILLITIDA	
FAMILY VERTICILLITIDAE	*Vaceletia*

After Bergquist 1968; Bergquist, Ayling, and Wilkinson 1988; Bergquist and Fromont 1988; Bergquist and Kelly-Borges 1991; Kelly-Borges and Bergquist 1994; Bergquist and Warne 1980; Desqueyroux-Faundez 1984, 1987; Diaz and van Soest 1994; Fromont 1991, 1993; Hadju, van Soest, and Hooper 1994; Hooper 1991; Hooper and Bergquist 1992; Hooper and Lévi 1993a, b, 1994; Hooper and Weidenmayer 1994; Kelly-Borges and Vacelet, in press; Kelly-Borges and Pomponi 1994; Lévi 1993; Lévi and Lévi 1983, 1988; Vacelet, pers. com.

RECOMMENDATIONS FOR INFORMATION MANAGEMENT

Sponge Distribution and Ecosystem Classification

As for most marine invertebrates, sponge species in the tropical island region of Oceania are predominately habitat-specific, with only a small number of species occurring in all reef environments. For the fringing reef surrounding Motupore Island in Bootless Inlet, Papua New Guinea, Kelly (1986) identified specific sponge assemblages rather than habitats, as sponge distribution did not correspond to the standard coral reef ecosystem classification system (e.g., reef ridge, reef flat, fore-reef slope, back reef lagoon, seagrass beds). Rather, patterns of distribution were more complicated, with species distributed along major physical oceanographic gradients so that the faunas differed on leeward and windward sides of the island. Similarly, the sponge assemblages in six of eighty marine lakes in Belau seemed to be determined by a complex combination of the lag-time between tidal cycle at the seawater source and within the lake, which is a function of the distance of the lake from the seawater source, the physico-chemical nature of the lake water, and the geomorphology of the surrounding terrain (Kelly-Borges, Bell, and Sharron 1994). Sponge distributions in both locations were further complicated by asexual reproductive mechanisms in many species, leading to dispersal of individuals with prevailing wind-driven currents (Kelly-Borges and Bergquist 1988). Recent evidence suggests that asexual reproduction is surprisingly common in a diversity of sponge morphologies and that the recruits generated have a remarkable behavioral and survivorship capability (Battershill and Bergquist 1990). Thus, biological and physical factors considerably modify sponge distribution within typical reef habitats. This fact and the presence of special sponge-dominated habitats, such as marine lakes, submarine caves, nutrient-rich environments, and areas of high current activity, require recognition when applying standard coral reef habitat classifications to sponge distribution data.

Geographic and habitat scale is also extremely important in the consideration of sponge distribution. In general, the sponge faunas of Oceania can be divided into four broad groups: (1) regionally endemic sponges that are habitat- or locality-specific (e.g., patch-reefs within nutrient-rich bays, specific marine lakes, or caves), (2) regionally endemic sponges that occur within a single atoll or island group, (3) species that are found in "super-regions" (e.g., Micronesia alone or Solomon Islands alone), and (4) species that are found throughout the broad Indo-West Pacific region (de Laubenfels 1954; Hooper 1994).

Biodiversity Database Structure for Resource Management

The minimum information required for management of sponge species data include SPECIES, SITE, and BIBLIOGRAPHIC information, managed in a relational database management system (DBMS) (B. E. Picton, pers. com. 1995). Additional information that might be considered for provision within the DBMS include protection status (CITES listing) and commercial use or application (aquaculture, mariculture, pharmacology). Suggested data fields including those employed in the Species Directory Marine Database (BioMar) of Trinity College, Ireland, currently being developed to serve the Porifera research community (B. E. Picton, pers. com. 1995) are given in Table 6.13. Although SPECIMEN information may seem optional for resource management, where species lists and distributional data are the primary goals, careful consideration should be given for the

Table 6.13 Minimum data requirements for management of sponge biodiversity information in a relational database management system

Species	Specimen	Site	Bibliographic	Addresses
⇔	**specimen number**	⇔		
	collection date	⇔		
	preservation method			
	status of specimen: cited, type, etc.			
⇔	collection depository			⇔ (museum or institution)
⇔	photographs			⇔ (photographer)
	collector			⇔
⇔	⇔	**site name**		
	⇔	site code		
	⇔	**site description**		
⇔	⇔	**latitude**		
⇔	⇔	**longitude**		
⇔	⇔	**habitat**		
⇔	**depth range**			
⇔	**substrate**			
⇔	**abundance at site**			
⇔	morphology			
⇔	size range			
⇔	surface			
⇔	texture			
⇔	external color			
⇔	internal color			
⇔	exudates			
⇔	odor			
phylum	⇔			
class	⇔			
order	⇔			
family	⇔			
genus	⇔			
species	⇔			
authority	⇔		⇔	⇔
⇔	taxonomist			⇔
	taxonomy updater			⇔

Fields useful for resource managers are in bold-face text, columns are DATA TABLES, rows are DATA FIELDS within tables, ⇔ = data that are accessed from other files.

provision of specimen information in the database structure. Individual specimens provide the fundamental data that derive the "species" name applied to that group of specimens. Collections and inventory initiatives are increasing within Oceania, and specimen data, if provided, will greatly facilitate revisionary systematics (taxonomic monography), and thus biodiversity inventory within the region. Indeed, electronic dissemination of specimen-based databases, via GOPHER, or a WORLDWIDE WEB (WWW) server, is a requirement of many recent biodiversity inventory initiatives such as the National Science Foundation's Biotic Surveys and Inventories Program.

The database should allow the storage of hierarchical Linnean taxonomy, enabling the sorting of species by taxa while taking into account changing taxonomies. Attention should also be given to the coding of synonymous species names. Literature references should be incorporated into the database and linked to the species names. The importance of employing a fully relational database is well illustrated within the SPECIES data fields in Table 6.13. For example, the same personnel information may be required within multiple fields, because a person may be a collector, a taxonomist, a donor, and an author. These personnel data, which might include the details of their employing organization, address, contact details, and research interests, for example, need only to be linked to the use of their name, rather than be stored multiple times in separate fields. The same applies to literature, photographic, and site information.

CONCLUSION

The sponge fauna of the tropical island region of Oceania has received little attention and is relatively unknown. With the exception of some Micronesian, Papua New Guinea, and New Caledonian localities, records consist of only the most common and obvious shallow-water sponges. There is an urgent need to expand sponge biodiversity initiatives within Oceania, several regional components of which are among and adjacent to regions of the world's greatest sponge biodiversity. Sponges are under considerable threat worldwide due to destructive fishing and recreational practices, siltation and pollutants, and exploitation for biomedical purposes. Understanding the mechanisms that promote biodiversity in marine ecosystems will enhance our ability to manage and conserve them. Documenting the biodiversity of sponges in Oceania and ultimately understanding the phylogenetic relationships between these organisms is a crucial first step to conserving them.

Acknowledgments—We thank John N. A. Hooper (Queensland Museum, Brisbane, Australia), Bernard Picton (Trinity College, Dublin, Ireland), Shirley A. Pomponi (Harbor Branch Oceanographic Institution, Fort Pierce, Florida), Rob van Soest (Zoologisch Museum, Amsterdam, The Netherlands), Ruth Desqueyroux-Faundez (Muséum d'Histoire Naturelle, Genève, Switzerland), Patricia R. Bergquist (University of Auckland, Auckland, New Zealand), and Jean Vacelet (Station Marine d'Endoume, Marseille, France) for information pertaining to this review.

REFERENCES

1. Adamczeski, M., E. Quinoa, and P. Crews. 1988. Unusual anthelminthic oxazoles from a marine sponge. *Journal of the American Chemical Society* 110(5): 1598–1608.

2. Agassiz, A. 1906. General Report of the Expedition. Report on the scientific results of the expedition to the Eastern Tropical Pacific, in charge of A. Agassiz, by the U. S. Fisheries Commission Steamer *Albatross*, 1904–1905 and 1888–1904. 21:*Memoirs of the Museum of Comparative Zoology, Harvard Collection* 41:261–323.

3. Alvi, K. A., and P. Crews. 1992. Homoscalarane sesterterpenes from *Lendenfeldia frondosa*. *Journal of Natural Products* 55(7): 859–865.

4. Alvi, K. A., M. C. Diaz, P. Crews, D. L. Slate, R. H. Lee, and R. Moretti. 1992. Evaluation of new sesquiterpene quinones from two *Dysidea* sponge species as inhibitors of protein tyrosine kinase. *Journal of Organic Chemistry* 57(24): 6604–6607.

5. Alvi, K. A., B. M. Peters, L. M. Hunter, and P. Crews. 1993. 2-Aminoimidazoles and their zinc

complexes from Indo-Pacific *Leucetta* sponges and *Notodoris* nudibranchs. *Tetrahedron Letters* 49(2): 329–336.

6. Amade, P., D. Pesando, and L. Chevolot. 1982. Antimicrobial activities of marine sponges from French Polynesia and Brittany. *Marine Biology* 70(3): 105–127.

7. Baar, R. 1903. Hornschwamme aus dem Pacific. Ergebnisse einer Reise nach dem Pacifc, Schauinsland, 1896–1897. *Zoologische Jahrbucher, Jena* 14:27–36.

8. Baba, K. 1983. *Spongicoloides hawaiiensis*, a new species of shrimp (Decapoda: Stenopodidea) from the Hawaiian Islands. *Journal of Crustacean Biology* 3(3): 477–481.

9. Baker, B. J., P. J. Scheuer, and J. N. Shoolery. 1988. Papuamine, an antifungal pentacyclic alkaloid from a marine sponge, *Haliclona* sp. *Journal of the American Chemical Society* 110(3): 965–966.

Bakus, G. J., and B. Ormesby. 1994. Comparison of species richness and dominance in Northern Papua New Guinea and Northern Gulf of Alaska. In *Sponges in Time and Space*, ed. R. W. M. van Soest, Th. M. G. van Kempen, and J. C. Braekman, 169–174. Rotterdam: Balkema.

10. Barraclough-Fell, H. 1950. The Kirk collection of sponges (Porifera) in the Zoology Museum, Victoria University College. *Zoological Publications of the Victoria University Collection, New Zealand* 4:1–12.

11. Basilie, L. L., R. J. Cuffey, and D. F. Kosich. 1984. Sclerosponges, pharetronids and sphinctozoans (relict cryptic hard-bodied Porifera) in the modern reefs of Enewetak Atoll. *Journal of Paleontology* 58(3): 636–650.

Battershill, C. N., and P. R. Bergquist. 1990. The influence of storms on asexual reproduction, recruitment, and suvivorship of sponges. In *New Perspectives in Sponge Biology*, ed. K. Rützler, V. V. Macintyre, and K. P. Smith, 397–403. Washington, D. C.: Smithsonian Institution Press.

12. Bergquist, P. R. 1965. The Sponges of Micronesia. I. The Palau Archipelago. *Pacific Science* 3(2): 123–204.

13. Bergquist, P. R. 1967. Additions to the sponge fauna of the Hawaiian Islands. *Micronesica* 3:159–173.

14. Bergquist, P. R. 1968. The Marine fauna of New Zealand Porifera, Demospongiae. Part I. Tetractinomorpha and Lithistida. *New Zealand Oceanographic Institute Memoir* 37:1–106.

15. Bergquist, P. R. 1979. Porifera. In *Reef and Shore Fauna of Hawaii. Section 1. Protozoa through Ctenophora*, ed. D. M. Devaney and L. G. Eldredge. *Bernice P. Bishop Museum Special Publications* 64(1): 53–70.

16. Bergquist, P. R. 1994. Preface: Onwards and upwards with sponges. In *Sponges in Time and Space*, ed. R. W. M. van Soest, Th. M. G. van Kempen, and J. C. Braekman, xiii–xviii. Rotterdam: Balkema.

17. Bergquist, P. R., A. M. Ayling, and C. R. Wilkinson. 1988. Foliose Dictyoceratida of the Australian Great Barrier Reef. I. Taxonomy and Phylogenetic Relationships. *Marine Ecology* 9(4): 291–319.

18. Bergquist, P. R., and P. J. Fromont. 1988. The Marine Fauna of New Zealand: Porifera, Demospongiae, Part 4 (Poecilosclerida). *New Zealand Oceanographic Institute Memoir* 96:1–197.

19. Bergquist, P. R., and M. Kelly-Borges. 1991. An evaluation of the genus *Tethya* (Porifera: Demospongiae: Hadromerida) with descriptions of new species from the southwest Pacific. *Beagle* 8(1): 37–72.

20. Bergquist, P. R., and M. Kelly-Borges. In press. Systematics and Biogeography of the Genus *Ianthella* (Demospongiae: Verongida: Ianthellidae) in the South Pacific. *Beagle*.

21. Bergquist, P. R., J. E. Morton, and C. A. Tizard. 1971. Some Demospongiae from the Solomon Islands with descriptive notes on major sponge habitats. *Micronesica* 7:99–121.

22. Bergquist, P. R., and K. P. Warne. 1980. The Marine Fauna of New Zealand: Porifera, Demospongiae, Part 3 (Haplosclerida and Nepheliospongida). *New Zealand Oceanographic Institute Memoir* 87:1–77.

23. Bobzin, S. C., and D. J. Faulkner. 1989. Novel

rearranged spongian diterpenes from the Palauan sponge *Dendrilla* sp.: reassessment of the structures of dendrilloide A and dendrilloide B. *Journal of Organic Chemistry* 54(24): 5727–5731.

24. Bobzin, S. C., and D. J. Faulkner. 1991. Aromatic alkaloids from the marine sponge *Chelonaplysilla* sp. *Journal of Organic Chemistry* 56(14): 4403–4407.

25. Bobzin, S. C., and D. J. Faulkner. 1991. Diterpenes from the Pohnpeian marine sponge *Chelonaplysilla* sp. *Journal of Natural Products* 54(1): 225–232.

26. Borojevic, R. 1967. Eponges calcaires receiullies en Nouvelle-Caledonie par la mission Singer-Polignac. *Expedition Francaise sur les recifs coralliens de la Nouvelle-Caledonie, Paris, Foundation Singer-Polignac* 2:1–12.

Boury-Esnault, N., A. M. Solé-Cava, and J. P. Thorpe. 1992. Genetic and cytological divergences between two colour morphs of the Mediterranean sponge *Oscarella lobularis* Schmidt (Porifera, Demospongiae, Oscarellidae). *Journal of Natural History* 26:271–284.

27. Bowerbank, J. S. 1873. Contributions to a general history of the Spongiadae. Part IV. *Proceedings of the Zoological Society of London* 4:3–25.

28. Bowerbank, J. S. 1875. Contributions to a general history of the Spongiadae. Part VII. *Proceedings of the Zoological Society of London* 7:281–296.

29. Bowerbank, J. S. 1877. Description of five new species of sponges discovered by A. B. Meyer on the Philippine Islands and New Guinea. Proceedings of the Zoological Society of London, 456–464.

30. Braekman, J. C., D. Daloze, P. Macedo de Abreu, C. Piccini-Leopardi, G. Germain, and M. van Meerssche. 1982. A novel type of bis-quinolizidine alkaloid from the sponge *Petrosia seriata*. *Tetrahedron Letters* 23(41): 4277–4280.

31. Braekman, J. C., D. Daloze, and B. Moussinaux. 1987. Jaspimide from the marine sponge *Jaspis johnstoni*. *Journal of Natural Products* 50(5): 994–995.

32. Breitfuss, L. L. 1898. Die Kalkschwamme der Sammlung Plate. *Zoologische Jahrbucher, Jena [Fauna Chilensis (3)]* 4:455–470.

33. Briggs, J. C. 1974. *Marine Zoogeography.* New York: McGraw Hill.

34. Briggs, J. C. 1987. *Biogeography and plate tectonics. Developments in Palaeontology and Stratigraphy.* Amsterdam: Elsevier.

35. Bruce, A. J. 1981. Pontoniine shrimps from Vitu Levu, Fijian Islands. *Micronesica* 17(1–2): 77–95.

36. Bruce, A. J. 1985. Some caridean associates of scleractinian corals in the Ryukyu Islands. *Galaxea* 4(1): 1–21.

37. Bruce, A. J. 1988. A redescription of *Periclimenaeus fimbriatus* Borradaile, 1915, with the designation of a new genus (Crustacea: Decapoda: Palaemonidae). *Zoological Journal of the Linnean Society* 94(3): 219–232.

38. Bruce, A. J. 1991. Crustacea Decapoda: further deep-sea palaemonid shrimps from New Caledonian waters. *Memoirs of the Museum of Natural History Series A Zoology* 152:299–411.

39. Bryan, P. G. 1973. Growth rate, toxicity and distribution of the encrusting sponge *Terpios* sp. (Hadromerida, Suberitidae) in Guam, Mariana Islands. *Micronesica* 9(2): 237–242.

40. Burton, M. 1934. Sponges. *Great Barrier Reef Expedition 1928–1929.* 4(14): 513–621.

41. Carney, J. R., P. J. Scheuer, and M. Kelly-Borges. 1993a. A new bastadin from the sponge *Psammaplysilla purpurea. Journal of Natural Products* 56(1): 153–157.

42. Carney, J. R., P. J. Scheuer, and M. Kelly-Borges. 1993b. Three unprecedented chlorosteroids from the Maui sponge *Strongylacidon* sp. kiheisterones C, D and E. *Journal of Organic Chemistry* 58(12): 3460–3462.

43. Carté, B., C. B. Rose, and D. J. Faulkner. 1985. 5-epi-ilimiquinone, a metabolite of the sponge *Fenestraspongia* sp. *Journal of Organic Chemistry* 50(15): 2785–2787.

44. Caspers, H. 1981. Fauna inhabiting the sponge *Damiriana hawaiiana* in Kaneohe Bay,

Hawaii. *International Symposium on Biological Management of Mangroves in Tropical Shallow-Water Communities* 2:1–20.

45. Caspers, H. 1985. The brittle star, *Ophiactis savignyi* (Muller and Troschel), an inhabitant of the Pacific sponge, *Damiriana hawaiiana* de Laubenfels. *Proceedings of the International Conference on Echinoderms* 1984:603–607.

46. Castilla, J. C., and N. Rozbaczylo. 1987. Invertebrados marinos de Isla de Pascua y Sala-y-Gomez. In *Islas Oceanicas Chilenas: Conocimientos cientifico y necesidades de investigaciones*, ed. J. C. Castilla, 191–215. Chile: Universidad Catolica de Chile.

47. Chan, G. W., S. Mong, M. E. Hemling, A. J. Freyer, P. H. Offen, C. W. DeBrosse, H. M. Sarau, and J. W. Westley. 1993. New leukotriene B_4 receptor antagonist: leucettamine A and related imidazole alkaloids from the marine sponge *Leucetta microrhapis*. *Journal of Natural Products* 56(1): 116–121.

Colin, P. L., and C. Arneson. In press. *Tropical Pacific Invertebrates*. San Diego: Coral Reef Press.

48. Copp, B. R., C. M. Ireland, and L. R. Barrows. 1992. Psammaplysin C: a new cytotoxic dibromotyrosine-derived metabolite from the marine sponge *Druinella (Psammaplysilla) purpurea*. *Journal of Natural Products* 55(6): 822–823.

49. Crews, P., C. Jimenez, and M. O'Neill-Johnson. 1991. Using spectroscopic and database strategies to unravel structures of polycyclic bioactive marine sponge sesterterpenes. *Tetrahedron Letters* 47(22): 3585–3600.

50. Davidson, B. S. 1991. Cytotoxic five-membered cyclic peroxides from a *Plakortis* sponge. *Journal of Organic Chemistry* 56(23): 6722–6724.

51. Davies-Coleman, M. T., D. J. Faulkner, G. M. Dubowchik, G. P. Roth, C. Polson, and C. Fairchild. 1993. A new EGF-active polymeric pyridinium alkaloid from the sponge

Callyspongia fibrosa. *Journal of Organic Chemistry* 58(22): 5925–5930.

52. D'Auria, M. V., L. G. Paloma, L. Minale, R. Riccio, and C. Debitus. 1992. Structure characterization by two-dimensional NMR spectroscopy, of two marine triterpene oligoglycosides from a Pacific sponge of the genus *Erylus*. *Tetrahedron Letters* 48(3): 491–498.

53. D'Auria, M. V., L. G. Paloma, L. Minale, R. Riccio, C. Debitus, and C. Lévi. 1992. Unique 3 ß-O-methyl sterols from the Pacific sponge *Jereicopsis graphidiophora*. *Journal of Natural Products* 55(3): 311–320.

54. D'Auria, M. V., L. G. Paloma, L. Minale, R. Riccio, A. Zampella, and C. Debitus. 1993. Metabolites of the New Caledonian sponge *Cladocroce incurvata*. *Journal of Natural Products* 56(3): 418–423.

55. de Guzman, F. S., and F. J. Schmitz. 1990. Peroxy aliphatic esters from the sponge *Plakortis lita*. *Journal of Natural Products* 53(4): 926–931.

56. de Guzman, F. S., and F. J. Schmitz. 1991. Naurol A and B, novel triterpene alcohols from a Pacific sponge. *Journal of Natural Products* 56(1): 55–58.

57. de Laubenfels, M. W. 1935. A collection of sponges from Puerto Galera, Mindoro, Philippine Islands. *Philippine Journal of Science* 156(3): 327–336.

58. de Laubenfels, M. W. 1943. Biological results of the last cruise of the *Carnegie*. *Scientific Results of the Last Cruise of the Carnegie VII (1928–1929), Biology* 4:91.

59. de Laubenfels, M. W. 1949. New sponges from Yap Archipelago. *Pacific Science* 3(2): 124–126.

60. de Laubenfels, M. W. 1950a. The sponges of Kaneohe Bay. *Pacific Science* 4(1): 3–36.

61. de Laubenfels, M. W. 1950b. Sponges in Hawaii. *Cooperative Fishery Investigations Circular* 9 (June): 1–2.

62. de Laubenfels, M. W. 1951. The sponges of the islands of Hawaii. *Pacific Science* 5(3): 256–271.

63. de Laubenfels, M. W. 1954. The Sponges of the

West Central Pacific. *Oregon State Monographs in Zoology* 7:1–306.

64. de Laubenfels, M. W. 1955. The sponges of Onotoa. *Pacific Science* 9(2): 137–143.

65. de Laubenfels, M. W. 1957. New species and records of Hawaiian sponges. *Pacific Science* 11:236–251.

66. de Riccardis, F., A. Rozanov, M. Lorizzi, C. Debitus, and C. Lévi. 1993. Marine sterols. Side-chain-oxygenated sterols, possibly of abiotic origin, from the New Caledonian sponge *Stelodoryx chlorophylla*. *Journal of Natural Products* 56(2): 282–287.

67. de Silva, E. D., R. J. Andersen, and T. M. Allen. 1990. Geodiamolides C to F, new cytotoxic cyclodepsipeptides from the marine sponge *Pseudaxynissa* sp. *Tetrahedron Letters* 31(4): 482–492.

68. de Silva, E. D., J. S. Racok, R. J. Andersen, T. M. Allen, L. S. Brinen, and J. Clardy. 1991. Neamphine, a sulfur-containing aromatic heterocycle isolated from the marine sponge *Neamphius huxleyi*. *Tetrahedron Letters* 32(24): 2707–2710.

69. Desqueyroux, R. 1972. Demospongiae (Porifera) de la Costa de Chile. *Gayana* 20:71.

70. Desqueyroux-Faundez, R. 1981. Revision de la collection d'eponges d'Amboine (Moluques, Indonesie) constituee par Bedot et Pictet et conservee au Museum d'Histoire Naturelle de Geneve. *Revue Suisse de Zoologie* 88(3): 723–764.

71. Desqueyroux-Faundez, R. 1984. Description de la faune des Haplosclerida (Porifera) de la Nouvelle-Caledonie. 1. Niphatidae - Callyspongiidae. *Revue Suisse de Zoologie* 91(3): 765–827.

72. Desqueyroux-Faundez, R. 1987. Description de la faune des Petrosida (Porifera) de la Nouvelle-Caledonie. 1. (Petrosidae - Oceanapiidae). *Revue Suisse de Zoologie* 94(1): 177–243.

73. Desqueyroux-Faundez, R. 1990. Spongiaires (Demospongiae) de l'Ile de Pacques (Isla de Pascua). *Revue Suisse de Zoologie* 97(2): 373–401.

74. Desqueyroux-Faundez, R. 1994. Biogeography of Chilean marine sponges (Demospongiae). In *Sponges in Time and Space*, ed. R. W. M. van Soest, Th. M. G. van Kempen, and J. C. Braekman, 183–189. Rotterdam: Balkema.

75. Diaz, M. C., and R. W. M. van Soest. 1994. The Plakinidae: a systematic review. In *Sponges in Time and Space*, ed. R. W. M. van Soest, Th. M. G. van Kempen, and J. C. Braekman, 93–109. Rotterdam: Balkema.

76. Di Salvo, L. H., J. E. Randall, and A. Cea. 1988. Ecological reconnaissance of the Easter Island sublittoral marine environment. *National Geographic Research* 4(4): 451–473.

77. Doi, Y., H. Shigemori, M. Shibashi, F. Mizobe, A. Kawashima, S. Nakaike, and J. Kobayashi. 1993. New sesterterpenes with the nerve growth factor synthesis-stimulating activity from the Okinawa sponge *Hyrtios* sp. *Chemical and Pharmaceutical Bulletin (Tokyo)* 41(12): 2190–2191.

78. Duffy, J. E., and V. J. Paul. 1992. Prey nutritional quality and the effectiveness of chemical defenses against tropical reef fishes. *Oecologia* 90(3): 333–339.

79. Edmondson, C. E. 1935. Reef and shore fauna of Hawaii. *Special publications of the Bernice P. Bishop Museum, Honolulu* 22:17–21.

80. Erdman, T. R., and P. J. Scheuer. 1976. 28-Isofucosterol: Major sterol of a marine sponge. *Journal of Natural Products* 38(4): 359–360.

81. Fahy, E., T. F. Molinski, M. K. Harper, B. W. Sullivan, D. J. Faulkner, L. Parkanyi, and J. Clardy. 1988. Haliclonadiamine, an antimicrobial alkaloid from the sponge *Haliclona* sp. *Tetrahedron Letters* 29(28): 3427–3428.

82. Fromont, P. J. 1991. Descriptions of species of Petrosida (Porifera: Demospongiae) occurring in the tropical waters of the Great Barrier Reef. *Beagle* 8(1): 73–96.

83. Fromont, P. J. 1993. Descriptions of species of Haplosclerida (Porifera: Demospongiae) occurring in the tropical waters of the Great Barrier Reef. *Beagle* 10(1): 7–40.

Fromont, P. J. 1994. Reproductive development and timing of tropical sponges (Order Haplosclerida) from the Great Barrier Reef. *Coral Reefs* 13:127–133.

Fromont, P. J., and P. R. Bergquist. 1994. Reproductive biology of three sponges from the genus *Xestospongia* (Porifera: Demospongiae: Petrosida). *Coral Reefs* 13:119–126.

84. Garson, M. J. 1994. The biosynthesis of sponge secondary metabolites: Why it is important. In *Sponges in Time and Space*, ed. R. W. M. van Soest, Th. M. G. van Kempen, and J. C. Braekman, 427–440. Rotterdam: Balkema.

85. Guella, G., I. Mancini, D. Duhet, B. Richer de Forges, and F. Pietra. 1989. Ethyl 6-bromo-3-indolcarboxylate and 3-hydroxyacetal-6-bromoindole, novel bromoindoles from the sponge *Pleroma menoui* of the Coral Sea. *Zeitschrift für Naturforschung* 44(11–12): 914–916.

86. Guerriero, A., M. d'Ambrosio, and F. Pietra. 1990. Hydroxyicosatetraenoic, hydroyicosapentaenoic, hydroxydocosapentaenoic, acids from the marine sponge *Echinochalina mollis* of the Coral Sea. *Journal of Natural Products* 53(1): 57–61.

87. Gulvita, N. K., S. P. Gunasekera, and S. A. Pomponi. 1992. Isolation of latrunculin A, 6,7-epoxylatrunculin A, fijian A and euryfuran from a new genus of the family Thorectidae. *Journal of Natural Products* 55(4): 506–508.

Hajdu, E., and R. Desqueyroux-Faundez. 1994. A synopsis of South American *Mycale* (*Mycale*) (Poecilosclerida, Demospongiae) with a description of three new species and a cladistic analysis of Mycalidae. *Revue Suisse de Zoologie* 101(3): 563–600.

Hajdu, E., R. W. M. van Soest, and J. N. A. Hooper. 1994. Proposal for a phylogenetic subordinal classification of Poecilosclerid sponges. In *Sponges in Time and Space*, ed. R. W. M. van Soest, Th. M. G. van Kempen, and J. C. Braekman, 123–140. Rotterdam: Balkema.

88. Hartman, W. D. 1976. A new ceratoporellid sponge (Porifera: Sclerospongiae) from the

Pacific. In *Aspects of Sponge Biology*, ed. F. W. Harrison and R. R. Cowden, 329–347. London: Academic Press.

89. He, H-Y., and D. J. Faulkner. 1989. Renieramycins E and F from the sponge *Reniera* sp.: reassignment of the stereochemistry of the renieramycins. *Journal of Organic Chemistry* 54(24): 5822–5824.

90. He, H-Y., J. Salva, R. F. Catalos, and D. J. Faulkner. 1992. Sesquiterpene thiocyanates and isothiocyanates from *Axinyssa aplysinoides*. *Journal of Organic Chemistry* 57(11): 3191–3194.

91. Hentschel, E. 1912. Kiesel - und Hornschwamme der Aru - und Kei - Inseln. *Abhandlungen Senckenbergiana naturforschende Gessellschaft* 34:293–448.

92. Hirayama, A. 1990. Two new caprellidean (n. gen.) and known gammeridean amphipods (Crustacea) collected from a sponge in Noumea, New Caledonia. *Beagle* 7(2): 21–29.

93. Hooper, J. N. A. 1991. Revision of the Family Raspailiidae (Porifera: Demospongiae), with description of Australian Species. *Invertebrate Taxonomy* 5:1179–1418.

94. Hooper, J. N. A. 1994. Coral reef sponges of the Sahul Shelf—A case for habitat preservation. *Memoirs of the Queensland Museum* 36(1): 93–106.

95. Hooper, J. N. A., and P. R. Bergquist. 1992. *Cymbastela*, a new genus of lamellate coral reef sponges. *Memoirs of the Queensland Museum* 32(1): 99–137.

Hooper, J. N. A., R. J. Capon, C. P. Keenan, D. L. Parry, and N. Smit. 1992. Chemotaxonomy of marine sponges: Families Microcionidae, Raspailiidae and Axinellidae and their relationships with other Families in the Orders Poecilosclerida and Axinellida (Porifera: Demospongiae). *Invertebrate Taxonomy* 6:261–301.

96. Hooper, J. N. A., and C. Lévi. 1993a. Poecilosclerida from the New Caledonian Lagoon (Porifera: Demospongidae). *Invertebrate Taxonomy* 7:1221–1302.

97. Hooper, J. N. A., and C. Lévi. 1993b. Axinellida from the New Caledonian Lagoon

(Porifera: Demospongidae). *Invertebrate Taxonomy* 7:1395–1472.

98. Hooper, J. N. A., and C. Lévi. 1994. Biogeography of Indo-Pacific Sponges: Microcionidae, Raspailiidae, Axinellidae. In *Sponges in Time and Space*, ed. R. W. M. van Soest, Th. M. G. van Kempen, and J. C. Braekman, 191–212. Rotterdam: Balkema.

99. Hooper, J. N. A., and F. Wiedenmayer. 1994. Porifera. In *Zoological Catalogue of Australia*, ed. A. Wells, 12:1–624. Melbourne: CSIRO.

100. Hoshino, T. 1985a. Description of two new species in the genus *Agelas* (Demospongiae) from Zamami Island, Ryukyus, Japan. *Proceedings of the Japanese Society for Systematic Zoology* 30:1–10.

101. Hoshino, T. 1985b. *Dysidea minna* sp. nov., a new demosponge from the Ryukyus. *Galaxea* 4(1): 49–52.

102. Hoshino, T. 1987. A preliminary report on demosponges from the coral reef of the Okinawa Islands. *Galaxea* 6(1): 25–29.

103. Hyatt, A. 1875. Revision of the North American Poriferae; with remarks upon the foreign species. Part 2. *Memoirs of the Boston Society of Natural History* 2:481–554.

104. Iguchi, K., A. Sahashi, J. Kohno, and Y. Yamada. 1990. New sesquiterpenoid hydroquinone and quinones from Okinawan marine sponge (*Dysidea* sp.). *Chemistry and Pharmaceutical Bulletin (Tokyo)* 38(5): 1121–1123.

105. Iwagawa, T., M. Nakatani, N. Balat, A. Murphy, J. Willet, and J. L. C. Wright. 1992. A steroidal saponin from a Papua New Guinean sponge. *Kagoshima University Research Center for the South Pacific Occasional Papers* 23:39–40.

106. Jimenez, C., and P. Crews. 1990. Novel marine sponge amino acids. 10. Xestoaminols from *Xestospongia* sp. *Journal of Natural Products* 53(4): 978–982.

107. Jimenez, C., E. Quinoa, M. Adamczeski, L. M. Hunter, and P. Crews. 1991. Novel sponge derived amino acids. 12. Tryptophan-derived pigments and accompanying sesterterpenes from *Fascaplysinopis reticulata. Journal of Organic Chemistry* 56(10): 3403–3410.

108. Johnson, S. 1981. Blue camouflage in a nudibranch. *Hawaiian Shell News* 29(5): 14.

109. Johnson, S. 1983. Distribution of two nudibranch species on a subtidal reef on the western shore of Oahu, Hawaii. *Veliger* 25(4): 356–364.

110. Jurek, J., W. Y. Yoshida, P. J. Scheuer, and M. Kelly-Borges. 1993. Three new bromotyrosine-derived metabolites from the sponge *Psammaplysilla purpurea. Journal of Natural Products* 59(6): 1609–1612.

111. Kakou, Y., P. Crews, and G. J. Bakus. 1987. Dendrolasin and latruculin A from the Fijian sponge *Spongia mycofijiensis* and an associated nudibranch *Chromodoris lochi. Journal of Natural Products* 50(3): 482–484.

112. Karuso, P., and P. J. Scheuer. 1987. Longchain, a, -bisisothiocyanates from a marine sponge. *Tetrahedron Letters* 28(4): 5461–5464.

113. Kelly, M. 1986. Systematics and ecology of the sponges of Motupore Island, Papua New Guinea. Master's Thesis, University of Auckland, New Zealand, 1–190.

114. Kelly-Borges, M., and P. R. Bergquist. 1988. Success in a shallow reef environment, sponge recruitment by fragmentation through predation. *Proceedings of the 6th International Coral Reef Symposium, Townsville, Australia*, 2:757–762.

Kelly-Borges, M., and P. R. Bergquist. 1994. A redescription of *Aaptos aaptos* with descriptions of new species of *Aaptos* (Hadromerida: Suberitidae) from northern New Zealand. *Journal of Zoology, London* 234:301–323.

Kelly-Borges, M. 1995. Sponges out of their depth. *Nature* 373:284.

115. Kelly-Borges, M. 1995. Paradise Regained (The cultivation of natural bath sponges, a green alternative). *European Superstore Decisions*. Winter Issue, 98–99.

116. Kelly-Borges, M., and J. Vacelet. In press. A revision of *Diacarnus* Burton and *Negombata* de Laubenfels (Demospongiae: Latrunculiidae) with

descriptions of new species from the West Central Pacific and Red Sea. *Memoirs of the Queensland Museum.*

117. Kelly-Borges, M., L. J. Bell, and L. Sharron. 1994. Sponge diversity in Palauan Marine Lakes. *Marine Biodiversity: Causes and Consequences*, University of York, UK (Conference abstract).

118. Kelly-Borges, M., and P. R. Bergquist. 1988. Sponges from Motupore Island, Papua New Guinea. *Indo-Malayan Zoology* 5(2): 121–159.

Kelly-Borges, M., P. R. Bergquist, and P. L. Bergquist. 1991. Phylogenetic relationships within the Order Hadromerida (Porifera, Demospongiae, Tetractinomorpha) as indicated by ribosomal RNA sequence comparisons. *Biochemical Systematics and Ecology* 19(2): 117–125.

Kelly-Borges, M., and S. A. Pomponi. 1994. Phylogeny and classification of lithistid sponges (Porifera: Demospongiae): a preliminary assessment using ribosomal RNA sequence comparisons. *Molecular Marine Biology and Biotechnology* 3(2): 87–103.

Kelly-Borges, M., E. V. Robinson, S. P. Gunasekera, M. Gunasekera, N. V. Gulvita, and S. A. Pomponi. 1994. Species differentiation in the marine sponge genus *Discodermia* (Demospongiae: Lithistida): the utility of ethanol extract profiles as species-specific chemotaxonomic markers. *Biochemical Systematics and Ecology* 2(4): 353–365.

119. Kernan, M. R., R. C. Cambie, and P. R. Bergquist. 1991. Chemistry of sponges. 12. A new dihydric phenol from the sponge *Fasciospongia* sp. *Journal of Natural Products* 54(1): 269–270.

Kerr, R. G., and M. Kelly-Borges. 1994. Biochemical and morphological heterogeneity in the Carribean sponge *Xestospongia muta* (Petrosida: Petrosiidae). In *Sponges in Time and Space*, ed. R. W. M. van Soest, Th. M. G. van Kempen, and J. C. Braekman, 65–74. Rotterdam: Balkema.

120. Kirk, H. B. 1911. Sponges collected at the Kermadec Islands by W. R. B. Oliver. *Transactions of the New Zealand Institute* 43:574–581.

121. Kirkpatrick, R. 1900. Description of sponges from Funafuti. *Annals and Magazines of Natural History, London* 7(6): 345–362.

122. Kirkpatrick, R. 1908. On two new genera of recent Pharetronid sponges. *Annals and Magazines of Natural History, London* 8(2): 503–515.

123. Kobayashi, J., K. Ishida, K. Naitoh, H. Shigemori, Y. Mikami, and T. Sasaki. 1993. Xestokerols A, B and C_{29} steroids with a cyclopropane ring from Okinawan marine sponge *Xestospongia* sp. *Journal of Natural Products* 56(8): 1350–1355.

124. Kobayashi, J., F. Itagaki, H. Shigemori, and T. Sasaki. 1993. Three new onnamide congeners from the Okinawan marine sponge *Theonella* sp. *Journal of Natural Products* 56(6): 976–981.

125. Kobayashi, J., K. Kondo, M. Ishibashi, M. R. Walchli, and T. Nakamura. 1993. Theonezolide A; a novel polyketide natural product from the Okinawan marine sponge *Theonella* sp. *Journal of the American Chemical Society* 115(15): 6661–6665.

126. Kobayashi, J., O. Murata, H. Shigemori, and T. Sasaki. 1993. Jaspisamides A-C, new cytotoxic macrolides from the Okinawan marine sponge *Jaspis* sp. *Journal of Natural Products* 56(5): 787–791.

127. Kobayashi, J., C.-M. Zeng, M. Ishibashi, and T. Sasaki. 1993. Luffariolides F and G, new manalide derivatives from the Okinawan marine sponge *Luffariella* sp. *Journal of Natural Products* 56(3): 436–469.

128. Kobayashi, M., K. Kawazoe, T. Okamoto, T. Sasaki, and I. Kitagawa. 1994. Marine natural products. 31. Structure-activity correlation of a potent cytotoxic dimeric macrolide swinholide A, from the Okinawan marine sponge *Theonella swinhoei* and its isomers. *Chemical and Pharmaceutical Bulletin (Tokyo)* 42(1): 19–26.

129. Kobayashi, M., K. Kondo, and I. Kitagawa. 1994. Antifungal peroxyketal acids from the Okinawan marine sponge of *Plakortis* sp. *Chemical and Pharmaceutical Bulletin (Tokyo)* 41(7): 1324–1326.

130. Kobayashi, M., T. Okamoto, K. Hayashi, N. Yokoyama, T. Sasaki, and I. Kitagawa. 1994. Marine natural products. 32. Absolute configurations of C-4 of the monoalide family, biologically active sesterterpenes from the marine sponge *Hyrtios erecta*. *Chemical and Pharmaceutical Bulletin (Tokyo)* 42(2): 265–270.

131. Kobayashi, M., Y. Okamoto, and I. Kitagawa. 1991. Marine natural products. 28. The structures of sarasinosides A₁, A₂, A₃, B₁, B₂, B₃, C₁, C₂ and C₃, nine new norlanostane-triterpenoidal oligoglycosides from the Palauan marine sponge *Asteropus sarasinorum*. *Chemical and Pharmaceutical Bulletin (Tokyo)* 39(11): 2867–2877.

132. Kong, F., and D. J. Faulkner. 1993. Leucettamols A and B, two antimicrobial lipids from the calcareous sponge *Leucetta microraphis*. *Journal of Organic Chemistry* 58(4): 970–971.

133. Kourany-Lefoll, E., M. Pais, T. Sevenet, E. Guittet, A. Montagnac, C. Fountaine, D. Guenard, M. T. Adeline, and C. Debitus. 1992. Phloeodictines A and B: new antibacterial and cytotoxic bicyclic amidinium salts from the New Caledonia sponge, *Phloeodictyon* sp. *Journal of Organic Chemistry* 57(14): 1244–1249.

134. Ksebati, M. B., F. J. Schmitz, and S. P. Gunasekera. 1988. Puosides A-E, novel triterpene galactosides from a marine sponge, *Asteropus* sp. *Journal of Organic Chemistry* 53(17): 3917–3921.

135. La Barre, S. C., D. Laurent, P. W. Sammarco, W. T. Williams, and J. C. Coll. 1988. Comparative ichthyotoxicity of shallow and deep water sponges of New Caledonia. *Proceedings of the 6th International Coral Reef Symposium, Townsville, Australia*, 3:5–59.

136. Lambe, L. M. 1900. Catalogue of recent marine sponges of Canada and Alaska. *Ottawa Natural History* 14(9): 153–172.

137. Lendenfeld, R. von. 1885a. The Homocoela hitherto described from Australia and the new family Homodermidae. *Proceedings of the Linnean Society of New South Wales* 9(4): 896–907.

138. Lendenfeld, R. von. 1885b. A monograph of the Australian sponges. Part III. Preliminary description and classification of the Australian Calcispongiae. *Proceedings of the Linnean Society of New South Wales* 9(4): 1083–1150.

139. Lendenfeld, R. von. 1886. A monograph of the Australian sponges. Part VI. The genus *Euspongia*. *Proceedings of the Linnean Society of New South Wales* 10(3): 481–553.

140. Lendenfeld, R. von. 1888. *Descriptive Catalogue of the Sponges in the Australian Museum*. Sydney, London: Taylor and Francis.

141. Lendenfeld, R. von. 1889. *Monograph of the Horny Sponges*. London: Royal Society.

142. Lendenfeld, R. von. 1906. Die Tetraxonia. *Wissenschaftliche Ergebnisse Tiefsee-Expedition Valdivia II* 11(2): 59–373.

143. Lendenfeld, R. von. 1910a. The sponges 1. The Erylidae. Reports on the scientific results of the expedition to the eastern Tropical Pacific by the *Albatross* 1904–1905 and 1888–1904. *Memoirs of the Museum of Comparative Zoology, Harvard Collection* 41:261–323.

144. Lendenfeld, R. von. 1910b. The sponges 1. The Geodiidae. Reports on the scientific results of the expedition to the eastern Tropical Pacific by the *Albatross* 1904–1905 and 1888–1904. *Memoirs of the Museum of Comparative Zoology, Harvard Collection* 41(1): 1–258.

145. Lendenfeld, R. von. 1915. The sponges: Hexactinellida. Reports on the scientific results of the expedition to the eastern Tropical Pacific by the *Albatross* 1904–1905 and 1891–1899. *Memoirs of the Museum of Comparative Zoology, Harvard Collection* 42:1–397.

146. Lévi, C. 1956. Etude de *Halisarca* de Roscoff. Embryologie et Systematique des Demosponges. *Archives de Zoologie Expérimentale et Générale* 93:1–181.

147. Lévi, C. 1961. Eponges intercodiales de Nha Trang (Viet Nam). *Archives de Zoologie Expérimentale et Générale* 100:127–141.

148. Lévi, C. 1964. Spongiaires des zones bathyale, abyssale et hadale. *Galathea Reports* 7:63–112.

149. Lévi, C. 1967. Demosponges récoltées en Nouvelle-Calédonie par la mission Singer-Polignac. *Expédition Française sur les récifs coralliens de la Nouvelle-Calédonie, Editions de la Foundation Singer-Polignac* 2:13–28.

150. Lévi, C. 1979. The Demosponge fauna from the New Caledonia area. *Proceedings of the International Symposium on Marine Biogeography and Evolution in the Southern Hemisphere. New Zealand Oceanographic Institute Special Volume,* 307–315.

151. Lévi, C. 1983. *Pseudaxinyssa canthrella* n. sp., demosponge Axinellida du lagon Noumea (Nouvelle-Caledonie). *Bulletin du Muséum National d'Histoire Naturelle* 5(3): 719–722.

152. Lévi, C. 1993. Porifera Demospongiae: Spongiaires bathyaux de Nouvelle-Caledonie recoltes par le *Jean Carcot* Campagne BIOCAL, 1985. In *Resultats des Campagnes MUSORSTOM,* ed. A. Crosnier, *Volume 11: Memoire du Muséum National d'Histoire Naturelle* 158:9–87.

153. Lévi, C., and P. Lévi. 1979. *Lepidosphaera,* nouveau genre de demosponges a spicules en ecailles. *Bulletin Societe Zoologie France* 103(4): 443–448.

154. Lévi, C., and P. Lévi. 1982. Spongiaires hexactinellides du Pacific sud-ouest (Nouvelle-Caledonie). *Bulletin du Muséum National d'Histoire Naturelle* 4(3–4): 283–317.

155. Lévi, C., and P. Lévi. 1983. Eponges tetractinellides et lithistides bathyales de Nouvelle-Caledonie. *Bulletin du Muséum National d'Histoire Naturelle* 5(3): 101–168.

156. Lévi, C., and P. Lévi. 1984. Demosponges bathyales recoltees par le N/O *Vauban* au sud de la Nouvelle-Caledonie. *Bulletin du Muséum National d'Histoire Naturelle* 5(4): 931–937.

157. Lévi, C., and P. Lévi. 1988. Nouveaux spongiaires lithistides bathyaux a affinites cretaceés de la Nouvelle-Caledonie. *Bulletin du Muséum National d'Histoire Naturelle* 10(2): 241–263.

158. Lindgren, N. 1897. Beitrag zur Kenntniss der Spongienfauna des Malaiischen Archipels und Chinesichen Meere. *Zoologische Anzeiger* 10:480–487.

159. Lister, J. J. 1900. *Asterosclera willeyana,* the Type of a new Family of Sponges. *A. Willey's Zoological Results* 4:459–482.

160. Manes, L. V., S. Naylor, P. Crews, and G. J. Bakus. 1985. Suvanine, a novel sesterpene from an *Ircinia* marine sponge. *Journal of Organic Chemistry* 50(2): 284–286.

161. McCauley, R. D., M. J. Riddle, S. J. Sorokin, P. T. Murphy, P. M. Goldsworthy, A. J. McKenna, J. T. Baker, and R. A. Kelley. 1992. Australian Institute of Marine Bioactivity Group Marine Invertebrate Collection VII: Papua New Guinea, Thailand and the Philippines. *Australian Institute of Marine Science, Townsville, Australia.*

162. McKee, T. C., and C. M. Ireland. 1987. Cytotoxic and antimicrobial alkaloids from the Fijian sponge *Xestospongia caycedoi. Journal of Natural Products* 50(4): 754–756.

Minelli, A. 1993. *Biological Systematics: the state of the art.* London: Chapman and Hall.

163. Moliniski, T. F., and C. M. Ireland. 1988. Dysidazirine, a cytotoxic azacyclopropene from the marine sponge *Dysidea fragilis. Journal of Organic Chemistry* 53(9): 3334–3337.

164. Mori, K., and M. Horiguchi. 1975. Discovery of a new recent sclerosponge from Ngargol, Palan Islands and its significance. *Journal of the Geological Society of Japan* 81(12): 787–789.

165. Mori, K. 1977. A calcite sclerosponge from the Ishigaki-Shima coast, Ryukyu Islands, Japan. *Science Reports of the Tohoku University (Geology)* 47(1–2): 1–5.

166. Oger, J-M., P. Richomme, J. Bruneton, H. Guinaudeau, T. Sevenet, and C. Debitus. 1991. Steroids from *Neosiphonia supertes,* a marine fossil sponge. *Journal of Natural Products* 54(1): 273–275.

167. Pettit, G. R., Z. Cichacz, J. Barkoczy, A.-C. Dorsaz, D. L. Herald, M. D. Williams, D. L. Doubek, J. M. Schmidt, and L. P. Tackett. 1993. Isolation and structure of the marine sponge cell

growth inhibitory cyclic peptide phakellistatin 1. *Journal of Natural Products* 56(2): 260–267.

Picton, B. E. In press. The Species Directory Marine Database: a hierarchical taxonomic database for species oriented biological recording in the marine environment. *Bulletin de l'Institut Royal des Sciences Naturelle de Belgique.*

168. Plucer-Rosario, G. 1987. The effect of substratum on the growth of *Terpios*, an encrusting sponge which kills coral. *Coral Reefs* 5(4): 197–200.

169. Polejaeff, N. de. 1884. Keratosa. Report on the scientific results of the Voyage of H. M. S. *Challenger. Zoology* 11:1–88.

170. Potts, B. C. M., R. J. Capon, and D. J. Faulkner. 1992. Luffalacetone and (4E, 6E) – dehydromanoalide from the sponge *Luffariella variabilis. Journal of Organic Chemistry* 57(10): 2965–2967.

171. Preiwisch, J. 1903. Kalkschwamme aus dem Pacific. Ergebnisse einer Reise nach dem Pacific, Schauinsland, 1896–1897. *Zoologische Jahrbucher, Jena* 14:9–26.

172. Quinoa, E., and P. Crews. 1987. Niphatynes, methyoxylamine pyridines from the marine sponge, *Niphates* sp. *Tetrahedron Letters* 28(22): 2467–2468.

173. Quinoa, E., and P. Crews. 1988. Phenolic constituents of *Psammaplysilla. Tetrahedron Letters* 28(28): 3229–3232.

174. Quinoa, E., and P. Crews. 1988. Melynes, polyacetylene constituents from a Vanuatu marine sponge. *Tetrahedron Letters* 29(17): 2037–2040.

175. Quiron, J-C., T. Sevenet, H-P. Husson, B. Weniger, and C. Debitus. 1992. Two new alkaloids from *Xestospongia* sp., a New Caledonia sponge. *Journal of Natural Products* 55(10): 1505–1508.

176. Ravi, B. N., T. R. Erdman, and P. J. Scheuer. 1974. An antimicrobial constituent from a sponge, *Chondrosia* sp. *Proceedings of a Conference on Food and Drugs from the Sea,* 258–262.

177. Raymundo, A. K., T. O. Zulaybar, and G. P. Corpuz. 1988. Antimicrobial activity of some marine sponges from San Juan, Batangas. *Philippine Journal of Science* 117(2): 193–209.

Reiswig, H. M., and G. O. Mackie. 1983. Studies on Hexactinellid sponges III. The taxonomic status of Hexactinellida within the Porifera. *Philosophical Transactions of the Royal Society (Biology)* 301:419–428.

Reitner, J. 1991. Phylogenetic aspects and new descriptions of spicule-bearing hadromerid sponges with secondary calcareous skeleton (Tetractinomorpha, Demospongiae). In *Fossil and Recent Sponges,* ed. J. Reitner and H. Kuepp, 179–211. Berlin: Springer-Verlag.

178. Reitner, J., and T. S. Engeser. 1987. Skeletal structures and habitat of recent and fossil *Acanthochaetetes* (subclass Tetractinomorpha, Demospongiae, Porifera). *Coral Reefs* 6(1): 13–18.

179. Ridley, S. O. 1884a. Collections from Melanesia. Spongida: Part I. *Report on the Zoological Collections made in the Indo-Pacific during the Voyage of H. M. S. Alert, 1881–1882:*366–482.

180. Ridley, S. O. 1884b. Collections from the West Indian Ocean. Part 2. *Report on the Zoological Collections made in the Indo-Pacific during the Voyage of H. M. S. Alert, 1881–1882:*582–630.

181. Ridley, S. O., and A. Dendy. 1886. Preliminary report on the Monaxonida collected by H. M. S. *Challenger. Annals and Magazines of Natural History* 18(1): 325–351, II:470–493.

182. Ridley, S. O., and A. Dendy. 1887. Monaxonida. Report on the scientific results of the Voyage of H. M. S. *Challenger. Zoology* 20(59): 1–275.

183. Rodgers, S. D., and V. J. Paul. 1991. Chemical defenses of three *Glossodoris* nudibranchs and their dietary *Hyrtios* sponges. *Marine Ecology Progress Series* 77(2–3): 221–232.

Rodrigo, A. G., P. R. Bergquist, P. L. Bergquist, and R. A. Reeves. 1994. Are sponges animals? An investigation into the vagaries of phylogenetic inference. In *Sponges in Time and Space,* ed. R. W. M. van Soest, Th. M. G. van Kempen, and J. C. Braekman, 47–54. Rotterdam: Balkema.

184. Rodriguez, A. D., W. Y. Yoshida, and P. J. Scheuer. 1990. Popolohuanone A and B. Two sesquiterpenoid aminoquinones from a Pacific sponge, *Dysidea* sp. *Tetrahedron Letters* 46(24): 8025–8030.

185. Roll, D. M., P. J. Scheuer, G. K. Matsumoto, and J. Clardy. 1983. Halenaquinone, a penta-cyclic polyketide from a marine sponge. *Journal of the American Chemical Society* 105(19): 6177–6178.

186. Roll, D. M., C. M. Ireland, H. S. M. Lu, and J. Clardy. 1988. Fascaplysin, an unusual antimicrobial pigment from the marine sponge *Fascaplysinopsis* sp. *Journal of Organic Chemistry* 53(14): 3276–3278.

187. Rosser, R. M., and D. J. Faulkner. 1984. Two steroidal alkaloids from a marine sponge, *Plakina* sp. *Journal of Organic Chemistry* 49(26): 5157–5160.

188. Rützler, K. 1970. Freshwater sponges from New Caledonia. *Cahiers de l'office de la Recherche Scientifique et Technique Outre-Mer, Paris* 2:57–66.

189. Rützler K., and K. Muzik. 1993. *Terpios hoshinota,* a new cyanobacteriosponge threaten-ing Pacific reefs. *Scientia Marina* 57(4): 273–432.

190. Salvat, B., and J. Renaud-Mornant. 1969. Étude ecologique du macrobenthos et du meiobenthos d'un fond sableux du lagon de Mururoa (Tuamotu-Polynesie). *Cahiers du Pacifique* 13:65–82.

191. Santavy, D. 1986. A blue-pigmented bacterium symbiotic with *Terpios granulosa,* a coral reef sponge. *Hawaii Institute of Marine Biology Technical Report* 37:380–393.

192. Sará, M. 1988. Two new species of *Tethya* (Porifera: Demospongiae) from New Caledonia. *Bulletin du Muséum National d'Histoire Naturelle* 10(4): 651–659.

193. Sará, M. 1992. New Guinean *Tethya* (Porifera: Demospongiae) from Laing Island with

description of three new species. *Cahiers de Biologie Marine* 33(4): 447–467.

194. Schmidt, F. J., S. K. Agarwal, S. P. Guna-sekera, P. G. Schmidt, and J. N. Schoolery. 1983. Amphimedine, new aromatic alkaloid from the Pacific sponge, *Amphimedon* sp. Carbon con-nectivity determination from natural abundance ^{13}C ^{13}C coupling constants. *Journal of the Amer-ican Chemical Society* 105(4): 4835–4836.

195. Schmidt, F. J., and S. J. Bloor. 1988. Xesto- and ahlenaquinone derivatives from a sponge, *Adocia* sp., from Truk Lagoon. *Journal of Organic Chemistry* 53(1): 3922–3925.

196. Schulte, B. A., and G. J. Bakus. 1992. Predation deterrence in marine sponges: laboratory versus field studies. *Bulletin of Marine Science* 50(1): 205–211.

197. Schulze, F. E. 1887. Hexactinellida. Report on the scientific results of the Voyage of H. M. S. *Challenger. Zoology* 21:1–513.

198. Selenka, E. 1867. Ueber einige neue Schwamme aus der Sudsee. *Zeitschrift für Wissenshaftliche Zoologie* 17:565–571.

199. Seurat, L. G. 1934. La fauna et le peuple-ment de la Polynesie francaise. *Memoires de la Société Biogéographie, Paris* 4:41–74.

200. Sichang, M., and R. J. Andersen. 1990. Cytotoxic metabolites from the sponge *Ianthella basta* collected in Papua New Guinea. *Journal of Natural Products* 53(6): 1441–1446.

201. Sichang, M., and R. J. Andersen. 1991. Callydiyne, a new diacetylenic hydrocarbon from the sponge *Callyspongia flammea. Journal of Natural Products* 54(5): 1433–1434.

202. Smith, L. C. 1986. Larval release in the sponge *Callyspongia diffusa. Hawaii Institute of Marine Biology Technical Reports* 37:394–400.

203. Smyth, M. J. 1989. Bioerosion of gastropod shells: with emphasis on effects of coralline algal cover and shell microstructure. *Coral Reefs* 8(3): 119–125.

204. van Soest, R. W. M. 1989. The Indonesian sponge fauna: A status report. *Netherlands Journal of Sea Research* 23(2): 223–230.

The Sponges of the Tropical Island Region of Oceania: A Taxonomic Status Review   119

205. van Soest, R. W. M. 1990. Shallow water reef sponges of Eastern Indonesia. In *New Perspectives in Sponge Biology*, ed. K. Rützler, 1–533. Washington, D.C.: Smithsonian Institution.

van Soest, R. W. M. 1994. Demosponge distribution patterns. In *Sponges in Time and Space*, ed. R. W. M. van Soest, Th. M. G. van Kempen, and J. C. Braekman, 213–224. Rotterdam: Balkema.

van Soest, R. W. M., and J. N. A. Hooper. 1993. Taxonomy, phylogeny and biogeography of the marine sponge genus *Rhabderemia* Topsent, 1890 (Demospongiae, Poecilosclerida). *Scientia Marina* 57(4): 319–351.

Solé-Cava, A. M., M. Klautau, N. Boury-Esnault, R. Borojevic, and J. P. Thorpe. 1991. Genetic evidence for cryptic speciation in allopatric populations of two cosmopolitan species of the calcareous sponge *Clathrina*. *Marine Biology* 111:381–386.

206. Sollas, W. J. 1888. Report on the Tetractinellida collected by H. M. S. *Challenger*, during the years 1873–1876. *Challenger Reports* 25:1–458.

207. Stierle, D. B., and D. J. Faulkner. 1991. Two new pyrroloquinoline alkaloids from the sponge *Damiria* sp. *Journal of Natural Products* 54(4): 1131–1133.

208. Sullivan, B. W., and D. J. Faulkner. 1982. An antimicrobial sesterpene from a Palauan sponge. *Tetrahedron Letters* 23(9): 907–910.

209. Sullivan, B. W., D. J. Faulkner, K. T. Okamoto, M. H. M. Chen, and J. Clardy. 1987. (6R, 7S)-7-amino-7,8-dihydro-a-bisabolene, an antimicrobial metabolite from the marine sponge *Halichondria* sp. *Journal of Organic Chemistry* 51(26): 5134–5136.

210. Tanita, S. 1943. Studies on the Calcarea of Japan. *Scientific reports of Tohoku Imperial University* 4(17): 353–490.

211. Tabachnik, K. R. 1990. Hexactinellid sponges from Nasca and Sala-y-Gomez. *Trudy Instituta Okeanologii Akademiya Nauk SSSR* 124:161–173.

212. Tendal, O. S. 1969. Demospongiae from the Fiji Islands. *Vidensk. Meddr. dansk. naturh. Foren* 132:31–44.

213. Tendal, O. S., D. Barthel, and K. R. Tabachnik. 1993. An enigmatic organicism disclosed—and some new enigma. *Deep-Sea Newsletter* 21:11–12.

214. Thiele, J. 1898. Studien über pacifische Spongien. I. *Zoologica* 24:1–72.

215. Thiele, J. 1899. Studien über pacifische Spongien. II. *Zoologica* 24:1–33.

216. Thiele, J. 1900. Kieselschwamme von Ternate. I. *Abhandlungen Senckenbergiana naturforschende Gessellschaft* 25(1): 17–80.

217. Thiele, J. 1903. Kieselschwämme von Ternate. II. *Abhandlugen der Senckenbergischen Naturforschenden Gesellschaft* 25(4): 933–968.

218. Thiele, J. 1905. Die Kiesel- und Hornschwamme der Sammlung Plate. *Zoologische Jahrbucher, Jena Supplement* 6 [Fauna Chilensis (3)]:407–496.

219. Thomas, P. A. 1982. Sponges from Papua and New Guinea. 1. Order Keratosida Grant. *Journal of the Marine Biological Association of India* 24(1–2): 15–22.

220. Thomas, P. A. 1991. Sponges from Papua and New Guinea. 2. Order Haplosclerida. *Journal of the Marine Biological Association of India* 33(1–2): 308–316.

221. Topsent, E. 1932. Note sur les Clionides. *Archives de Zoologie Expérimentale et Générale* 47:549–579.

222. Topsent, E. 1934. Sur quelques Clionides (eponges) du Pacifique. *Memoires de la Société Biogéographie, Paris (Contribution a l'Etude du Peuplement zoologique et botanique des Iles du Pacifique)* 4:235–236.

223. Uchio, Y. 1992. A search for antineoplastic compounds from marine organisms. *Kagoshima University Research Center for the South Pacific Occasional Papers* 23:37–38.

224. Unson, M. D., C. B. Rose, D. J. Faulkner, L. S. Brinen, J. R. Steiner, and J. Clardy. 1993. New

polychlorinated amino acid derivatives from the marine sponge *Dysidea herbacea. Journal of Organic Chemistry* 58(23): 6336–6343.

225. Vacelet, J. 1977. Eponges pharetronids actuelles et sclerosponges de Polynesie francaise de Madagscar et de la Reunion. *Bulletin du Muséum National d'Histoire Naturelle* 307:345–367.

226. Vacelet, J. 1979. Description et affinities d'une eponge sphinctozoaire actuelle. *Colloque international CNRS number 291: Biologie des Spongiaires, Paris* 291:259–270.

227. Vacelet, J. 1979. Description et affinities d'une eponge sphinctozoaire actuelle. *Colloque international CNRS number 291: Biologie des Spongiaires, Paris* 291:483–493.

228. Vacelet, J. 1981. Eponges hypercalcifiees ('pharetronides', sclerosponges') des cavities des recifs coralliens de Nouvelles-Caledonie. *Bulletin du Muséum National d'Histoire Naturelle* 3(2): 313–351.

229. Vacelet, J. 1982. Algal-sponge symbioses in the coral reefs of New Caledonia: a morphological study. *Proceedings of the International Coral Reef Symposium* 4(2): 713–719.

230. Vacelet, J., J-P. Cuif, M. Massot, B. Richer de Forges, and H. Zibrowius. 1992. Un spongiaire sphinctozoaire colonail apparente aux constructeurs de recifs triasiques survivant dans le bathyal de Nouvelle-Caledonie. *Compte Rendu de l'Académie des Sciences* 314(9): 379–385.

231. Wicksten, M. K. 1981. Sponges and murex: a mutualistic relationship. *Of Sea and Shore* 12(1): 59.

232. Willey, A. 1902. Contribution to the Natural History of the Pearly Nautilus. I. Personal Narrative. *Willey's Zoological Results* 6:691–734. Cambridge University Press.

233. Whitelegge, T. 1897. The sponges of Funafuti. *Memoirs of the Australian Museum* 3:323–332.

234. Williams, D. E., D. L. Burgoyne, S. J. Rettig, R. J. Andersen, Z. R. Fathi-Afshar, and T. M. Allen. 1993. The isolation of majusculamide C from the sponge *Ptilocaulis trachys* collected in Enewetak and determination of the absolute configuration of the 2-methyl-3-aminopentanoic acid residue. *Journal of Natural Products* 56(4): 545–551.

235. Wilson, H. V. 1925. Siliceous and horny sponges collected by the U.S. Fisheries steamer *Albatross* during the Philippine expedition, 1907–10. *Bulletin of the United States National Museum of Washington* 100, 2(4): 273–506.

236. Wilson, B. R., and G. R. Allen. 1987. Major components and distribution of marine fauna. In *Fauna of Australia. General Articles*, ed. G. R. Dyne and D. W. Dalton. Volume 1A, chapter 3. Canberra: Australian Government Publishing Services.

237. Zabriskie, T. M., and C. M. Ireland. 1989. The isolation and structure of modified bioactive nucleosides from *Jaspis johnstoni. Journal of Natural Products* 52(6): 1353–1356.

CHAPTER 7

Polychaetes of Western Pacific Islands: A Review of Their Systematics and Ecology

J. H. Bailey-Brock

ABSTRACT

Tropical islands in the central and west Pacific region have speciose polychaete assemblages that are mostly cryptic and endolithic or infaunal. The most well known coral associated polychaetes are the brightly colored Christmas tree worms and the infamous fire worms. A few polychaetes play a role in Pacific Island cultures: the Palolo worm is collected and consumed during its annual spawning events, and a terebellid was valued by the ancient Hawaiians for medicinal properties. Large worms of many families are used for bait, and for fish and shrimp feeds.

Polychaetes are important components of the diet of fishes and reef gastropods, which in turn feed on algae, small invertebrates, detritus-laden sediments, and particulate organic matter. The diversity of polychaete communities may be as diverse as the habitats and ecological niches that reefs provide. Carbonate rocks and rubble, sands and muds, live corals, algal turf, and sponges all provide habitats for different assemblages of polychaetes.

Polychaetes may be found in tide pools of rocky intertidal areas, on terraced benches, reef flats, and subtidal reef slopes with extensive coral cover. Some tubicolous families are well represented in man-made harbors and natural embayments; burrowing species are typical of most sandy reef flat habitats and in lagoonal sediments. These habitats are briefly described, and the most conspicuous polychaetes are listed for each.

The characteristics of seventeen families commonly present on reefs are described. A table shows the distribution of known genera at six Pacific Island locations (Hawaii, Fiji, Tonga, the Solomon Islands, the Cook Islands, and Enewetak in the Marshall Islands). The number of genera is a good indication of the sampling effort at each location and does not necessarily imply a limited polychaete fauna for those islands with only a few

recorded genera. The consistency in generic number is similar for those families with few

genera (e.g., Arabellidae, Lumbrineridae, and Chaetopteridae) at all islands. Greater

collecting effort usually produces more genera and species for families with many genera

(e.g., Syllidae and Spionidae), especially if the specialists are available to work on identifi-

cations and generic revisions.

INTRODUCTION

Although polychaetes are not the most conspicuous fauna of coral reefs, they are some of the most ecologically important of the invertebrates. Their role in the trophic dynamics of reefs has only been understood in the last forty years and their impact on the geological structure of reefs in the last few decades. Polychaetes even contribute to the diet and economy of Pacific islanders, and some have antitumor properties (Tabrah, Kashiwagi, and Norton 1970). Most are small and easily overlooked, but they occur abundantly in all reef habitats and in live corals.

Polychaetes of tropical islands are not well known compared to those from more temperate regions. Expeditions to coral reefs in the western Pacific and Indian oceans have resulted in some taxonomic works that serve as foundations to the specialists who have added to them over the last 100 years. These publications are faunal lists and taxonomic revisions of polychaete taxa from a few island groups and coral reef systems but do not represent a very complete picture of the polychaete fauna of this extensive region. Another reason for this lack of information is that polychaetes are not a high priority taxon compared to fish and corals that are more conspicuous components of reefs and coastal waters. The results of collections made during expeditions with the *Challenger* (McIntosh 1885), *Eugenies* (Kinberg 1855, 1857, 1865, 1866, 1867), and the *Albatross* (Treadwell 1906) in the western Pacific and throughout the Hawaiian chain formed a basis for the tropical polychaete fauna. French scientists worked extensively in New Caledonia (Rullier 1972) and Tahiti (Fauvel 1919); Japanese researchers have collected in the Ryukus and nearby islands (Imajima and Hartman 1964); and polychaete researchers from many nations have collected from the Australian Great Barrier Reef (Monro 1931; Day and Hutchings 1979), Fiji (Bailey-Brock 1985), the Solomon Islands (Gibbs 1971), Cook Islands (Gibbs 1972), Tonga (Bailey-Brock 1987), the Marshalls (Devaney and Bailey-Brock 1987; Bailey-Brock, White, and Ward 1980), and Marianas (Bailey-Brock in prep.).

Polychaete collecting was a means of forming a faunal inventory to give a more complete picture of the diversity of coral reef assemblages at various localities. Efforts to understand community relationships came later when the question of reef metabolism and the trophic dynamics of reef dwellers were being investigated. The ecological analysis of Japtan reef (Odum and Odum 1955) included abundance and biomass estimates of polychaetes from the various zones across the reef. This generated an interest in their population dynamics and distribution. Researchers went on to examine the role of polychaetes on reefs, which added considerably to the information on their ecology and life-history events.

Polychaetes are important components of the diet of fishes. Analyses of the gut contents of coral reef fishes in the Marshall Islands (Hiatt and Strasburg 1960) and Hawaiian Islands (Hobson 1974; Dee and Parrish 1994) revealed that polychaetes contributed to the diet of many reef-inhabiting species. Polychaetes are selected as appropriate prey by goat fishes that have specialized structures, the chin barbels, for detecting small invertebrates in the sand. Parrot fish are not very discriminating, but scrape coral rock and digest the crypto-faunal community, including the worms, which make them both carnivores and herbivores (Brock 1991). Puffers also have coral-rock-crushing mouth parts, and *Canthigaster jactator* picks at tube worms on rocks (Hobson 1974).

Other fishes have predatory behaviors that enable them to feed on polychaetes. Chaetodontids at Enewetak bite off the opercula of serpulids. These fish have been collected with guts full of Christmas tree worm opercula (Motta 1980; Reese, pers. com.).

The importance of polychaetes in the diet of invertebrates is well illustrated by coniid gastropods. This has been carefully documented from the east and west Pacific and on Indian Ocean reefs (Kohn 1959; Kohn and Lloyd 1973a, 1973b). Kohn's work emphasized the availability of various species of worms on reef flats and benches, and the predatory specificity of the gastropods found in those habitats. Leviten (1976, 1978) showed that juvenile and small *Conus* spp. are specific in their dietary preferences and that they feed on small errant polychaetes, including nereids, eunicids, and many syllid species.

Polychaetes have little apparent effect on the geological framework of reefs or as predators. They are readily overlooked because of their small size and cryptic behavior, and their impact is assumed to be slight or negligible compared to larger invertebrates and reef fish. Some worms stabilize sand on reef flats by their tube-building activities (Bailey-Brock 1979, 1984a); others bore into coral rock contributing to the erosion of reef materials (White 1980). Many burrow through sediments; a few species are commensals living among the tube feet of echinoderms (Devaney 1967); and others reside on live corals, sponges, and among thalloid algae. The impact worms have in the reef community depends on their diversity and abundance, and the ability of the various species present to maintain high population densities in a particular habitat or location. The diet of worms and method of food procurement vary and are not always interpretable from the presence of jaws, and the nature of the pharyngeal and head structures. The most useful reference to polychaete diets gives a general perspective at the family level based on an extensive literature review and more specific information where this is known (Fauchald and Jumars 1979). These authors develop a guild approach to combine polychaete feeding and motility characteristics. Worms are tubicolous, motile, or discretely motile, and their trophic profile includes raptorial predators, omnivorous scavengers, filter or suspension feeders, deposit feeders, and selective deposit feeders. They can also be described by scale (e.g., macro- or microherbivores) and according to the presence or absence of jaws and kind of pharyngeal development.

Life histories include larvae that are pelagic or brooded in tubes or jelly cocoons for relatively short periods ranging from a few hours to a number of days, and direct development of juveniles that may be demersal for all or a portion of their dispersal phase.

Reproductive and gravid individuals may become temporary components of the plankton just prior to spawning. These worms frequently have a different morphology from non-reproductive adults, with enlarged eyes, modified setae for swimming, and segments swollen with gametes. The most well known example is the Palolo worm, which is the epitoke of *Palola siciliensis*, a rock-dwelling species. Ripe epitokes swim to the surface and spawn on moonlit nights in the fall. In Fiji, the Solomons, Cook Islands, Vanuatu, and American Samoa, the swarming of the Palolo worms is a celebrated event and the worms are scooped up and eaten raw or cooked (Miller and Pen 1959; Bailey-Brock 1985). They may be difficult to identify as they no longer have the normal parapodial structures and setae necessary for species diagnosis. Holly (1935) described a number of nereid species that were later recognized as the reproductive stages of known species. Coral reef benthic polychaetes occupy three types of habitat; they are associated with "hard" or "soft" bottom materials or live among marine vegetation (seagrasses, mangroves, and fleshy or thalloid algae). Truely pelagic polychaetes are restricted to a few families characterized by well-developed parapodia and lensed eyes. They are not well known on coral reefs, but about thirty species are known from subtropical waters across the northern Pacific Ocean (Tebble 1962).

COLLECTING METHODS

Hard Substrata

Basalt rocks, live coral, dead coral such as rock or rubble, coralline algal encrusted rocks, mollusc shells, and crustacean carapaces all become encrusted or bored by a diverse polychaete fauna, which is removed by scraping or dissolving the carbonate substrate (Bailey-Brock 1972, 1976). Endolithic worms will often leave their burrows if the water becomes anaerobic or if chilled for a few hours in a refrigerator. Coral rocks can be broken with a hammer and chisel, and worms extracted with forceps while they are being viewed with a dissecting microscope (Kohn and White 1977). Worms may be fragmented and damaged by this method, but it can be the best method at field stations. Worms are extracted in better condition if the rubble samples are fixed in 5–10 percent formalin for a minimum of forty-eight hours and dissolved in a nitric acid bath (Brock and Brock 1977). This method is used to obtain quantitative estimates of the fauna in the rock. Although calcareous opercula may soften, setae and parapodial features remain in good condition.

Soft Sediments

Sands and muds at the sediment-water interface and between the surface to a depth of 10 cm contain most of the polychaetes, except for a few large species that burrow to greater depths. Reef sediments are usually derived from coral rock and are calcium carbonate in origin. Samples can be sieved gently before or after fixation to remove the infauna, or the sand is elutriated by swirling with water in a shallow pan and pouring the water through a sieve of an appropriate mesh size (Sanders, Hessler, and Hampson 1965). The latter is usually better for soft-bodied organisms including polychaetes, sipunculans, and echiurans. Elutriation of formalin-fixed sediment samples produces worms in the best condition. In order to compare samples, the elutriation and sieving routine should be repeated six to eight times for each replicate and the same protocol followed throughout the study. Polychaetes trapped on the sieve (0.5 mm recommended) should be rinsed into a container using a wash bottle, then formalin-fixed, or, if previously preserved, a container of 70 percent ethanol. The acid dissolution method (Brock and Brock 1977) works well for quantitative sand samples and is most appropriate when sand and rubble are mixed, which is often the case on reef flats and among coral outcrops. Mixed sediments can be processed both by elutriation and then acid dissolution of the rubble fraction.

Algal Mat and Sponge Communities

Algae that form dense mats or turfs and encrusting or massive growths of sponge often have speciose polychaete communities. These can be processed by the refrigeration and manual-picking methods or by letting the water become anaerobic and removing worms as they swim to the surface. Formalin fixation of the entire algal or sponge sample is a sure method to retain all the associated fauna.

TAXONOMY OF POLYCHAETES

Most researchers categorize polychaetes at the family level (Table 7.1), although they are sometimes grouped in orders in reference works to emphasize morphological adaptations. Some families that were previously at the subfamily level are discussed as complexes of related families (e.g., the eunicid complex, the polynoid group). The eunicid complex is comprised of six families, which are all well represented on coral reefs, whereas the polynoid group is comprised of four families of scale-bearing worms, with two of these commonly found in coral reef environments. Ideally, generic and specific diagnoses are preferable for taxonomic studies, but much can be done to characterize an area or habitat by grouping worm assemblages at the family level (e.g., sand flats may be dominated by the sand tubes of chaetopterids and spionids, and boat harbors by sabellid and calcareous tube worms). The former conveys a habitat with sand and shallow depths; the latter indicates the availability of hard substrata in calm, protected waters suitable for suspension feeders. Assemblages of burrowing and tubicolous worms (e.g., capitellids and onuphids) indicate fine sediments with high organic content and the possibility of anaerobic conditions occurring just below the sediment surface.

Characteristics of the Phylum Annelida and Class Polychaeta are briefly presented here; they are fully described in most invertebrate texts. The phylum exemplifies the origin of segmental organization, with paired appendages best developed in the Polychaeta. The diagnoses of polychaete families are fully described in the references cited, and only a few conspicuous features most typical of each are given here.

Phylum Annelida—the segmented worms, triploblastic coelomate worms.

Class Polychaeta—bristle worms, literally means many setae. Worms have well-developed, segmentally arranged parapodia, each bearing a dorsal (notopodial) and ventral (neuropodial) tuft of setae. The majority of polychaetes are marine, but freshwater and brackish worms live in streams and estuaries of tropical regions. The taxonomy of these groups is being reviewed, but little is known about their life histories or diet. The classi-

Table 7.1. Number of recorded genera for the commonly present polychaete families from central and west Pacific Islands

Family	Hawaii	Fiji	Tonga	Solomon Islands	Cook Islands	Enewetak
Polynoidae	7	1	1	8	2	5
Sigalionidae	2		1	5		
Chrysopetalidae	2			2	1	3
Amphinomidae	8		1	7	1	2
Phyllodocidae	6		1	5	3	5
Syllidae	22	1	3	10	1	15
Nereididae	9	3	4	8	3	8
Eunicidae	5			5	2	5
Dorvilleidae	5			1		2
Lumbrineridae	1			1		1
Arabellidae	1		1	2		2
Onuphidae	3			1		
Spionidae	19	2	1	9	5	10
Cirratulidae	5		2	3		3
Chaetopteridae	4		3	4	3	3
Capitellidae	4		2	6	1	7
Arenicolidae	1			1		1
Sabellariidae	4			1		
Terebellidae	12		4	9		4
Sabellidae	13		5	6	2	4
Serpulidae	13	5	7	6		7
Spirorbidae	10	5	8			2
No. families	51[a]	7[b]	19[c]	34[d]	16[e]	35[f]

a. Bailey-Brock 1976; Bailey-Brock and Hartman 1987; Bailey-Brock in prep.

b. Bailey-Brock 1985; Treadwell 1906

c. Bailey-Brock 1987

d. Gibbs 1971

e. Gibbs 1972

f. Devaney and Bailey-Brock 1987

cal division of polychaetes into two groups, the Errantia and Sedentaria, is convenient and suggests that free living and burrowing or tubicolous lifestyles are consistent within the families in each group. This is not always the case, however. There are tubicolous families in the Errantia and free-living examples in the Sedentaria, but the basic grouping is still useful in general terms.

Polychaete Families Typically Found on Pacific Reefs

A key to forty-three families of Hawaiian polychaetes is published in the *Reef and Shore Fauna of Hawaii* (Bailey-Brock and Hartman 1987). Two families are entirely pelagic; the rest are benthic for all or part of their life cycle. Approximately twenty-two are typical of reef environments either by the abundance of a few species (e.g., chaetopterids and some fanworm species) or by the diverse group of species within a family that are found in proximity to each other (e.g., serpulid and spirorbid tube worms). Another useful reference is *A Monograph of the Polychaeta of Southern Africa* (Day 1967), which includes an illustrated key to families. Fauchald's key (1977) includes taxonomy at the order and family levels and contains keys to genera. A few of the families are briefly described here.

F. Polynoidae—Scaleworms. Occur under rocks and among rubble on reef flats and as commensals with echinoderms in deeper waters. Scaleworms are active jawed carnivores with well-developed eyes and sensory structures. Keys focus on the number of segments with scales (if the elytra have fallen off during handling and fixation, the elytrophores should be counted) and the insertion position of antennae on the head.

F. Chrysopetalidae—Chrysopetalids are small cryptic worms that occur under corals and among rubble. The dorsum is covered with transverse rows of golden paleae (broad, flat setae) that often cover the head. Specific identification is best left to the specialist.

F. Amphinomidae—Fireworms. Amphinomids are restricted to tropical latitudes. They occur in rubble habitats in shallow and deeper waters. Some species are found on floating debris such as glass floats and cargo nets; others are demersal but spawn at the surface and become stranded on beaches. Amphinomids are named for the itching caused by tufts of calcareous setae that break off when touched, releasing a toxin into the wound. Head with a chemosensory lobe, the caruncle. They are jawless omnivores and carnivores feeding on attached invertebrates, coral tissue, fish, and detrital laden sediment.

F. Phyllodocidae—Active, colorful worms with heart-shaped heads and uniramous parapodia with well-developed dorsal cirri that may be broad and leaflike or more slender lobes. Found in crevices and among algae, these worms have well-developed eyes and a jawless eversible proboscis to capture prey.

F. Syllidae—Probably the most abundant and speciose family of very small worms. The unique pharynx and anterior gut are diagnostic and facilitate feeding by piercing and sucking the tissues and cell contents of sessile invertebrates and algae. Parapodia are uniramous with dorsal and ventral cirri and short compound setae. Budding and epitoky are typical reproductive modes.

F. Nereididae—The nereids (or nereidids). Motile, crawling worms with well-developed head structures and parapodia. The pharynx is eversible and bears a pair of jaws and many small teeth. The number, shape, and arrangement of the paragnaths are diagnostic at generic and specific levels. Most marine nereids have bi-lobed parapodia with dorsal and ventral cirri, and compound setae. Reproductive worms produce specialized setae for swimming, and spawn at the surface at night.

F. Eunicidae—(and the eunicid complex: Dorvilleidae, Lysaretidae, Lumbrineridae, Arabellidae, Onuphidae). Motile worms that occupy the many types of habitat in coastal areas that burrow in coral rock, sand and mud, and crawl among algae and mangrove roots. The Onuphidae are tubicolous and may be locally abundant on reef flats. Head region with antennae, palps, paired eyes, jawed and scraping mouth parts. Parapodia appear uniramous and bear heavy setae, acicula and acicular setae; many segments have dorsal branchiae associated with the parapodia. Includes the Palolo worm, an epitokous stage of a coral burrower with predictable seasonal spawning in the west Pacific.

F. Spionidae—Tubicolous worms associated with sand, mixed sediments, and rubble. They are selective particulate feeders that may be abundant on reef flats and in deeper parts of coral reefs. Some species in the *Polydora* complex are borers in shells and coral rock. Anterior region with paired extensile palps, parapodial branchiae, and biramous parapodia; posterior with cuff-shaped pygidium or anal cirri.

F. Cirratulidae—Burrowing worms living just below the sediment surface with long filamentous gills and tentacles that extend above the sand as bright red or purple filaments. Worms are deposit feeders that selectively remove food particles with the grooved tentacular filaments. Taxonomy is based on the setae, which are nearly all simple, and the segmental origin of the tentacular filaments. Worms are found in all types of sediment and as borers in calcareous algal encrusted coral rock.

F. Chaetopteridae—Tubicolous worms with the body divided into three regions with specialized parapodia modified for filter feeding with mucus sacs. Chaetopterids may not preserve well in their tubes due to the presence of mucus in the tubes and their narrow diameter. Worms removed from their tubes prior to fixation would be more useful for identification than those that have been preserved in bulk. The presence or absence of tentacular cirri and details of the 4th setiger setae are necessary to follow the keys.

F. Capitellidae—Burrowing worms that resemble earthworms because they lack head appendages, and parapodia are reduced to two bundles of setae. Diagnostic features include the number of thoracic setigers and structure of the hooded hook setae. *Capitella capitata* is sexually dimorphic; the males have a pair of thoracic setae that aid copulation. This species is found in habitats receiving high levels of organic enrichment.

F. Arenicolidae—Large, soft-bodied worms that burrow in sand on reef flats and form U-shaped burrows. Worms have a protrusible proboscis that excavates and engulfs sediment while maintaining the burrow. The anterior region bears branching gills on some segments and has a broader diameter than the posterior region, which lacks setae. Fecal casts often mark the position of burrows, but removing intact specimens can be difficult because the worms are often in the rubble underlying the sand. Worms lay eggs in jelly cocoons or in their burrows (Bailey-Brock 1984b).

F. Sabellariidae—Tubicolous worms that construct strong sand-grain tubes on rocks, shells, and other firm substrates. The head has an operculum of modified setae (paleae) that close the tube to reduce desiccation at low tide and to exclude predators. The head region has filamentous tentacles for suspension feeding and manipulating tube-building materials. The number of rows of opercular paleae, number of thoracic setigers, and presence of oar and capillary setae are diagnostic. Some species are gregarious, forming rocky reefs as dense populations.

F. Terebellidae—Spaghetti worms. Tubicolous, deposit-feeding worms that remain mostly buried in the sediments or attached to rocks and extend their feeding tentacles across the substrate to selectively remove particles of detritus. Anterior region typically with 2–3 pairs of gills and lateral lobes, parapodia with neurosetae as uncini.

F. Sabellidae—Fan worms. Filter-feeding worms that form mud- or sand-covered tubes on docks, floats, pier walls, and among coral rocks. Branchial crown may bear eyes and stylodes. Thorax of eight segments with uncinigerous neuropodia, abdomen usually longer than the thorax with segments inverted.

F. Serpulidae—Calcareous tube worms. Filter-feeding worms that secrete calcareous tubes and occupy similar protected habitats as the fan worms. Serpulids are typically smaller than the conspicuous reef flat and harbor sabellids; most are attached to rocks and solid substrates; a few species are associated with live coral. The branchial crown includes an operculum, which closes the tube when the worm is retracted. Opercula may be lost to predaceous fish and another will grow, usually on the opposite side of the branchial crown. Regenerated opercula may have different morphologies as well as point of origin, which leads to separation as a new species or confusion at the generic or specific level. Setal and opercular characteristics are diagnostic features for these worms and the spirorbids, and except for a few easily identified species, they should be referred to specialists.

F. Spirorbidae—Spirally coiled, calcareous tube worms found on the undersides of coral rock, stones, and shells, and attached to seaweeds. Frequently occurring in boat harbors, tide pools, and in open coral reef habitats where they can be very abundant. Comprised of a branchial crown, an operculum, four or fewer thoracic segments, and a longer abdomen with both male and female reproductive organs. Embryos are brooded in the tube or in a chamber that forms below the operculum, and larvae are released at an advanced stage of development. Changes in the structure of the operculum in juvenile, prereproductive and reproductive worms can lead to incorrect specific diagnoses and difficulty with following keys.

DISTRIBUTION OF POLYCHAETES IN CORAL REEF HABITATS

Rocky Intertidal Areas

Tide pools, algal turf, and shallow sand-filled depressions associated with lava rocks, basalt, and limestone benches contain diverse polychaete assemblages. Depending on the amount of physical exposure to surf and wave action and the tidal movement of water level, pools are flushed on a regular basis and sand scour may be predictable. Island groups that receive seasonal heavy rains may have a reduced bench or tide pool fauna but one that is tolerant of reduced salinities following rains. The supralittoral pools of Talafofo, Guam, experienced extreme dilution in August 1981, and dead polychaetes, shrimp, and fish rotted with decomposing algae until the pools were flushed with sea water. The windward bench at Enewetak is exposed to a gradient of temperature and periods of immersion from shore to reef crest. Polychaete abundance and species richness was much higher in refuges such as shallow pools with sand bottoms and among algal turf than on exposed areas of the bench (Bailey-Brock, White, and Ward 1980). Tide pools often contain tubicolous worms, for example, spirorbid and serpulid tube worms, sabellids, spionids, terebellids, sabellariids and chaetopterids, and small syllids and capitellids.

The terraced benches at Talafofo, Guam, had a dominant algal cover of *Padina* sp., and the fireworm *Eurythoe complanata* occurred abundantly across the extensive bench and crest.

Reef Flats

Reef flats and the sandy tops of patch reefs and sand cays are shallow subtidal or intertidal habitats that may be impacted by storms or changes to the coastline and adjacent watershed. These areas are often characterized by sediment-stabilizing worms—the chaetopterids, *Phyllochaetopterus* spp., *Mesochaetopterus sagittarius*, and *Spiochaetopterus* spp., and spionids in Tonga (Bailey-Brock 1987) and the Solomons (Gibbs 1971). They form sand-encrusted tubes and occur in dense populations. In Hawaii these worms are often found with the hemichordate *Ptychodera flava* and the seagrass *Halophila hawaiiana*.

Reef flats often have hard substrate covered with an algal turf or bare reef rock that provides habitat for carnivorous and grazing gastropod molluscs, oysters, sponges, tunicates, anemones, zoanthids, corals, and polychaetes. The rocks may be encrusted with an intertidally restricted serpulid *Pomatoleios kraussii* and provide anchorage for the large fanworm *Sabellastarte sanctijosephi*. Errant worms (e.g., polynoids, phyllodocids, fireworms, syllids, and nereids) are often collected from the undersides of rocks and among turf algae. The endolithic fauna (from acid-dissolved or broken-up rocks) would include eunicids, the boring cirratulid *Dodecaceria laddi* if coralline algae are present, boring spionids, and sabellids and sipunculans.

Living Corals

Only a few species of invertebrates live with live corals compared to the diverse community that can be found in dead coral. Stinging nematocysts and mucus production by corals are deterrents to larvae that may attempt to settle on corals. The Christmas tree worm, *Spirobranchus giganteus corniculatus*, is the most well known and visually conspicuous in shallow coral-rich habitats. The multicolored branchial crowns of this serpulid stand out against the coral unless disturbed by swimmers or predaceous fish, which causes them to retract into their tubes. Trochophore larvae are thought to settle in open areas that may have been scraped by a fish or urchin to expose the coral skeleton or adjacent to the growing edge of the coral (Smith 1984). The tubes of the worms are overgrown by the coral so that coral and tube cannot be separated, and the only tube feature not overgrown is the median spine that projects over the opening. The operculum of this serpulid is elaborately branched and spinous and considered to deter predaceous fish.

Boat Harbors

Harbors and anchorages usually provide protected conditions and calm waters that allow filter-feeding tube worms to flourish. They colonize vessel hulls, floats, docks, and harbor walls and are important components of the harbor community. Many species are transported worldwide in ballast water and on ship hulls and have a cosmopolitan distribution. Worms that were introduced by shipping some time ago may have become established outside the harbor and spread to other protected locations. Recent introductions may be restricted to a single harbor or enclosed boatway. The serpulid *Hydroides elegans* has been dispersed by shipping, and a number of spirorbids, including *Pileolaria militaris*, may have the same history. As spirorbids are hermaphroditic, a single worm on a ship's hull could start a population in each harbor visited.

Shallow reef flats of Suva Harbor have dense populations of sabellariids forming masses of cemented tubes of sand and mud on firm substrates. Nereids, spionids, and a polynoid species were collected in the vicinity of the sabellariids (Bailey-Brock 1985).

Mangroves and Seagrasses

Mangroves are typically associated with the landward parts of reefs such as on atolls, estuaries, and river mouths, but flourish along coasts with normal marine conditions. Mangrove forest follows the shoreline and extends out onto the reef flat as groves or small copses of trees. The diversity of trees and associated fauna increase across the Pacific to Southeast Asia where mangrove forests are best developed despite commercial logging operations. Mangroves on the Klang River in W. Malaysia have an invertebrate fauna on their trunks that is zoned like the intertidal with barnacles and mussels in discrete horizontal bands. The prop roots of *Rhizophora mangle*, which was intentionally intro-duced to Hawaii within the last century (see Ellison, this volume), have few epifaunal organisms compared to the Malaysian trees. Palau, Fiji, Tonga, and other west Pacific locations may have more diverse communities, especially if oysters are present. Oysters thrive in estuarine habitats and provide a suitable substrate for tube worms and cryptic errants. The shells may be bored by polydorids (Spionidae) and fouled by sponges and tunicates, which provide another habitat for polychaetes. Spirorbids and serpulids attach to prop roots and floating fruits, and the sediments beneath the trees may have an abundant infaunal community (Bailey-Brock 1979).

Seagrasses often become encrusted with spirorbids but may not be very long lived sub-strates. *Janua (Dexiospira) foraminosa* was found on seagrasses at Pago Bay, Guam. Seagrasses in the lagoon at Saipan were free of tube worms and the same is true for the only species, *Halophila hawaiiana*, found in Hawaii (see Coles and Kuo, this volume). Blades of seagrasses may be broken off or lost periodically as they are fouled, so sampling may need to be more frequent to check for tube worms. The proximity of seagrass beds to hard substrates that may have many tube worms may be relevant to the recruitment of worms to the plant.

Algae

Seaweeds on reefs and in protected lagoons, as well as algal crusts and turfs, support diverse polychaete assemblages. *Janua (D.) foraminosa* has been collected on *Dictyo-sphaeria cavernosa*, *Caulerpa* sp., and *Acanthophora spicifera* in Hawaii. The same spirorbid was collected on *Halimeda* and *Padina* on Guam. This type of substrate specific-ity is typical of some spirorbid species and indicative of larval selection behavior. The algae forming the turf on the intertidal bench at Enewetak provided a three-dimensional habitat or refuge to amphinomids, syllids, nereids, chaetopterids, and cirratulids (Bailey-Brock, White, and Ward 1980).

Mariculture Farms

Commercially cultured bivalve molluscs often become infected with the shell-boring spionid *Polydora websteri*. Oysters raised for food and pearl culture are appropriate habitats for the worms that produce advanced planktonic larvae that readily infect adjacent oysters. Naturally occurring populations of boring spionids inhabitating coral rock can be particularly difficult to control in extensive reef-culture programs.

Floating Slippers, Glass Floats, and Debris

This category of "habitat" has gained importance as corals are known to be rafted across oceans in this manner (Jokiel 1990). A large polychaete identified as *Eunice aphroditois* was collected on a transoceanic slipper in the Seychelles by Dr. Jokiel. Some amphinomids are collected from the netting around glass balls and fishing floats. *Hipponoe gaudichaudi*, which is associated with stalked barnacles, and *Amphinome rostrata* are only known from this habitat in Hawaiian waters.

Acknowledgments—I thank R. Brock for always keeping my interests in mind and collecting polychaetes across the Pacific whenever he could. I am grateful to R. Brock and E. A. Kay for reading the manuscript, P. Jokiel for the floating slipper and worm geographic information, L. Ward and K. Fauchald of the United States National Museum, Smithsonian Institution, for providing information and advice on polychaetes, and to L. Eldredge and C. Birkland for hosting a visit to the Marine Laboratory at Mangilao, Guam.

REFERENCES

Bailey-Brock, J. H. 1972. Deep water tube worms (Polychaeta: Serpulidae) from the Hawaiian Islands. *Pac. Sci.* 26(4): 405–8.

———. 1976. Habitats of tubicolous polychaetes from the Hawaiian Islands and Johnston Atoll. *Pac. Sci.* 30:690–81.

———. 1979. Sediment trapping by chaetopterid polychaetes on a Hawaiian fringing reef. *J. Mar. Res.* 37:643–56.

———. 1984a. Ecology of the tube building polychaete *Diopatra leuckarti Kinberg*, 1865 (Onuphidae) in Hawaii: Community structure, and sediment stabilizing properties. *J. Linn. Soc. Lond. Zool.* 80:191–99.

———. 1984b. Spawning and development of *Arenicola brasiliensis* (Nonato) in Hawaii. In *Proc. First International Polychaete Conference*, Sydney, 1983, ed. P. A. Hutchings, 439–49. New South Wales: J. Linn. Soc.

———. 1985. Polychaetes from Fijian coral reefs. *Pac. Sci.* 39(2): 195–220.

———. 1987. The polychaetes of Fanga'uta lagoon and coral reefs of Tongatapu, Tonga, with discussion of the Serpulidae and Spirorbidae. *Proc. Biol. Soc. Wash.* 7:280–94.

———. In prep. Polychaetes of shallow coastal habitats of Guam, Mariana Islands.

Bailey-Brock, J. H., and O. Hartman. 1987. Polychaeta (Annelida) of the Hawaiian Islands. Revision of C. H. Edmondson, *Reef and shore fauna of Hawaii*. Honolulu: Bernice P. Bishop Museum Press.

Bailey-Brock, J. H., J. White, and L. Ward. 1980. Effects of refuges on polychaete assemblages of a windward reef bench at Enewetak Atoll. *Micronesica* 16(1): 43-58.

Brock, R. E. 1991. Species profiles: Life histories and environmental requirements of coastal vertebrates and invertebrates, Pacific Ocean Region; 5. The Parrotfishes, Family Scaridae. Technical Report EL-89-10. U.S. Army Engineer Waterways Experiment Station, Vicksburg, Mississippi.

Brock, R. E., and J. H. Brock. 1977. A method for quantitatively assessing the infaunal community in coral rock. *Limnol. Oceanogr.* 22:948–51.

Day, J. H. 1967. *A monograph of the Polychaeta of Southern Africa*. Part 1. *Errantia*; Part 2. *Sedentaria*. British Museum (Natural History), London. Vols. 1 and 2.

Day, J. H., and P. A. Hutchings. 1979. An annotated check-list of Australian and New Zealand Polychaeta, Archiannelida, and Myzostomida. *Rec. Aust. Mus.* 32(3): 80–161.

Dee, A. J., and J. D. Parrish. 1994. Reproductive and trophic ecology of the soldierfish *Myripristis amaena* in tropical fisheries. *Fishery Bulletin* 92:516–30.

Devaney, D. M. 1967. An ectocommensal polynoid associated with Indo-Pacific echinoderms, primarily ophiuroids. *Occas. Pap. Bernice P. Bishop Mus.* 23(13): 287–304.

Devaney, D., and J. H. Bailey-Brock. 1987. Polychaetes of Enewetak Atoll. In Vol. 2, *The Natural History of Enewetak Atoll.* U.S. Dept. of Energy, Office of Scientific and Technical Information.

Fauchald, K. 1970. Polychaetous annelids of the families Eunicidae, Lumbrineridae, Iphitimidae, Arabellidae, Lysaretidae, and Dorvilleidae from western Mexico. *Allan Hancock Monogr. Mar. Biol.* 5:1–335.

———. 1977. *The polychaete worms: Definitions and keys to the orders, families, and genera.* Nat. Hist. Mus. Los Ang. Cty. Sci. Ser. 28:1–190.

Fauchald, K., and P. A. Jumars. 1979. The diet of worms: A study of polychaete feeding guilds. *Oceanogr. Mar. Biol. Ann. Rev.* 17:193–284.

Fauvel, P. 1919. Annelides polychetes des Iles Gambier et Touamotou. *Paris Mus. Nat. d'Hist. Natur., Bull.* 25:336–43.

Gibbs, P. E. 1971. Polychaete fauna of the Solomon Islands. *Bull. Br. Mus. (Nat. Hist) Zool.* 21:101–211.

———. 1972. Polychaete annelids from the Cook Islands. *J. Zool. (Lond.)* 168:199–220.

Hiatt, R. W., and D. W. Strasburg. 1960. Ecological relationships of the fish fauna on coral reefs of the Marshall Islands. *Ecol. Mongr.* 30:65–127.

Hobson, E. S. 1974. Feeding relationships of teleostean fishes on coral reefs in Kona, Hawaii. *Fishery Bulletin* 72:915–1031.

Holly, M. 1935. *Polychaeta from Hawaii.* Bernice P. Bishop Mus. Bull. 129.

Imajima, M., and O. Hartman. 1964. The polychaetous annelids of Japan. *Allan Hancock Found. Occas. Pap.* 26:1–452.

Jokiel, P. L. 1990. Long distance dispersal by rafting: Reemergence of an old hypothesis. *Endeavour* 14(2): 66–73.

Kinberg, J. 1855. Nya slagter och arter af Annelider. *Ofver. Svenska Vetensk. Akad. Forh.* 12:381–88.

———. 1857. Nya slagter och arter of Annelider. *Ofver. Svenska Vetensk. Akad. Forh.* 14:11–14.

———. 1865. Annulata Nova. *Ofver. Svenska Vetensk. Akad. Forh.* 21:559–74.

———. 1866. Annulata Nova. *Ofver. Svenska Vetensk. Akad. Forh.* 22:239–58.

———. 1867. Annulata Nova. *Ofver. Svenska Vetensk. Akad. Forh.* 23:337–57.

Kohn, A. J. 1959. The ecology of *Conus* in Hawaii. *Ecol. Monogr.* 29:47–90.

Kohn, A. J., and M. C. Lloyd. 1973a. Marine polychaete annelids of Easter Island. *Int. Revue Ges. Hydrobiol.* 58(5): 691–712.

———. 1973b. Polychaetes of truncates reef limestone substrates on eastern Indian Ocean coral reefs: Diversity, abundance, and taxonomy. *Int. Revue Ges. Hydrobiol.* 58(3): 369–99.

Kohn, A. J., and J. K. White. 1977. Polychaete annelids of an intertidal reef limestone platform at Tanguisson, Guam. *Micronesica* 13(2): 199–215.

Leviten, P. J. 1976. The foraging strategy of vermivorous conid gastropods. *Ecol. Monogr.* 46(2): 157–78.

Leviten, P. J. 1978. Resource partitioning by predatory gastropods of the genus *Conus* on subtidal Indo-Pacific coral reefs: The significance of prey size. *Ecology* 59(3): 614–31.

McIntosh, W. 1885. Reports on the Annelida Polychaeta collected by H.M.S. *Challenger* during the years 1873–76. *Rep. Sci. Results of the Voyage of H.M.S. Challenger.* 12:1–554.

Miller, C. D., and F. Pen. 1959. Composition and

nutritive value of Palolo (*Palola siciliensis* Grube).
Pac. Sci. 13:191–94.

Monro, C. C. A. 1931. Polychaeta, Oligochaeta,
Echiuroidea, and Sipunculoidea. Great Barrier
Reef Expedition 1928–29. *Sci. Rep.* 4(1): 1–37.

Motta, P. J. 1980. Functional anatomy of the jaw
apparatus and the related feeding behavior of the
butterflyfishes (Chaetodontidae) including a
review of jaw protrusion in fishes. Ph.D. diss.,
University of Hawaii.

Odum, H. T., and E. P. Odum. 1955. Trophic
structure and productivity of a windward coral
reef community on Eniwetok Atoll. *Ecol. Monogr.*
25(3): 291–320.

Rullier, F. 1972. Annelides polychetes de
Nouvelle-Caledonie recueillies par Y. Plessis et B.
Salvat.-Expedition Francaise sur les recifs
coralliens de la Nouvelle-Caledonie 16:1–169.

Sanders, H. L., R. R. Hessler, and G. R. Hampson. 1965. An introduction to the deep sea faunal
assemblages along the Gay Head–Bermuda
transect. *Deep Sea Res.* 12:845–67.

Smith, R. 1984. Development and settling of
Spirobranchus giganteus (Polychaeta: Serpulidae).
In *Proc. First International Polychaete Confer-
ence*, Sydney, 1983, ed. P. A. Hutchings, 461–83.
New South Wales: J. Linn. Soc.

Tabrah, F. L., M. Kashiwagi, and T. R. Norton.
1970. Antitumor activity in mice of tentacles of
two tropical sea annelids. *Science* 170:181–83.

Tebble, N. 1962. The distribution of pelagic
polychaetes across the north Pacific Ocean. *Bull.
Br. Mus. (Nat. Hist) Zool.* 7(9): 373–492.

Treadwell, A. L. 1906. Polychaetous annelids of
the Hawaiian Islands, collected by the steamer
Albatross in 1902. *Bull. U.S. Fish. Comm.*
23:1145–81.

White, J. K. F. 1980. Distribution, recruitment,
and development of the borer community in dead
coral on shallow Hawaiian reefs. Ph.D. diss.,
University of Hawaii.

CHAPTER 8

Pacific Island Marine Mollusks: Systematics

E. Alison Kay

ABSTRACT

This first approximation of marine molluscan species diversity among the islands of the oceanic Pacific is approached by listing families, genera, and number of species known from Guam and the Northern Mariana Islands; Kwajalein and Enewetak, Marshall Islands; the Hawaiian Islands; the Society and Tuamotu islands; the Pitcairn Islands; and Easter Island. The accounting includes about 2,600 species associated with these islands and demonstrates several aspects of the distribution of diversity: attenuation both in species numbers, drop-out of major family and generic groups, and changes in species composition from west to east across the Pacific. Explanations for anomalies in the distribution patterns of diversity are proposed.

INTRODUCTION

Sven Ekman (1935, 1953) points to a "centre and focus" of the wealth of animal life that occurs within the compass of what he terms the Indo-West Pacific, the great oceanic realm "which contains the tropical and subtropical portions of the whole Indian Ocean and the western and central part of the Pacific." That "centre and focus" is the Indo-Malayan region from which, as Ekman visualizes it, other faunas, such as that of the oceanic Pacific, recruit their biota. Ekman estimates the molluscan fauna of the region as 6,000 species; Faustino (1928) cites 5,000 species in the Philippines alone. Ekman also suggests that "The further one moves from this centre . . . the more the fauna appears as a progressively impoverished Indo-Malayan fauna." In this review of the diversity of the marine molluscan faunas of the oceanic islands of the Pacific, the faunas of five island groups for which there are reasonable census data are assessed in terms of the patterns of diversity proposed by Ekman.

MARINE MOLLUSKS IN THE OCEANIC PACIFIC

Several thousand names are now associated with species of marine mollusks in the tropical Pacific, and new species continue to be described—an average of twenty-four new species each year during the past five years. The three volumes by Cernohorsky (1967, 1972, 1978) list 1,200 species as found from Indonesia and the Philippines to Hawaii and

Easter Island. Census data for three island groups report approximately 1,000 species each for the Hawaiian Islands (Kay 1979) and Enewetak, Marshall Islands (Kay and Johnson 1987); 1,495 species from Kwajalein, Marshall Islands (Johnson, pers. com.); 400 species from the Pitcairn Islands—Henderson, Pitcairn, Oeno, and Ducie islands (Preese in press); and 130 species from Easter Island (Rehder 1980). In this summary, based on numbers from Guam and the Northern Mariana Islands; Kwajalein and Enewetak, Marshall Islands; the Hawaiian Islands; the Society and Tuamotu islands; the Pitcairn Islands; and Easter Island, approximately 2,600 species are assessed.

Families, genera, and numbers of species of marine mollusks for the islands noted in the previous paragraph are listed in the Appendix to this paper. Although the families and genera cited are generally familiar, several unfamiliar placements reflect recent changes in the scheme of the higher classification (see, for example, Ponder and Warén 1988; Hickman and McLean 1990). Superfamilies, although not always used in invertebrate classifications, appear in accounts of mollusks as they indicate some useful relationships.

The listing and the summary (see Appendix) clearly demonstrate that the marine mollusks of the oceanic islands of the Pacific are a subset of the fauna of the Indo-Malayan region, including as they do genera such as *Hippopus*, *Tridacna*, and *Nautilus* that occur only in the Indo-West Pacific. The data also graphically demonstrate Ekman's thesis of increasing impoverishment from west to east across the Pacific—from 1,200 species associated with islands in the western Pacific, the total drops to 121 at Easter Island, the farthest reaches of the eastern edge of the Indo-West Pacific.

The distribution patterns of diversity are not without anomalies. Several of those anomalies can be accounted for, simply because some island faunas are better known than others and because specialists in certain taxonomic groups have contributed to knowledge of those groups on some islands but not on others. The Hawaiian Islands, for example, have the broadest representation of species, reflecting a collection history going back to the voyage of Captain Cook in 1778; intense collection and interest by malacologists such as W. H. Pease, William H. Dall, Harald Rehder, Henry A. Pilsbry, Charles H. Edmondson, and Jens M. Ostergaard from the mid-nineteenth century to the mid-twentieth century; and more recently by avid scuba divers and other collectors. In the Hawaiian Islands, too, considerable attention has been given to some unfamiliar groups and "difficult" groups such as micromollusks (the families Orbitestellidae, Triphoridae, and Cerithiopsidae, often with shells less than 5 mm in maximum dimension); the Eulimidae, parasites on echinoderms (Warén 1983; Ponder and Gooding 1978); Epitoniidae, which are associated with coelenterates (Du Shane 1987, 1988); the notoriously difficult Turridae (Kay 1990b); and opisthobranchs (Kay 1979; Kay and Young 1969; Gosliner 1979, pers. com.). Data for the Hawaiian Islands prior to 1979 are summarized in Kay (1979). But even in Hawaii, hundreds of species remain unreported: Kay (1979) reported about 130 opisthobranchs; Gosliner (pers. com.) now has recorded more than 430 species of opisthobranchs in Hawaii. Hochberg (pers. com.) has recorded fifteen species of benthic octopods, but nearly half of them are undescribed.

PATTERNS OF DISTRIBUTION

Numbers themselves, although indicative of more species, genera, or families associated with one island from another, are meaningless without context. Context in the oceanic Pacific lies in the nature of islands themselves.

Attenuation

The listing of the numbers of marine mollusks known from different islands in the Pacific basin, although not without anomalies, reflects the basic rule of distribution in the Pacific cited by Ekman (1935, 1953): there are fewer species from west to east across the Pacific (see Appendix). Attenuation, however, is not solely a matter of numbers; for example, sixty-three species of Cypraeidae are recorded from Guam, but only thirty-seven species are present in the Hawaiian Islands. Not only species, but major groups of mollusks fall out from west to east: the chambered nautilus *(Nautilus),* abalones *(Haliotis),* and the bailer shell *Melo* are not found east of the Caroline Islands and Samoa (Kay in press). The question of interest thus becomes that of finding a pattern to the attenuation and in the changes in species composition—as the cowrie genera *Erronea* and *Purpuradusta* drop out from west to east, *Erosaria* and *Lyncina* become increasingly prominent (Kay 1990a).

Restricted Distributions

While some species reach remote parts of islands, others are restricted in their distribution. At least three factors may be involved: (1) Dispersal ability—that is, the distribution of mollusks with long-lived larval stages compared with those that are direct developers or with lecithotrophic larva—was recognized by Hedley (1899) as a prime determinant of distribution. Scheltema (1986) continues to find patterns associated with larval dispersal. (2) Some mollusks are restricted to certain habitats and are thereby restricted in their occurrence—*Cellana* and *Hipponix* are associated with basalt substrates; *Smaragdia* is found only on seagrasses. (3) Endemism is clearly associated with isolation—the more isolated the island, the higher the degree of endemism. Kay (1984) identifies two foci of endemism in the Pacific, the Hawaiian Islands, and the islands of southeastern Polynesia where endemism in the marine mollusks approaches more than 25 percent. In the Marshall Islands, which lie in the midst of a stream of biological traffic, no recent endemic marine mollusks have been identified.

Topography

Atolls and volcanic islands are very different, one from the other. And atolls differ one from another: two extremes among atolls are those that have a single pass into the lagoon, and those that have several passes. At both Fanning Island (Kay and Switzer 1974) and Canton Island (Kay 1978), a single deepwater pass leads into the lagoon consisting of an outer, clear water lagoon and an inner, turbid water lagoon. The molluscan assemblages of the seaward reef slopes of both atolls are characterized by high species diversity, low abundance, and numerous faunal grazers represented by triphorids, cerithiopsids, and marginellids (Kay and Switzer 1974). Inside the lagoon, mollusks are divisible into an outer assemblage with lower abundance than in the inner lagoon assemblage, more species, and a preponderance of epifaunal microherbivores compared with a high proportion of suspension feeders in the inner reaches. At Funafuti, the lagoon is open to the ocean through several deepwater passes; the molluscan species composition of the seaward reefs and the channels are similar to that of the seaward reefs at Fanning and Canton, and the lagoon resembles that of the outer lagoon at Fanning and Canton.

Molluscan assemblages around the coastlines of high islands such as Oahu in the Hawaiian Islands are "patchy": species composition from similar benthic substrates at similar depths can be strikingly different on windward and leeward coasts, and even more striking differences can be seen when assemblages from bays such as Pearl Harbor are compared with those from open coastal areas. Trophic structure in assemblages from the coastlines is dominated by micrograzers; inside Pearl Harbor, virtually every mollusk in the sediments is a suspension feeder or pyramidellid.

Chance

When all else fails, there is one final resort as explanation for distribution patterns: chance. The dominant cone on the seaward reef flats at Enewetak in the Marshall Islands is not necessarily the dominant cone on the seaward reef flats at Kwajalein or at Majuro in the same island archipelago; the dominant gastropod in the lagoon at Fanning is *Tricolia variabilis*, of which only thirty-one shells were seen at Canton within a few hundred miles away (Kay and Switzer 1974; Kay 1978). All other things being equal (similar reef flats, approximately same location with respect to equator, etc.), chance may also play a role, if only because dispersal—the distribution of planktonic larvae, the primary vehicle for the dispersal of marine mollusks from Ekman's "centre and focus" (1935, 1953)—must be a matter of chance.

CONCLUSIONS

The accounting here is a first approximation toward a count of the numbers of species found among the oceanic islands of the Pacific. The account confirms in several respects generalizations that have long been cited as characteristic of the distribution of marine organisms in the Pacific. But it is apparent that the numbers will change. Gosliner's astonishing figure (pers. com.), for what is undoubtedly the more accurate count of Hawaiian opisthobranchs in an island group for which mollusks have been collected for more than 200 years, indicates the current assessment of marine molluscan diversity is far from complete.

Acknowledgments—I am particularly indebted to Scott Johnson of Kwajalein, Marshall Islands, for his continuing work on the Marshallese molluscan fauna, and I thank Terence Gosliner and Eric Hochberg for allowing me to use some unpublished data.

APPENDIX

Families, Genera, and Number of Species of Reef and Shore Mollusks Recorded from Each Locality

Abbreviations—Gum = Guam and the Northern Marianas; Kwa and Ene = Kwajalein and Enewetak, Marshall Islands; Haw = Hawaiian Islands; Tah = Society and Tuamotu islands; PI = Pitcairn, Henderson, Ducie, and Oeno islands; and Eas = Easter Island

	Genera	Gum	Kwa	Ene	Haw	Tah	PI	Eas
ORDER ARCHAEOGASTROPODA								
SUPERFAMILY PLEUROTOMARIOIDEA								
HALIOTIDAE								
	Haliotis	5	1	4		1	1	
SCISSURELLIDAE								
	Scissurella	3	1	5	3	2		1
FISSURELLIDAE								
	Diodora	4	6	3	3	1	1	
	Emarginula	7	1	6	2		1	2
	Montfortista	1						
	Montfortulana	1						
	Rimula			1				
	Scutus	1	1					
	Tugali				1			
	Zeidora							1
SUPERFAMILY PATELLOIDEA								
ACMAEIDAE								
	Acmaea	1						
	Colisella	1						
	Nipponacmea	1						
	Notoacmaea	1						
	Yayoiacmea	1						
PATELLIDAE								
	Cellana	2	?		4	2	2	
	Patella	2	1	1		1	1	
	Patelloida	1		1		1	2	1
SUPERFAMILY TROCHOIDEA								
TROCHIDAE								
	Astele			1				
	Broderipia			1		1	1	
	Calliostoma	1		1	1			
	Calliotropis				?3			
	Cantharidus	2	2				1	
	Chrysostoma	1						
	Clanculus	4	3	3	1			
	Danilia	1			1			
	Diloma	1						
	Ethalia	1						
	Euchelus	4	4	5	2	3	1	1
	Eurytrochus	1	2					
	Fautor	1						
	Hybochelus			1				
	Isanda			1				
	Mesoclanculus	1						
	Monilea	3	1	3				
	Monodonta	2						
	Pseudostomatella	1	1	3			1	
	Stomatella	4	2	3		2	1	1
	Stomatellina	3				1		

Appendix *(continued)*

	Genera	Gum	Kwa	Ene	Haw	Tah	PI	Eas
	Stomatia		4					1
	Synaptocochlea	1	1	1	1		1	
	Talopina	1						
	Tectus	1	3	3				
	Thalotia	1		1	2	1		
	Tosatrochus			1				
	Tristichotrochus		3		1			
	Trochus	9	3	3	1	1		
	Turcica			1				
	Umbonium	2						
TURBINIDAE								
	Angaria	1	1	1				
	Astraea	2	3	1		1		
	Astralium	1				2	2	
	Collonista	2						
	Gabrielona			1				
	Galeoastrea	1			1			
	Leptothyra	3	4	5	3	2		
	Phasianella	1						
	Tricolia	1	1	1	1		1	
	Turbo	7	3	3	1	4	2	
SKENEIDAE								
	Brookula				1			
	Collonista	1						
	Cyclostrema					1		
	Cyclostremiscus				2		1	
	Leucorhynchia	1					1	
	Lophocochlias	1		1	1			
SUPERFAMILY NERITOIDEA								
NERITIDAE								
	Clithon	1						
	Nerita	13	8	5	4	10	2	2
	Neritilia				1			
	Neritina	2						
	Pisulina	1						
	Puperita	1		1		1		
	Septaria	3						
	Smaragdia	2			1			
	Theodoxus	1			3			
NERITOPSIDAE								
	Neritopsis	1	1	1	1	1	1	
PHENACOLEPADIDAE								
	Phenacolepas	2		2	2	3	1	
ORDER NEOTAENIOGLOSSA SUBORDER DISCOPODA								
SUPERFAMILY CERITHIOIDEA OBTORTIONIDAE								
	Finella	1			1	1	1	
SCALIOLIDAE								
	Scaliola	1		2	3			
DIALIDAE								
	cf. *Alaba*	1						

Appendix *(continued)*

	Genera	Gum	Kwa	Ene	Haw	Tah	PI	Eas
	Cerithidium	2			2	1	1	1
	Diala	1	2		2	3		
	Mellitestea				1			1
LITIOPIDAE								
	Styliferina			1	1	1		
CERITHIIDAE								
	Argyropeza				1			
	Cerithium	22	21	16	13	18	10	6
	Clypeomorus	6	4	2		2		1
	Itibittium	1			1	1		
	Plesiotrochus	1	1		1	1	2	
	Pseudovertagus	2	1		1	1		
	Rhinoclavis	8	5	5	4	5	3	
	Royella		1	1		1	1	
POTAMIDIDAE								
	Terebralia					1		
PLANAXIDAE								
	Fissilabia	1						
	Hinea						1	1
	Planaxis	3	2		2	4	1	
FOSSARIDAE								
	Fossarus				2		3	
MODULIDAE								
	Modulus	1	1	1	1	2		
TURRITELLIDAE								
	Turritella	1	2	1				
	Vermicularia	2	1	1				
SILIQUARIDAE								
	Siliquaria	1						
	Stephopoma			1				
SUPERFAMILY LITTORINOIDEA								
LITTORINIDAE								
	Echininus	1				1		
	Littoraria	4	4	4	4	2	1	
	Littorina					2		
	Nodilittorina	2			1	3		1
	Peasiella	2		2	1	1		
	Tectarius					1		
SUPERFAMILY CINGULOPSOIDEA								
EATONIELLIDAE								
	Eatoniella				2		1	
	Eatonina						1	
RASTODENTIDAE								
	Rastodens				2			
	Rufodardanula				2			
SUPERFAMILY TRUNCATELLOIDEA								
RISSOIDAE								
	Alvania	1						
	Alvinia	1		1	1		1	
	Barleeia				1			
	Elachisina				1			
	Isseliella			1	1	1	2	
	Merelina				2	1		

Appendix *(continued)*

	Genera	Gum	Kwa	Ene	Haw	Tah	PI	Eas
	Parashiela	1		1	1		1	
	Pelycidion				1			
	Phosinella	1						
	Powellsetia				1			
	Pusillina					1	1	
	Rissoa					2		
	Rissoina	10	20	18	7	13	8	
	Schwartziella	2	1	1	2	1	1	
	Stoiscia					1	2	1
	Vitricithna		1		1	1		
	Zebina	2	5	3	2	2	4	
TRUNCATELLIDAE								
	Truncatella	1	1			1		
ASSIMINEIDAE								
	Assiminea	2	1	1	11	2	1	1
CAECIDAE								
	Caecum	4	1	2	7	2	1	
	Meoceras				1			
	Strebloceras				1			
PICKWORTHIIDAE								
	Marelepetopoma			11	3	2		
	Sansonia							
VITRINELLIDAE								
	Elacorbis				1			
	Leucorhyncha	1	1	2				
	Microliotia			1				
	Munditella			1				
	Solariorbis			1				
	Teinostoma			5	1			
SUPERFAMILY STROMBOIDEA								
STROMBIDAE								
	Lambis	5	5	5		4	2	
	Strombus	15	15	15	7	10	2	1
	Terebellum	1	1	1				
SUPERFAMILY VANIKOROIDEA								
HIPPONICIDAE								
	Hipponix		1		3	2		2
	Malluvium	1						
	Pilosabia							1
	Sabia	1	1	1	1	1	1	
VANIKORIDAE								
	Vanikoro	4	3	3	3	4	3	
SUPERFAMILY CALYPTRAEOIDEA								
CALYPTRAEIDAE								
	Cheilea	2	1	1	1	1	1	1
	Crepidula		1	1	1	1		
	Crucibulum				1			
CAPULIDAE								
	Capulus		4		1	3		
SUPERFAMILY XENOPHOROIDEA								
XENOPHORIDAE								
	Xenophora			1	1			

Appendix *(continued)*

Genera	Gum	Kwa	Ene	Haw	Tah	PI	Eas
SUPERFAMILY VERMETOIDEA							
VERMETIDAE							
Bivona	1						
Dendropoma	5	4	5	5			2
Petaloconchus	2	1	2	1	1		
Serpulorbis	3	2	2	2		1	1
Thylaeodus						1	
Vermetus					1		
SUPERFAMILY CYPRAEOIDEA							
CYPRAEIDAE							
Bistolida	6	7	6	3	4	1	
Chelycypraea	1	1	1		1	1	
Cribrarula	1	1	2	1	3	1	
Cypraea	2	2	2	1	2	1	
Erosaria	11	12	10	8	10	8	2
Erronea	6	2	2				
Ipsa	1	1	1	1	1	1	
Lyncina	7	8	7	7	6	4	
Mauritia	7	7	3	7	2		
Notadusta	2	2	2				
Nucleolaria	1	1	1	1	1		1
Ovatipsa	1	1	1	1	2		
Palmadusta	5	3	2		2		
Purpuradusta	3	3	1		4	1	
Pustularia	4	4	4	4	4	3	
Staphylaea	2	2	2	1	1		
Talparia	1	1	1	2	1		
OVULIDAE							
Alcyvolva	1						
Calpurnus	1	2					
Cymbovula		2					
Lunovula	1			1			
Margovula				1			
Ovula	1						
Phenacovolva		1		3	1		
Primovula		1					
Prosimnia		1					
Spicula				1			
SUPERFAMILY LAMELLAROIDEA							
ERATOIDAE							
Erato				1	1	1	
Pedicularia		1			1		
Proterato	1	1			1		
Pseudocypraea	1	1			1		
TRIVIIDAE							
Trivia	5	7	1	5	7	2	
LAMELLARIIDAE							
Lamellaria			1				
Lamellaria berghi		1					
SUPERFAMILY NATICOIDEA							
NATICIDAE							
Eunaticina				1			
Euspira				1			

Appendix *(continued)*

	Genera	Gum	Kwa	Ene	Haw	Tah	PI	Eas
	Mammilla	2	2	1	2	2	1	
	Natica	4	9	1	4	5	3	1
	Notocochlis	1						
	Polinices	2	6	1	3	3		
	Sinum					1		
	Tectonatica			1	1		1	
SUPERFAMILY TONNOIDEA								
CASSIDAE								
	Casmaria	2	2	2	2	2	1	1
	Cassis							
	Cypraecassis		1			1		
	Phalium		2		2	1		
TONNIDAE								
	Malea	1	1	1	1	1	1	
	Tonna	1	2	1	2	2	2	
RANELLIDAE								
	Charonia	1	1	1	1	1	1	
	Cymatium	14	18	10	16	13	3	
	Distorsio	2	2	2	3	2		
	Gyrineum	2	2		2	2		
BURSIDAE								
	Bursa	8	11	5	6	5	3	
	Tutufa		1		1			
SUBORDER PTENOGLOSSA								
SUPERFAMILY TRIPHOROIDEA								
CERITHIOPSIDAE								
	Ataxocerithium	1						
	Cerithiopsis		1	2	1	1		2
	Cyrbasia							1
	Horologica	2						
	Joculator	2	2	6	5			
TRIPHORIDAE								
	Bouchetriphora			1	1			
	Cautor				3			
	Euthymella			3	4			
	Inella			9				
	Iniforis	3		3	6			1
	Liniphora						1	
	Litharium				2			
	Mastonia			2	3			
	Mesophora				2			
	Metaxia			2	3		1	1
	Nanophora			3				
	Subulophora				2			
	Tetraphora			3				
	Triphora	2		1	16			6
	Viriola	3		5	4			
	Viriolopsis			1	1			
SUPERFAMILY JANTHINOIDEA								
EPITONIIDAE								
	Amaea				1			
	Asperiscala				1			

Appendix *(continued)*

	Genera	Gum	Kwa	Ene	Haw	Tah	PI	Eas
	Cirsotrema				1			
	Cycloscala			1	4	1		
	Depressiscala			1	1			1
	Epitonium			5	8		1	
	Fragilopalia				1			
	Gyroscala						1	
	Laeviscala				3	1		
	Opalia			1	2			
	Parviscala			2	6	3		
	Stylapex							
JANTHINIDAE								
	Janthina	1						
	Recluzia							
SUPERFAMILY EULIMOIDEA								
EULIMIDAE								
	Apicalis				1			
	Balcis	1	3	4	11	10	2	4
	Echineulima	1			3			
	Eulima				1			
	Goubinia			1	1			
	Hemileiostraca			1	1			1
	Lutzenia							1
	Melanella			1	1			
	Mucronalia							1
	Parvioris				1			
	Peasistilifer				1	1		
	Pelseneeria				1			
	Pseudoretusa				1	1	1	
	Pulocochlea				1	2	1	
	Pyramidelloides	1		1	3		1	
	Scalenostoma				1	1		
	Stilifer	1	1	1	1	1	1	1
	Stylapex			1				
	Thyca		1	1	1			
	Trocostilifer				1			
	Vitreolina				2			1
ORDER NEOGASTROPODA								
SUPERFAMILY MURICOIDEA								
MURICIDAE								
	Aspella	2	3	1	1	1		
	Attiliosa		2					
	Boreotrophon	2						
	Chicoreus	6	6	5	2	8		
	Drupa	10	6	6	5	6	5	2
	Drupella	6	5	4	2	4	2	
	Favartia		5	1	1	1		
	Homolocantha	1	2	1	2	2		
	Maculotriton	3	2	1	1	3	1	
	Mancinella	1				1		
	Morula	12	20	7	9	11	5	2
	Muricodrupa	1						
	Naquetia	1	2	2		2		
	Nassa	1	1	1	1	1	1	
	Neothais			1	1			1

Appendix *(continued)*

	Genera	Gum	Kwa	Ene	Haw	Tah	PI	Eas
	Pascula							1
	Phyllocoma	1	2	1		1	1	
	Phyllonotus		1	1				
	Pinaxia			1	1			
	Pterynotus		4	2	2	2		
	Purpura	1	1	1	1	1		
	Thais	6	4	3	3	2	2	
	Vexilla	1	3	1	3	1	2	
	Vitularia	1		1	1			
CORALLIOPHILIDAE								
	Babelomurex				1			
	Coralliobia		1		1			
	Coralliophila	5	7	3	4	4	2	1
	Hirtomurex				1			
	Latiaxis			1	3			
	Magilopsis	1			1	1	1	
	Quoyula	1		1	1	1	1	1
	Rapa	1	1	1			1	
	Rhizochilus			1	1			
TURBINELLIDAE								
	Vasum	2	2	2		2	1	
BUCCINIDAE								
	Caducifer	3	2	2	3	2	1	1
	Cantharus	5	5	2	3	7		
	Clivipollia	1	1	1	1			
	Colubraria		8	4	3		2	
	Cyrtulus				1			
	Dolicholatirus		1		2			
	Engina	4	8	5	4	6	2	
	Enginopsis	1						
	Fasciolaria		1		1			
	Fusinus		1	2	3	1	1	
	Fusolatirus				1			
	Latirulus				1			
	Latirus	6	9	4	5	4	1	
	Nassaria		1					
	Nassarius	27	18	7	7	19	3	1
	Peristernia	3	3	6	3	6	1	
	Phos	1		1	1			
	Pisania		2			2	1	
	Pleuroploca	1						
	Prodotia	2	5	2	2	2	2	
	Zafrona			1				
	Zeuis	1						
COLUMBELLIDAE								
	Aesopus	1	1	1	1			
	Anachis				1			
	Columbella		1					
	Euplica	4	10	4	7	1	3	1
	Graphicomassa	1						
	Mitrella	7	5	5	2	6	1	
	Nodochila							1
	Pyrene	9	4	7	1	7	2	

Appendix *(continued)*

	Genera	Gum	Kwa	Ene	Haw	Tah	PI	Eas
	Seminella	2	9	5	1		1	1
	Zafra	5	1	2		3		
	Zafrona							1
OLIVIDAE								
	Ancilla	1						
	Belloliva				1			
	Oliva	8	7	5	4	8		
	Olivella			1	1			
HARPIDAE								
	Harpa	3	3	4	4	4	1	
	Morum		1					
MARGINELLIDAE								
	Crithe	1						
	Cystiscus	3			1		1	
	Dentimargo				1			
	Gibberulina	1						
	Granula				1		1	1
	Granulina	1	1		1			
	Haloginella		2	2				
	Kogomea	2	3					
	Marginella		1			7		
	Persicula			1				
	Tringinella			1				
	Volvarina			4	1			
	Volvarinella	1						
	Marginellid spp.	4	7	8	5	7	2	1
COSTELLARIIDAE								
	Thala	3	4		3	2		
	Vexillum	71	90	53	45	40	4	1
MITRIDAE								
	Cancilla	4	5	4	2	4		
	Imbricaria	5	5	5	3	4		1
	Mitra	45	55	47	38	45	12	2
	Nebularia	2						
	Neocancilla	3	2	2	4	3		
	Pterygia	5	5	4	3	5		
	Scabricola	6	4	3	1	3		
	Subcancilla	2	1	2	2	2		
	Ziba						1	
VOLUTOMITRIDAE								
	Volutomitra				1			
SUPERFAMILY CANCELLARIOIDEA								
CANCELLARIIDAE								
	Plesiotriton		1					
	Scalptia	2						
	Tritonoharpa		1					
SUPERFAMILY CONOIDEA								
TURRIDAE								
	Agathotoma		1					
	Anacithara					1	2	
	Carinapex	2	3	1	2	2		
	Ceritoturris				1			
	Clathurella					1		1

Appendix *(continued)*

	Genera	Gum	Kwa	Ene	Haw	Tah	PI	Eas
	Clavus	7	15	3	6	8		1
	Comitas		1					
	Daphnella	5	7	2	3	7	3	
	Drillia					1		
	Elaiocyma	1						
	Etrema		4		1	1	1	
	Eucithara	1	18	5	2	9	3	
	Eucyclotoma	1	4		1	3	2	
	Gemmula	1	4	2	6	2		
	Glyphostoma				1	1		
	Hemilienardia		3					
	Iredalea	2	2					
	Kermia		14	10	9	3	6	2
	Lienardia	6	29	11	6	11	4	1
	Lophiotoma	2	7	2				
	Lovellona	1	1	2	2	1	1	
	Macteola	2	5	3	1	1	1	
	Mangelia					2	1	
	Microdaphne	1	1	1	1		1	
	Mitrolumna	2	4	4	3	2	4	
	Mitromorpha	1			1	1		
	Naudedrillia		1					
	Nodotoma			1				
	Paradrillia	1						
	Paramontana				1	1		
	Philbertia	1		3		3		
	Pleurotomella				1	1		
	Pseudodaphnella	3	5	2	1	3	1	
	Spergo			2	2			
	Splendrillia		6			1		
	Thatcheriasyrinx				1	1		
	Tritonoturris	2	8	4	4	3	2	
	Turridrupa	6	12	10	4	4		
	Turris	1	1	1	1			
	Tylotia	1						
	Tylotilla		1					
	Veprecula				1	1		
	Vexitoma	1						
	Xenuroturris	3	5	2	4	1	1	
CONIDAE								
	Conus	89	79	34	33	71	21	3
TEREBRIDAE								
	Acuminata							1
	Duplicaria	3			2			
	Egenteleria							1
	Hastula	6	4	4	10	1		
	Terebra	34	26	19	29	37	4	
ORDER HETEROSTROPHA SUPERFAMILY VALVATOIDEA ORBITESTELLIDAE								
	Orbitestella				2		1	

Appendix *(continued)*

	Genera	Gum	Kwa	Ene	Haw	Tah	PI	Eas
SUPERFAMILY ARCHITECTONICOIDEA								
ARCHITECTONICDAE								
	Architectonica	2			1	1		
	Heliacus	1	4	2	4	8		1
	Philippia	1	2	2	2	1		
SUPERFAMILY RISSOELLOIDEA								
RISSOELLIDAE								
	Rissoella				1			
SUPERFAMILY OMALOGYROIDEA								
OMALOGYRIDAE								
	Omalogyra				1			
SUPERFAMILY PYRAMIDELLOIDEA								
PYRAMIDELLOIDEA								
PYRAMIDELLIDAE								
	Chemnitzia			1				
	Chrysalida			1		1		
	Contraxiala					1		
	Costabielta			1				
	Evalea		1	3	2			
	Herviera			1	2		1	
	Hinemoa				1			
	Koolonella				1			
	Miralda	2		1	2		1	
	Nesiodostomia				3			
	Odostomella	1						
	Odostomia			3	5	3		1
	Otopleura	1	1	2	1	1		
	Pyramidella	3	2	2	3	3		
	Pyrgulina	2		2	2			
	Styloptygma	2						
	Syrnola				2	1	1	
	Turbonilla	1			3	3	4	2
	Pyramidellid spp.		2					
SUBCLASS OPISTHOBRANCHIA								
ORDER CEPHALASPIDEA								
SUPERFAMILY ATYOIDEA								
OPHTHALMIDAE								
	Phanerophthalmus	1	2	1	2	2	1	2
SMARAGDINELLIDAE								
	Smaragdinella	1	1	1	1	1	1	1
BULLIDAE								
	Bulla	3	1	1	1	4	2	
SUPERFAMILY ACTEONOIDEA								
ACTEONIDAE								
	Cylichnatys			1				
	Pupa	3	1	3	2	3	1	
BULLINIDAE								
	Bullina				4			
RINGICULIDAE								
	Ringicula			1				
HYDATINIDAE								
	Hydatina	1	2	3	3	2		
	Micromelo	1	1	1	1			

Appendix *(continued)*

	Genera	Gum	Kwa	Ene	Haw	Tah	PI	Eas
SUPERFAMILY HAMINOIDEA								
HAMINEIDAE								
	Atys	2	3	5	3	4	2	
	Diniatys		1	1				
	Haminoea	1	3	6	3	4	1	
RETUSIDAE								
	Coleophysis			1				
	Retusa						1	
SUPERFAMILY PHILINOIDEA								
AGLAJIDAE								
	Aglaja	1				1		
	Chelidonura	2	2	2	3			
	Odontoglaja	1	1	1				
	Philinopsis	3	2	1	3			
SCAPHANDRIDAE								
	Acteocina	1		1	2	1		
	Adamnesia			1				
	Cylichna				1	2	1	1
	Nakamigawaia			1				
	Punctacteon	1						
	Scaphander				2			
PHILINIDAE								
	Cryptophthalmus					2		
	Philine				1		1	
GASTROPTERIDAE								
	Gastropteron	3	1	1	1			
	Sagaminopteron	3	1					
	Siphopteron	1	1					
DIAPHANIDAE								
	Colpodaspis	1	1					
SUPERFAMILY RUNCINOIDEA								
RUNCINIDAE	*Ilia*	1						
	Metaruncina	1						
ANASPIDEA								
SUPERFAMILY APLYSIOIDEA								
APLYSIIDAE								
	Aplysia	1	1	3	5	5		
	Dolabella	1	1	1	1	1		1
	Petalifera				2	1		
	Phyllaplysia	1				1		
NOTARCHIDAE								
	Dolabrifera	1	1	1	1	1	1	1
	Stylocheilus	1	1	1	1	1		
NOTASPIDEA								
SUPERFAMILY PLEUROBRANCHIOIDEA								
PLEUROBRANCHIDA								
	Berthella		5	2	2			
	Berthellina	1	1	1	1			1
	Euselenops			1				
	Pleurobranchus	1			2	3		
	Pleurohedra		1	1				

Appendix *(continued)*

	Genera	Gum	Kwa	Ene	Haw	Tah	PI	Eas
VOLVATELLIDAE								
	Volvatella		2	1	2			
UMBRACULIDAE								
	Umbraculum				1	1		1
SACCOGLOSSA								
SUPERFAMILY OXYNOIDEA								
OXYNOEIDAE								
	Lobiger	2			1			
	Oxynoe	1	1			2		
CYLINDROBULLIDAE								
	Volvatella	1	2		2	2		
SUPERFAMILY JULIOIDEA								
JULIIDAE								
	Berthelinia				1			1
ELYSIIDAE								
	Elysia	9	7	8	7	4		
	Julia	1	1		1		1	1
	Lobifera					1		
	Plakobranchus	2	1	1	1	3		
STILIGERIDAE								
	Costasiella	1						
	Hermaea	1						
HERMAEIDAE								
	Branchophyllum	2	1	1	1			
	Cyerce	2	3	1	3			
	Hermaea	1						
	Phyllobranchillus				1		1	
	Styliger	1	2	1	6			
NUDIBRANCHIA								
SUPERFAMILY DORIDOIDEA								
DORIDIDAE	*Casella*	2				1		
	Doriopsis	3	1	2	2	2		
	Doriorbis				1			
	Miamira	2			1			
	Pusikus					1		
	Rosodoris					1		
ROSTANGIDAE								
	Rostanga		2	1				
ARCHIDORIDIDAE								
	Archidoris		1		2	2		
PLATYDORIDIDAE								
	Platydoris	2	4	2	2	1		
CADLINIDAE								
	Cadlinella			1				
KENTRODORIDIDAE								
	Asteronotus	1			1			
	Jorunna		2	2	1			
	Kentrodoris	1	1					
DISCODORIDIDAE								
	Carminodoris	1	3	2	3			
TRIPPIDAE								
	Trippa	2	3	2	2			

Appendix *(continued)*

	Genera	Gum	Kwa	Ene	Haw	Tah	PI	Eas
HALGERDIDAE								
	Halgerda	10	6	2	1			
	Orodoris					1		
	Sclerodoris		3	2	3			
DORIDIDAE								
	Doris	1						
	Peltodoris				1		2	
	Thordisa				2	1		
CHROMODORIDIDAE								
	Ceratosoma		2		1			
	Chromodoris	11	14	13	8	5		
	Durvilledoris		2	1		1		
	Glossodoris	6	3	2	3	1		
	Hypselodoris	6	5	7	5	1		
	Miamira		1	1				
	Noumea	2	10	5	2	1		
	Risbecia	2	1	2	1	2		
	Thorunna	1	3	4	1			
HEXABRANCHIDAE								
	Hexabranchus	1	2	1	3			
ACTINOCYCLIDAE								
	Actinocyclus		1		1			
	Hallaxa		1					
ALDISIDAE								
	Aldisa		1		1			
DENDRODORIDIDAE								
	Dendrodoris	5	9	6	5	1	1	
GYMNODORIDIDAE								
	Gymnodoris	5	16	5	5	2		
SUPERFAMILY DENDRONOTOIDEA								
POLYCERIDAE								
	Nembrotha	3	5	1				
	Paloplocamus			1				
	Polycera	1	3					
	Robastra	1	2	1				
	Tambja		1		1			
TRIOPHIDAE								
	Plocamopherus		3	1	2			
VAYSSIEREIDAE								
	Okadaia	1		1	1			
BONIODORIDAE								
	Goniodoridiella			1				
	Goniodoris	1	1	1	1			
	Okenia		1		1			
	Trapania		3					
AEGIRIDAE								
	Aegires	1	4	3				
PHYLLIDIIDAE								
	Freyeria	2			1			
	Phyllidia	7	4	2	2	4	2	
TETHYIDAE								
	Melibe				2			

Appendix *(continued)*

	Genera	Gum	Kwa	Ene	Haw	Tah	PI	Eas
TRITONIIDAE								
	Aruncus	1						
	Marianina	1	1	1				
	Tritonia				1			
	Tritoniopsilla		2					
	Tritonid spp.		2					
ARMINIDAE								
	Dermatobranchus	1	1	1	1			
DORIDOMORPHIDAE								
	Doridomorpha		1	1				
JANOLIDAE								
	Doto	1						
	Janolus		2		1			
BORNELLIDAE								
	Bornella	1	3	2		1	1	
SCYLLAEIDAE								
	Notobryon				1			
SUPERFAMILY AEOLIDIOIDEA								
CORYPHELLIDAE								
	Flabellina		3	1	1			
TERGEPEDIDAE								
	Cuthona		1		2			
	Phestilla	1	2	1	4			
FACELINIDAE								
	Caloria				1			
	Facelinella				1			
GLAUCIDAE								
	Cratena	1	1					
	Favorinus	1	2	2	1			
	Glaucus	1	1	1	1		1	
	Godiva		1					
	Herviella		1	2	1			
	Phidiana	3	2	1				
	Phyllodesmium		1	1				
	Pteraeolidia	1	1	1	1			
AEOLIDIDIDAE								
	Aeolidiella			2	2			
	Baeolidia		2	1	1			
	Berghia				1			
	Cerberilla			1				
	Embletonia	1	1	1	1			
	Noumeaella			1	1			
	Protaeolidiella		1					
	Spurilla			1	1			
SUBCLASS PULMONATA								
ORDER SYSTELLOMMATOPHORA								
SUPERFAMILY ONCHIDIOIDEA								
ONCHIDIIDAE								
	Onchidiella			1				
	Onchidium					1	1	
	Peronia		1	1	1			

Appendix *(continued)*

Genera	Gum	Kwa	Ene	Haw	Tah	PI	Eas
ORDER BASOMMATOPHORA							
SUPERFAMILY MELAMPOIDEA							
MELAMPIDAE							
Allochroa				1	1		
Auriculastra	1						
Blauneria				1			
Laemodonta				1	3		
Leuconopsis							1
Melampus	3	1	1	2	3		1
Pedipes				1			
Pira				1			
Plectotrema					1		
Pythia	1						
Rangitotoa							1
SIPHONARIIDAE							
Siphonaria	2	1	1	1	1		1
Williamia	1			1			
TRIMUSCULIDAE							
Trimusculus							1
CLASS BIVALVIA							
SUPERFAMILY MYTILOIDA							
MYTILIDAE							
Amygdalum				2			
Botula			1	1	1	1	
Brachidontes				1			
Lithophaga			1	1	4		
Modiolus	1		2	2	2	1	1
Musculus				1			
Septifer	4		1	2	3	1	1
Terua				1			
SUPERFAMILY ARCOIDA							
ARCIDAE							
Anadara		1	1		1		
Arca	3	6	2	2	2	1	
Barbatia	10	6	6	8	8	1	1
Bathyarca				1			
Bentharca				1			
GLYCYMERIDAE							
Glycymeris	1		1	3	2		
SUPERFAMILY PTERIOIDA							
PTERIIDAE							
Electroma			1				
Pinctada	2	1	1	2	2	2	
Pteria		2	1	1			
MALLEIDAE							
Malleus	1		1	1	1		1
ISOGNOMONIDAE							
Isognomon	3	2	4	4	4	3	2
Melina			1				
PINNIDAE							
Atrina		2	2	2	1		
Pinna		1	1	2	2		
Streptopinna			1	1	1	1	

Appendix *(continued)*

	Genera	Gum	Kwa	Ene	Haw	Tah	PI	Eas
SUPERFAMILY LIMOIDA								
LIMIDAE								
	Ctenoides		1					
	Lima	2	3	3	4	2	2	1
	Limaria					1		1
SUPERFAMILY OSTREOIDA								
OSTREIDAE								
	Dendrostrea					1		
	Lopha		1					
	Neopycnodonta				1			
	Ostrea		2	2	2			
	Saxostrea	1						
	Striostrea					1		
GRYPHAEIDAE								
	Hyotissa						1	
	Parahyotissa					1		
DIMYIDAE								
	Dimya				1			
PLICATULIDAE								
	Plicatula	1		1		1		
PECTINIDAE								
	Amusium		1					
	Anguipecten		1		1			
	Bractechlamys		1					
	Chlamys	4	8	2	1	6	2	1
	Comptopallium		2					
	Cryptopecten				1			
	Decatopecten				1			
	Excellichlamys	1	1			1	1	
	Gloripallium		2	1	1	1		
	Haumea				1	2		
	Juxtamusium		1	1				
	Mirapecten		3	2	1	2	2	
	Pecten				2			
	Pedum		1	1		1		
	Semipallium		2			3	1	
SPONDYLIDAE								
	Spondylus	4	3	4	3	5	2	
ANOMIIDAE								
	Anomia				1			
SUPERFAMILY VENEROIDA								
CHAMIDAE								
	Chama	4	3	4	2	8	4	1
FIMBRIIDAE								
	Fimbria	1		1				
LUCINIDAE								
	Anodontia			1	1	1	1	
	Codakia		2	1	1	2	1	
	Ctena			1	2	1	1	1
	Lucina			2				1
	Pillucina				2			

Appendix *(continued)*

	Genera	Gum	Kwa	Ene	Haw	Tah	PI	Eas
SUPERFAMILY MYOIDA								
GASTROCHAENIDAE								
	Gastrochaena				3			
HIATELLIDAE								
	Hiatella				1			
PHOLADIDAE								
	Martesia				1			
SUPERFAMILY VENEROIDA								
LASAEIDAE								
	Lasaea				1			
GALEOMMATIDAE								
	Galeomma						1	
	Leiochasmea				1			
	Scintilla			1	1			
	Scintillona				1			
SPORTELLIDAE								
	Anisodonta				2			
CONDYLOCARDIIDAE								
	Carditella				1			
CARDIIDAE								
	Cardium		1					
	Corculum			1		1	1	
	Discors			1				
	Fragum		1	2	1	4	1	
	Laevicardium		1	1				
	Microfragum			1				
	Nemocardium					1	1	
	Trachycardium		1	1	1	2	1	
CARDITIDAE								
	Cardita	1			2			
TRIDACNIDAE								
	Hippopus		1	1				
	Tridacna	2	3	3		2	2	
MESODESMATIDAE								
	Atactodea			1				
	Ervilia			1	2	1	2	
TELLINIDAE								
	Macoma		1	3	2	3	1	
	Pinguitellina	1		1	1	1	1	
	Tellina	3	5	13	6	10	5	2
PSAMMOBIIDAE								
	Aspahis			1	1	1	1	
	Gari			1	2	2		
	Grammatomya				1			
	Solecurus				1			
SEMELIDAE								
	Abra					1		
	Lonoa				1	1	1	
	Semelangulus			1	1	1	1	
	Semele			1	2	1	2	1
TRAPEZIIDAE								
	Trapezium	1	1	2	1	2	1	

Appendix *(continued)*

	Genera	Gum	Kwa	Ene	Haw	Tah	PI	Eas
VENERIDAE								
	Dosinia			1				
	Gafrarium	2	1		1	1		
	Glycodonta		1	1				
	Lioconcha	1	5	1	1	2		
	Periglypta	2	3	3	1	3	1	
	Pitar			1		1		
	Timoclea					1		
CLASS CEPHALOPODA								
ORDER OCTOPODA								
ARGONAUTIDAE								
	Argonauta				2	2		
OCTOPODIDAE								
	Berrya				1			
	Octopus			1	4	4	1	1
	Scaergus				1			
FAMILY OMMASTREPHIDAE								
	Hyaloteuthis				1		1	
ORDER SEPIOIDEA								
SEPIOLIDAE								
	Euprymna				1			
ORDER TETHOIDEA								
LOLIGINIDAE								
	Sepioteuthis				1			
	"*Octopus*"					2		
SUBCLASS NAUTILOIDEA								
NAUTILIDAE								
	Nautilus			2				
CLASS POLYPLACOPHORA								
ORDER CHITONID								
ACANTHOCHITONIDAE								
	Acanthochitona	1			1			
	Cryptoplax		1	1				
CHITONIDAE								
	Acanthopleura	1						
	Rhyssoplax sp.	1			1			
ISCHNOCHITONIDAE								
	Ischnochiton			1	1			
LEPIDOPLEURIDAE								
	Leptochiton			1				
MOPALIIDAE								
	Plaxiphora				1			1
	Mopaliid sp.	1						
	Chiton spp.	1	1					

SUMMARY: Number of species of reef and shore mollusks recorded from each locality

	Gum	Kwa	Ene	Haw	Tah	PI	Eas
Archaeogastropoda	127	61	76	46	48	25	10
Discopoda	242	251	199	215	206	99	26
Ptenoglossa	18	8	62	114	22	10	21
Neogastropoda	509	645	388	365	437	126	30
Heterostropha	16	12	26	37	23	6	2
Cephalaspidea	63	55	57	75	52	14	10
Nudibranchia	95	155	97	93	29	7	0
Pulmonata	8	2	2	8	8	0	5
Bivalvia	56	88	101	104	118	49	15
Cephalopoda	0	0	10	10	8	2	1
Polyplacophora	5	2	3	4	8	0	1
Total	1139	1279	1021	1071	959	338	121

Abbreviations—Gum = Guam and the Northern Marianas; Kwa and Ene = Kwajalein and Enewetak, Marshall Islands; Haw = Hawaiian Islands; Tah = Society and Tuamotu islands; PI = Pitcairn, Henderson, Ducie, and Oeno islands; and Eas = Easter Island

REFERENCES

Cernohorsky, W. O. 1967. *Marine shells of the Pacific.* Sydney, Australia: Pacific Publications.

———. 1972. *Marine shells of the Pacific.* Vol. 2. Sydney, Australia: Pacific Publications.

———. 1978. *Tropical marine shells.* Sydney, Australia: Pacific Publications.

Du Shane, H. 1987. Classification of three species of Epitoniidae found in Hawaiian waters. *Hawaiian Shell News* 35(6): 1, 4.

———. 1988. New Hawaiian species of Epitoniidae (Mollusca: Gastropoda). *Veliger* 31(3/4): 267–71.

Ekman, S. 1935. *Tiergeographie des Meeres.* Akademische Verlagsgesellschaft. Leipzig.

———. 1953. *Zoogeography of the sea.* London: Sidgwick and Jackson.

Faustino, L. A. 1928. *Summary of Philippine marine and freshwater mollusks.* Manila Bureau of Science Monograph No. 25, Philippines.

Gosliner, T. 1979. The systematics of the Aeolidacea (Nudibranchia: Mollusca) of the Hawaiian Islands, with description of two new species. *Pacific Science* 33(1): 37–77.

Hedley, C. 1899. A zoogeographic scheme for the mid-Pacific. *Proceedings of the Linnean Society of New South Wales* 24:391–423.

Hickman, C., and J. McLean. 1990. *Systematic revision and supra-generic classification of Trochacean Gastropods.* Science Series No. 35. Natural History Museum of Los Angeles County.

Kay, E. A. 1978. Molluscan distribution patterns at Canton Atoll. In *Phoenix Islands report,* ed. S. V. Smith and R. S. Henderson. *Atoll Research Bulletin* 221:159–69.

———. 1979. *Hawaiian marine shells: Reef and shore fauna of Hawaii.* Section 4: Mollusca. Bernice P. Bishop Museum Special Publication 64(4), Honolulu.

———. 1984. Patterns of speciation in the Indo-West Pacific. In *Biogeography of the tropical Pacific,* ed. P. H. Raven, F. J. Radovsky, and S. H. Sohmer, 15–31. Association of Systematics Collections and the Bernice P. Bishop Museum, Honolulu.

———. 1990a. Cypraeidae of the Indo-Pacific: Cenozoic fossil history and biogeography. *Bulletin of Marine Science* 47:23–34.

Kay, E. A. 1990b. Turrid faunas of Pacific islands. *Malacologia* 32:79–87.

———. In press. Diversification and differentiation: Two evolutionary patterns in the molluscan fauna of Pacific islands with consequences for conservation.

Kay, E. A., and S. Johnson. 1987. Mollusca of Enewetak Atoll. In *The natural history of Enewetak Atoll,* ed. D. M. Devaney, E. S. Reese, B. L. Burch, and P. Helfrich. Vol. 2, Biogeography and systematics. U.S. Dept. of Energy.

Kay, E. A., and M. F. Switzer. 1974. Molluscan distribution patterns in Fanning Island lagoon and a comparison of mollusks of the lagoon and the seaward reefs. *Pacific Science* 28:275–95.

Kay, E. A., and D. K. Young. 1969. The Doridacea (Opisthobranchia: Mollusca) of the Hawaiian Islands. *Pacific Science* 23:172–231.

Ponder, R. W., and R. U. Gooding. 1978. Four new Eulimid Gastropods associated with shallow-water Diadematid Echinoids in the western Pacific. *Pacific Science* 32(2): 157–81.

Ponder, W. F., and A. Warén. 1988. Classification of the Caenogastropoda and Heterostropha: A list of the family-group names and higher taxa. *Malacological Review,* Suppl. 4:288–326.

Preese, R. In press. The composition and relationships of the marine molluscan fauna of the Pitcairn Islands. *Biological Journal of the Linnean Society.*

Rehder, H. A. 1980. The marine mollusks of Easter Island (Isla de Pascua) and Sala y Gómez. *Smithsonian Contributions to Zoology,* No. 289.

Scheltema, R. S. 1986. Long-distance dispersal by planktonic larvae of shoal-water benthic invertebrates among central Pacific islands. *Bulletin of Marine Science* 39:241–56.

Warén, A. 1983. A generic revision of the family Eulimidae (Gastropoda, Prosobranchia). *Journal of Molluscan Studies,* Suppl. 13:1–96.

CHAPTER 9

Status of Crustacean Systematics

L. G. Eldredge

Crustaceans are among the most abundant and diverse animals in the waters of the tropical/subtropical islands of the Pacific. The larger crustaceans, the shrimp and the crab, have been widely sought after both commercially and for subsistence and traditional purposes. Unfortunately, there are virtually no region-wide monographs or reviews of any of the crustaceans.

STATUS

More than 2,000 species of crabs, alone, are estimated to occur within the tropical and subtropical waters of the Indo-west Pacific. No estimates have been made for other crustacean groups. Systematic endeavors conclude that species numbers are estimates. From 14 years of *Zoological Record* data, an average of 220 new genera and subgenera have been described annually (for the same period, 210 mollusk genera and subgenera have been described each year). No estimates are available for descriptions of new species annually.

Few of the 113 families of the decapod crustaceans have been monographed. (An additional 543 crustacean families are recognized among the other taxonomic groups.) Some examples of recent works follow. The spider crabs (Family Majidae, exclusive of one boreal subfamily and members of two boreal genera) of the Indo-west Pacific were reviewed by Griffin and Tranter (1986). Based largely on collections obtained through the Siboga Expedition (1899–1900) and examination of specimens from several other museums, the Griffin and Tranter monograph is a good example of what ought to be done; they found the following—384 species; Indian Ocean only, 83 (22 percent) species; Indo-west Pacific, 125 (32 percent) species; and west Pacific only, 176 (46 percent) species. For the sponge crabs (Family Dromiidae), McLay (1993) reported on species from New Caledonia and the Philippines. He also included a review of all the genera of the family, as well as records of numerous other species found in the Indo-west Pacific. A total of 99 species in 28 genera are noted. Where previously only six species were known from New Caledonia and the Philippines, McLay (1993) reported 29 species in 13 genera. In their report on the shrimp (Superfamily Palaemonoidea) of the *Albatross* Expedition, 1907–1910, Chace and Bruce (1993) included information on additional collections and listed 408 species (70 genera) as belonging to the Subfamily Pontoniinae and 194 species in the genus *Macro-*

brachium, alone. At the other extreme, the Family Parthenopidae has not been reviewed since 1927, and some families have never been monographed.

There are virtually no complete crustacean studies in the area of the tropical island Pacific. Comparative comments on distribution and diversity are impossible to make at this time. Publications pertaining to Hawaii (Eldredge and Miller 1995) and Enewetak (Devaney et al. 1987) are the only ones that might be considered somewhat complete for aquatic species of the island region, although there are gaps in the information from both places.

Number of species	Hawaii	Enewetak
Ostracods	32	10+
Cirripeds	58	9
Copepods	100	157
Stomatopods	17	12
Mysids	14	16
Amphipods	172	4
Isopods	27	?
Tanaidaceans	8	?
Euphausiaceans	27	1
Decapods		
Caridean	149	133
Penaeidean	30	5
Stenopodidean	5	7
Palinura	13	4
Anomurans	43	76
Brachyurans	189	291
	884	725

Outside the tropical island Pacific, the crustacean fauna of Japan and China Seas is relatively well known. For Japan, Sakai (1976) reported more than 900 crabs in 328 genera and 31 families. In his recent two-volume color-photograph series, Miyake (1991) lists 951 species of brachyuran decapods (true crabs) from 351 genera; 485 species (151 genera) of penaeid through palinurids; 332 species (87 genera) of anomurans; and 48 (27 genera) of stomatopods. In the China Seas, Dai and Yang (1991) provide descriptions of more than 800 species.

CLASSIFICATION

Specialists cannot agree on whether the Crustacea should be classified as a Phylum, Subphylum, or Superclass. The classification proposed by Bowman and Abele (1982), with modifications by Chace (1992), Holthuis (1991), Kropp and Manning (1985), and Holthuis (1993), is appended and classify crustaceans as a separate phylum. Of the six classes, the Malacostraca contain the most familiar and the majority of species:

Class Cephalocarida (cephalocarids)
    Branchiopoda (conchostrans, cladocerans, anostracans)
    Remipedia (remipeds)
    Maxillopoda (barnacles, copepods, branchiurans)

Ostracoda (ostracods)

Malacostraca (stomatopods, mysids, amphipods, isopods, tanaidaceans, cumaceans, euphausiacean, decapods [shrimps, lobsters, hermit crabs, true crabs])

It should be pointed out that the level of Superfamily is generally not used in taxonomic works, although it is in the classification scheme where it can be used for convenience to group families. Crustacean taxonomists do not entirely agree with the Bowman and Abele list; many have their own listings.

In the decapod crustaceans, there are as many as 13 taxonomic levels:

Phylum
  Class
    Subclass
      Superorder
        Order
          Suborder
            Infraorder
              Section
                Superfamily
                  Family
                    Subfamily
                      Genus
                        Species

This degree of complexity seems unwieldy for database purposes. As with other phyla, this should be reduced to four or five levels: Class, Order, Family, Genus, and species with an appropriate authority file, so that intervening taxonomic levels could be searched efficiently. The "simplified" taxonomy avoids the ever-changing higher classification changes.

In addition to the lack of monographs and review articles, there is a severe lack of systematists. In the United States, the only group of crustacean taxonomists is at the National Museum of Natural History, and that group is getting smaller as members retire or leave. Positions are not being refilled. The future of crustacean systematics is in dire trouble.

As can be seen from the preceding examples, there are great opportunities, as there are in all the invertebrate groups.

APPENDIX

Classification of the Recent Crustacea
(modified from Bowman and Abele 1982)

Phylum, Subphylum, or Superclass Crustacea Pennant, 1777 [subordinal levels given for Peracarida and subsequent groups]

Class Cephalocarida Sanders, 1955

Class Branchiopoda Latreille, 1817
    Subclass Calmanostraca Tasch, 1969
    Subclass Diplostraca Gerstaecker, 1866
        Order Conchostraca Sars, 1867
        Order Cladocera Latreille, 1829
    Subclass Sarsostraca Tasch, 1969
        Order Anostraca Sars, 1867

Class Remipedia Yager, 1981

Class Maxillopoda Dahl, 1956
    Subclass Mystacocarida Pennak and Zinn, 1943
    Subclass Cirripedia Burmeister, 1834
        Order Ascothroacica Lacaze-Duthiers, 1880
        Order Thoracica Darwin, 1854
        Order Acrothoracica Gruvel, 1905
        Order Rhizocephala F. Müller, 1862
    Subclass Copepoda Milne Edwards, 1840
        Order Calanoida Sars, 1903
        Order Harpacticoida Sars, 1903
        Order Cyclopoida Burmeister, 1834
        Order Poecilostomatoida Thorell, 1859
        Order Siphonostomatoida Thorell, 1859
        Order Monstrilloida Sars, 1903
        Order Misophrioida Gurney, 1933
        Order Mormonilloida Boxshall, 1979
        Order uncertain [four families]
    Subclass Branchiura Thorell, 1864
        Order Arguloida Rafinesque, 1815

Class Ostracoda Latreille, 1806
    Subclass Myodocopa Sars, 1866
        Order Mydocopida Sars, 1866
        Order Halocyprida Dana, 1853
    Subclass Podocopa Müller, 1894
        Order Platycopida Sars, 1866
        Order Podocopida Sars, 1866
    Subclass Palaeocopa Henningsmoen, 1953
        Order Palaeocopida Henningsmoen, 1953

Class Malacostraca Latreille, 1806
    Subclass Phyllocarida Packard, 1879
    Subclass Hoplocarida Calman, 1904
        Order Stomatopoda Latreille, 1817
    Subclass Eumalacostraca Grobben, 1892
      Superorder Syncarida Packard, 1885
        Order Bathynellacea Chappuis, 1915
        Order Anaspidacea Calman, 1904
      Superorder Pancarida Siewing, 1958
        Order Thermosbaenacea Monod, 1927
      Superorder Peracarida Calman, 1904

Order Spelaeogriphacea Gordon, 1957
Order Mysidacea Boas, 1883
   Suborder Lophogastrida Boas, 1883
Order Amphipoda Latreille, 1816
   Suborder Gammaridea Latreille, 1803
   Suborder Caprellidea Leach, 1814
   Suborder Hyeriidea H. Milne Edwards, 1830
Order Isopoda Latreille, 1817
   Suborder Gnathiidea Leach, 1814
   Suborder Anthuridea Leach, 1814
   Suborder Microcerberidea Lang, 1961
   Suborder Flabellifera Sars, 1882
   Suborder Asellota Latreille, 1803
   Suborder Valvifera Sars, 1882
   Suborder Phreatocidea Stebbing, 1957
   Suborder Epicaridea Latreille, 1831
   Suborder Oniscidea Latreille, 1803
Order Tanaidacea Hansen, 1895
   Suborder Tanaidomorpha Sieg, 1980
   Suborder Neotanaidomorpha Sieg, 1980
   Suborder Apseudomorpha Sieg, 1980
Order Cumacea Kryer, 1846
Superorder Eucarida Calman, 1904
Order Euphausiacea Dana, 1852
Order Amphionidacea Williamson, 1973
       Family Amphionididae Holthuis, 1955
Order Decapoda Latreille, 1803
   [Suborder Natantia Boas, 1880]
   Suborder Dendrobranchiata Bate, 1888
      Superfamily Penaeoidea Rafinesque, 1815
        Family Aristeidae Wood-Mason, 1891
          Penaeidae Rafinesque, 1815
          Solenoceridae Wood-Mason and Alcock, 1891
          Sicyoniidae Ortmann, 1898
      Superfamily Sergestoidea Dana, 1852
        Family Sergestidae Dana, 1852
   Suborder Pleocyemata Burkenroad, 1963
    Infraorder Stenopodidea Claus, 1872
        Family Stenopodidae Claus, 1872
          Spongicolidae Schram, 1986
    Infraorder Caridea Dana, 1852 [modified, Chace (1992)]
      Superfamily Procaridoidea Chace and Manning, 1972
        Family Procarididae Chace and Manning, 1972
      Superfamily Atyoidea De Haan, 1849
        Family Atyidae De Haan, 1852
      Superfamily Nematocarcinoidea Smith, 1884
        Family Nematocarcinidae Smith, 1884
      Superfamily Oplophoroidea Dana, 1852
        Family Oplophoridae Dana, 1852
      Superfamily Stylodactyloidea Bate, 1888
        Family Stylodactylidae Bate, 1888
      Superfamily Pasiphaeoidea Dana, 1852
        Family Pasiphaeidae Dana, 1852
      Superfamily Bresilioidea Calman, 1896
        Family Bresiliidae Calman, 1896
      Superfamily Nemataocarcinoidea

Family Eugonatonotidae Chace, 1936
Nematocarcinidae Smith, 1884
Rhynchocinetidae Ortmann, 1890
Xiphocarididae Ortmann, 1895
Superfamily Campylonotoidea Sollaud, 1913
Family Campylonotidae Sollaud, 1913
Bathypalaemonellidae de Saint Laurent, 1985
Superfamily Palaemonoidea Rafinesque, 1815
Family Anchistioididae Gurney, 1938
Desmocarididae Borradaile, 1915
Gnathophyllidae Dana, 1852
Hymenoceridae Ortmann, 1890
Palaemonidae Rafinesque, 1815
Subfamily Palaemoninae Rafinesque, 1815
Pontoniinae Kingsley, 1878
Family Typhlocarididae Annandale and Kemp, 1913
Subfamily Euryrhnchinae Holthuis, 1950
Typhlocaridinae Annandale and Kemp, 1913
Superfamily Psalidopodoidea Wood-Mason and Alcock, 1892
Family Psalidopodidae Wood-Mason and Alcock, 1892
Superfamily Stylodactyloidea Bate, 1888
Family Stylodactylidae Bate, 1888
Superfamily Alpheoidea Rafinesque, 1815
Family Alpheidae Rafinesque, 1815
Hippolytidae Dana, 1852
Ogrididiae Hay and Shore, 1918
Superfamily Processoidea Ortmann, 1896
Family Processidae Ortmann, 1896
Superfamily Pandaloidea Haworth, 1825
Family Pandalidae Haworth, 1825
Thalassocarididae Bate, 1888
Superfamily Physetocaridoidea Chace, 1940
Family Physetocarididae Chace, 1940
Superfamily Crangonoidea Haworth, 1825
Family Crangonidae Haworth, 1825
Glyphocrangonidae Smith, 1884
[Suborder Macrura Reptantia—modified, Holthuis (1991)]
Infraorder Astacidea Latreille, 1802
Superfamily Nephropoidea Dana, 1852
Family Nephropidae Dana, 1852
Thaumastochelidae Bate, 1888
Superfamily Astacoidea Latreille, 1802
Family Astacidae Latreille, 1802
Cambaridae Hobbs, 1942
Superfamily Parastacoidea Huxley, 1879
Family Parastacidae Huxley, 1879
Infraorder Thalassinidea Latreille, 1831
Superfamily Thalassinoidea Latreille, 1831
Family Axianassidae Schmitt, 1924
Axiidae Huxley, 1879
Callianassidae Dana, 1852
Callianideidae Kossmann, 1880
Laomediidae Latreille, 1831
Thalassinidae Latreille, 1831
Upogebiidae Borradaile, 1903

Infraorder Palinuridea Latreille, 1802
    Superfamily Glypheoidea Zittel, 1885
        Family Glypheidae Zittel, 1885
    Superfamily Eryonoidea De Haan, 1841
        Family Polychelidae Wood-Mason, 1874
    Superfamily Palinuroidea Latreille, 1802
        Family Palinuridae Latreille, 1802
            Scyllaridae Latreille, 1825
            Synaxidae Bate, 1881
Infraorder Anomura H. Milne Edwards, 1832
    Superfamily Coenobitoidea Dana, 1851
        Family Coenobitidae Dana, 1851
            Diogenidae Ortmann, 1892
            Lomisidae Bouvier, 1895
            Pomatochelidae Miers, 1879
    Superfamily Paguroidea Latreille, 1803
        Family Lithodidae Samouelle, 1819
            Paguridae Latreille, 1803
            Parapaguridae Smith, 1882
    Superfamily Galatheoidea Samouelle, 1819
        Family Aeglidae Dana, 1852
            Chirostylidae Ortmann, 1892
            Galatheidae Samouelle, 1819
            Porcellanidae Haworth, 1825
    Superfamily Hippoidea Latreille, 1825
        Family Albuneidae Stimpson, 1858
            Hippidae Latreille, 1825
Infraorder Brachyura Latreile, 1803
    Section Dromiacea Da Haan, 1833
        Superfamily Dromioidea De Haan, 1833
            Family Dromiidae De Haan, 1833
                 Dynomenidae Ortmann, 1892
                 Homolodromiidae Alcock, 1899
    Section Archaeobrachyura Guinot, 1977
        Superfamily Tymoloidea Alcock, 1896
            Family Cymonomidae Bouvier, 1897
                 Tymolidae Alcock, 1896
        Superfamily Homoloidea De Haan, 1839
            Family Homolidae De Haan, 1839
                 Latreilliidae Stimpson, 1858
        Superfamily Raninoidea De Haan, 1839
            Family Raninidae De Haan, 1839
    Section Oxystomata H. Milne Edwards, 1834
        Superfamily Dorippoidea MacLeay, 1838
            Family Dorippidae MacLeay, 1838
        Superfamily Leucosioidea Samouelle, 1819
            Family Calappidae De Haan, 1833
                 Leucosiidae Samouelle, 1819
    Section Oxyrhyncha Latreille, 1803
        Superfamily Majoidea Samouelle, 1819
            Family Majidae Samouelle, 1819
        Superfamily Hymenosomatoidea MacLeay, 1838
            Family Hymenosomatidae MacLeay, 1838
        Superfamily Mimilambroidea Williams, 1979
            Family Mimilambroidae Williams, 1979
        Superfamily Parthenopoidea MacLeay, 1838

Family Parthenopidae MacLeay, 1838
Section Cancridea Latreille, 1803
    Superfamily Cancroidea Latreille, 1893
        Family Atelecyclidae Ortmann, 1893
                Cancridae Latreille, 1803
                Corystidae Samouelle, 1819
                Pirimelidae Alcock, 1899
                Thiidae Dana, 1852
Section Brachyrhyncha Borradaile, 1907
    Superfamily Portunoidea Rafinesque, 1815
        Family Geryonidae Colosi, 1923
                Portunidae Rafinesque, 1815
    Superfamily Bythograeoidea Williams, 1980
        Family Bythograeidae Williams, 1980
    Superfamily Xanthoidea MacLeay, 1838
        Family Goneplacidae MacLeay, 1838
                Hexopodidae Miers, 1886
                Platyxanthidae Guinot, 1977
                Xanthidae MacLeay, 1838
    Superfamily Bellioidea Dana, 1852
        Family Belliidae Dana, 1852
    Superfamily Grapsidoidea MacLeay, 1838
        Family Gecarcinidae MacLeay, 1838
                Grapsidae MacLeay, 1838
                Mictyridae Dana, 1851
    Superfamily Pinnotheroidea De Haan, 1833
        Family Pinnotheridae De Haan, 1833
    Superfamily Potamoidea Ortmann, 1896
        Family Deckeniidae Bott, 1970
                Gecarcinucidae Rathbun, 1904
                Isolapotamidae Bott, 1970
                Parathelphusidae Alcock, 1910
                Potamidae Ortmann, 1896
                Potamocarcinidae Ortmann, 1899
                Potamonautidae Bott, 1970
                Pseudothelphusidae Ortmann, 1893
                Sinopotamidae Bott, 1970
                Sundathelphusidae Bott, 1969
                Trichodactylidae H. Milne Edwards, 1853
    Superfamily Ocypodoidea Rafinesque, 1815
        Family Ocypodidae Rafinesque, 1815
                Palicidae Rathbun, 1898
                Retroplumidae Gill, 1894
    Superfamily Cryptochiroidea Paulson, 1875
        Family Cryptochiridae Paulson, 1875

REFERENCES

Bowman, T. E., and L. G. Abele. 1982. Classification of the recent Crustacea. In *The biology of Crustacea, vol. 1. Systematics, the fossil record, and biogeography*, ed. L. G. Abele, 1–27. New York: Academic Press.

Chace, F. A., Jr. 1992. On the classification of the Caridea (Decapoda). *Crustaceana* 63(1): 70–80.

Chace, F. A., Jr., and A. J. Bruce. 1993. The caridean shrimps (Crustacea: Decapoda) of the *Albatross* Philippine Expedition, 1907–1910. Part 6: Superfamily Palaemonoidea. *Smithsonian Contrib. Zool.* 543:1–152.

Dai, A., and S. Yang. 1991. *Crabs of the China Seas.* Beijing: China Ocean Press.

Devaney, D. M., E. S. Reese, B. L. Burch, and P. Helfrich. 1987. *The natural history of Enewetak Atoll. Vol. 2. Biogeography and systematics.* U.S. Department of Energy, Office of Scientific and Technical Information.

Eldredge, L. G., and S. E. Miller. 1995. How many species are there in Hawaii? *Bishop Mus. Occas. Pap.* 41:3–18.

Griffin, D. J. G., and H. A. Tranter. 1986. The Decapoda Brachyura of the Siboga Expedition. Part 8. Majidae. *Siboga-Expeditie Monogr. 39, C4 (Livr. 138).*

Holthuis, L. B. 1991. Marine lobsters of the world. FAO Species Catalogue 13. *FAO Fish. Synop.* 125, 13:1–292.

———. 1993. *The recent genera of the caridean and stenopodidean shrimps (Crustacea, Decapoda) with an appendix on the Order Amphionidacea.* Nationaal Natuurhistorisch Museum, Leiden.

Kropp, R. K., and R. B. Manning. 1985. Cryptochiridae, the correct name for the family containing the gall crabs (Crustacea: Decapoda: Brachyura). *Proc. Biol. Soc. Wash.* 98(4): 954–55.

McLay, C. L. 1993. Crustacea Decapoda: The sponge crabs (Dromiidae) of New Caledonia and the Philippines with a review of the genera. In *Résultats des Campagnes MUSORSTOM 10*, ed. A. Crosnier. *Mem. Mus. Nat. Hist. Nat.* 156:111–251.

Miyake, S. 1991. *Japanese crustacean decapods and stomatopods in color.* 2 vols. Osaka: Hoikusha Publishing.

Sakai, T. 1976. *Crabs of Japan and the adjacent seas.* 3 vols. Tokyo: Kodansha.

CHAPTER 10

Echinoderms of the Tropical Island Pacific: Status of Their Systematics and Notes on Their Ecology and Biogeography

David L. Pawson

ABSTRACT

The Phylum Echinodermata, comprising more than 7,000 extant species, is represented in all seas and at all depths. Six classes are recognized today. For the most part, the systematic characters at the family level and above are reasonably stable and accessible. At the genus and species level, the situation is less satisfactory, and several major groups urgently require substantial revision. Despite some shortcomings in classification at various levels, the phylum is one of the best-known taxonomically in the marine realm.

The tropical Indo-Pacific marine region supports a nearshore/inshore echinoderm fauna of approximately 1,300 species. In a recent monograph (Clark and Rowe 1971), a total of 327 species were listed from the tropical island Pacific. Today, twenty-five years on, the known tropical island Pacific fauna now comprises 452 species, an increase of approximately 38 percent over the 1971 total. There are 8 families, 21 genera, and 40 species of crinoids; 19 families, 47 genera, and 105 species of asteroids; 12 families, 39 genera, and 103 species of ophiuroids; 20 families, 50 genera, and 90 species of echinoids; and 6 families, 24 genera, and 114 species of holothuroids. The recent increase in the known fauna is due mostly to exploration of formerly poorly researched areas, in particular New Caledonia, and discovery there of known Indo-Pacific species until now unrecorded from the tropical island Pacific.

The twenty-three recently described new tropical island Pacific species represent approximately 18 percent of the new records. The new data indicate that more intensive exploration virtually anywhere in the Pacific will result in numerous additions to the faunal list for the general area. It is recommended that (1) any new field exploration be coupled with the study of existing collections in major museums, where many unidentified or unpublished collections are housed; (2) as the island Pacific echinoderm fauna remains poorly known, systematic and ecological studies of Pacific echinoderms must be encouraged whenever possible; and (3) in light of the ecological and growing economic importance of the echinoderms, production of a computerized island Pacific echinoderm database is a necessary next step.

INTRODUCTION

In the tropical island Pacific (here regarded as comprising Melanesia, Micronesia, and Polynesia, including Hawaii and the Kermadec Islands [but excluding mainland New Zealand], Easter and Sala-y-Gómez Islands, Marcus, Bonins, and Ryukyus), the shallow-water echinoderms are usually conspicuous, and they are usually ecologically important. There are some famous and some infamous species—the crown-of-thorns starfish *Acanthaster planci* would qualify for both titles! The echinoderms of the area are also of some economic importance, particularly the sea cucumber species that are prized as beche-de-mer or trepang, and some sea urchins whose gonads are edible. Sea cucumbers and sea urchins have been subjected to intense fishing pressure in many easily accessible areas (Conand and Byrne 1993; Régis and Pérès 1994). Of negative economic importance are the species that can devastate coral reefs, chiefly the crown-of-thorns starfish (Birkeland and Lucas 1990). Because of their beauty, symmetry, and their long fossil record, the echinoderms have received much attention from scientists; worldwide, in all depths, they are one of the better-known groups of marine invertebrates. Approximately 7,000 species in total are known today, and another 13,000 fossil species have been described.

While as a group the echinoderms can be characterized as "well-known," difficulties arise when one studies the classification at all levels. Contemporary experts disagree about higher levels of classification (see, for example, Paul and Smith 1988): should there be three subphyla or none, six classes or five, and within the classes should there be super-orders, superfamilies, and so forth? Then within families and genera there are difficulties in all groups. On the face of it, echinoderms present a plethora of useful systematic characters—for example, the typical brittlestar, with its upper, lower, and lateral arm-plates, armspines and tentacle scales, radial shields, disc with plates, granules or spines, oral area with oral shields, jaws, oral papillae, and teeth papillae, seems to offer endless possibilities for the systematist. However, there are great difficulties involved in studying brittlestars—problems of variability within and between populations and, in many cases, profound growth changes contribute to the confusion. Smith, Paterson, and Lafay (1994) found some value in a combined morphological and molecular approach to classification of brittlestars, but there are still many major difficulties to overcome.

Our knowledge of the systematics of the echinoderms of the tropical island Pacific can best be described as fragmentary, and probably in this respect the echinoderms compare with most other invertebrate groups. Some individual specialists have contributed greatly to our knowledge of the Pacific fauna. The older generation of A. H. Clark, H. L. Clark, and Theodore Mortensen, who were active during the first half of this century, were prolific publishers, and they established a sound basis for our knowledge of the Pacific fauna. Coming closer to the present day, I pay tribute to the late and deeply lamented Dennis Devaney, who published several important papers on Pacific brittlestars. Loisette Marsh of the Western Australian Museum has contributed greatly to our knowledge of Pacific seastars. And then there are Ailsa Clark, formerly of the British Museum (Natural History), and Frank Rowe, formerly of the British Museum and then of the Australian Museum. Clark and Rowe published in 1971 their definitive volume, *Shallow Water Indo–West Pacific Echinoderms*, with annotated keys and distribution data for the approximately 1,100 species then known from this vast area.

METHODS

The results presented here have been extracted from the scientific literature. For the purposes of this paper, I have used the splendid, meticulously researched, and authoritative book by Clark and Rowe (1971) as the baseline for all pre-1969 data, and I have attempted to bring the faunal lists up-to-date by adding published reports of species from the study area between 1969 and 1994. The list (see appendix to this paper) includes the shallow-water species only (i.e., those reported from a depth of 20 meters or less, the same depth criterion used by Clark and Rowe 1971).

CURRENT KNOWLEDGE OF THE ECHINODERM FAUNA

Today, we can list approximately 1,300 shallow-water echinoderm species from the general Indo-Pacific. From the tropical island Pacific study area that forms the topic of this paper, 452 species are currently known; this comprises approximately 35 percent of the total Indo-Pacific fauna. These 452 species are arranged into 65 families and 187 genera. The composition of the tropical island Pacific echinoderm fauna is analyzed for families and genera in Table 10.1, and for species in Table 10.2.

HOW WELL DO WE KNOW THE BIOLOGY OF TROPICAL ISLAND PACIFIC ECHINODERMS?

In a review article on biology of echinoderms of coral reefs, Clark (1976b) commented that, despite great advances in our knowledge obtained during the twenty years prior to 1976, there remained large gaps in our knowledge of reproduction, early life histories, food and feeding habits, migrations, and predator-prey relationships of echinoderms. Over the past twenty years, several important papers have helped to close some gaps and open others.

Table 10.1 Families and genera of echinoderms in the tropical island Pacific

Class	Families/genera, 1971		Pacific added fam./gen. 1969–94	Pacific % increase 1969–94	Pacific totals 1994
	Indo-Pacific	Pacific			
Crinoids	9/36	7/16	1/5	14/31	8/21
Asteroids	17/59	14/30	5/17	36/57	19/47
Ophiuroids	11/59	9/30	3/9	33/30	12/39
Echinoids	21/59	18/39	2/11	11/28	20/50
Holothuroids	9/58	6/20	0/4	0/20	6/24
Total	67/271	54/135	11/46	20/34	65/181

Table 10.2 Species of echinoderms in the tropical island Pacific

Class	Species, 1971		Pacific species added 1969–94	Pacific % increase 1969–94	Pacific new totals 1994
	Indo-Pacific	Pacific			
Crinoids	138	30	10 (0 new spp.)	33	40
Asteroids	294	63	42 (6)	67	105
Ophiuroids	231	81	22 (3)	27	103
Echinoids	146	65	25 (1)	38	90
Holothuroids	287	88	26 (13)	30	114
Total	1,096	327	125 (23)	38	452

Acanthaster planci, the Crown-of-Thorns Starfish

The study of the general biology of the echinoderms has been dominated by research on this species. The crown-of-thorns starfish is surely the best-known coral reef–associated animal in the Indo-Pacific, at least in terms of its biology. Since the late 1960s, well in excess of 1,000 articles have been published on the crown-of-thorns; 30 percent have been in popular or semipopular magazines and journals, and 70 percent are in the primary scientific literature (Birkeland and Lucas 1990).

In our knowledge of systematics and distribution, there have been some important changes over the past twenty-five years. It is now almost universally believed that the eastern Pacific crown-of-thorns starfish, formerly known as *A. ellisi*, is merely an outlier of *A. planci*. The known geographic range of *planci* has increased eastward to the west coast of Central America, and southward to include Norfolk and Lord Howe islands and the Kermadec Islands. In suitable habitats, therefore, the species ranges the entire Pacific and Indian oceans reaching approximately 33° N latitude and 33° S latitude. Many aspects of the biology of this seastar are discussed in recent publications—occurrence, population size, fluctuations in populations, reproductive biology, indeed most aspects of autecology and synecology. For a superb summary of what we know and do not know about this species, see Birkeland and Lucas (1990). Also, two sets of collected papers (Wilkinson 1990; Wilkinson and MacIntyre 1992) add further valuable information on this fascinating animal.

Despite the intensive study of this species, some larger questions have yet to be definitively answered. An active debate still continues about whether or not the *Acanthaster* infestations are related in some way to activities of the human species. Authors are also debating historical aspects: Are the infestations a cyclic phenomenon, and were there *Acanthaster* infestations hundreds and thousands of years ago? How useful are fossil and subfossil remains in indicating population densities in the past? The matter of control is still under discussion; .control programs (removal or killing of individuals in some reef areas) are still in effect on some islands. By thinning populations of seastars, we may actually be increasing the reproductive potential of the species by maintaining small breeding stocks. Reefs around high islands, where devastating plagues occur most commonly, are slightly more diverse than around low islands. It has been suggested that control programs in these areas may be reducing the potential diversity of coral reefs. This supposed connection with diversity requires further investigation.

The Echinoderms of New Caledonia

Since the landmark Indo-Pacific work of Clark and Rowe (1971), the most comprehensive publication on the tropical island Pacific is Guille, Laboute, and Menou (1986) on the echinoderms of New Caledonia, a beautiful work with *in situ* color photographs of virtually all of the 246 species they discuss. The book represents the result of extensive diving, collecting, and photographing of echinoderms over a period of approximately ten years. Their study added seventy-seven new records to the New Caledonia fauna, of which seventeen were new species (see Table 10.3). This work has created a major change in our knowledge of the New Caledonia fauna and its relationships. The new records that were not new species are mostly taxa that were previously known from the Australian region.

The most striking conclusion to be gained from Guille, Laboute, and Menou (1986) is that a thorough investigation of a circumscribed area can reveal a large number of taxa that are new to the fauna of the area. These new additions can dramatically affect our concepts of the composition and relationships of the fauna of whatever larger area may be under study. Guille, Laboute, and Menou (1986) added approximately 46 percent to the echinoderm faunal list of New Caledonia. One might assume that a thorough study of other Pacific islands could yield similarly impressive results.

The Common Sea Urchin *Echinometra mathaei*—One Species or Four?

Tsuchiya and Nishihira (1984) and subsequent authors have suggested that the very common reef-associated rock-burrowing urchin *Echinometra mathaei* might actually be a complex of up to four species. In recent years Japanese biologists in particular have studied this problem intensively and have produced evidence from morphology, reproductive biology, ecology, and mitochondrial DNA studies that lends some support to the idea that four species or incipient species are living in relative proximity to one another, in Okinawa and elsewhere. Results of recent investigations by a variety of authors can be found in Yanagisawa et al. (1991, 89–139). The *Echinometra* complex, while lagging far behind *Acanthaster* in terms of attention from biologists, is today one of the better-known echinoids in the Indo-Pacific.

Table 10.3 New Caledonia echinoderms: Impact of Guille et al., 1986

Class	No. species prior to 1986	New records Guille et al., 1986	New total spp. Guille et al., 1986	% increase
Crinoids	31	6 (0 new spp.)	37	20
Asteroids	35	19 (2)	54	54
Ophiuroids	38	19 (2)	57	50
Echinoids	32	11 (0)	43	34
Holothuroids	33	22 (13)	55	67
Total	169	77 (17)	246	46

Recent Advances in Knowledge of Ecology and Reproductive Biology

While I do not intend to cover this topic in any detail here, it is important to point out that over the past twenty-five years, several important and informative papers have been published on the biology of echinoderms, and these have a direct bearing on ideas about distribution and dispersal of taxa. Papers dealing with the tropical island Pacific include Yamaguchi (1973, 1974, 1975, 1977), Yamaguchi and Lucas (1984), Glynn and Krupp (1986), Meyer and Macurda (1980), Kerr, Stoffel, and Yoon (1993), Kerr (1994), Ogden, Ogden, and Abbott (1989), Rowe and Doty (1977), and many others too numerous to mention here.

PATTERNS OF DISTRIBUTION OF PACIFIC ISLAND ECHINODERMS

Several authors have commented in the past twenty-five years on broad distribution patterns of Pacific echinoderms: Clark and Rowe (1971), Clark (1976a), Yamaguchi (1977), Yamaguchi and Lucas (1984), Devaney (1973), Marsh (1977), and others. Species of echinoderms, in general, appear to be widely distributed in the Indo-Pacific. Apart from the twenty-three newly described species, which are so far known from very few records in circumscribed areas, endemics comprise a virtually nonexistent component of the Pacific island echinoderm fauna. Most tropical island Pacific species are known from several scattered localities across the entire Indo-Pacific region. No distribution patterns appear to be related to the configuration or characteristics of the Pacific lithospheric plate (i.e., the Pacific Plate of Springer 1982 and later authors), or even to existing ocean surface current patterns. The echinoderm fauna as we know it today is different from the fauna we knew twenty-five years ago. Kay (1984) made an excellent analysis of distribution patterns of the Indo-West Pacific echinoderms, along with mollusks and fishes. She concluded that there are three major areas of endemism—the Indian Ocean, the western Pacific, and the Pacific Basin. She regarded the western Pacific not as a center of origin as envisaged by Ekman (1953), but as an area of accumulation of taxa. Perhaps the new echinoderm records might cause a slight shift in Dr. Kay's results—for the echinoderms anyway, but these new data have yet to be analyzed.

Of special interest in this context is the new information available on reproductive biology of some echinoderms, particularly seastars (e.g., Yamaguchi 1973, 1974, 1977; Yamaguchi and Lucas 1984). Yamaguchi (1977) invokes differences in behavior patterns of planktonic larvae to account for differences in present-day distribution patterns of adult seastars. Larvae that display negative geotaxis represent widely distributed species, such as *Acanthaster planci*, while positively geotactic larvae represent species that are less widely distributed, such as *Protoreaster nodosus*. This is an interesting idea that might reward further investigation. There are exceptions that perhaps "prove the rule," such as the viviparous brittlestar *Amphipholis squamata* which, while it apparently has no active means of dispersal, inexplicably has a cosmopolitan distribution pattern.

In broader discussions of Pacific island biogeography (e.g., Springer 1982; Springer and Williams 1990; Kay 1984), the conflicting or complementary (depending upon one's preferences) ideas of larval dispersal versus vicariance theory are invoked, and the reader is referred to these works for further information. Suffice it to say here that, for the echinoderms, we understand their biology a little better today, and we also have some more details on their distribution. Further analyses based on current data might yield something closer to the real picture.

CONCLUSIONS AND RECOMMENDATIONS

Our knowledge of the composition and relationships of the island Pacific echinoderm fauna has increased steadily over the past twenty-five years. The surprising results of the New Caledonia study (Guille, Laboute, and Menou 1986) suggest that numerous new records and new taxa await the careful collector almost anywhere in the Pacific. In many of the world's museums, large Pacific island collections, either unidentified or identified but unpublished, await the attention of specialists. These collections will be a valuable source of systematic and biogeographic information, just as the British Museum (Natural History) collections provided much new data for Clark and Rowe (1971). The number of active specialists studying island Pacific faunas has decreased in recent years, and it is recommended that, wherever possible, study of systematics and ecology of echinoderms from these areas be actively encouraged. In light of the ecological and economic importance of the echinoderms, a computerized database for this phylum should be instituted at the earliest opportunity.

Acknowledgements—I am most grateful to Doris Vance for her careful and thorough assistance in searches of the literature on echinoderms. Gordon Hendler kindly reviewed an early draft of this paper. I thank the Ocean Policy Institute of the Pacific Forum and the Program on Environment of the East-West Center for inviting me to participate in the Marine and Coastal Biodiversity Workshop.

APPENDIX

Checklist of Pacific Islands Shallow-Water Echinodermata

The following checklist is intended to be comprehensive for the tropical island Pacific to depths of 20 meters. Data for the period 1758–1969 were derived from Clark and Rowe (1971); post-1969 data were obtained from the Zoological Record and the primary literature. For convenience, the taxonomic arrangement closely follows that in Clark and Rowe (1971). In the list, taxa marked with an asterisk (*) are new records for the region; these taxa are accompanied by a literature citation and a geographic name.

PHYLUM ECHINODERMATA

CLASS CRINOIDEA

Family Comasteridae
Capillaster multiradiatus (Linnaeus)

*Cenolia spanoschistum (H. L. Clark)   Rowe 1989, Norfolk Is.

Comantheria polycnemis (A. H. Clark)
**C. briareus* (Bell)   Meyer and Macurda 1980, Guam; Guille, Laboute, and Menou 1986, New Caledonia

Comanthina schlegeli (Carpenter)

Comanthus bennetti (Müller)
C. parvicirrus (Müller)
C. samoanus A. H. Clark

Comaster gracilis (Hartlaub)
C. multifidus (Müller)
*C. distinctus (Carpenter)   Guille, Laboute, and Menou 1986, New Caledonia

Comatella maculata (Carpenter)
*C. nigra (Carpenter)   Meyer and Macurda 1980, Guam
C. stelligera (Carpenter)

*Comissia pectinifera A. H. Clark   Guille, Laboute, and Menou 1986, New Caledonia

Family Eudiocrinidae
Eudiocrinus tenuissimus Gislén

Family Himerometridae
Amphimetra tessellata (Müller)
Himerometra magnipinna A. H. Clark
H. robustipinna (Carpenter)

Family Miriametridae
Dichrometra flagellata (Müller)

Lamprometra palmata (Müller)

Liparometra regalis (Carpenter)

Oxymetra finschi (Hartlaub)

*Stephanometra echinus (A. H. Clark)   Meyer and Macurda 1980, Guam; Guille, Laboute, and Menou 1986, New Caledonia

S. indica (Smith)
S. oxyacantha (Hartlaub)
S. spicata (Carpenter)
S. tenuipinna (Hartlaub)

Family Colobometridae
Cenometra bella (Hartlaub)

*Cyllometra manca (Carpenter)   Meyer and Macurda 1980, Guam

Colobometra diadema A. H. Clark
C. perspinosa (Carpenter)

Oligometra serripinna (Carpenter)

Pontiometra andersoni (Carpenter)

****Family Tropiometridae**
*Tropiometra afra (Hartlaub)   Guille, Laboute, and Menou 1986, New Caledonia

Family Antedonidae
*Antedon incommoda Bell   McKnight 1977a, Norfolk Is.
*A. detonna McKnight   McKnight 1977a, Norfolk Is.

Dorometra nana (Hartlaub)

Euantedon tahitiensis A. H. Clark

Mastigometra pacifica A. H. Clark

CLASS ASTEROIDEA

Family Luidiidae
Luidia aspera Sladen
L. avicularia Fisher
*L. maculata Müller and Troschel   Guille, Laboute, and Menou 1986, New Caledonia
L. magnifica Fisher
L. savignyi (Audouin)

Family Astropectinidae
Astropecten polyacanthus Müller and Troschel
A. triseriatus Müller and Troschel
*A. validispinosus   Oguro 1982, Palau and Yap

Family Archasteridae
Archaster angulatus Müller and Troschel
A. typicus Müller and Troschel

*Family Goniasteridae
*Tosia queenslandensis Livingstone   Guille, Laboute, and Menou 1986, New Caledonia

*Family Asteridiscididae
*Asterodiscides helonotus (Fisher)   Guille, Laboute, and Menou 1986, New Caledonia

Family Oreasteridae
Choriaster granulatus Lütken

Culcita grex Müller and Troschel
C. novaeguineae Müller and Troschel

*Halityle regularis Fisher   Guille, Laboute, and Menou 1986, New Caledonia

Pentaceraster alveolatus (Perrier)
P. regulus (Müller and Troschel)

Pentaster hybridus Döderlein
P. obtusatus (Bory de St. Vincent)

*Poraster superbus (Moebius)   Guille, Laboute, and Menou 1986, New Caledonia

Protoreaster nodosus (Linnaeus)

Family Ophidiasteridae
Celerina heffernani (Livingston)

*Cistina columbiae Gray   Guille, Laboute, and Menou 1986, New Caledonia

Dactylosaster cylindricus (Lamarck)

Fromia balansae Perrier
F. hemiopla Fisher
F. indica (Perrier)
F. milleporella (Lamarck)
F. monilis Perrier
**F. nodosa*   Oguro and Sasayama 1984, Marshall Is.
F. pacifica H. L. Clark
**F. polypora* H. L. Clark   Rowe 1989, Norfolk Is.

Gomophia egyptiaca Gray
**G. watsoni* (Livingstone)   Guille, Laboute, and Menou 1986, New Caledonia

**Heteronardoa carinata* (Koehler) Guille, Laboute, and Menou 1986, New Caledonia

Leiaster brevispinus H. L. Clark
L. coriaceus Peters
L. glaber Peters
L. leachi (Gray)
**L. speciosus* von Martens   Guille, Laboute, and Menou 1986, New Caledonia

Linckia guildingi Gray
L. laevigata (Linnaeus)
L. multifora (Lamarck)

*Nardoa frianti Koehler   Guille, Laboute, and Menou 1986, New Caledonia
N. gomophia (Perrier)
N. lemonnieri Koehler
N. mollis de Loriol
N. novaecaledoniae (Perrier)
N. pauciforis (v. Martens)
N. tuberculata Gray
N. tumulosa Fisher

Neoferdina cancellata (Grube)
**N. cumingi* (Gray)   Guille, Laboute, and Menou 1986, New Caledonia
N. ocellata (H. L. Clark)
**N. offreti* (Koehler)   Marsh 1977, Palau

**Ophidiaster confertus* H. L. Clark McKnight 1967, Norfolk and Lord Howe Is.
O. cribrarius Lütken
**O. easterensis* (Ziesenhenne)   DiSalvo, Randall, and Cea 1988, Easter Is.
O. granifer Lütken
O. hemprichi Müller and Troschel
**O. kermadecensis* Benham   H. E. S. Clark 1970, Kermadec Is.
O. lorioli Fisher
O. perplexus A. H. Clark
**O. robillardi* de Loriol   Marsh 1977, Palau
O. squameus Fisher

*Tamaria fusca (Gray)   Guille, Laboute, and Menou 1986, New Caledonia
T. pusilla Müller and Troschel

*Family Chaetasteridae
**Chaetaster moorei* Bell   Guille, Laboute, and Menou 1986, New Caledonia

Family Asteropseidae
Asteropsis carinifera (Lamarck)

*Petricia vernicina (Lamarck)   Rowe 1989, Kermadec Is.
*P. imperialis (Farquhar)   H. E. S. Clark 1970, Kermadec Is.

Family Asterinidae
*Asterina alba H. L. Clark   Rowe 1989, Norfolk and Lord Howe Is.
*A. anomala H. L. Clark   Marsh 1977, Palau
A. burtoni Gray
*A. corallicola Marsh   1977, Palau
A. coronata v. Martens
A. granulosa Perrier

Disasterina abnormalis (Perrier)

*Nepanthia briareus (Bell)   Guille, Laboute, and Menou 1986, New Caledonia

Patiriella exigua (Lamarck)
*P. oliveri (Benham)   H. E. S. Clark 1970, Kermadec Is.
*P. regularis (Verrill)   H. E. S. Clark 1970, Kermadec Is.

*Family Poraniidae
*Marginaster sp.   H. E. S. Clark 1970, Kermadec Is.

*Family Solasteridae
*Seriaster regularis Jangoux   Guille, Laboute, and Menou 1986, New Caledonia

Family Acanthasteridae
Acanthaster planci (Linnaeus)

Family Valvasteridae
Valvaster striatus (Lamarck)

Family Pterasteridae
Euretaster insignis (Sladen)
*E. attenuatus Jangoux   Guille, Laboute, and Menou 1986, New Caledonia

Family Mithrodiidae
Mithrodia clavigera (Lamarck)
M. fisheri Holly

*Thromidia catalai Pope and Rowe   Guille, Laboute, and Menou 1986, New Caledonia

Family Echinasteridae
Echinaster callosus v. Marenzeller
*E. colemani Rowe and Albertson   1987, Norfolk Is.
E. luzonicus (Gray)
*E. varicolor H. L. Clark   Guille, Laboute, and Menou 1986, New Caledonia

Family Sphaerasteridae
Podosphaeraster polyplax A. M. Clark
*P. pulvinatus Rowe and Nichols   1980, Guam

Family Asteriidae
*Allostichaster polyplax Müller and Troschel   H. E. S. Clark 1970, Kermadec Is.

Astrostole paschae (A. M. Clark)   DiSalvo, Randall, and Cea 1988, Easter Is.
A. rodolphi (Perrier)   A. M. Clark 1950, Kermadec Is.

Coronaster pauciporis Jangoux   Guille, Laboute, and Menou 1986, New Caledonia

Coscinasterias calamaria (Gray)

Distolasterias edmondi (Benham)   H. E. S. Clark 1970, Kermadec Is.

Stichaster australis (Verrill)   H. E. S. Clark 1970, Kermadec Is.

CLASS OPHIUROIDEA

Family Ophiomyxidae
Ophiomyxa australis Lütken

*Family Euryalidae
Euryale aspera Lamarck   Guille, Laboute, and Menou 1986, New Caledonia

*Family Asteroschematidae
Astrobrachion constrictum (Farquhar)   Guille, Laboute, and Menou 1986, New Caledonia

Family Gorgonocephalidae
Astrocladus tonganus Döderlein

Astroboa granulatus (H. L. Clark)   Guille, Laboute, and Menou 1986, New Caledonia
A. nuda (Lyman)   Guille, Laboute, and Menou 1986, New Caledonia

*Family Ophiacanthidae
Amphilimna tanyodes Devaney   1974, Pitcairn Is.

Family Amphiuridae
Amphioplus (Amphioplus) iuxtus Murakami
A. (Amphioplus) platycanthus Murakami
A. (Lymanella) caelatus Ely
A. (Lymanella) depressus (Ljungman)
A. (Lymanella) laevis (Lyman)

Amphipholis misera (Koehler)
A. squamata (D. Chiaje)

Amphiura bountyia Devaney   1974, Pitcairn Is.
A. (Amphiura) crossota Murakami
A. (Amphiura) immira Ely
A. (Amphiura) luetkeni Duncan

Ophiodaphne formata (Koehler)   Guille, Laboute, and Menou 1986, New Caledonia

Ophiocentrus asper (Koehler)   Guille, Laboute, and Menou 1986, New Caledonia
O. dilatatus (Koehler)   Guille, Laboute, and Menou 1986, New Caledonia

Family Ophiactidae
Ophiactis modesta Brock
O. savignyi Müller and Troschel

Family Ophiotrichidae
*Macrophiothrix belli (Döderlein)   Guille, Laboute, and Menou 1986, New Caledonia
*M. demessa (*Lyman)
M. expedita (Koehler)
M. galateae (Lütken)
M. koehleri A. M. Clark
M. longipeda (Lamarck)
M. lorioli A. M. Clark
*M. rugosa H. L. Clark   Guille, Laboute, and Menou 1986, New Caledonia

Ophiogymna elegans Ljungman

Ophiomaza cacaotica Lyman

Ophiopteron vitiensis Koehler

Ophiothela danae Verrill

Ophiothrix ciliaris (Lamarck)
O. elegans Lütken
O. foveolata Marktanner-Turneretscher
*O. picteti de Loriol   Guille, Laboute, and Menou 1986, New Caledonia
O. trilineata Lütken
O. (Acanthophiothrix) proteus Koehler
O. (Acanthophiothrix) purpurea v. Martens
*O. (Acanthophiotrix) vigelandi A. M. Clark   Guille, Laboute, and Menou 1986, New Caledonia
O. (Acanthophiothrix) scotiosa Murakami
O. (Keystonea) martensi Lyman
O. (Keystonea) nereidina Lamarck
O. (Keystonea) propinqua Lyman
O. (Placophiothrix) hybrida H. L. Clark
O. (Placophiothrix) virgata Lyman

Family Ophiocomidae
Ophiarthrum elegans Peters
O. pictum Müller and Troschel

Ophiocoma anaglyptica Ely
O. coma brevipes Peters
O. ocoma dentata Müller and Troschel
O. coma erinaceus Müller and Troschel
O. coma latilanxa Murakami
*O. coma longispina (H. L. Clark)   Devaney 1973, Easter Is.
O. ocoma pica Müller and Troschel
*O. coma pusilla (Brock)   Devaney 1973, Australs, Pitcairn
O. coma schoenleini Müller and Troschel
O. coma scolopendrina (Lamarck)
O. coma wendti: sensu Koehler

Ophiocomella sexradia (Duncan)

Ophiomastix annulosa (Lamarck)
O. asperula Lütken
O. bispinosa H. L. Clark
O. caryophyllata Lütken
O. palaoensis Murakami

O. sexradiata A. H. Clark
O. variabilis Koehler

Ophiopsila multipapillata Guille   Guille, Laboute, and Menou 1986, New Caledonia

Family Ophionereidae
Ophionereis degeneri (A. H. Clark)
**O. dubia* (Müller and Troschel)   Guille, Laboute, and Menou 1986, New Caledonia
O. fusca Brock
O. porrecta Lyman
O. thryptica Murakami

Family Ophiodermatidae
Cryptopelta longibrachialis Koehler   Guille, Laboute, and Menou 1986, New Caledonia

Distichophis clarki Ely

Ophiarachna affinis Lütken
O. incrassata (Lamarck)
**O. megacantha erythema* Devaney   1974, Pitcairn Is.

Ophiarachnella gorgonia (Müller and Troschel)
O. infernalis (Müller and Troschel)
O. macracantha H. L. Clark
O. parvispina H. L. Clark
O. septemspinosa (Müller and Troschel)
**O. snelliusi* A. H. Clark   Guille, Laboute, and Menou 1986, New Caledonia

Ophiochaeta hirsuta Lütken

Ophioclastus hataii Murakami

Ophioconis cincta Brock
O. cupida Koehler

Ophioderma tonganum Lütken

Ophiodyscrita pacifica (Murakami)

Ophiopeza dubiosa de Loriol
**O. kingi* Devaney   1974, Pitcairn Is.
O. spinosa (Ljungman)

Ophiostegastus novaecaledoniae Guille   Guille, Laboute, and Menou 1986, New Caledonia
  and Vadon

Family Ophiuridae
Ophiolepis cardioplax Murakami
O. cincta Müller and Troschel
O. superba H. L. Clark

Ophioplocus imbricatus Müller and Troschel

Ophiotylos leucus Murakami

Ophiura kinbergi (Ljungman)

CLASS ECHINOIDEA

Family Cidaridae
Chondrocidaris brevispina (H. L. Clark)
*C. gigantea A. Agassiz   Guille, Laboute, and Menou 1986, New Caledonia

*Compsocidaris pyrsacantha   Shigei 1970, Bonin Is.

Eucidaris metularia (Lamarck)

*Phyllacanthus dubius Brandt   Shigei 1970, Bonin Is.
P. imperialis (Lamarck)

*Prionocidaris australis (Ramsay)   Guille, Laboute, and Menou 1986, New Caledonia
*P. baculosa (Lamarck)   Shigei 1970, Bonin Is.
P. verticillata (Lamarck)

*Family Echinothuriidae
*Asthenosoma varium Grube   Guille, Laboute, and Menou 1986, New Caledonia

Family Diadematidae
Astropyga radiata (Leske)

*Centrostephanus rodgersi (A. Agassiz)   Pawson 1965, Kermadec Is.; Rowe 1989, Norfolk Is.

*Diadema palmeri Baker   Rowe 1989, Norfolk Is.
D. paucispinum A. Agassiz
D. savignyi Michelin
D. setosum (Leske)

Echinothrix calamaris (Pallas)
E. diadema (Linnaeus)

Family Stomopneustidae
Stomopneustes variolaris (Lamarck)

Family Temnopleuridae
*Holopneustes inflatus Lütken   Rowe 1989, Norfolk Is.

Mespilia globulus (Linnaeus)

Paratrema doederleini (Mortensen)

*Salmacis belli Döderlein   Guille, Laboute, and Menou 1986, New Caledonia
S. sphaeroides (Linnaeus)

*Temnopleurus reevesi (Gray)   Shigei 1970, Bonin Is.
T. toreumaticus (Leske)

Temnotrema hawaiiense (A. Agassiz and
   H. L. Clark)

Family Toxopneustidae
Cyrtechinus verruculatus (Lütken)

*Gymnechinus epistichus H. L. Clark   Guille, Laboute, and Menou 1986, New Caledonia

Nudechinus ambonensis Mortensen
N. inconspicuus (Mortensen)
N. multicolor (Yoshiwara)

Pseudoboletia indiana (Michelin)

Toxopneustes maculatus (Lamarck)
T. pileolus (Lamarck)

Tripneustes gratilla (Linnaeus)

Family Parasaleniidae
Parasalenia gratiosa A. Agassiz
P. poehli Pfeffer

Family Echinometridae
*   *Anthocidaris crassispina* (A. Agassiz)   Shigei 1970, Bonin Is.

Colobocentrotus atratus (Linnaeus)
C. mertensi Brandt
C. pedifer (de Blainville)

Echinometra mathaei (de Blainville)

Echinostrephus aciculatus A. Agassiz
E. molaris (de Blainville)

*   *Heliocidaris tuberculata* (Lamarck)   Rowe 1989, Norfolk Is.

Heterocentrotus mammillatus (Linnaeus)
H. trigonarius (Lamarck)

Zenocentrotus kellersi A. H. Clark
Z. paradoxus A. H. Clark

Family Echinoneidae
Echinoneus abnormis de Loriol
E. cyclostomus Leske

Family Clypeasteridae
*   *Clypeaster australasiae* (Gray)   Pawson 1965, Norfolk Is.
C. eurypetalus H. L. Clark
C. humilis (Leske)
C. ohshimensis Ikeda
C. reticulatus (Linnaeus)

Family Arachnoididae
Arachnoides placenta (Linnaeus)

Family Fibulariidae
Echinocyamus australis (Desmoulins)
E. crispus Mazetti
E. incertus H. L. Clark
E. megapetalus H. L. Clark
*   *E. polyporus* Mortensen   Pawson 1965, Norfolk Is.

Fibularia acuta Yoshiwara
F. ovulum Lamarck
F. volva L. Agassiz

Family Laganidae
Laganum centrale H. L. Clark
*L. decagonale (Blainville)   Rowe 1989, Norfolk Is.
L. depressum Lesson
L. L. (Leske)

*Peronella hinemoae Mortensen   Pawson 1965, Norfolk Is.

Family Scutellidae
Echinodiscus tenuissimus (L. Agassiz)
*E. bisperforatus truncatus (L. Agassiz)   Guille, Laboute, and Menou 1986, New Caledonia

Family Echinolampadidae
Echinolampas alexandri de Loriol

Family Spatangidae
Maretia carinata Bolau
M. planulata (Lamarck)

*Family Loveniidae
*Lovenia elongata (Gray)   Guille, Laboute, and Menou 1986, New Caledonia

Family Palaeostomatidae
Palaeostoma mirabile (Gray)

Family Schizasteridae
Protenaster rostratus (Smith)

*Schizaster lacunosus (Linnaeus)   Guille, Laboute, and Menou 1986, New Caledonia

Family Brissidae
Brissopsis luzonica (Gray)
*B. oldhami Alcock   Pawson 1965, Norfolk Is.

*Brissus agassizi Döderlein   Baker 1967, Norfolk Is.
B. latecarinatus (Leske)

*Eupatagus rubellus Mortensen   Guille, Laboute, and Menou 1986, New Caledonia

*Metalia angustus De Ridder   Guille, Laboute, and Menou 1986, New Caledonia
M. dicrana H. L. Clark
M. spatagus (Linnaeus)
M. sternalis (Lamarck)

Rhynobrissus hemiasteroides A. Agassiz

CLASS HOLOTHUROIDEA

Family Holothuriidae
*Actinopyga albonigra Cherbonnier and Féral   Guille, Laboute, and Menou 1986, New Caledonia
*A. crassa Panning   Guille, Laboute, and Menou 1986, New Caledonia
A. echinites (Jaeger)

A. flammea Cherbonnier   Guille, Laboute, and Menou 1986, New Caledonia
A. lecanora (Jaeger)
A. mauritiana (Quoy and Gaimard)
A. miliaris (Quoy and Gaimard)
A. obesa Selenka
A. palauensis Panning
*A. spinea Cherbonnier   Guille, Laboute, and Menou 1986, New Caledonia

Bohadschia argus Jaeger
B. bivitata (Mitsukuri)
B. graeffei (Semper)
B. koellikeri (Semper)
*B. maculisparsa Cherbonnier and Féral   Guille, Laboute, and Menou 1986, New Caledonia
B. marmorata Jaeger
B. paradoxa (Selenka)
B. similis (Semper)
B. tenuissima (Semper)
B. vitiensis (Semper)

Labidodemas rugosum (Ludwig)
L. semperianum Selenka

*Holothuria (Acanthotrapeza) coluber Semper   Guille, Laboute, and Menou 1986, New Caledonia
H. (Acanthotrapeza) kubaryi Ludwig
H. (Cystipus) inhabilis Selenka
H. (Cystipus) rigida (Selenka)
H. (Halodeima) atra Jaeger
H. (Halodeima) edulis Lesson
H. (Halodeima) pulla Selenka
*H. (Halodeima) signata Ludwig   Paulay 1989, Pitcairn Is.
H. (Lessonothuria) glandifera Cherbonnier
H. (Lessonothuria) insignis Ludwig
H. (Lessonothuria) pardalis Selenka
H. (Lessonothuria) verrucosa Selenka
H. (Mertensiothuria) fuscocinerea Jaeger
H. (Mertensiothuria) leucospilota Bandt
H. (Mertensiothuria) pervicax Selenka
H. (Metriatyla) albiventer Semper
H. (Metriatyla) bowensis Ludwig
*H. (Metriatyla) ocellata (Jaeger)   Guille, Laboute, and Menou 1986, New Caledonia
H. (Metriatyla) scabra Jaeger
H. (Microthele) nobilis (Selenka)
*H. (Microthele) axiologa H. L. Clark   Rowe and Doty 1977, Guam
*H. (Microthele) fuscogilva Cherbonnier   Guille, Laboute, and Menou 1986, New Caledonia
*H. (Microthele) fuscopunctata Jaeger   Guille, Laboute, and Menou 1986, New Caledonia
H. (Platyperona) difficilis Semper
H. (Selenkothuria) erinaceus Semper
H. (Selenkothuria) moebii Ludwig
H. (Semperothuria) cinerascens (Brandt)
H. (Semperothuria) flavomaculata Semper
H. (Semperothuria) imitans Ludwig
H. (Stauropora) discrepans Semper
H. (Stauropora) fuscoolivacea Fisher
H. (Stauropora) hawaiiensis Fisher
H. (Stauropora) ludwigi Lampert
*H. (Stichothuria) coronopertusa Cherbonnier   Guille, Laboute, and Menou 1986, New Caledonia

H. (Theelothuria) maculosa Pearson   Guille, Laboute, and Menou 1986, New Caledonia
H. (Thelothuria) samoana Ludwig
H. (Theelothuria) turriscelsa Cherbonnier Guille, Laboute, and Menou 1986, New Caledonia
H. (Theelothuria) squamifera Semper
H. (Thymiosycia) arenicola Semper
H. (Thymiosycia) altaturricula Cherbonnier Guille, Laboute, and Menou 1986, New Caledonia
 and Féral
H. (Thymiosycia) conusalba Cherbonnier Guille, Laboute, and Menou 1986, New Caledonia
 and Féral
H. (Thymiosycia) hilla Lesson
H. (Thymiosycia) impatiens (Forskål)
H. (Thymiosycia) gracilis Semper

Family Stichopodidae
Stichopus chloronotus Brandt
S. horrens Selenka
*S. noctivagus Cherbonnier   Guille, Laboute, and Menou 1986, New Caledonia
*S. pseudhorrens Cherbonnier   Guille, Laboute, and Menou 1986, New Caledonia
S. variegatus Semper

Thelenota ananas (Jaeger)
*T. anax H. L. Clark   Rowe and Doty 1977, Guam

Family Cucumariidae
Pentacta anceps (Selenka)
*P. australis (Ludwig)   Guille, Laboute, and Menou 1986, New Caledonia

*Thyone okeni Bell   Rowe and Doty 1977, Guam

Family Phyllophoridae
Afrocucumis africana (Semper)

Cladolabes acicula (Semper)

*Neothyonidium magnum (Ludwig)   Guille, Laboute, and Menou 1986, New Caledonia

*Ohshimella castanea Cherbonnier   Guille, Laboute, and Menou 1986, New Caledonia

Thyonidiella oceana Heding and Panning

Family Synaptidae
Anapta gracilis Semper

Euapta godeffroyi (Semper)
E. magna Heding
E. tahitiensis Cherbonnier

*Leptosynapta dolabrifera (Stimpson) Rowe 1989, Norfolk Is.

Ophiodesoma australiensis Heding
O. glabra (Semper)
O. grisea (Semper)
O. spectabilis Fisher
O. variabilis Heding

Patinapta laevis (Bedford)
P. ooplax (von Marenzeller)

Pendekaplectana grisea Heding

Polyplectana galatheae Heding
P. kefersteini (Selenka)
P. longogranula Heding
P. samoae Heding
P. unispicula Heding

Synapta maculata (Chamisso and Eysenhardt)
⁺S. media Cherbonnier and Féral   Guille, Laboute, and Menou 1986, New Caledonia
S. oceanica (Lesson)
S. reticulata (Semper)

Synaptula macra (H. L. Clark)
S. purpurea Heding
S. recta (Semper)

Family Chiridotidae
⁺Chiridota exuga Cherbonnier   1986, Moorea
C. hawaiiensis Fisher
C. intermedia Bedford
C. liberata Sluiter
C. rigida Semper
C. stuhlmanni Lampert

Polycheira rufescens (Brandt)
P. vitiensis (Semper)

REFERENCES

Baker, A. N. 1967. Two new echinoids from northern New Zealand, including a new species of *Diadema*. Transactions of the Royal Society of New Zealand. *Zoology* 8(23): 239–45.

Birkeland, C., and J. Lucas. 1990. Acanthaster planci: *Major management problem of coral reefs.* Boca Raton, Fla.: CRC Press.

Cherbonnier, G. 1986.*Chiritoda exuga,* nouvelle espece d'Holothurie apode de Polynesie francaise. *Bulletin Museum d'Histoire Naturelle Paris,* 4th Series 8:39–41.

Clark, A. M. 1950. A new species of sea-star from Norfolk Island. *Annals and Magazine of Natural History* (12)3: 808–11.

———. 1976a. Echinoderms of coral reefs. In *Biology and geology of coral reefs,* ed. O. A. Jones and R. Endean, 95–123. Vol. 3: Biology 2. New York: AcademicPress.

———. 1976b. Review of present status of knowledge of Pacific echinoderms. *Micronesica* 12(1): 193–95.

Clark, A .M., and F. W. E. Rowe. 1971. *Shallow water Indo–West Pacific echinoderms.* London: British Museum (Natural History).

Clark, H. E. S. 1970. Sea-stars (Echinodermata: Asteroidea) from "Eltanin" Cruise 26, with a review of the New Zealand asteroid fauna. *Zoology Publications* (Victoria University of Wellington) 52:1–34.

Codoceo, M. 1974. Equinodermos de la Isla de Pascua. *Boletin del Museo Nacional de Historia Natural, Chile* 33:53–63.

Conand, C., and M. Byrne. 1993. A review of recent developments in the world sea cucumber fisheries. *Marine Fisheries Review* 55(4): 1–13.

Devaney, D. M. 1973. Zoogeography and faunal composition of south-eastern Polynesian asterozoan echinoderms. In *Oceanography of the South Pacific 1972,* ed. R. Fraser, 357–66. New Zealand National Commission for UNESCO, Wellington, New Zealand.

Devaney, D. M. 1974. Shallow-water asterozoans of southeastern Polynesia. II. Ophiuroidea. *Micronesica* 10(1): 105–204.

DiSalvo, L. H., J. E. Randall, and A. Cea. 1988. Ecological reconnaissance of the Easter Island sublitoral marine environment. *National Geographic Research* 4(4) :451–73.

Ekman, S. 1953. *Zoogeography of the sea.* London: Sidgwick and Jackson.

Fell, F. J. 1974. The echinoids of Easter Island (Rapa Nui). *Pacific Science* 28(2): 147–58.

Glynn, P. W., and D. A. Krupp. 1986. Feeding biology of a Hawaiian sea star corallivore, *Culcita novaeguineae* Müller and Troschel. *Journal of Experimental Marine Biology and Ecology* 96:75–96.

Guille, A., P. Laboute, and J.-L. Menou. 1986. *Guide des étoiles de mer, oursins et autres échinodermes du lagon de Nouvelle-Calédonie,* 1–238. Paris: Éditions de l'ORSTOM.

Hoggett, A. K., and F. W. E. Rowe. 1986. Southwest Pacific cidarid echinoids (Echinodermata), including two new species. *Indo-Malayan Zoology* 3:1–13.

———. 1987. Zoogeography of echinoderms on the world's most southern coral reefs. In *Echinoderm biology,* ed. R. D. Burke, P. V. Mladenov, P. Lambert, and R. L. Parsley, 379–87. Rotterdam: A. A. Balkema.

Kay, E. A. 1984. Patterns of speciation in the Indo–West Pacific. In *Biogeography of the tropical Pacific,* ed. F. J. Radovsky, P. H. Raven, and S. H. Sohmer, 15–31. Bernice P. Bishop Museum Special Publication No. 72, Honolulu.

Kerr, A. M. 1994. Shallow-water holothuroids (Echinodermata) of Kosrae, Eastern Caroline Islands. *Pacific Science* 48(2): 161–74.

Kerr, A. M., E. M. Stoffel, and R. L. Yoon. 1993. Abundance distribution of holothuroids (Echinodermata: Holothuroidea) on a windward and leeward fringing coral reef, Guam, Mariana Islands. *Bulletin of Marine Science* 52(2): 780–91.

Marsh, L. M. 1977. Coral reef asteroids of Palau, Caroline Islands. *Micronesica* 13(2): 251–81.

McKnight, D. G. 1967. Some asterozoans from Norfolk Island. *New Zealand Journal of Marine and Freshwater Research* 1(3): 324–26.

———. 1977a. Crinoids from Norfolk Island and Wanganella Bank. *NZOI Records* 3(14): 129–37.

———. 1977b. Some crinoids from the Kermadec Islands. *NZOI Records* 3(13): 121–28.

———. 1978. *Acanthaster planci* (Linnaeus) (Asteroidea: Echinodermata) at the Kermadec Islands. *NZOI Records* 4(3): 17–19.

———. 1979. *Acanthaster planci* (Linnaeus) (Asteroidea: Echinodermata) in the northern Tasman Sea. *NZOI Records* 4(4): 21–23.

Meyer, D. L., and D. B. Macurda. 1980. Ecology and distribution of the shallow-water crinoids of Palau and Guam. *Micronesica* 16(1): 59–99.

Ogden, N. B., J. C. Ogden, and I. A. Abbott. 1989. Distribution, abundance, and food of sea urchins on a leeward Hawaiian reef. *Bulletin of Marine Science* 45(2): 539–49.

Oguro, C. 1982. Notes on the sea-star fauna and a new species of *Astropecten* in Palau and Yap islands. *Proceedings of the Japanese Society of Systematic Zoology* 23:71–79.

———. 1983. Supplementary notes on the sea-stars from the Palau and Yap islands. 1. *Annotationes Zoologicae Japonensis* 56(3): 221–26.

Oguro, C., and Y. Sasayama. 1984. Occurrence of *Fromia nodosa* A. M. Clark (Asteroidea: Ophidiasteridae) from the Marshall Islands, the western Pacific. *Proceedings of the Japanese Society of Systematic Zoology* 27:101–6.

Paul, C. R. C., and A. B. Smith. 1988. *Echinoderm phylogeny and evolutionary biology.* Oxford: Clarendon Press.

Paulay, G. 1989. Marine invertebrates of the Pitcairn Islands: Species composition and biogeography of corals, molluscs, and echinoderms. *Atoll Research Bulletin,* no. 326:1–28.

Pawson, D. L. 1965. Some echinozoans from north of New Zealand. Transactions of the Royal Society of New Zealand. *Zoology* 5(15): 197–224.

Régis, M.-B., and J. M. Pérès. 1994. L'Avenir du marché international de la rogue d'oursin dans l'Indo-Pacifique. In *Echinoderms through time*, ed. B. David, A. Guille, J.-P. Féral, and M. Roux, 845–48. Rotterdam: A. A. Balkema.

Rowe, F. W. E. 1981. A new genus and species in the Family Ophidiasteridae (Echinodermata: Asteroidea) from the vicinity of Lord Howe Island, Tasman Sea. *Proceedings of the Linnaean Society of New South Wales* 105(2): 90–94.

———. 1985. Six new species of *Asterodiscides* A. M. Clark (Echinodermata: Asteroidea), with a discussion of the origin and distribution of the Asterodiscididae and other "amphi-Pacific" echinoderms. *Bulletin Museum d'Histoire Naturelle Paris*, 4th series 3:531–77.

———. 1989. Nine new deep-water species of Echinodermata from Norfolk Island and Wanganella Bank, northeastern Tasman Sea, with a checklist of the echinoderm fauna. *Proceedings of the Linnaean Society of New South Wales* 111(4): 257–91.

Rowe, F. W. E., and L. Albertson. 1987. A new species in the echinasterid genus *Echinaster* Müller and Troschel, 1840 (Echinodermata: Asteroidea) from southeastern Australia and Norfolk Island. *Proceedings of the Linnaean Society of New South Wales* 109(3): 195–202.

Rowe, F. W. E., and J. E. Doty. 1977. The shallow-water holothurians of Guam. *Micronesica* 13(2): 217–50.

Rowe, F. W. E., and D. Nichols. 1980. A new species of *Podosphaeraster* Clark and Wright, 1962. (Echinodermata: Asteroidea) from the Pacific. *Micronesica* 16(2): 289–95.

Shigei, M. 1970. Echinoids of the Bonin Islands. *Journal of the Faculty of Science*, University of Tokyo (4)12: 1–25.

Smith, A. B., G. Paterson, and B. Lafay. 1994. Morphological and molecular phylogeny of ophiuroids. In *Echinoderms through time*, ed. B. David, A. Guille, J.-P. Féral, and M. Roux, 489. Rotterdam: A. A. Balkema.

Springer, V. G. 1982. Pacific Plate biogeography, with special reference to shorefishes. Smithsonian Contributions to Zoology No. 367:1–182.

Springer, V. G., and J. T. Williams. 1990. Widely distributed Pacific Plate endemics and lowered sea level. *Bulletin of Marine Science* 47(3): 631–40.

Tsuchiya, M., and M. Nishihira. 1984. Ecological distribution of two types of the sea-urchin *Echinometra mathaei* (Blainville), on Okinawan reef flat. *Galaxea* 4:37–48.

Wilkinson, C. R., ed. 1990. Special issue *Acanthaster planci*. *Coral Reefs* 9(3): 93–171.

Wilkinson, C. R., and I. G. MacIntyre, eds. 1992. Special issue: The *Acanthaster* debate. *Coral Reefs* 11(2): 51–124.

Yamaguchi, M. 1973. Recruitment of coral reef asteroids, with emphasis on *Acanthaster planci*. *Micronesica* 9(2): 207–12.

———. 1974. Larval life span of the coral reef asteroid *Gomophia egyptiaca* Gray. *Micronesica* 10(1): 57–64.

———. 1975. Coral-reef asteroids of Guam. *Biotropica* 7(1): 12–23.

———. 1977. Larval behavior and geographic distribution of coral reef asteroids in the Indo-West Pacific. *Micronesica* 13(2): 283–96.

Yamaguchi, M., and J. S. Lucas. 1984. Natural parthenogenesis, larval and juvenile development, and geographical distribution of the coral reef asteroid *Ophidiaster granifer*. *Marine Biology* 83:33–42.

Yanagisawa, T., I. Yasumasu, C. Oguro, N. Suzuki, and T. Motokawa, eds. 1991. *Biology of Echinodermata*. Rotterdam: A. A. Balkema.

CHAPTER 11

Zoogeographic Analysis of the Inshore Hawaiian Fish Fauna

John E. Randall

In 1905 Jordan and Evermann documented 441 inshore and surface-dwelling species of fishes in the Hawaiian Islands. Gosline and Brock (1960) recorded 448 species of native inshore and surface-dwelling fishes for the islands, hence an apparent increase of only seven species. However, the real increase was far greater because many of the species listed by Jordan and Evermann proved to be synonyms (due especially to their not recognizing juvenile stages or sexually dichromatic phases) as well as some erroneous records.

For this report, only the reef and shore fishes of Hawaii occurring on the insular shelf above a depth of 200 m will be considered. The offshore pelagic species such as the billfishes, tunas (except for *Euthynnus affinis*, which is often observed inshore), mahi-mahi, flyingfishes, and lanternfishes are not treated.

The number of species of reef and shore fishes now known in Hawaii is 557, disregarding three snappers, a grouper, two sardines, a mullet, a goatfish, a goby, and two species of tilapia (*Oreochromis*) that have become established in the Hawaiian marine environment (marginally for the tilapia) as a result of introductions (Maciolek 1984; Randall 1987; Randall et al. 1993). In addition, a blenny appears to be an unintentional introduction (Springer 1991). Not considered here as Hawaiian species of fishes are those found at Johnston Island (which has a predominantly Hawaiian fish fauna) that do not range north to the Hawaiian Islands.

These 557 Hawaiian species of fishes are representatives of 99 families (Springer 1982 listed 98, but one specimen of an ephippid fish has since been reported, as noted below). The most speciose is the wrasse family (Labridae) with 42 species in Hawaii, followed by the moray eels (Muraenidae) with 38 species, the gobies (Gobiidae) with 27, the surgeonfishes (Acanthuridae) with 24, the jacks (Carangidae) and butterflyfishes (Chaetodontidae) each with 23, and the scorpionfishes (Scorpaenidae) with 22.

The number of species of the various families of fishes in the Hawaiian Islands is often out of proportion to the number of species of these families in the rest of the Indo-Pacific region in general. Away from Hawaii, the Gobiidae nearly always has far more species than any other family; thus, 27 species in Hawaii is a poor representation for the group. Other large families represented by relatively few species in Hawaii are the damselfishes (Pomacentridae) with 17 species, the blennies (Blenniidae) with 13 species, and the cardinalfishes (Apogonidae) with 10. The fishes of the first three of these families lay demersal eggs, and those of the fourth are mouth-brooders. Therefore, they may generally be expected to have a relatively short planktonic larval life. Considering the vast distance a larval fish must travel in an ocean current to reach the Hawaiian Islands from any locality, the paucity of species of some of these large families is more understandable. Limited data on the length of larval life of gobies and cardinalfishes (Brothers and Thresher 1985) suggest a short period. That of damselfishes is variable but generally short; Wellington and Victor (1989) and Thresher, Colin, and Bell (1989) reported that it varied from 7 to 47 days. As noted by Brothers and Thresher (1985), the nonendemic Hawaiian damselfishes are those with the longest larval life.

The Labridae, with an estimated 500 species worldwide, is second only to the Gobiidae among marine fishes in family size, so its having the largest number of species in Hawaii might be expected. However, from the size of the family in the tropical Indo-Pacific region, one would expect more species to occur in Hawaiian waters. The wrasses are among those families exhibiting a wide range in the duration of larval life. Victor (1986) showed that it varied from 15 to 120 days for the 100 species he studied. The species of the genera *Thalassoma* and *Xyrichtys* have the longest larval life. These two genera are the best represented in Hawaii with six and five species, respectively.

Among those families of fishes with short larval life, there are species with unusually broad distributions. An example from the cardinalfishes is *Apogon kallopterus*, which is found throughout most of the Indo-Pacific region. Bruce C. Mundy (pers. com.) has identified small juveniles of this species in midwater trawls 5 to 15 nautical miles from land, so it may be able to transform and still survive in the pelagic realm. Even more surprising is the distribution of *Apogon evermanni*, which is known from Hawaii to Mauritius and the West Indies; it may be a relic from Miocene time.

Mention should be made of some shore fishes such as carangids, *Lobotes surinamensis*, and species of *Kyphosus* and *Abudefduf* that are sometimes found associated as juveniles with floating *Sargassum*, logs, or other drifting objects on the surface of the sea (also in the case of carangids as commensals with jellyfishes). Therefore, they may attain a broader distribution by rafting than would be possible as larvae.

Conversely, some families of fishes such as the moray eels and surgeonfishes have more species in Hawaii than would be expected from the size of these families within the Indo-Pacific region. The fishes of these two families attain a large larval size, and most that have been studied have a long sojourn in the pelagic realm (Brothers and Thresher 1985).

Also unique and noteworthy for the Hawaiian fish fauna is the absence of many large, widespread, shallow-water families and genera of bony fishes: the toadfishes (Batrachoididae); the flatheads (Platycephalidae); groupers of the subfamily Epinephelinae, except for two species of *Epinephelus* (the very large and rare *E. lanceolatus* and the deepwater *E. quernus*); the glassfishes (Ambassidae); the dottybacks (Pseudochromidae); the longfins (Plesiopidae); the terapons (Teraponidae); snappers of the genera *Lutjanus*, *Macolor*, and *Paracaesio*; the fusiliers (Caesionidae); the grunts (Haemulidae); the emperors of the genus *Lethrinus*; the breams (Nemipteridae); the porgies (Sparidae); the whitings

(Sillaginidae); the drums and croakers (Sciaenidae); the mojarras (Gerreidae); the pony-fishes (Leiognathidae); the sweepers (Pempheridae); the rabbitfishes (Siganidae); the jawfishes (Opistognathidae); and the clingfishes (Gobiescocidae). Also there are no skates (Rajidae) in Hawaii and only five inshore representatives of 13 families of rays.

Some of these groups are entirely or primarily continental in their distribution; they are absent from other oceanic islands of the Pacific as well as Hawaii. For stenothermal tropical species, the cool Hawaiian winter and early spring sea temperature may be limiting. As discussed earlier, a short larval life could also be the explanation for some groups. There is more to be considered, however, with respect to larval development than duration in the plankton. As noted by Leis (1986), all the ecological requirements of the larvae of fishes need to be addressed when assessing the dispersal capabilities of species. There may be a need for development in water of lower salinity or in plankton-rich inshore water. The wide-ranging and abundant small grouper *Epinephelus merra* was introduced to the Hawaiian Islands from the Society Islands in 1956; 469 fish were released into Kaneohe Bay, Oahu, an environment approximating their usual shallow-water lagoon habitat, and another 132 at the island of Kauai. The introduced fish lived out their lives in Hawaiian waters, but no juveniles were ever found. Because this fish occurs in cooler seas than Hawaii in southern Japan and Rapa, temperature would not seem to be the limiting factor. Something lacking among the requirements for developing larvae may have been the reason for its failure to become established.

There are eight zoogeographic categories of the native inshore Hawaiian fish fauna: circumglobal (including those fishes absent from the eastern Pacific; they may have been present when the western Atlantic and eastern Pacific were confluent), wide-ranging Indo-Pacific, eastern tropical Pacific, Japan to Taiwan, antitropical, central Pacific, waifs, and endemics.

The circumglobal species are those of tropical and/or subtropical seas of the Indo-Pacific and Atlantic (and for some, the eastern Pacific as well). The species of this category, which occur in Hawaiian waters, are the moray eel *Monopenchelys acuta*; the lizardfish *Trachinocephalus myops*; the round herring *Etrumeus teres*; the frogfishes *Antennarius nummifer* and *Histrio histrio* (though this species is usually pelagic); the needlefishes *Ablennes hians*, *Platybelone argalus* (regarded as subspecifically different in the Atlantic), *Tylosurus acus*, and *T. crocodilus*; the cornetfish *Fistularia petimba*; the boarfish *Antigonia capros*; the tripletail *Lobotes surinamensis*; the bigeyes *Cookeolus japonicus* and *Heteropriacanthus cruentatus*; the cardinalfish *Apogon evermanni*; the jacks *Alectis ciliaris*, *Caranx lugubris*, *Decapterus macarellus*, *D. tabl*, *Elagatis bipinnulata*, *Pseudocaranx dentex*, *Selar crumenophthalmus*, *Seriola lalandi*, *S. rivoliana*, and *Uraspis secunda* (*P. dentex* and *S. lalandi* are also antitropical); the mullet *Mugil cephalus*; the triggerfish *Melichthys niger*; the filefishes *Aluterus monoceros* and *A. scriptus*; and the porcupinefishes *Chilomycterus reticulatus*, *Diodon holocanthus*, and *D. hystrix*. Also in this category are the spotted eagle ray *Aetobatis narinari* and eight sharks that may be found in inshore waters, but their distribution depends on their ability to move long distances in the pelagic realm as adults or juveniles (though such open-ocean movements may occur only rarely).

The Indo-Pacific component represents those Hawaiian fishes that occur broadly in the tropical and subtropical Indo-Pacific region. Strictly speaking, this category should include species that occur in the Indian Ocean as well as the Pacific, but with respect to the occurrence of a widely distributed Pacific fish in Hawaii, it does not matter whether it also extends its distribution to the Indian Ocean; therefore, Indo-Pacific as here understood includes species broadly distributed in the Pacific, whether they occur in the Indian Ocean or not (or whether they extend their range to the eastern Pacific as well). This category of 343 species—hence 61.3 percent of the total inshore fish fauna of Hawaii—is much the largest component.

There seem to be only three species of shore fishes common to both the eastern Pacific and the Hawaiian Islands as breeding populations—the stingray *Dasyatis brevis* (reported by Nishida and Nakaya 1990), the bigeye *Priacanthus alalaua* (family revision by Starnes 1988), and the pearlfish *Encheliophis dubius* (distribution from Markle and Olney 1990). Two other eastern Pacific species appear to reach Hawaii only as strays (see below).

The following inshore species of fishes are shared by the Hawaiian Islands and the region from southern Japan to Taiwan, but not areas to the south: the dogfish shark *Squalus mitsukurii*; the stingray *Dasyatis latus*; the snake eel *Ophichthus erabo*; the lizardfishes *Synodus ulae* and *S. usitatus*; the scorpionfish *Rhinopias xenops*; the armored searobin *Satyrichthys engyceros*; the roughy *Paratrachichthys prosthemius*; the serranids *Caprodon schlegelii, Liopropoma maculatum, Plectranthias helenae,* and *Pseudanthias thompsoni*; the jack *Seriola quinqueradiata*; the slopefish *Symphysanodon maunaloae*; the angelfish *Centropyge interruptus*; the armored boarfish *Pseudopentaceros wheeleri*; the knifejaws *Oplegnathus fasciatus* and *O. punctatus*; the wrasse *Novaculichthys woodi*; the sand perch *Parapercis roseoviridis*; the dragonets *Callionymus corallinus* and *Synchiropus rubrovinctus*; the flatfish *Parabothus coarctatus*; the trunkfish *Kentrocapros aculeatus*; and the puffer *Canthigaster inframacula*. The moray *Gymnothorax ypsilon* and the wrasse *Bodianus cylindriatus* are recorded only from Japan and Hawaii, but they are known only at depths greater than 200 m in the Hawaiian Islands.

Antitropical or antiequatorial species of fishes are here regarded as a separate category of the Hawaiian Islands fish fauna, though all the species are shared with other categories such as the two circumglobal jacks mentioned earlier. Most are tallied as Indo-Pacific species. Antitropical fishes found in Hawaii include the moray *Gymnothorax eurostus*; the lizardfish *Synodus capricornis*; the soldierfishes *Ostichthys archiepiscopus, O. sandix,* and *Pristilepis oligolepis*; the basslet *Plectranthias kelloggi*; the morwong *Cheilodactylus vittatus*; the snapper *Randallichthys filamentosus*; the jacks *Carangoides dasson, Pseudocaranx dentex,* and *Seriola lalandi*; the rovers *Emmelichthys karnellai, E. struhsakeri,* and *Erythrocles scintillans*; the stripey *Microcanthus strigatus*; the bearded boarfish *Evistias acutirostris*; the sea chub *Kyphosus bigibbus*; the butterflyfish *Hemitaurichthys thompsoni*; the bothid flatfish *Engyprosopon arenicola*; the gobies *Kelloggella oligolepis* and *Trimma unisquamis*; the surgeonfishes *Acanthurus leucopareius* and *Naso maculatus*; and the triggerfish *Xanthichthys mento*. Three more antitropical species found in Hawaii are unique in having both north-south and east-west disjunct populations: the moray eel *Gymnothorax berndti*, reported from Hawaii, Japan, Taiwan, Mauritius, and Réunion (Kailola 1975); the cardinalfish *Lachneratus phasmaticus*, described from the Hawaiian Islands, Fiji, and Comoro Islands by Fraser and Struhsaker (1991); and the hawkfish *Cirrhitops fasciatus*, known only from the Hawaiian Islands, Mauritius, and Madagascar.

Antitropical distributions can be explained in three ways: the tropical zone may have been transgressed by subtropical or temperate organisms at the time of the ice ages when the seas were cooler (most recently during the Pleistocene). Although recent studies have shown that the cooling of sea temperatures in the Pacific during the Pleistocene may have been only 1–2°C, this could be enough to result in a break in the distribution of those subtropical species barely able to extend their range across the tropical zone during an ice age (Randall 1982). The second explanation involves species descending into the cooler, deeper water of the tropics, thus resulting in an apparent north-south break of the shallow-water distribution. This may take place for some pelagic species such as the blue shark (*Prionace glauca*) and the albacore (*Thunnus alalunga*), but it is not likely for Pacific shore fishes. The third explanation is the occurrence of shallow-water reefs and islands forming north-south stepping stones across the tropics in the past. Because of subsidence, plate movement, or sea-level rise, these islands and reefs are no longer present or are too deep to support shallow-water organisms, resulting in the north-south discontinuities in some of the distributions of today.

The aforementioned antitropical species *Synodus capricornis*, *Emmelichthys karnellai*, *Engyprosopon arenicola*, and *Kelloggella oligolepis* have their north-south disjunct distributions confined to eastern Oceania. *S. capricornis* is known only from the Hawaiian Islands, Easter Island, and Pitcairn; *E. karnellai* from Hawaii, Easter, and Rapa, and the flatfish and goby only from Hawaii and Easter. These are also representatives of the next component of the Hawaiian fish fauna that is termed central Pacific. Springer (1967) regarded the Hawaiian endemic blenny *Entomacrodus strasburgi* to be most closely related to the Easter Island endemic *E. chapmani*, so it may be included, though not at the species level. There are also north-south distributions confined to the eastern part of the central Pacific that are not antitropical. These include the snake eel *Apterichthys flavicaudus*, recorded from Hawaii, Marquesas, Society Islands, and Rapa; the tiny goby *Eviota epiphanes* known only from Hawaii, Johnston, and the Line Islands (Lachner and Karnella 1980); the scorpionfish *Sebastapistes coniorta* with the same distribution (Eschmeyer and Randall 1975); the trunkfish *Ostracion whitleyi* from Hawaii, Johnston, Marquesas, Tuamotus, and Society Islands (Randall 1972); and the halfbeak *Hemiramphus depauperatus*, which occurs only in the Hawaiian, Line, and Marquesas islands (Parin, Collette, and Shcherbachev 1980). Rehder (1980) noted that ten of the 67 nonendemic species of mollusks of Easter Island are otherwise known only from the Hawaiian Islands. Based on present island locations and current patterns, these distributions are difficult to explain. We may postulate that there were more shallow-water areas in intermediate locations in the central Pacific than today or that some island groups were closer together than they are today. Rehder (1980) wrote, "This could well have been when many of the submarine mountain ranges, seamounts, and guyots present now in the central Pacific were at or near the surface and some 25 degrees to the southeast of their present location" (Ladd 1960, 148; Ladd, Newman, and Sohl 1974, 518).

Rotondo et al. (1981) proposed "island integration" to explain the north-south disjunct distributions such as those of the marine animals restricted to Easter Island and Hawaii. According to this hypothesis, "Other island groups formed well to the SE on the Pacific plate close to the East Pacific Rise. These latter islands shared a biota that was not present at/on the emergent Hawaiian Islands of the time. . . . Ultimately, these emergent islands integrated with the emergent Hawaiian Islands, thus combining non-Hawaiian biotas with the Hawaiian." Newman and Foster (1983) and Newman (1986) rejected this hypothesis on both geological and biological grounds, stating, "There are no early Tertiary, much less Cretaceous, high islands on the Pacific Plate today." And the similarities of the Hawaiian fauna with the eastern South Pacific "are at specific rather than generic or higher taxonomic levels."

Chave and Mundy (1994) remarked on the similarity of the composition of the deep-water fish faunas of Hawaii and the Nazca and Sala y Gómez Ridges in the southeastern Pacific, particularly in the Macrouridae. There are, of course, different patterns of dispersal for deep-water fishes from those of shallow-water fishes, the most obvious of which is the availability of the deeper seamounts for dispersal. Also, as pointed out by Chave and Mundy, the larvae of many deep-sea benthic fishes, including macrourids, are rarely found above the thermocline. The currents transporting these larvae may be very different in direction and strength from the wind-driven currents of the surface.

The Hawaiian category called waifs is for those species that occur so rarely that they are probably not represented by breeding populations in the islands. These include two species from the eastern Pacific: the rudderfish *Sectator ocyurus* and the triggerfish *Balistes polylepis*. The soldierfish *Myripristis occidentalis* (= *M. leiognathus*) has been listed as a possible waif from the eastern Pacific by Gosline and Brock (1960, 26, 321) based on 12 specimens, allegedly from Hawaii, described as *M. sealei* by Jenkins (1903). Gosline and Brock suspected that the 12 specimens were not collected in Hawaii, and Greenfield (1965) agreed. Gosline and Brock also listed the tripletail *Lobotes surinamensis* as a possible immigrant from Panama, but it could more easily have drifted in from southern Japan (Springer 1982). The following Indo-Pacific species of fishes appear to be waifs: the spadefish *Platax boersii* (reported from Midway Atoll by Randall et al. 1993), the butterflyfish *Chaetodon ulietensis* (two individuals reported by Randall et al. 1993), the angelfish *Pomacanthus imperator* (first mentioned by Gosline and Brock 1960, 26, 197), the wrasse *Halichoeres marginatus* (one individual from the island of Hawaii reported by Randall 1981), the surgeonfish *Acanthurus lineatus* (one listed from vicinity of South Point, Hawaii, by Randall 1981, and a second individual recently photographed by John P. Hoover off Hanauma Bay, Oahu), the trunkfish *Ostracion cubicus* (one specimen reported by Fowler 1923, and a juvenile observed by E. H. Chave, pers. com.), the toby *Canthigaster solandri* (two specimens reported from the Hawaiian Islands by Allen and Randall 1977), and *Arothron manilensis* (one individual reported from Oahu by Randall et al. 1993).

Some Indo-Pacific fishes such as the grouper *Cromileptes altivelis*; the spadefish *Platax orbicularis*; the angelfishes *Apolemichthys xanthopunctatus*, *Centropyge flavissimus*, and *C. multicolor*; and the surgeonfishes *Paracanthurus hepatus* and *Zebrasoma rostratum* have been found in Hawaiian waters, but these are most likely aquarium releases (Richard L. Pyle, pers. com.). One individual of *Platax orbicularis* was recently discovered in the Florida Keys—hence an obvious aquarium release. The emperor angelfish, which was listed as a waif earlier, is a popular aquarium fish, especially when young. However, its record from Hawaii would seem to be valid as a natural occurrence because it was based on a fish discovered on the Kona coast of Hawaii in 1948, hence before there was a flourishing trade in marine aquarium fishes.

Gosline and Brock (1960) reported 34 percent of the reef fishes of the Hawaiian Islands as endemics. Randall (1992) revised the percentage of endemism in Hawaii to 25 percent after finding some of these endemics at localities away from Hawaii and including the many new records of Indo-Pacific fishes for the islands. With the same trends continuing, the current level is now 24.3 percent, still slightly higher than the 23 percent level of endemic shore fishes at Easter Island. The third highest area of endemism within the Indo-Pacific region is South Africa with 13 percent (Smith and Heemstra 1986).

It is possible to find a closely related species for many of the endemic Hawaiian fishes from which one might assume they evolved (or that the two had a common ancestor). Examples of such apparent sister species are as follows (the Hawaiian member of each pair is given last): the morays *Gymnothorax kidako* and *G. mucifer*; the bigeyes *Priacanthus hamrur* and *P. meeki*; the butterflyfishes *Chaetodon guentheri* and *C. miliaris*; the damselfishes *Dascyllus trimaculatus* and *D. albisella*; the wrasses *Anampses caeruleopunctatus* and *A. cuvier*; the parrotfishes *Calotomus japonicus* and *C. zonarchus*; the blennies *Cirripectes variolosus* and *C. obscurus*, and *Istiblennius bellus* and *I. zebra*; two pairs of sabertooth blennies: *Plagiotremus rhynorhynchus* and *P. ewaensis*, and *P. tapeinosoma* and *P. goslinei*; the filefishes *Cantherhines pardalis* and *C. sandwichiensis*; and the puffers *Canthigaster janthinoptera* and *C. jactator*. Systematists sometimes differ whether to regard a Hawaiian member of such pairs as a species or a subspecies. For tallying endemics, subspecies are here given the same rank as species.

Other Hawaiian endemic fishes have no apparent close relatives and may be relics. Perhaps they were more broadly distributed in the past but now survive only in Hawaii. Or they may have been very early colonizers to the islands and diverged in Hawaii from the progenitor stock that may no longer be extant or recognizable. The Hawaiian Islands have been forming intermittently over a relatively fixed site of volcanism for more than 70 million years (Rotondo et al. 1981). As the Pacific Plate moved to the northwest, the islands of the Hawaiian-Emperor chain eroded to reefs and ultimately subducted as seamounts under the Asian continent, but the resident fishes were able to disperse the short distance to the east to the newly emerging islands. Examples of apparent relic Hawaiian fishes are the scorpionfish *Pterois sphex*, the grouper *Epinephelus quernus* (whose closest relatives are *E. niveatus* of the western Atlantic and *E. niphobles* of the eastern Pacific), the butterflyfish *Chaetodon fremblii*, the angelfishes *Centropyge potteri* and *Genicanthus personatus* (both appear to be the most primitive members of their genera and have no close relatives), the wrasse *Coris flavovittata*, and the parrotfish *Chlorurus perspicillatus* (the placement of this species in the genus *Chlorurus* is based on a study by Bellwood 1994).

As noted by Gosline and Brock (1960) and Randall (1976), the endemic fishes of Hawaii are usually among the most abundant within their genera. The squirrelfish *Sargocentron xantherythrum*, the butterflyfish *Chaetodon miliaris*, the angelfish *Centropyge potteri*, the damselfish *Chromis ovalis*, the wrasse *Thalassoma duperrey*, the surgeonfish *Acanthurus triostegus sandvicensis*, and the puffer *Canthigaster jactator* are obvious examples. Whether they are derivatives of widespread Indo-Pacific species or are relics, these successful fishes have probably existed in the Hawaiian Islands for a long time and hence have had ample opportunity to become well adapted to the environment.

The great majority of the native reef and shore fishes (disregarding sharks and rays) of Hawaii must have arrived in the islands as a result of passive transport as larvae in ocean currents. It would be interesting to speculate by what route or routes these immigrants made their initial journey. The prevailing ocean current that now bathes the Hawaiian chain is the westward-moving North Equatorial Current. Any Hawaii-bound larvae of shallow-water fishes in this current system would have to come from California or northern Mexico. Two major problems face such larvae. First, they would have to survive transport for a distance well over 4,000 km, and second, the great majority would find the sea in Hawaii too warm. That only three resident species of fishes are common to both the eastern Pacific and the Hawaiian Islands is indicative that this immigration route is of negligible importance.

The closest island group to Hawaii today is the Line Islands, which lie directly south of the main Hawaiian Islands between latitudes 2° and 6° N; however, the surface currents between these islands and Hawaii would result in transport to the east (Counterequatorial Current) or to the west (North Equatorial Current). There would be a much greater chance of fish larvae from the Line Islands reaching Hawaii if Johnston Island, a biogeographic outlier of the Hawaiian Islands to the northwest of the Line Islands at 16°45' N, were a stepping stone. Gosline (1955) favored the Line Islands as the major source of Indo-Pacific fishes immigrating to the Hawaiian Islands, with Johnston Island serving as a way station or, in his term, "filter bridge." For assessing this role of Johnston, he divided the fishes of the atoll into three groups: those Central Pacific species reaching Johnston but not Hawaii (14 species), those Hawaiian species reaching south to Johnston but not beyond (16 species), and those found both in Hawaii and in the Central Pacific (86 species). He wrote, "Of those tropical fishes that have reached Johnston, the great majority seem to have passed on through to Hawaii." Randall, Lobel, and Chave (1985) and Kosaki et al. (1991), however, have pointed out that our expanding knowledge of the systematics of Indo-Pacific fishes and their distributions substantially modified Gosline's three categories. More important, as noted by Randall, Lobel, and Chave (1985), the present-day current patterns clearly indicate that the most likely route of immigration of shallow-water fishes to Hawaii would be from the islands of southern Japan, beginning with the Kuroshio

Current and arriving via eddies at shallow seamounts, low islands, and reefs at the northwestern end of the Hawaiian chain.

It is difficult to say by what route any of the wide-ranging Indo-Pacific species of fishes arrived in Hawaii. That some may have originated in the Line Islands seems possible for a few species that are known today from these islands and Hawaii but not from southern Japan, such as the angelfish *Centropyge loriculus*, the surgeonfish *Acanthurus achilles*, the blenny *Cirripectes quagga*, and the aforementioned goby *Eviota epiphanes*, and scorpionfish *Sebastapistes coniorta*. Unfortunately, the fish fauna of the Line Islands is not as well known as that of Japan, Hawaii, and most island groups of the Pacific, so it is premature to list species that are missing from the fauna but present at localities such as islands of the western Pacific. Even if one could confidently demonstrate that a certain species of fish is not present in the Line Islands today, it could have been present in the past.

Whereas most of the Hawaiian fishes may well have arrived in the islands via present-day or very similar current patterns, some of the distributions can be explained only in terms of the geological history of the Pacific Plate, as stressed by Kay (1980), Rehder (1980), Rotondo et al. (1981), Springer (1982), and Newman (1986).

Acknowledgments—Thanks are due E. H. Chave, John L. Earle, David W. Greenfield, Richard L. Pyle, Victor G. Springer, Arnold Y. Suzumoto, Jeffrey M. Williams, and especially Bruce C. Mundy for their critical review of the manuscript.

REFERENCES

Allen, G. R., and J. E. Randall. 1977. Review of the sharpnose pufferfishes (subfamily Canthigasterinae) of the Indo-Pacific. *Records of the Australian Museum* 30(17): 475–517.

Bellwood, D. R. 1994. A phylogenetic study of the parrotfishes family Scaridae (Pisces: Labroidei), with a revision of genera. *Records of the Australian Museum*, Suppl. 20:1–86.

Brothers, E. B., and R. E. Thresher. 1985. Pelagic duration, dispersal and the distribution of Indo-Pacific coral-reef fishes. Vol. 2, *The ecology of deep and shallow coral reefs*, ed. M. Reaka, 53–93. NOAA Symposium Series Undersea Res., U.S. Department of Commerce, Washington, D.C.

Chave, E. C., and B. C. Mundy. 1994. Deep-sea benthic fish of the Hawaiian Archipelago, Cross Seamount, and Johnston Atoll. *Pacific Science* 48(4): 367–409.

Eschmeyer, W. N., and J. E. Randall. 1975. The scorpaenid fishes of the Hawaiian Islands, including new species and new records (Pisces: Scorpaenidae). *Proceedings of the California Academy of Science* 40(11): 265–334.

Fowler, H. W. 1923. New or little-known Hawaiian fishes. *Bernice P. Bishop Museum Occasional Paper* 8:375–92, Honolulu.

Fraser, T. H., and P. J. Struhsaker. 1991. A new genus and species of cardinalfish (Apogonidae) from the Indo-West Pacific, with a key to the Apogonine genera. *Copeia*, no. 3:718–22.

Gosline, W. A. 1955. The inshore fish fauna of Johnston Island, a central Pacific atoll. *Pacific Science* 9(4): 442–80.

Gosline, W. A., and V. E. Brock. 1960. *Handbook of Hawaiian fishes*. Honolulu: University of Hawaii Press.

Greenfield, D. W. 1965. Systematics and zoogeography of *Myripristis* in the eastern tropical Pacific. *California Fish and Game* 51(4): 229–47.

Jenkins, O. P. 1903. Report on collections of fishes made in the Hawaiian Islands, with descriptions of new species. *Bull. U.S. Fish Commission* 22 (1902): 417–511.

Jordan, D. S., and B. W. Evermann. 1905. The aquatic resources of the Hawaiian Islands. Part I—The shore fishes. *Bull. U.S. Fish Commission* 23:xxvii + 574 pp.

Kailola, P. J. 1975. The rare moray eel *Gymnothorax pikei* Bliss recorded from Papua New Guinea. *Pacific Science* 29(2): 165–70.

Kay, E. A. 1980. *Little worlds of the Pacific: An essay on Pacific Basin biogeography.* University of Hawaii, Harold L. Lyon Arboretum Lecture 9:1–40.

Kosaki, R. K., R. L. Pyle, J. E. Randall, and D. K. Irons. 1991. New records of fishes from Johnston Atoll, with notes on biogeography. *Pacific Science* 45(2): 186–203.

Lachner, E. A., and S. J. Karnella. 1980. Fishes of the Indo-Pacific genus *Eviota* with descriptions of eight new species (Teleostei: Gobiidae). *Smithsonian Contributions to Zoology*, No. 315.

Ladd, H. S. 1960. Origin of the Pacific island molluscan fauna. *American Journal of Science, Bradley Volume* 258-A:137–50.

Ladd, H. S., W. A. Newman, and N. F. Sohl. 1974. Darwin Guyot, the Pacific's oldest atoll. *Proceedings of the Second International Coral Reef Symposium* 2:513–22.

Leis, J. M. 1986. Ecological requirements of Indo-Pacific larval fishes: A neglected zoogeographic factor. In *Indo-Pacific fish biology: Proceedings of the Second International Conference on Indo-Pacific Fishes*, ed. T. Uyeno, R. Arai, T. Taniuchi, and K. Matsuura, 759–66. Tokyo: Ichthyological Society of Japan.

Maciolek, J. A. 1984. Exotic fishes in Hawaii and other islands of Oceania. In *Distribution, biology and management of exotic fishes*, ed. W. R. Courtenay Jr. and J. R. Stauffer Jr., 131–61. Baltimore: Johns Hopkins University Press.

Markle, D. F., and J. E. Olney. 1990. Systematics of the pearlfishes (Pisces: Carapidae). *Bull. Marine Science* 47(2): 269–410.

Newman, W. A. 1986. Origin of the Hawaiian marine fauna: Dispersal and vicariance as indicated by barnacles and other organisms. In *Crustacean biogeography*, ed. R. H. Gore and K. L. Heck, 21–49. Rotterdam: A. A. Balkema.

Newman, W. A., and B. A. Foster. 1983. The Rapanuian fauna district (Easter and Sala y Gómez): In search of ancient archipelagos. *Bull. Marine Science* 33:633–44.

Nishida, K., and K. Nakaya. 1990. Taxonomy of the genus *Dasyatis* (Elasmobranchii, Dasyatididae) from the North Pacific. *NOAA Technical Report*, NMFS 90, 327–46.

Parin, N. V., B. B. Collette, and Y. N. Shcherbachev. 1980. Preliminary review of the marine halfbeaks (Hemirhamphidae, Beloniformes) of the tropical Indo-West-Pacific. *Transactions of the P. P. Shirshov Institute of Oceanology* 97:7–173 (in Russian with English summary).

Randall, J. E. 1972. The Hawaiian trunkfishes of the genus *Ostracion*. *Copeia*, no. 4:756–68.

———. 1976 . The endemic shore fishes of the Hawaiian Islands, Lord Howe Island and Easter Island. *Trav. Doc. ORSTOM*, no. 47:49–73.

———. 1981. New records of fishes from the Hawaiian Islands. *Pacific Science* (1980) 34(3): 211–32.

———. 1982. Examples of antitropical and antiequatorial distribution of Indo-West-Pacific fishes. *Pacific Science* (1981) 35(3): 197–209.

———. 1987. Introduction of marine fishes to the Hawaiian Islands. *Bull. Marine Science* 41(2): 490–502.

———. 1992. Endemism of fishes in Oceania. In *Coastal resources and systems of the Pacific Basin: Investigation and steps toward protective management*, 55–67. UNEP Regional Seas Reports and Studies, No. 147.

Randall, J. E., J. L. Earle, T. Hayes, C. Pittman, M. Severns, and R. J. F. Smith. 1993. Eleven new records and validations of shore fishes from the Hawaiian Islands. *Pacific Science* 47(3): 222–39.

Randall, J. E., P. S. Lobel, and E. H. Chave. 1985. Annotated checklist of the fishes of Johnston Island. *Pacific Science* 39(1): 24–80.

Rehder, H. A. 1980. The marine mollusks of Easter Island (Isla de Pascua) and Sala y Gómez. *Smithsonian Contributions to Zoology,* No. 289.

Rotondo, G. M., V. G. Springer, G. A. F. Scott, and S. O. Schlanger. 1981. Plate movement and island integration: A possible mechanism in the formation of endemic biotas, with special reference to the Hawaiian Islands. *Syst. Zool.* 30(1): 12–21.

Smith, M. M., and P. C. Heemstra, eds. 1986. *Smiths' sea fishes.* Johannesburg, South Africa: Macmillan.

Springer, V. G. 1967. Revision of the circum-tropical shorefish genus *Entomacrodus* (Blenni-idae: Salariinae). *Proceedings of the U.S. National Museum* 122(3582): 1–150.

———. 1982. Pacific Plate biogeography, with special reference to shore fishes. *Smithsonian Contributions to Zoology,* No. 367.

———. 1991. Documentation of the blenniid fish *Parablennius thysanius* from the Hawaiian Islands. *Pacific Science* 45(1): 72–75.

Starnes, W. C. 1988. Revision, phylogeny and biogeographic comments on the circumtropical marine percoid fish family Priacanthidae. *Bull. Marine Science* 43(2): 117–203.

Thresher, R. E., P. L. Colin, and L. J. Bell. 1989. Planktonic duration, distribution and population structure of western and central Pacific damselfishes (Pomadentridae). *Copeia,* no. 2:420–34.

Victor, B. C. 1986. Duration of the planktonic larval stage of one hundred species of Pacific and Atlantic wrasses (family Labridae). *Marine Biology* 90:317–26.

Wellington, G. M., and B. C. Victor. 1989. Planktonic larval duration of one hundred species of Pacific and Atlantic damselfishes (Pomacentridae). *Marine Biology* 101:557–67.

CHAPTER 12

Pacific Reef and Shore Fishes

Richard L. Pyle

ABSTRACT

Published and unpublished data on Indo-Pacific reef and shore fish species were entered on a computer database and analyzed for patterns of biodiversity. At least 3,392 species classified among 839 genera, 166 families, 30 suborders, 30 orders, and 2 classes have been reported from Pacific island localities within the region defined as Oceania. This suggests that previous estimates of the total number of Indo-Pacific reef and shore fish species are too low. The most frequently used taxonomic levels in marine ichthyological literature are family, genus, and species. The taxonomy of reef and shore fishes at these levels is relatively complete and stable compared with higher levels of taxonomic distinction, and with similar taxonomic levels of marine invertebrates. A declining rate of new species descriptions in recent years does not seem to be a result of fewer new species discoveries, because 13 percent of the known species remain undescribed. To utilize reef and shore fish species as indicators of patterns of biodiversity in marine environments, existing published and unpublished information needs to be combined in a single database management system that allows for analyzing species diversity across defined geographic regions. In addition, comprehensive species checklists of certain regions within Oceania are needed, as is further investigation of the fish fauna inhabiting the deep reefs.

INTRODUCTION

Reef and shore fish species are those that regularly inhabit or are usually associated with shoreline, lagoonal, and/or coral reef habitats at depths of less than about 150–200 meters (Figure 12.1). Excluded are species that are primarily pelagic or abyssal in their distribution or are restricted to freshwater environments. At depths in excess of 150–200 m, there is insufficient light penetration to sustain photosynthetic activity, and there appears to be a corresponding demarcation in the composition of species assemblages; very few non-pelagic species frequent depths both significantly less than, and significantly greater than,

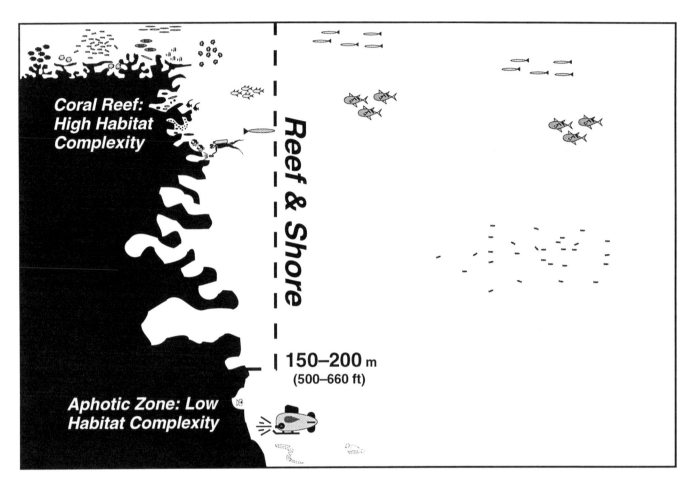

Figure 12.1 Reef and shore habitat

this transition depth zone (Strasburg, Jones, and Iverson 1968; Thresher and Colin 1986; Chave and Mundy 1994).

Estimates of the total number of reef and shore fish species worldwide have not been calculated with any degree of confidence. Springer (1982) estimated that more than 4,000 species occur within the tropical Indo-Pacific region, an area spanning more than half the circumference of the Earth from the Red Sea and western Indian Ocean eastward to Polynesia. He suggested that this may be an underestimate, and the findings presented in this chapter support this suggestion. Based on a reappraisal of these estimates in conjunction with discussions with colleagues, it appears that the total number of reef and shore fish species worldwide (including both the Indo-Pacific and Atlantic Ocean) is on the order of about 6,000, or one-quarter of the estimated 24,000 species of all fishes.

The islands of Oceania (Figure 12.2) represent the largest biogeographical province for marine reef and shore fishes within the Indo-Pacific. No single published work has listed or summarized all known valid species of reef and shore fishes in Oceania. Springer (1982) examined biogeographic patterns of reef and shore fishes on the Pacific Plate and adjacent regions, and estimated that 1,312 species among 461 genera were then known to occur

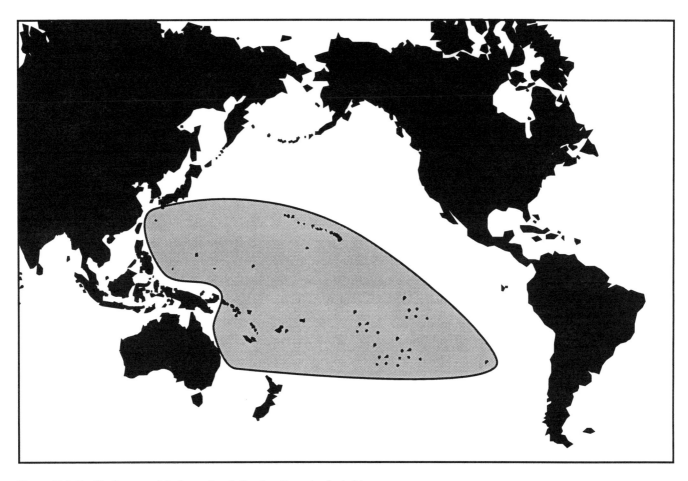

Figure 12.2 Pacific Ocean, with the region defined as Oceania shaded in gray

nonmarginally on the Pacific Plate. Myers (1991) provides a checklist of just over 1,400 species of inshore and epipelagic fishes known to occur in Micronesia, which includes most of the western half (and the most speciose region) of Oceania. Comprehensive regional checklists of reef and shore fish faunas of several other localities within Oceania, including the Ryukyu Islands, Ogasawara Islands, New Caledonia, Lord Howe Island, Norfolk Island, Kermadec Islands, Midway Atoll, main Hawaiian Islands, Johnston Atoll, Samoa, Society Islands, Marquesas Islands, Rapa, and Easter Island, have either been published or are currently in preparation. Additionally, information on species occurrence within the region can be extracted from the numerous taxonomic revisions that have been published in recent years. This chapter represents the first comprehensive summary of the systematics of reef and shore fishes of Oceania.

METHODS

A database of Pacific Ocean marine fish species was developed by the author using Microsoft Access® software. The data structure includes several linked tables with fields for both systematic and biogeographic data pertaining to almost 4,500 tropical and subtropical species, of which nearly 4,300 are considered to be reef and shore. Although the database is not complete for all of the Indo-Pacific region, it currently includes an estimated 98 percent of the known, valid species of reef and shore fish species (both described and undescribed) occurring at the islands within Oceania.

Data on fish species of Oceania were obtained from a variety of sources, but primarily from systematic revisions of higher-level taxa (e.g., Lachner and Karnella 1980; Randall 1980; Randall and Bruce 1983; Compagno 1984; Allen 1985a, 1985b; Bruce and Randall 1985; Dawson 1985; Murdy and Hoese 1985; Randall and Hoese 1985; Hutchins 1986; Pietsch and Grobecker 1987; Carpenter 1988; Springer 1988; Williams 1988; Carpenter and Allen 1989; Palsson and Pietsch 1989; Larson 1990; Markle and Olney 1990; Williams 1990; Woodland 1990; Holleman 1991; Randall and Heemstra 1991; Randall and McCosker 1992; Gon 1993; McCosker and Rosenblatt 1993; Orr and Fritzsche 1993; Hoese and Larson 1994; Springer and Williams 1994); published regional checklists of fishes (e.g., Zama and Yasuda 1979; Randall and Egaña 1984; Randall, Lobel, and Chave 1985; Randall 1985, 1986; DiSalvo, Randall, and Cea 1988; Masuda et al. 1988; Myers 1988, 1991; Kosaki et al. 1989; Randall, Allen, and Steene 1990; Francis 1993; Randall et al. 1993; Kulbicki, Randall, and Rivaton 1994); unpublished regional checklists of fishes (particularly of New Caledonia, Palau, Ogasawara Islands, Hawaiian Islands, Cook Islands, and Fiji); specimen holdings of the Bernice P. Bishop Museum Ichthyology collection; and unpublished data obtained from personal observations by the author and other researchers.

Subjective interpretations of taxonomic-level usage and taxonomic certainty among different reef and shore fish families were made by the author after an extensive review of literature, and are consistent with similar interpretations of other researchers.

RESULTS

At least 3,392 distinct species of reef and shore fishes are known to occur within the tropical Pacific islands of Oceania, of which an estimated 442 (13 percent) are currently undescribed. Thus, slightly more than half of the estimated 6,000 total worldwide species of reef and shore fishes occur within Oceania. The species of Pacific island reef and shore fishes are classified among 839 genera, 166 families, 30 suborders (not all orders of fishes are subdivided into suborders), 30 orders, and 2 classes, all within the phylum Chordata (Table 12.1).

A complete listing of all reef and shore fish species is far beyond the scope of this chapter. A listing of all genera, arranged under existing classes, orders, suborders, and families, including numbers of described and undescribed species, is given in the Appendix to this chapter. More than two-thirds of the species, more than half of the genera, and about half of the families of Pacific reef and shore fishes are within the order Perciformes. The twenty most speciose families are the Gobiidae, Labridae, Serranidae, Pomacentridae, Apogonidae, Blenniidae, Muraenidae, Scorpaenidae, Carangidae, Syngnathidae, Chaetodontidae, Ophichthidae, Lutjanidae, Acanthuridae, Tripterygiidae, Pomacanthidae, Tetraodontidae, Holocentridae, Scaridae, and Monacanthidae. Together, these families comprise 65 percent of all Pacific reef and shore fish species (Figure 12.3).

Table 12.1 Summary of reef and shore fish systematics

Taxonomic level	Frequency of use	Relative stability	Reef and shore Pacific	Reef and shore Total	All fishes
Phylum	Never	Very High	1	1	1
Subphyplum	Never	Very High	1	1	1
Superclass	Never	Very High	1	1	1
Class	Occasional	Moderate	2	2	5
Subclass	Rare	Low	?	?	?
Superorder	Very Rare	—	?	?	?
Order	Frequent	Moderate	30	40	56
Suborder	Occasional	Moderate	30	38	62
Superfamily	Never	—	—	—	—
Family	Very Frequent	High	166	~200	450
Subfamily	Frequent	Moderate	~20	~30	76
Tribe	Occasional	High	?	?	?
Genus	Very Frequent	Moderate	839	~1,500	~4,500
Subgenus	Frequent	Moderate	?	?	?
Species complex	Occasional	Moderate/High	?	?	?
Species	Very Frequent	Moderate/High	3,392	~6,000	~24,000
Subspecies	Very Rare	Low	<20	<100	?

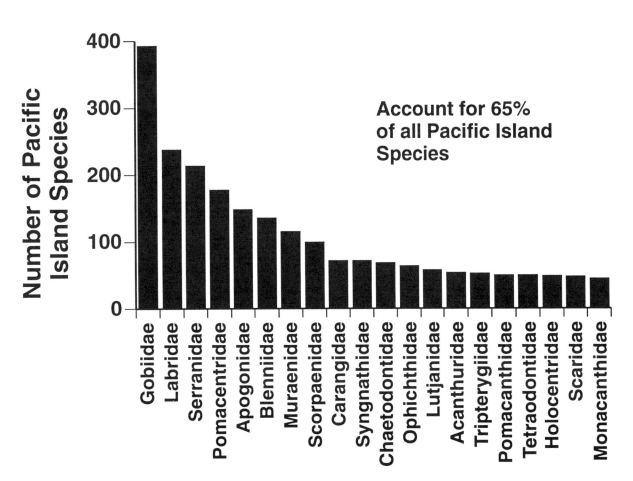

Figure 12.3 Twenty most speciose reef and shore fish families of Oceania

Figures 12.4a and 12.4b summarize the numbers of currently recognized species described in each decade since the time of Linnaeus. It should be noted that the values presented in these figures only reflect species currently recognized as valid; they do not include species names that have subsequently been synonymized or otherwise considered invalid.

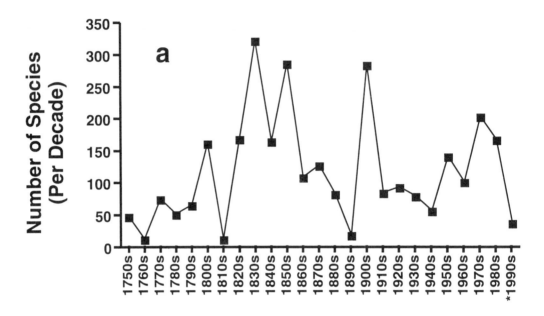

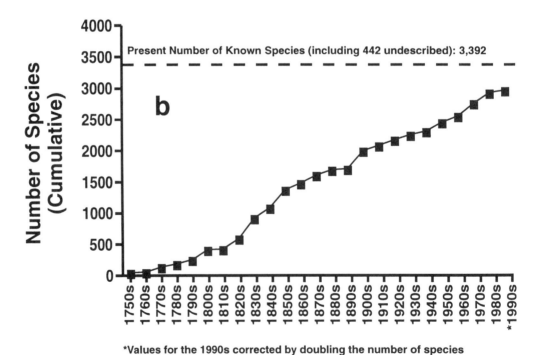

*Values for the 1990s corrected by doubling the number of species described from 1990-1994.

Figure 12.4 Number of described reef and shore fish species by decade

Based on the results of nearly 3,400 distinct species of fishes known from Oceania, it appears that Springer (1982) was correct in suggesting that his estimate of 4,000 species for the entire Indo-Pacific region is too low. Because Cohen (1973) estimated that there are 3,000–4,000 reef and shore fish species in the Indian Ocean, many of which do not occur in Oceania, it seems likely that the total number of Indo-Pacific reef and shore fish species is considerably higher.

The classification of lower-level taxa (family and below) of Pacific island reef and shore fishes is relatively stable and well resolved when compared to classifications of many marine invertebrates. Most publications that discuss reef and shore fish systematics and taxonomy (e.g., species checklists, taxonomic monographs) classify fishes by family, genus, and species. Some works make references to higher taxonomic levels, particularly orders and suborders; and occasionally divisions at the level of class are discussed (especially for the class Elasmobranchii, the sharks and rays). However, higher-level taxonomy for marine fishes has been less stable than the levels of family and below, so many recent authors have avoided including the higher taxonomy in their works.

Most families of reef and shore fishes are not subdivided into subfamilies; however, for those that are, the subfamily level is often very important to classification schemes (e.g., Serranidae). Some groups, for example the blennies (family Blenniidae), are subdivided into tribes. Superspecies are seldom utilized in marine fishes; instead, systematists have informally designated "species complexes" in many cases. Subspecies and other taxonomic levels below species are seldom incorporated; however, some subspecies have been described and are consistently recognized as such.

Taxonomic descriptions of Pacific island reef and shore fishes reached a peak in the early-to-mid-nineteenth century, primarily as a result of the efforts of the systematists Bleeker, Cuvier, Valenciennes, and Günther. Species descriptions experienced a renaissance of sorts in the early twentieth century, mostly as a result of large-scale exploratory expeditions by research vessels such as the *Albatross*, through the work of numerous ichthyologists including Jordan and Evermann. Another resurgence of species descriptions is evident during the 1950s to 1980s. This likely reflects the large numbers of species discovered with the aid of SCUBA. It is noteworthy that a precipitous drop in species descriptions has occurred in recent years. Fewer than twenty new species from Oceania have been described since 1990, compared with more than 200 in the 1970s. Given that at least 442 species remain to be described from this region, the decline in new species descriptions seems more likely a reflection of reduced available funding for alpha-level taxonomic research, rather than a drop in new species discoveries. Nevertheless, subjective interpretations by experienced reef-fish systematists indicate that probably more than 90 percent of all reef and shore fish species in the Pacific island region have already been discovered (but see discussion of deep reef species below).

As previously mentioned, at least 442 species of reef and shore fish species in the Pacific island region remain to be described. Nearly a third of these undescribed species are within the family Gobiidae. Indeed, 34 percent of all gobiids in Oceania are undescribed. Not surprisingly, all but one of the top ten families with the most undescribed species are also among the most speciose (Table 12.2). For the majority of these families, the rate of undescribed species is about 12–16 percent. However, in addition to the gobiids, an unusually large percentage of species in the families Tripterygiidae and Bothidae (57 percent and 26 percent, respectively) are undescribed.

Despite these undescribed species, the overall level of taxonomic certainty for species of fishes in the Pacific island region is relatively high. About 34 percent of these fishes have been investigated and the findings have appeared in recent or soon-to-be-published systematic revisions, and thus their level of taxonomic certainty is high. An additional 46 percent of Pacific marine fishes are among genera and families that are moderately well known and studied. Only 20 percent of the reef and shore fishes of this region are poorly known or are in need of taxonomic revision (Table 12.3).

Table 12.2 Summary of undescribed Pacific reef and shore fish species

Family	Number of undescribed species[a]	Total species in family	Percent of species in family
Gobiidae	134	393	34
Labridae	33	238	14
Tripterygiidae	30	53	57
Serranidae	30	213	14
Apogonidae	19	149	13
Muraenidae	18	116	16
Scorpaenidae	12	98	12
Bothidae	11	43	26
Pomacentridae	9	178	5
Ophichthidae	8	64	13
Other (55 families)	138	—	—
Total	442		

a. As of November 1994.

Table 12.3 Taxonomic certainty of Pacific reef and shore fishes

Level of taxonomic certainty	Description	Pacific island reef and shore fishes (%)
High	Recent or imminent revision	34
Moderate	Revision in progress or partially known	46
Low	In need of revision or poorly known	20

CONCLUSIONS AND RECOMMENDATIONS

Information Management

Perhaps the greatest need for more accurate assessment of the biodiversity and taxonomic status of reef and shore fishes in the Pacific is the combining and organizing of all the information already existing in published or unpublished form. Eschmeyer (1990) provides an exhaustive catalog of all genera of fishes (both freshwater and marine), and is currently preparing a similar catalog of all nominal fish species (see Eschmeyer, this volume). What is needed is a database of larger scope that includes information about the biology, ecology, and distribution of reef and shore fish species. Because much misinformation persists in published literature, extensive documentation of information sources must be included with such data.

The International Center for Living Aquatic Resources Management (ICLARM, Philippines) has developed a sophisticated database management system for information pertaining to fishes called "FishBase" (Froese and Pauly 1994; Froese and Palomares, this volume). This system should be adopted as the standard repository of data pertaining to fishes and should receive full cooperation from ichthyologists worldwide. To meet the needs of reef fish biogeographers, the FishBase Project should include mechanisms within its database structure for allowing detailed biogeographic analysis of marine fish distributions. Localities should be standardized and categorized in logical, hierarchical geographic sections (not necessarily along political borders, although political affiliation of localities should be included as additional data). Localities should be defined to a resolution of at least one order of magnitude greater than the scope of typical species endemicity. In the case of reef and shore fishes, few if any species are endemic to regions smaller than an "island group," so localities should be defined to the resolution of "island" within an island group. Much of the database structure required for this form of analysis is already extant in the FishBase system.

Required Research

In consideration of the apparent drastic decline in rates of species descriptions in recent decades, combined with the fact that 13 percent of all known Pacific island reef and shore fish species remain undescribed, it is clear that more emphasis on alpha-level taxonomy of fishes within this region is needed. Particular emphasis should be placed on the species of families listed in Table 12.2. Although much taxonomic work needs to be done, the relatively high diversity and otherwise well-resolved taxonomy of reef and shore fish species makes them ideal for testing various biogeographic models and for providing insight into modes of distribution and evolution.

Significant holes also persist in our knowledge of species distributions, both geographically and among habitats. With regard to the former, there are several regions in Oceania, including the Marquesas Islands, Cook Islands, Line Islands, Phoenix Islands, Gilbert Islands, and Solomon Islands, for which insufficient data on reef fish biodiversity have been accumulated. Comprehensive, accurate, annotated checklists of fishes from these areas are required to develop a clearer picture of total Pacific island species diversity and patterns of biogeography.

Perhaps even more significant is a near-complete lack of information regarding species that inhabit the deeper regions of the reef and shore environment. Figure 12.5 delineates the main habitat regions of the reef and shore environment. Of these, all but the deep reefs have received large-scale investigation by ichthyologists. By contrast, virtually nothing is known of the species diversity inhabiting the apparently complex habitat of deep coral reefs. The primary reason this habitat remains unexplored is that it extends below the depth attainable by divers using conventional SCUBA, and is shallower than

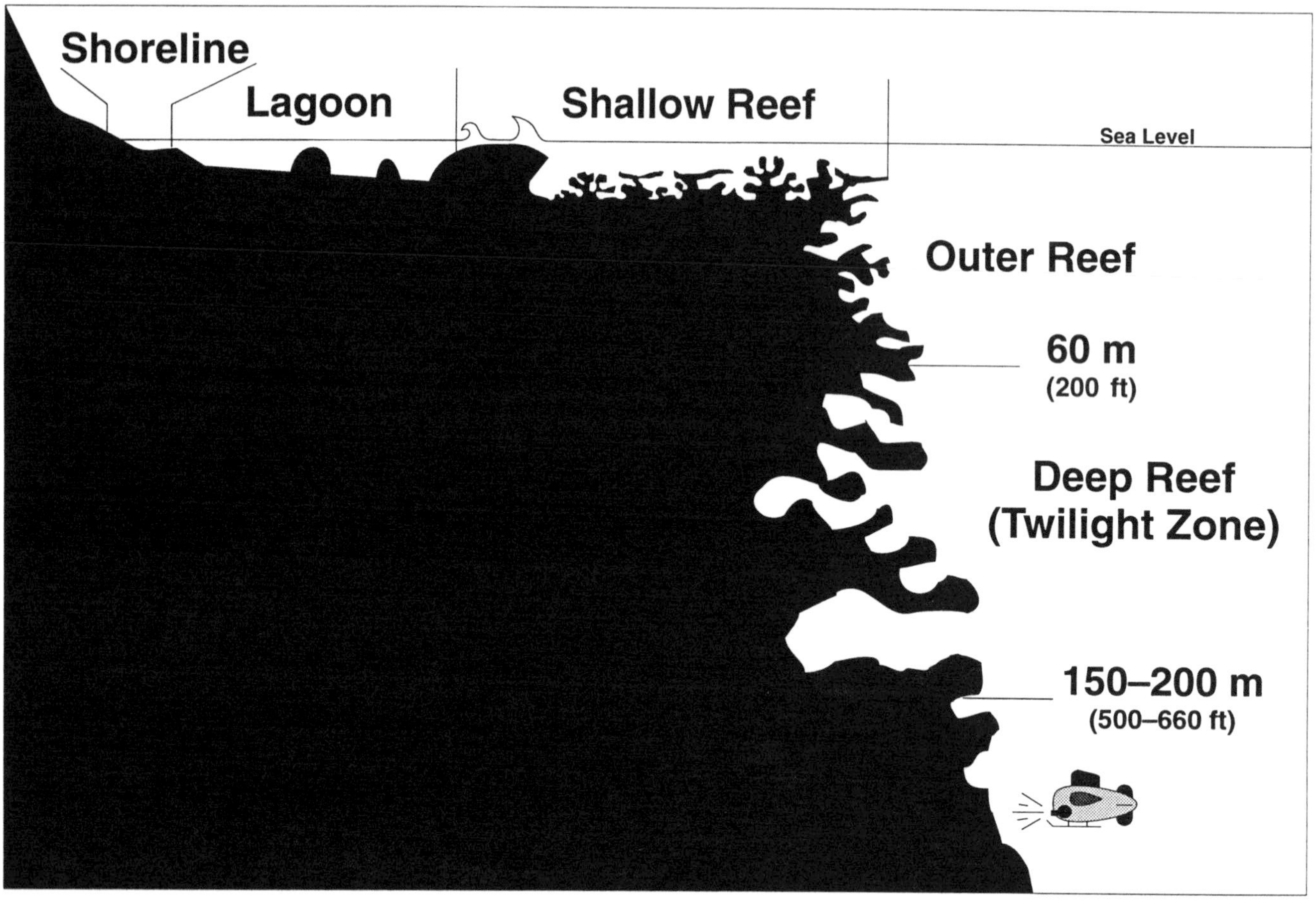

Figure 12.5 Reef and shore habitat zones

the depth of the majority of scientific deep-sea submersible activity (Pyle in press). Recent advances in mixed-gas SCUBA technology have allowed for preliminary investigations of the deep reefs by a few individuals (Sharkey and Pyle 1993). The scant information on fish species diversity at these depths that has become available through the use of this new diving technology indicates that a wealth of new species remain undiscovered. For example, more new fish species were discovered during the first sixty minutes of exploration of deep (100–120 m) reefs in Rarotonga than have been described in the past five years from all of Oceania (e.g., Pyle and Randall 1992). Because the species diversity at Rarotonga is about one-third the diversity of western-Pacific localities, and because no ichthyocides (necessary for collecting the many cryptic reef species) have yet been used on the deep reefs, it seems likely that a substantial number of species remain to be discovered.

Acknowledgements—I wish to thank John E. Randall for his extensive contributions to the information included in this chapter. Valuable contributions were also provided by William N. Eschmeyer and Lisa A. Privitera. Unpublished checklists of fishes compiled by M. Kulbicki (New Caledonia), T. Donaldson and R. Myers (Palau), and N. Simms (Cook Islands) were useful as sources of information on species distributions.

APPENDIX

All Known Genera of Reef and Shore Fishes in Oceania, Including Numbers of Described and Undescribed Species

Classification	Described species	Undescribed species[a]	Total species
Class: ELASMOBRANCHII	**101**	**3**	**104**
Order: HETERODONTIFORMES	**2**	—	**2**
HETERODONTIDAE	2	—	2
Heterodontus	2	—	2
Order: ORECTOLOBIFORMES	**14**	—	**14**
ORECTOLOBIDAE	5	—	5
Cirrhoscyllium	1	—	1
Eucrossohrinus	1	—	1
Orectolobus	3	—	3
HEMISCYLLIIDAE	5	—	5
Chiloscyllium	4	—	4
Hemiscyllium	1	—	1
STEGOSTOMATIDAE	2	—	2
Stegostoma	2	—	2
GINGLYMOSTOMATIDAE	2	—	2
Nebrius	2	—	2
Order: LAMNIFORMES	**3**	—	**3**
ODONTASPIDIDAE	2	—	2
Eugomphodus	2	—	2
ALOPIIDAE	1	—	1
Alopias	1	—	1
Order: CARCHARHINIFORMES	**32**	—	**32**
SCYLIORHINIDAE	4	—	4
Atelomycterus	1	—	1
Cephaloscyllium	1	—	1
Halaelurus	1	—	1
Scyliorhinus	1	—	1
TRIAKIDAE	2	—	2
Proscyllium	1	—	1
Triakis	1	—	1
HEMIGALEIDAE	1	—	1
Triaenodon	1	—	1
CARCHARHINIDAE	22	—	22
Carcharhinus	16	—	16
Galeocerdo	1	—	1
Loxodon	1	—	1
Negaprion	1	—	1
Rhizoprionodon	2	—	2
Scoliodon	1	—	1
SPHYRNIDAE	3	—	3
Sphyrna	3	—	3
Order: PRISTIOPHORIFORMES	**1**	—	**1**
PRISTIOPHORIDAE	1	—	1
Pristiophorus	1	—	1

Classification	Described species	Undescribed species[a]	Total species
Order: PRISTIFORMES	**1**	**—**	**1**
PRISTIDAE	1	—	1
Pristis	1	—	1
Order: TORPEDINIFORMES	**5**	**—**	**5**
TORPEDINIDAE	2	—	2
Crassinarke	1	—	1
Hypnos	1	—	1
NARKIDAE	3	—	3
Narcine	1	—	1
Narke	2	—	2
Order: RAJIFORMES	**9**	**—**	**9**
RHINOBATIDAE	5	—	5
Rhina	1	—	1
Rhinobatos	3	—	3
Rhynchobatus	1	—	1
PLATYRHINIDAE	1	—	1
Platyrhina	1	—	1
RAJIDAE	2	—	2
Raja	2	—	2
ANACANTHOBATIDIDAE	1	—	1
Anacanthobatis	1	—	1
Order: MYLIOBATIFORMES	**34**	**3**	**37**
DASYATIDAE	21	1	22
Dasyatis	11	1	12
Gymnura	2	—	2
Himantura	3	—	3
Taeniura	3	—	3
Urogymnus	2	—	2
MYLIOBATIDAE	6	—	6
Aetobatus	1	—	1
Aetomylaeus	2	—	2
Myliobatis	1	—	1
Myliobatus	1	—	1
Rhinoptera	1	—	1
UROLOPHIDAE	2	2	4
Urolophus	2	2	4
MOBULIDAE	5	—	5
Manta	2	—	2
Mobula	3	—	3
Class: ACTINOPTERYGII	**2,849**	**439**	**3,288**
Order: ELOPIFORMES	**3**	**1**	**4**
ELOPIDAE	2	1	3
Elops	2	1	3
MEGALOPIDAE	1	—	1
Megalops	1	—	1

Classification	Described species	Undescribed species[a]	Total species
Order: ALBULIFORMES	**2**	—	**2**
ALBULIDAE	2	—	2
Albula	2	—	2
Order: ANGUILLIFORMES	**192**	**35**	**227**
Suborder: ANGUILLOIDEI	**7**	—	**7**
MORINGUIDAE	7	—	7
Moringua	7	—	7
Suborder: MURAENOIDEI	**104**	**22**	**126**
CHLOPSIDAE	6	4	10
Chilorhinus	1	—	1
Kaupichthys	4	4	8
Xenoconger	1	—	1
MURAENIDAE	98	18	116
Anarchias	4	6	10
Channomuraena	1	—	1
Echidna	5	—	5
Enchelycore	5	—	5
Enchelynassa	1	—	1
Gymnomuraena	1	—	1
Gymnothorax	56	9	65
Monopenchelys	1	—	1
Rhinomuraena	1	—	1
Scuticaria	3	—	3
Siderea	3	1	4
Strophidon	2	—	2
Thyrsoidea	1	2	3
Uropterygius	14	—	14
Suborder: CONGROIDEI	**81**	**13**	**94**
OPHICHTHIDAE	56	8	64
Apterichtus	3	1	4
Bascanichthys	1	—	1
Brachysomophis	4	1	5
Caecula	1	—	1
Callechelys	4	—	4
Cirrhimuraena	1	—	1
Cirricaecula	1	—	1
Elapsopsis	1	—	1
Evipes	1	1	2
Glenoglossa	1	—	1
Ichthyapus	1	—	1
Lamnostoma	2	—	2
Leiuranus	2	—	2
Malvoliophis	1	—	1
Muraenichthys	10	—	10
Myrichthys	3	—	3
Myrophis	2	—	2
Ophichthus	8	2	10
Ophisurus	1	—	1
Phaenomonas	1	—	1
Phyllophichthus	1	1	2
Pisodonophis	1	—	1
Schismorhynchus	1	—	1

Classification	Described species	Undescribed species[a]	Total species
Shultzidia	2	—	2
Yirrkala	2	2	4
CONGRIDAE	22	5	27
Anago	1	—	1
Ariosoma	4	2	6
Conger	5	1	6
Gnathophis	2	—	2
Gorgasia	4	1	5
Heteroconger	3	—	3
Poeciloconger	1	1	2
Rhechias	2	—	2
MURAENESOCIDAE	3	—	3
Muraenesox	2	—	2
Oxyconger	1	—	1
Order: CLUPEIFORMES	**34**	**3**	**37**
CLUPEIDAE	22	2	24
Amblygaster	3	—	3
Anodontostoma	1	—	1
Dussumieria	2	2	4
Etrumeus	1	—	1
Herklotsichthys	1	—	1
Ilisha	1	—	1
Konosirus	1	—	1
Nematalosa	2	—	2
Sardinella	7	—	7
Spratelloides	3	—	3
ENGRAULIDAE	11	1	12
Encrasicholina	3	—	3
Engraulis	1	1	2
Stolephorus	6	—	6
Thrissa	1	—	1
CHIROCENTRIDAE	1	—	1
Chirocentrus	1	—	1
Order: GONORYNCHIFORMES	**4**	**—**	**4**
CHANIDAE	1	—	1
Chanos	1	—	1
GONORYNCHIDAE	3	—	3
Gonorynchus	3	—	3
Order: SILURIFORMES	**3**	**—**	**3**
PLOTOSIDAE	3	—	3
Paraplotosus	1	—	1
Plotosus	2	—	2
Order: AULOPIFORMES	**26**	**4**	**30**
Suborder: ALEPISAUROIDEI	**26**	**4**	**30**
SYNODONTIDAE	26	4	30
Saurida	6	—	6
Synodus	19	4	23
Trachinocephalus	1	—	1

Classification	Described species	Undescribed species[a]	Total species
Order: GADIFORMES	**6**	**—**	**6**
BREGMACEROTIDAE	4	—	4
Bregmaceros	4	—	4
MORIDAE	2	—	2
Lotella	2	—	2
Order: OPHIDIIFORMES	**25**	**16**	**41**
OPHIDIIDAE	3	4	7
Brotula	2	1	3
Ophidion	1	2	3
Otophidium	—	1	1
CARAPODIDAE	12	5	17
Carapus	3	5	8
Encheliophis	4	—	4
Onuxodon	3	—	3
Pyramodon	2	—	2
BYTHITIDAE	10	7	17
Brosmophyciops	1	—	1
Brotulina	2	—	2
Dermatopsis	1	—	1
Diancistrus	1	—	1
Dinematichthys	1	5	6
Microbrotula	2	—	2
Ogilbia	—	2	2
Oligopus	2	—	2
Order: BATRACHOIDIFORMES	**2**	**—**	**2**
BATRACHOIDIDAE	2	—	2
Halophryne	2	—	2
Order: LOPHIIFORMES	**17**	**—**	**17**
Suborder: ANTENNARIOIDEI	**16**	**—**	**16**
ANTENNARIIDAE	16	—	16
Antennarius	14	—	14
Antennatus	1	—	1
Histrio	1	—	1
Suborder: CHAUNACIOIDEI	**1**	**—**	**1**
CHAUNACIDAE	1	—	1
Chaunax	1	—	1
Order: GOBIESOCIFORMES	**12**	**1**	**13**
GOBIESOCIDAE	12	1	13
Alabes	1	—	1
Conidens	2	—	2
Diademichthys	1	—	1
Discotrema	2	—	2
Lepadichthys	4	—	4
Liobranchia	1	—	1
Pherallodus	1	—	1
Unidentified genus	—	1	1

Classification	Described species	Undescribed species[a]	Total species
Order: ATHERINIFORMES	**14**	**1**	**15**
ATHERINIDAE	14	1	15
Atherinomorus	3	—	3
Atherion	2	—	2
Hypoatherina	6	—	6
Iso	2	1	3
Stenatherina	1	—	1
Order: BELONIFORMES	**22**	**1**	**23**
Suborder: EXOCOETOIDEI	**22**	**1**	**23**
BELONIDAE	8	—	8
Ablennes	1	—	1
Platybelone	1	—	1
Strongylura	4	—	4
Tylosurus	2	—	2
HEMIRAMPHIDAE	14	1	15
Hemiramphus	4	1	5
Hyporamphus	4	—	4
Hyporhamphus	4	—	4
Oxyporhamphus	1	—	1
Zenarchopterus	1	—	1
Order: BERYCIFORMES	**52**	**3**	**55**
Suborder: BERCOIDEI	**51**	**3**	**54**
MONOCENTRIDAE	2	—	2
Cleidopus	1	—	1
Monocentris	1	—	1
TRACHICHTHYIDAE	1	—	1
Optivus	1	—	1
ANOMALOPIDAE	2	—	2
Anomalops	1	—	1
Photoblepharon	1	—	1
HOLOCENTRIDAE	46	3	49
Myripristis	14	1	15
Neoniphon	4	—	4
Ostichthys	6	—	6
Plectrypops	1	—	1
Pristilepis	1	—	1
Sargocentron	20	2	22
Suborder: POLYMIXIOIDEI	**1**	**—**	**1**
POLYMIXIIDAE	1	—	1
Polymixia	1	—	1
Order: GASTEROSTEIFORMES	**3**	**—**	**3**
PEGASIDAE	3	—	3
Eurypegasus	3	—	3
Order: SYNGNATHIFORMES	**75**	**6**	**81**
Suborder: AULOSTOMOIDEI	**6**	**—**	**6**
AULOSTOMIDAE	1	—	1
Aulostomus	1	—	1
FISTULARIIDAE	2	—	2
Fistularia	2	—	2

Classification	Described species	Undescribed species[a]	Total species
CENTRISCIDAE	3	—	3
Aeoliscus	1	—	1
Centriscus	1	—	1
Macroramphosus	1	—	1
Suborder: SYNGNATHOIDEI	**69**	**6**	**75**
SOLENOSTOMIDAE	2	1	3
Solenostomus	2	1	3
SYNGNATHIDAE	67	5	72
Acentronura	2	—	2
Bhanotia	1	—	1
Bulbonaricus	1	—	1
Campichthys	—	2	2
Choeroichthys	3	—	3
Coelonotus	1	—	1
Corythoichthys	9	—	9
Cosmocampus	5	—	5
Doryrhamphus	6	—	6
Dunckerocampus	1	—	1
Festucalex	3	—	3
Halicampus	8	—	8
Hippichthys	3	—	3
Hippocampus	4	3	7
Micrognathus	3	—	3
Microphis	4	—	4
Minyichthys	2	—	2
Oostethus	1	—	1
Parasyngnathus	1	—	1
Phoxocampus	2	—	2
Solegnathus	1	—	1
Sygnathus	1	—	1
Syngnathoides	1	—	1
Trachyrhamphus	3	—	3
Urocampus	1	—	1
Order: SCORPAENIFORMES	**134**	**20**	**154**
Suborder: SCORPAENOIDEI	**110**	**16**	**126**
SCORPAENIDAE	88	12	100
Ablabys	1	—	1
Apistus	1	—	1
Brachypterois	1	—	1
Dendrochirus	5	—	5
Ebosia	1	—	1
Ectreposebastes	1	—	1
Erosa	1	—	1
Inimicus	3	1	4
Iracundus	1	—	1
Lioscorpius	1	—	1
Maxillicosta	1	—	1
Minous	3	—	3
Neomerinthe	2	1	3
Neosebastes	1	—	1
Paracentropogon	1	—	1
Parapterois	1	—	1

Classification	Described species	Undescribed species[a]	Total species
Parascorpaena	2	—	2
Phenacoscorpius	1	1	2
Plectrogenium	1	—	1
Pontinus	1	1	2
Pteroidichthys	1	—	1
Pterois	4	—	4
Rhinopias	3	1	4
Richardsonichthys	1	1	2
Scorpaena	11	—	11
Scorpaenodes	13	—	13
Scorpaenopsis	10	4	14
Sebastapistes	7	—	7
Sebasticus	1	—	1
Setarches	2	—	2
Synanceia	2	—	2
Taenionotus	1	—	1
Tetraroge	2	1	3
Trachyscorpia	—	1	1
CARACANTHIDAE	3	—	3
Caracanthus	3	—	3
APLOACTINIDAE	6	3	9
Cocotropis	3	3	6
Neoaploactis	1	—	1
Paraploactis	2	—	2
CONGIOPODIDAE	4	1	5
Amblyapistus	1	—	1
Neocentropogon	1	1	2
Ocosia	1	—	1
Snyderina	1	—	1
TRIGLIDAE	9	—	9
Chelidonichthys	2	—	2
Lepidotrigla	5	—	5
Pterygotrigla	2	—	2
Suborder: DACTYLOPTEROIDEI	2	—	2
DACTYLOPTERIDAE	2	—	2
Dactyloptena	2	—	2
Suborder: PLATYCEPHALOIDEI	22	4	26
PLATYCEPHALIDAE	19	4	23
Cociella	1	—	1
Cymbacephalus	1	—	1
Inegocia	1	—	1
Kumococcius	1	—	1
Onigocia	2	1	3
Papilloculiceps	1	—	1
Parabembras	1	—	1
Platycephalus	2	2	4
Rogadius	2	—	2
Sorsogona	1	—	1
Suggrundus	1	1	2
Thysanophrys	5	—	5
BEMBRIDAE	2	—	2
Bembradium	2	—	2
HOPLICHTHYIDAE	1	—	1
Hoplichthys	1	—	1

Classification	Described species	Undescribed species[a]	Total species
Order: PERCIFORMES	**1,996**	**324**	**2,320**
Suborder: PERCOIDEI	**1,110**	**93**	**1,203**
CENTROPOMIDAE	1	—	1
Psammoperca	1	—	1
PERCICHTHYIDAE	3	—	3
Lateolabrax	1	—	1
Malakichthys	1	—	1
Niphon	1	—	1
ACROPOMATIDAE	5	—	5
Acropoma	3	—	3
Doederleinia	1	—	1
Neoscombrops	1	—	1
POLYPRIONIDAE	1	—	1
Polyprion	1	—	1
SERRANIDAE	184	29	213
Acanthistius	3	—	3
Aetheloperca	1	—	1
Anyperodon	1	—	1
Aporops	2	—	2
Aulacocephalus	1	—	1
Belonoperca	1	1	2
Caprodon	3	—	3
Cephalopholis	16	—	16
Chelidoperca	1	1	2
Cromileptes	1	—	1
Diploprion	1	—	1
Elerkeldia	1	—	1
Epinephelus	55	—	55
Giganthias	1	—	1
Gracila	1	—	1
Gramistes	1	—	1
Grammistops	1	—	1
Hapalogenys	1	—	1
Holanthias	7	—	7
Hypoplectrodes	—	1	1
Liopropoma	14	3	17
Luzonichthys	3	—	3
Plectranthias	17	2	19
Plectropomus	7	—	7
Pogonoperca	1	—	1
Pseudanthias	31	15	46
Pseudogramma	1	4	5
Rainfordia	1	—	1
Sacura	1	—	1
Saloptia	1	—	1
Selenanthias	1	1	2
Serranocirrhitus	1	—	1
Suttonia	1	2	3
Tosana	1	—	1
Trachypoma	1	—	1
Triso	1	—	1
Variola	2	—	2
CENTROGENIIDAE	1	—	1
Centrogenys	1	—	1

Classification	Described species	Undescribed species[a]	Total species
PSEUDOCHROMIDAE	28	4	32
Chlidichthys	—	1	1
Congrogadus	1	—	1
Cypho	1	—	1
Ogilbyina	4	—	4
Pseudochromis	16	—	16
Pseudoplesiops	6	3	9
CALLANTHIIDAE	1	—	1
Grammatonotus	1	—	1
PLESIOPIDAE	12	1	13
Assessor	2	—	2
Calloplesiops	1	—	1
Ernogrammoides	1	—	1
Paraplesiops	1	—	1
Plesiops	6	1	7
Steenichthys	1	—	1
ACANTHOCLINIDAE	2	—	2
Acanthoplesiops	1	—	1
Belonepterygion	1	—	1
GLAUCOSOMATIDAE	1	—	1
Galucosoma	1	—	1
TERAPONIDAE	5	—	5
Pelates	1	—	1
Rhyncopelates	1	—	1
Terapon	3	—	3
KUHLIIDAE	7	—	7
Kuhlia	7	—	7
PRIACANTHIDAE	10	3	13
Cookeolus	2	—	2
Heteropriacanthus	1	—	1
Priacanthus	5	3	8
Pristigenys	2	—	2
APOGONIDAE	130	19	149
Apogon	78	11	89
Apogonichthys	2	—	2
Archamia	8	—	8
Brephostoma	1	—	1
Cercamia	2	—	2
Cheilodipterus	7	—	7
Coranthias	1	—	1
Foa	3	1	4
Fowleria	6	1	7
Gymnapogon	3	2	5
Lachneratus	1	—	1
Neamia	1	—	1
Pseudamia	6	—	6
Pseudamiops	1	2	3
Rhabdamia	2	1	3
Siphamia	5	1	6
Sphaeramia	2	—	2
Vincentia	1	—	1
SILLAGINIDAE	3	—	3
Sillago	3	—	3

Classification	Described species	Undescribed species[a]	Total species
MALACANTHIDAE	6	—	6
Hoplolatilus	4	—	4
Malacanthus	2	—	2
BRANCHIOSTEGIDAE	5	1	6
Branchiostegus	5	1	6
RACHYCENTRIDAE	1	—	1
Rachycentron	1	—	1
ECHENEIDAE	1	—	1
Echeneis	1	—	1
CARANGIDAE	70	2	72
Absalom	1	—	1
Alectis	2	—	2
Alepes	2	1	3
Atropus	1	—	1
Atule	1	—	1
Carangichthys	1	—	1
Carangoides	16	—	16
Caranx	8	—	8
Decapterus	8	—	8
Elagatis	1	—	1
Gnathanodon	1	—	1
Megalaspis	1	—	1
Naucrates	1	—	1
Parastromateus	1	—	1
Pseudocaranx	1	—	1
Scomberoides	4	—	4
Selar	2	—	2
Selaroides	1	—	1
Seriola	5	—	5
Seriolina	1	—	1
Trachinotus	4	1	5
Trachurus	3	—	3
Ulua	1	—	1
Uraspis	3	—	3
LEIOGNATHIDAE	12	—	12
Gazza	2	—	2
Leiognathus	10	—	10
ARRIPIDAE	1	1	2
Arripis	1	1	2
EMMELICHTHYIDAE	3	2	5
Emmelichthys	2	1	3
Erythrocles	1	1	2
LUTJANIDAE	57	1	58
Aphareus	2	—	2
Aprion	1	—	1
Etelis	3	—	3
Lipocheilus	1	—	1
Lutjanus	28	—	28
Macolor	2	—	2
Paracaesio	6	1	7
Parapristipomoides	1	—	1
Pinjalo	1	—	1
Pristipomoides	9	—	9

Classification	Described species	Undescribed species[a]	Total species
Randallichthys	1	—	1
Symphorichthys	1	—	1
Symphorus	1	—	1
CAESIONIDAE	15	1	16
Caesio	7	1	8
Pterocaesio	8	—	8
LOBOTIDAE	1	—	1
Lobotes	1	—	1
GERREIDAE	13	—	13
Gerres	12	—	12
Pentaprion	1	—	1
HAEMULIDAE	26	1	27
Diagramma	2	—	2
Haplogenys	1	—	1
Parapristipoma	1	—	1
Plectorhynchus	18	1	19
Pomadasys	4	—	4
SPARIDAE	7	3	10
Acanthopagrus	4	1	5
Cheimerius	1	—	1
Chrysophrys	1	—	1
Dentex	1	2	3
LETHRINIDAE	40	4	44
Gnathodentex	1	—	1
Gymnocranius	11	4	15
Lethrinus	26	—	26
Monotaxis	1	—	1
Wattsia	1	—	1
NEMIPTERIDAE	29	3	32
Nemipterus	8	1	9
Parascolopsis	1	—	1
Pentapodus	6	2	8
Scolopsis	14	—	14
SCIAENIDAE	4	—	4
Argyrosomus	1	—	1
Collichthys	1	—	1
Johnius	1	—	1
Nibea	1	—	1
MULLIDAE	37	2	39
Mulloidichthys	4	—	4
Parupeneus	21	—	21
Upeneichthys	1	—	1
Upeneus	11	2	13
PEMPHERIDAE	12	1	13
Parapriacanthus	3	—	3
Pempheris	9	1	10
TOXOTIDAE	1	—	1
Toxotes	1	—	1
KYPHOSIDAE	13	—	13
Girella	6	—	6
Girellops	1	—	1
Kyphosus	5	—	5
Sector	1	—	1

Classification	Described species	Undescribed species[a]	Total species
SCORPIDIDAE	5	—	5
Atypichthys	2	—	2
Microcanthus	1	—	1
Scorpis	2	—	2
LABRACOGLOSSIDAE	4	—	4
Bathystethus	2	—	2
Labracoglossa	2	—	2
DREPANIDAE	2	—	2
Drepane	2	—	2
EPHIPPIDAE	7	—	7
Ephippus	1	—	1
Platax	5	—	5
Zabidius	1	—	1
MONODACTYLIDAE	2	—	2
Monodactylus	2	—	2
SCATOPHAGIDAE	2	—	2
Scatophagus	2	—	2
CHAETODONTIDAE	69	—	69
Amphichaetodon	2	—	2
Chaetodon	50	—	50
Chelmon	2	—	2
Coradion	2	—	2
Forcipiger	2	—	2
Hemitaurichthys	2	—	2
Heniochus	6	—	6
Parachaetodon	1	—	1
Prognathodes	1	—	1
Roa	1	—	1
POMACANTHIDAE	49	1	50
Apolemichthys	4	—	4
Centropyge	24	—	24
Chaetodontoplus	7	—	7
Genicanthus	8	1	9
Pomacanthus	5	—	5
Pygoplites	1	—	1
PENTACEROTIDAE	6	—	6
Evistias	1	—	1
Histiopterus	1	—	1
Pentaceros	2	—	2
Pseudopentaceros	2	—	2
OPLEGNATHIDAE	2	—	2
Oplegnathus	2	—	2
POMACENTRIDAE	169	9	178
Abudefduf	10	—	10
Acanthochromis	1	—	1
Amblyglyphidodon	4	—	4
Amblypomacentrus	1	—	1
Amphiprion	13	—	13
Cheiloprion	1	—	1
Chromis	38	6	44
Chrysiptera	19	—	19
Dascyllus	7	—	7
Dischistodus	5	—	5

Classification	Described species	Undescribed species[a]	Total species
Hemiglyphidodon	1	—	1
Lepidozygus	1	—	1
Neoglyphidodon	3	—	3
Neopomacentrus	8	—	8
Parma	4	—	4
Plectroglyphidodon	8	—	8
Pomacentrus	27	2	29
Pomachromis	4	—	4
Premnas	1	—	1
Pristotis	1	—	1
Stegastes	10	—	10
Teixeirichthys	2	1	3
CIRRHITIDAE	21	—	21
Amblycirrhitus	3	—	3
Cirrhitichthys	4	—	4
Cirrhitops	2	—	2
Cirrhitus	2	—	2
Cyprinocirrhitus	1	—	1
Isocirrhitus	1	—	1
Neocirrhites	1	—	1
Oxycirrhites	1	—	1
Paracirrhites	6	—	6
APLODACTYLIDAE	1	—	1
Aplodactylus	1	—	1
CHEILODACTYLIDAE	10	—	10
Cheilodactylus	7	—	7
Goniistius	1	—	1
Nemadactylus	2	—	2
LATRIDIDAE	1	—	1
Latridopsis	1	—	1
CEPOLIDAE	4	—	4
Acanthocepola	2	—	2
Cepola	1	—	1
Owstonia	1	—	1
OPISTOGNATHIDAE	8	4	12
Opistognathus	7	4	11
Stalix	1	—	1
Suborder: MUGILOIDEI	**16**	**—**	**16**
MUGILIDAE	16	—	16
Crenimugil	1	—	1
Ellochelon	1	—	1
Liza	5	—	5
Moolgarda	4	—	4
Mugil	1	—	1
Myxus	1	—	1
Neomyxus	1	—	1
Oedalechilus	1	—	1
Valamugil	1	—	1
Suborder: POLYNEMOIDEI	**3**	**—**	**3**
POLYNEMIDAE	3	—	3
Polydactylus	3	—	3
Suborder: LABROIDEI	**251**	**35**	**286**
LABRIDAE	205	33	238
Anampses	11	—	11

Classification		Described species	Undescribed species[a]	Total species
	Bodianus	17	1	18
	Cheilinus	9	1	10
	Cheilio	1	—	1
	Choerodon	12	—	12
	Cirrhilabrus	12	9	21
	Coris	15	4	19
	Cymolutes	3	—	3
	Diproctacanthus	1	—	1
	Epibulus	1	1	2
	Gomphosus	1	—	1
	Halichoeres	22	2	24
	Hemigymnus	2	—	2
	Hologymnosus	4	—	4
	Labrichthys	1	—	1
	Labroides	5	—	5
	Labropsis	6	—	6
	Leptojulis	1	2	3
	Macropharyngodon	5	—	5
	Notolabrus	2	—	2
	Novaculichthys	2	—	2
	Novaculops	1	1	2
	Oxycheilinus	4	1	5
	Paracheilinus	—	2	2
	Parajulis	1	—	1
	Pseudocheilinops	1	—	1
	Pseudocheilinus	4	2	6
	Pseudocoris	4	—	4
	Pseudodax	1	—	1
	Pseudojuloides	4	—	4
	Pseudolabrus	7	—	7
	Pteragogus	4	1	5
	Semicossyphus	1	—	1
	Stethojulis	6	1	7
	Suezichthys	5	—	5
	Thalassoma	11	1	12
	Wetmorella	3	—	3
	Xiphocheilus	1	1	2
	Xyrichtys	14	3	17
SCARIDAE		46	2	48
	Bolbometapon	1	—	1
	Calotomus	4	—	4
	Cetoscarus	1	—	1
	Hipposcarus	1	—	1
	Leptoscarus	1	—	1
	Scarus	38	2	40
Suborder: TRACHINOIDEI		**44**	**14**	**58**
AMMODYTIDAE		4	2	6
	Ammodytoides	3	—	3
	Embolichthys	1	2	3
URANOSCOPIDAE		3	2	5
	Kathetostoma	—	1	1
	Uranoscopus	3	1	4
TRICHONOTIDAE		3	1	4
	Trichonotus	3	1	4

Classification	Described species	Undescribed species[a]	Total species
CREEDIIDAE	5	1	6
Chalyixodytes	1	—	1
Crystallodytes	2	—	2
Limnichthys	2	1	3
PERCOPHIDAE	8	2	10
Acanthaphrites	1	1	2
Bembrops	2	—	2
Chironema	1	—	1
Chrionema	2	—	2
Enigmapercis	—	1	1
Pteropsaron	1	—	1
Spinapsaron	1	—	1
PINGUIPEDIDAE	21	6	27
Parapercis	21	6	27
Suborder: BLENNIOIDEI	**160**	**35**	**195**
TRIPTERYGIIDAE	23	30	53
Enneapterygius	12	20	32
Helcogramma	7	6	13
Lepidoblennius	—	1	1
Norfolkia	3	1	4
Tripterygion	1	1	2
Ucla	—	1	1
CLINIDAE	5	—	5
Cristiceps	1	—	1
Heteroclinus	1	—	1
Petraites	1	—	1
Springeratus	1	—	1
Springerichthys	1	—	1
CHAENOPSIDAE	1	—	1
Neoclinus	1	—	1
BLENNIIDAE	131	5	136
Alticus	4	1	5
Andamia	2	—	2
Aspidontus	2	—	2
Atrosalarius	1	—	1
Blenniella	1	—	1
Cirripectes	14	—	14
Cirrisalarias	1	—	1
Crossosalarias	1	—	1
Ecsenius	16	—	16
Enchelyurus	3	—	3
Entomacrodus	16	—	16
Exallias	1	—	1
Glyptoparus	1	—	1
Istiblennius	16	—	16
Litobranchus	1	—	1
Medusablennius	1	—	1
Meiacanthus	6	—	6
Mimoblennius	1	—	1
Nannosalarias	1	—	1
Omobranchus	6	—	6
Omox	1	—	1
Parablennius	4	—	4
Parenchelyurus	1	—	1

Classification	Described species	Undescribed species[a]	Total species
Petroscirtes	7	—	7
Plagiotremus	5	—	5
Praealticus	5	3	8
Rhabdoblennius	3	—	3
Salarias	6	1	7
Stanulus	2	—	2
Xiphasia	2	—	2
Suborder: CALLIONYMOIDEI	**32**	**6**	**38**
CALLIONYMIDAE	30	6	36
Anaora	1	—	1
Callionymus	9	2	11
Calliurichthys	1	—	1
Dactylopus	1	—	1
Diplogrammus	2	—	2
Draculo	1	—	1
Eleutherochir	1	—	1
Neosynchiropus	1	—	1
Paradiplogrammus	1	1	2
Pseudocalliurichthys	—	1	1
Repomucenus	1	—	1
Synchiropus	11	2	13
DRACONETTIDAE	2	—	2
Draconetta	2	—	2
Suborder: GOBIOIDEI	**295**	**139**	**434**
ELEOTRIDAE	2	1	3
Calumia	2	—	2
Eleotris	—	1	1
GOBIIDAE	259	134	393
Acentrogobius	7	—	7
Amblyeleotris	13	4	17
Amblygobius	9	1	10
Asterropteryx	2	4	6
Austrolethops	1	—	1
Barbuligobius	1	—	1
Bathygobius	9	2	11
Bostrychus	1	—	1
Brachyamblyopus	1	—	1
Bryaninops	8	2	10
Cabillus	2	4	6
Callogobius	9	4	13
Chaenogobius	1	—	1
Coryphopterus	4	14	18
Cristatogobius	3	2	5
Cryptocentroides	1	—	1
Cryptocentrus	8	5	13
Ctenogobiops	5	2	7
Discordipinna	1	—	1
Drombus	—	2	2
Eutaeniichthys	1	—	1
Eviota	31	14	45
Exyrias	2	—	2
Favonigobius	2	—	2
Feia	—	1	1
Gladigobius	1	—	1

Classification	Described species	Undescribed species[a]	Total species
Glossogobius	5	—	5
Gnatholepis	4	6	10
Gobiodon	10	5	15
Gobiopsis	4	—	4
Hetereleotris	1	3	4
Istigobius	6	2	8
Kelloggella	4	—	4
Lotilia	1	—	1
Lucigobius	1	—	1
Macrodontogobius	1	—	1
Mahidolia	1	1	2
Mangarinus	1	—	1
Mugiligobius	7	—	7
Myersina	2	1	3
Oligolepis	1	1	2
Oplopomops	2	—	2
Oplopomus	1	—	1
Opua	1	—	1
Oxyurichthys	6	2	8
Palutris	1	—	1
Pandaka	2	—	2
Papillogobius	1	—	1
Parachaeturichthys	1	—	1
Paragobiodon	5	—	5
Periophthalmus	6	—	6
Pleurosicya	6	3	9
Priolepis	10	7	17
Pseudogobius	2	—	2
Psilogobius	1	—	1
Redigobius	3	—	3
Scartelaos	1	—	1
Schismatogobius	1	—	1
Sicyopterus	2	1	3
Signigobius	1	—	1
Silhouettea	1	1	2
Stonogobiops	2	2	4
Sueviota	1	—	1
Taenioides	2	—	2
Tomiyamaichthys	1	1	2
Trimma	10	27	37
Trimmatom	1	2	3
Valenciennea	7	5	12
Vanderhorstia	5	3	8
Waitea	1	—	1
Yongeichthys	3	—	3
KRAEMERIIDAE	6	—	6
Gobitrichinotus	1	—	1
Kraemeria	4	—	4
Parkraemeria	1	—	1
MICRODESMIDAE	25	3	28
Gunnellichthys	4	—	4
Nemateleotris	3	—	3
Paragunnelichthys	1	—	1
Parioglossus	8	1	9

Classification	Described species	Undescribed species[a]	Total species
Ptereleotris	9	2	11
XENISTHMIDAE	3	1	4
Allomicrodesmis	1	—	1
Xenisthmus	2	1	3
Suborder: ACANTHUROIDEI	**74**	**2**	**76**
SIGANIDAE	20	1	21
Siganus	20	1	21
ZANCLIDAE	1	—	1
Zanclus	1	—	1
ACANTHURIDAE	53	1	54
Acanthurus	26	—	26
Ctenochaetus	6	—	6
Naso	13	1	14
Paracanthurus	1	—	1
Prionurus	3	—	3
Zebrasoma	4	—	4
Suborder: SPHYRAENOIDEI	**13**	**—**	**13**
SPHYRAENIDAE	13	—	13
Sphyraena	13	—	13
Suborder: SCOMBROIDEI	**2**	**—**	**2**
GEMPYLIDAE	2	—	2
Epinnula	1	—	1
Rexea	1	—	1
Suborder: STROMATEOIDEI	**2**	**—**	**2**
STROMATEIDAE	2	—	2
Pampus	2	—	2
Order: PLEURONECTIFORMES	**78**	**20**	**98**
Suborder: PLEURONECTOIDEI	**78**	**20**	**98**
CITHARIDAE	2	—	2
Citharoides	1	—	1
Lepidoblepharon	1	—	1
PARALICHTHYIDAE	7	—	7
Paralichthys	1	—	1
Pseudorhombus	5	—	5
Tarphops	1	—	1
BOTHIDAE	32	11	43
Arnoglossus	4	1	5
Asterorhombus	1	3	4
Bothus	4	—	4
Crossorhombus	2	1	3
Engyprosopon	7	6	13
Japonolaeops	1	—	1
Kamoharaia	1	—	1
Laeops	2	—	2
Lophonectes	1	—	1
Neolaeops	1	—	1
Parabothus	2	—	2
Psettina	3	—	3
Taeniopsetta	2	—	2
Tosarhombus	1	—	1
PLEURONECTIDAE	6	2	8
Peltorhamphus	1	—	1
Poecilopsetta	2	1	3

Classification	Described species	Undescribed species[a]	Total species
Samaris	1	1	2
Samariscus	2	—	2
SOLEIDAE	18	7	25
Aesopia	1	1	2
Aseraggodes	8	6	14
Heteromycteris	1	—	1
Liachirus	1	—	1
Parachirus	1	—	1
Pardachirus	1	—	1
Rhinosolea	1	—	1
Soleichthys	1	—	1
Synaptura	2	—	2
Zebrias	1	—	1
CYNOGLOSSIDAE	13	—	13
Arelia	1	—	1
Cynoglossus	8	—	8
Paraplagusia	4	—	4
Order: TETRAODONTIFORMES	**143**	**4**	**147**
Suborder: BALISTOIDEI	**86**	**2**	**88**
TRIACANTHODIDAE	6	—	6
Atrophacanthus	1	—	1
Halimochirugus	1	—	1
Paratriacanthodes	1	—	1
Triacanthodes	2	—	2
Tydemania	1	—	1
TRIACANTHIDAE	1	—	1
Triacanthus	1	—	1
BALISTIDAE	23	—	23
Abalistes	1	—	1
Balistapus	1	—	1
Balistes	1	—	1
Balistoides	2	—	2
Canthidermis	1	—	1
Melichthys	2	—	2
Odonus	1	—	1
Pseudobalistes	2	—	2
Rhinecanthus	4	—	4
Sufflamen	3	—	3
Xanthichthys	4	—	4
Xenobalistes	1	—	1
MONACANTHIDAE	43	2	45
Acreichthys	3	—	3
Aluterus	2	—	2
Amanses	1	—	1
Brachaluteres	2	—	2
Cantherhines	8	—	8
Cantheschenia	2	—	2
Chaetoderma	1	—	1
Laputa	—	1	1
Monacanthus	1	—	1
Oxymonacanthus	1	—	1
Paraluteres	1	—	1

Classification	Described species	Undescribed species[a]	Total species
Paramonacanthus	2	—	2
Parika	1	—	1
Pervagor	6	1	7
Pseudaluteres	1	—	1
Pseudomonacanthus	1	—	1
Rudarius	2	—	2
Stephanolepis	1	—	1
Thamnaconus	7	—	7
OSTRACIIDAE	13	—	13
Kentrocapros	1	—	1
Lactoria	3	—	3
Ostracion	5	—	5
Rachynchostracion	2	—	2
Tetrosomus	2	—	2
Suborder: TETRAODONTOIDEI	**57**	**2**	**59**
TETRAODONTIDAE	48	2	50
Amblyrhinchotus	1	1	2
Arothron	10	—	10
Canthigaster	17	—	17
Chelonodon	1	—	1
Fugu	1	—	1
Lagocephalus	5	—	5
Sphoeroides	3	—	3
Takifugu	2	—	2
Tetractenos	1	—	1
Torquigener	7	1	8
DIODONTIDAE	9	—	9
Chilomycterus	1	—	1
Cyclichthys	2	—	2
Diodon	4	—	4
Lophodiodon	1	—	1
Tragulichthys	1	—	1
TOTAL	**2,950**	**442**	**3,392**

Note: Phylogeny follows Eschmeyer (1990).

a. Numbers of undescribed species are approximate, based on published records and personal communication with various investigators.

REFERENCES

Allen, G. R. 1985a. *Butterfly and angelfishes of the world.* Vol. 2. MERGUS Publishers Hans A. Baensch, Melle, Germany.

———. 1985b. FAO species catalogue. Vol. 6. Snappers of the World. *FAO Fish. Synop.* 125.

Bruce, R. W., and J. E. Randall. 1985. Revision of the Indo-Pacific parrotfish genera *Calotomus* and *Leptoscarus*. *Indo-Pacific Fishes* 5:1–32.

Carpenter, K. E. 1988. FAO species catalogue. Vol. 8. Fusilier Fishes of the World. *FAO Fish. Synop.* 125.

Carpenter, K. E., and G. R. Allen. 1989. FAO species catalogue. Vol. 9. Emperor Fishes and Large-eye Breams of the World. *FAO Fish. Synop.* 125.

Chave, E. H., and B. C. Mundy. 1994. Deep-sea benthic fish of the Hawaiian Archipelago, Cross Seamount, and Johnston Atoll. *Pac. Sci.* 48(4): 367–409.

Cohen, D. M. 1973. Zoogeography of the fishes of the Indian Ocean. In *Ecological Studies, Analysis and Synthesis,* ed. B. Zeitzschel, 3:451–63.

Compagno, L. J. V. 1984. FAO species catalogue. Vol. 4: Sharks of the World. *FAO Fish. Synop.* 125.

Dawson, C. E. 1985. Indo-Pacific pipefishes (Red Sea to Americas). Gulf Coast Res. Lab., Ocean Springs, MS.

DiSalvo, L. H., J. E. Randall, and A. Cea. 1988. Ecological reconnaissance of the Easter Island sublittoral marine environment. *Nat. Geog. Res.* 4(4): 451–73.

Eschmeyer, W. N. 1990. *Catalog of the genera of recent fishes.* San Francisco: California Academy of Sciences.

Francis, M. P. 1993. Checklist of the coastal fishes of Lord Howe, Norfolk, and Kermadec Islands, Southwest Pacific Ocean. *Pac. Sci.* 47(2): 136–70.

Froese, R., and D. Pauly. 1994. A strategy and a structure for a database on aquatic biodiversity. In *Data sources in Asian-Oceanic countries,* ed. J.-L. Wu, Y. Hu, and E. F. Westrum Jr., 209–20. Taipei: DSAO.

Gon, O. 1993. Revision of the cardinalfish genus *Cheilodipterus* (Perciformes: Apogonidae), with description of five new species. *Indo-Pacific Fishes* 22:1–59.

Hoese, D. F., and H. K. Larson. 1994. Revision of the Indo-Pacific gobiid fish genus *Valenciennea,* with descriptions of seven new species. *Indo-Pacific Fishes* 23:1–71.

Holleman, W. 1991. A revision of the tripterygiid fish genus *Norfolkia* Fowler, 1953 (Perciformes: Blennioidei). *Ann. Cape Prov. Mus. (Nat. Hist.).* 18:227–43.

Hutchins, J. B. 1986. Review of the monacanthid fish genus *Pervagor,* with descriptions of two new species. *Indo-Pacific Fishes* 12:1–35.

Kosaki, R. K., R. L. Pyle, J. E. Randall, and D. K. Irons. 1989. New records of fishes from Johnston Atoll, with notes on biogeography. *Pac. Sci.* 45(2): 186–203.

Kulbicki, M., J. E. Randall, and J. Rivaton. 1994. Checklist of the fishes of the Chesterfield Islands (Coral Sea). *Micronesica* 27(1/2): 1–43.

Lachner, E. A., and S. J. Karnella. 1980. Fishes of the Indo-Pacific genus *Eviota* with descriptions of eight new species (Teleostei: Gobiidae). *Scartichthys. Smithsonian Contrib. to Zool.* 315:1–127.

Larson, H. K. 1990. A revision of the commensal gobiid fish genera *Pleurosicya* and *Luposicya* (Gobiidae), with descriptions of eight new species of *Pleurosicya* and discussion of related genera. *Beagle, Rec. Northern Territory Mus. Arts Sciences* 7:1–53.

Markle, D. F., and J. E. Olney. 1990. Systematics of the pearlfishes (Pisces: Carapidae). *Bull. Mar. Sci.* 47:269–410.

Masuda, H., K. Amaoka, C. Araga, T. Uyeno, and T. Yoshino, eds. 1988. *The Fishes of the Japanese Archipelago.* Tokyo: Tokai University Press.

McCosker, J. E., and R. H. Rosenblatt. 1993. A revision of the snake eel genus *Myrichthys* (Anguilliformes: Ophichthidae) with the description of a new eastern Pacific species. *Proc. Calif. Acad. Sci.* 48(8): 153–69.

Murdy, E. O., and D. F. Hoese. 1985. Revision of the gobiid fish genus *Istigobius*. *Indo-Pacific Fishes* 4:1–41.

Myers, R. F. 1988. An annotated checklist of the fishes of the Mariana Islands. *Micronesica* 21(1–2).

———. 1991. *Micronesian reef fishes*. Guam: Coral Graphics.

Orr, J. W., and R. A. Fritzsche. 1993. Revision of the ghost pipefishes, Family Solenostomidae (Teleostei: Syngnathoidei). *Copeia*, no. 1:168–82.

Palsson, W. A., and T. W. Pietsch. 1989. Revision of the acanthopterygian fish family Pegasidae (Order Gasterosteiformes). *Indo-Pacific Fishes* 18:1–38.

Pietsch, T. W., and D. B. Grobecker. 1987. *Frogfishes of the world*. Stanford: Stanford University Press.

Pyle, R. L. In press. Life on the edge of darkness: A wealth of diversity awaits discovery on the deep coral reefs. *Nat. Hist.*

Pyle, R. L., and J. E. Randall. 1992. A new species of *Centropyge* from the Cook Islands, with a redescription of *Centropyge boylei*. *Revue Fr. Aquariol.* 19(4): 115–24.

Randall, J. E. 1980. Revision of the fish genus *Plectranthias* (Serranidae: Anthiinae) with descriptions of 13 new species. *Micronesica* 16(1): 101–87.

———. 1985. Fishes. In *French Polynesian coral reefs*, ed. B. Delesalle, R. Galzin, and B. Salvat, 379–520. Vol. 1, *Fifth Intl. Coral Reef Congress*, 27 May–1 June, Tahiti.

———. 1986. 106 new records of fishes from the Marshall Islands. *Bull. Mar. Sci.* 38(1): 170–252.

Randall, J. E., G. R. Allen, and R. C. Steene. 1990. *Fishes of the Great Barrier Reef and Coral Sea*. Bathurst, NSW: Crawford House Press.

Randall, J. E., and R. W. Bruce. 1983. The parrotfishes of the subfamily Scaridae of the Western Indian Ocean, with descriptions of three new species. *Bull. J.L.B. Smith Inst. Ichthyol., Rhodes Univ.*, no. 47:1–39.

Randall, J. E., J. L. Earle, R. L. Pyle, J. D. Parrish, and T. Hayes. 1993. Annotated checklist of the fishes of Midway Atoll, Northwestern Hawaiian Islands. *Pac. Sci.* 47(4): 356–400.

Randall, J. E., and A. C. Egaña. 1984. Native fish names of Easter Island fishes, with comments on the origin of the Rapanui people. *Occas. Pap. B.P. Bishop Mus.* 25(12): 1–16.

Randall, J. E., and P. C. Heemstra. 1991. Revision of the Indo-Pacific groupers (Perciformes: Serranidae: Epinephelinae), with descriptions of five new species. *Indo-Pacific Fishes* 20:1–332.

Randall, J. E., and D. F. Hoese. 1985. Revision of the Indo-Pacific dartfishes, genus *Ptereleotris* (Perciformes: Gobiodei). *Indo-Pacific Fishes* 7:1–36.

Randall, J. E., P. S. Lobel, and E. H. Chave. 1985. Annotated checklist of the fishes of Johnston Island. *Pac. Sci.* 39(1): 24–80.

Randall, J. E., and J. E. McCosker. 1992. Revision of the fish genus *Luzonichthys* (Perciformes: Serranidae: Anthiinae), with descriptions of two new species. *Indo-Pacific Fishes* 21:1–21.

Sharkey, P., and R. L. Pyle. 1993. The twilight zone: The potential, problems, and theory behind using mixed gas, surface-based scuba for research diving between 200 and 500 feet. In *Diving for Science . . . 1992, Proceedings of the American Academy of Underwater Sciences Twelfth Annual Scientific Diving Symposium*. American Academy of Underwater Sciences, Costa Mesa, California.

Springer, V. G. 1982. Pacific Plate biogeography, with special reference to shorefishes. *Smithsonian Contr. Zool.*, no. 367:1–182.

———. 1988. The Indo-Pacific blenniid fish genus *Ecsenius*. *Smithsonian Contr. Zool.*, no. 465:1–134.

Springer, V. G., and J. T. Williams. 1994. The Indo-Pacific blenniid fish genus *Istiblennius* reappraised: A revision of *Istiblennius, Blenniella,* and *Paralticus*, new genus. *Smithsonian Contr. Zool.*, no. 565:1–193.

Strasburg, D. W., E. C. Jones, and R. T. B. Iverson. 1968. Use of a small submarine for biological and oceanographic research. *J. Cons. perm. int. Explor. Mer.* 31(3): 410–26.

Thresher, R. E., and P. L. Colin. 1986. Trophic structure, diversity, and abundance of fishes of the deep reef (30–300 m) at Enewetak, Marshall Islands. *Bull. Mar. Sci.* 38(1): 253–72.

Williams, J. T. 1988. Revision and phylogenetic relationships of the blenniid fish genus *Cirripectes. Indo-Pacific Fishes* 17:1–78.

———. 1990. Phylogenetic relationships and revision of the blenniid fish genus *Scartichthys. Smithsonian Contrib. to Zool.* 492:1–30.

Woodland, D. J. 1990. Revision of the fish family Siganidae with descriptions of two new species and comments on distribution and biology. *Indo-Pacific Fishes* 19:1–136.

Zama, A., and F. Yasuda. 1979. An annotated checklist of fishes from the Ogasawara Islands— Suppl. I, with zoogeographical notes on the fish fauna. *J. Tokyo. Univ. Fish.* 65(2): 139–63.

CHAPTER 13

Marine Ecosystem Classification for the Tropical Island Pacific

Paul F. Holthus and James E. Maragos

ABSTRACT

A comprehensive, hierarchical classification system for the coastal and marine ecosystems of the tropical island Pacific, including the region's extensive coral reefs, has been developed. The classification system was developed with input from a wide range of coastal/marine scientists and resource managers and allows ecosystems to be inventoried on a systematic, regionally valid and comparable basis. Such inventories can be conducted as part of field surveys or by the use of existing information (field survey reports, maps, air photos) and are valuable for determining conservation priorities on a local, national, or regional scale. A preliminary inventory of ecosystems in U.S.-affiliated islands of the tropical Pacific was undertaken to test and refine the ecosystem classification system. A revised coastal/marine ecosystem classification system that is applicable to the broader tropical Indo-Pacific region is now presented. The revised classification system may be useful in inventorying and classifying coastal and marine ecosystems in other tropical coastal areas.

INTRODUCTION

The island countries and territories of the South Pacific region consist of 550,000 km² of land with 5.2 million inhabitants scattered across some 29 million km² of the Pacific Ocean. However, the figures drop to merely 87,587 km² of land area and 2.2 million people if Papua New Guinea (PNG) is excluded. Among these twenty-seven island states in the region, there are four with land areas of less than 100 km², eleven with 100 to 1,000 km², and another seven with land areas ranging from 2,935 to 27,556 km².

There are four principal types of islands in the South Pacific region: continental, volcanic, raised limestone, and low islands (Thomas 1963; Dahl 1980). The first three types are all "high" islands, whereas low islands are built of coral reef sediment and rubble that accumulate to elevations of only a few meters on reef platforms, such as the ring of reef that forms an atoll. Virtually all the islands in the region are entirely coastal in character.

That is, all parts of the island influence, or are influenced by, processes and activities occurring on coastal lands and in nearshore waters.

Coastal and marine ecosystems are critically important to Pacific island peoples, cultures, and economies. The coastal and nearshore marine areas of all islands in the tropical Pacific are the location of the vast majority of human habitation, the focus of subsistence and commercial agricultural and fisheries activity, and the target of most economic development (Connell 1984; Dahl 1984). This combination of factors is accelerating the destruction or degradation of coastal habitats, the overexploitation of natural resources, and conflicts in coastal resource use. Common problems in the nearshore environment include terrestrial sedimentation, sewage pollution, waste dumping, overharvesting of living resources, destructive fishing techniques, and dredging or filling nearshore areas (Brodie et al. 1990). The coastal areas of the Pacific, especially low elevation islands, are also subject to the destructive effects of natural hazards from extreme events such as cyclones. Global warming now threatens to exacerbate these hazards through accelerated sea-level rise and other possible changes to oceanographic conditions.

The island Pacific region contains a range of oceanic and nearshore marine environments. The term "coastal and marine environments" (hereinafter "marine environments") is used here to generally refer to the components, functions, and processes of the shoreline and intertidal area and the nearshore waters and benthos around small islands. Marine biological diversity generally decreases from west to east across the Pacific Ocean, with the regional center of high diversity in the southwest Pacific (Holthus and Maragos 1992). Within this broad pattern are ecosystems important to rare, threatened, or endangered species and a few known centers of endemism. Relatively little is known about the large ocean water masses of the Pacific and their deep-sea features and benthic communities. Shallow and nearshore marine areas support coral reef, lagoon, mangrove, and seagrass systems and contain a diverse flora and fauna that are critical to the cultural, subsistence, and economic life of Pacific islands. Marine protected areas (MPAs) are an important tool for conserving marine biological diversity; however, there are few MPAs in the island Pacific region, and the marine biological diversity of no part of the region can be considered effectively protected (Thomas, Holthus, and Reti 1992).

Coastal management problems are widespread in the region and in some areas require urgent attention. The potential for sustainable development of coastal and marine resources is being permanently lost or compromised at the same time as their value to the rest of the world (e.g., for biomedical purposes) is only beginning to be explored. Much of the degradation of coastal habitats and depletion of resources in the island Pacific could be avoided, reduced, or mitigated through Integrated Coastal Zone Management (ICZM) (i.e., comprehensive, multisectoral, integrated planning and management for the sustainable development, multiple use, and conservation of coastal and marine environments [Holthus in press]). ICZM is an essential approach for addressing the special environmental and development situation of small islands. The need for ICZM has been recognized globally and at the regional and national level in the Pacific.

NEED FOR SCIENTIFIC INFORMATION ON MARINE ENVIRONMENTS

Information on the biological, ecological, geological, and oceanographic characteristics and processes of marine environments is an essential component of ICZM. Unfortunately, despite the importance of coastal and marine environments to the region, this kind of information is usually lacking for most of the small islands of the island Pacific. Much research is needed to provide a sound scientific basis for managing marine environments in the Pacific. Scientific research is often a slow and expensive process, and is more so in the island Pacific region due to the complexity of tropical marine environments and the scattered and isolated location of the islands. In addition, the economic realities facing the developing countries in the region mean very little of the needed research is being undertaken.

The marine environments of small islands in the Pacific will not be able to be managed with scientific rigor for some time and will have to rely on less than adequate information for many years to come at the same time as development pressures and impacts are increasing. However, a substantial amount of information important to ICZM—the management of marine resources and sustainable island development—can be readily obtained through the inventory ecosystems using a regional ecosystem classification. Thus marine ecosystems can be evaluated, and conservation priorities set, on a systematic, comparable basis for any country, or for the region as a whole.

CLASSIFICATION AND INVENTORY OF MARINE ECOSYSTEMS

A regionally accepted framework for systematically identifying the presence of ecosystems, and inventorying their occurrence, is necessary as a basis for identifying areas of priority for conservation on a national, regional, or global scale.

An Ecosystem Classification for the Tropical Island Pacific

It has not been possible to determine the full range of ecosystems in each country or territory of the island Pacific, or in the region as a whole, without a complete ecosystem checklist based on a systematic classification. In spite of this, there have been efforts to inventory terrestrial biodiversity and recommend areas for conservation (Dahl 1988), but there has been no comparable effort for the vast majority of the tropical island Pacific—the marine ecosystems.

In response to the need to inventory ecosystems on a systematic, regionally valid, and comparable basis, the South Pacific Ecosystem Classification System (SPECS) was developed as a comprehensive, hierarchical, and scientific classification of the ecosystems of the tropical island Pacific region. Scientific and island country experts familiar with classifying natural systems and/or familiar with the ecosystems of the tropical island Pacific participated in workshops to develop the initial classification system, building on existing efforts. This process has resulted in:

- a revised biogeography of the region (Stoddart 1992),

- a freshwater ecosystem classification system (Polhemus, Maciolek, and Ford 1992),

- a review of past classifications and initial development of a marine classification system (Maragos 1992),

- the preliminary development of a terrestrial classification system (Fosberg and Pearsall 1993), and

- an integrated status report on the entire ecosystem classification initiative (Pearsall et al. 1992).

The Marine Ecosystem Classification System (MECS) component of the classification system combines biogeographical, morphological, and geophysical characteristics. The MECS has a simplified biogeographic component, with only continental and/or oceanic subdivisions, following Stoddart (1992). The basic structure of the marine ecosystem classification is presented in Figure 13.1. At its higher levels, the classification system divides the marine environment into ocean and benthic components. The ocean area is subdivided into the open ocean (pelagic) and the nearshore ocean (neritic), each of which has several types of water mass. The bottom (benthic) area is subdivided into continental and noncontinental (oceanic) classes. The next level divides islands into either high-island and atoll/table reef/low-island earthform types. The oceanic area also has submerged earthform types.

Further levels of classification of marine ecosystems are outlined in Appendix 1, which contains the full classification. Diagrams to illustrate the upper levels of the classification are found in Appendix 2. Ecological units, the lowest level of classification, are not listed in the classification. Examples of ecological units are listed in Table 13.1. Further work is required to fully develop the lowest levels of MECS because field surveys and ground truthing will generally be needed to identify dominant organisms in the ecological units. A glossary has been developed and continues to be updated as the MECS is refined (Appendix 3). French and German versions of the glossary are also being compiled.

Table 13.1 Examples of ecological units to be added to the lowest level of the marine component of the South Pacific Ecosystem Classification System

Classification element	Ecological unit	Dominant or conspicuous organisms
Open ocean, divergence, epipelagic	Yellowfin tuna aggregations	*Thunnus albacares*
Open ocean, boundary current, epipelagic	Albacore schools	*Thunnus alalunga*
Nearshore ocean, coastal ocean, epipelagic	Skipjack tuna schools	*Katsuwonus pelamis*
Open ocean, water mass, epipelagic	Bigeye tuna aggregations	*Thunnus obesus*
Open ocean, coastal ocean, epipelagic	Eastern little tuna	*Euthynnuis affinis*
Bottom continental, high island, sand beach	Ghost crab community	*Ocypode* sp.
Bottom continental, high island, boulder beach	Littorine snail community	*Littorina* sp., *Nerita* sp.
Bottom noncontinental, marine bench	Boring urchin community	*Echinometra mathaei*
Bottom, continental, lagoon fringing reef	Mangrove forest	*Brugiera gymnorhiza*
Bottom, noncontinental, reef area	Microatoll corals	*Porites lutea*

Source: Maragos (1992).

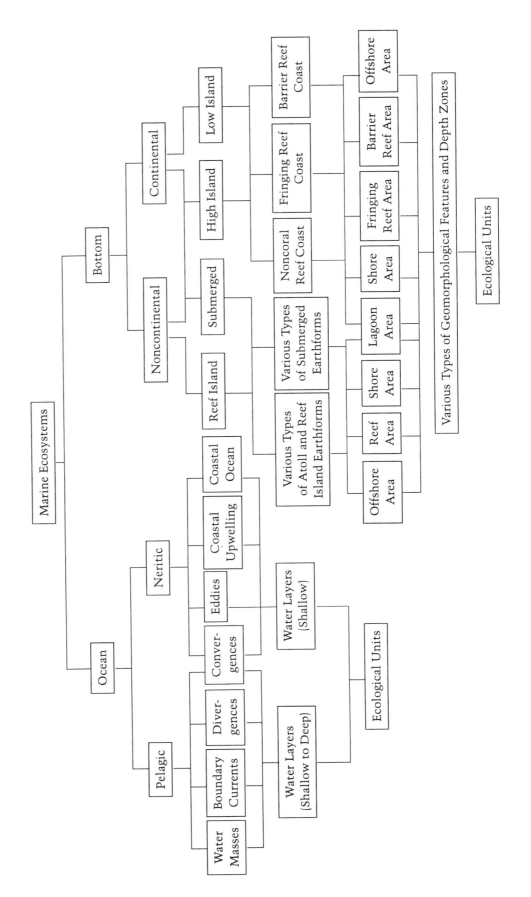

Figure 13.1 Basic structure of the marine component of the South Pacific Ecosystem Classification System (Source: Maragos 1992)

Preliminary Inventory of Pacific Island Ecosystems

It was anticipated that SPECS would be tested and refined through the preliminary inventory of ecosystems in a subset of Pacific island areas. Such a preliminary inventory was undertaken for the U.S.-affiliated islands of the tropical Pacific, including the Republic of the Marshall Islands, Federated States of Micronesia (FSM), Territory of Guam, Commonwealth of the Northern Mariana Islands, Republic of Palau, and Territory of American Samoa (Holthus et al. 1993).

A total of 127 main islands or island clusters were evaluated by applying SPECS to an analysis of the literature, maps, and aerial photographs available for each of the islands. The number of principal islands in each country or territory, and their island type, is summarized in Table 13.2. The actual number of islands covered by the inventory was higher due to the lumping of island clusters (e.g., the Rock Islands of Palau; the high and low islets surrounding Pohnpei Island) under a single island database record.

The kinds and level of information available for each island varied greatly, and for most of the islands information was very limited. The best information base available for an island was a combination of (1) large-scale maps (usually 1:25,000); (2) recent low altitude, color air photos; and (3) recent field survey reports with extensive coverage of the island. This combination of sources was rarely available. For many islands, the only information source was a small-scale hydrographic chart. The following aspects of each island were systematically evaluated using SPECS during the ecosystem inventory:

- island shape and dimensions;

Table 13.2 Island type and number of islands inventoried

	FSM Total	Kosrae State	Pohnpei State	Chuuk State	Yap State	Marshall Islands	Palau	Guam	Northern Mariana Islands	American Samoa	Total
High islands											
Continental	0	0	0	0	0	0	2	0	0	0	2
Volcanic (basalt)	16	1	1[a]	14[b]	0	0	0	0	9	5	30
Raised limestone	1	0	0	1	0	0	3[c]	0	3	0	7
Mixed	1	0	0	0	1[d]	0	2	1	2	0	6
Atoll/reef islands											
Atoll (many deep passes/open)	9	0	1	4	4	17	0	0	0	0	26
Atoll (few/one deep passes)	11	0	3	4	4[e]	6	2	0	1	1	21
Atoll (no deep passes)	13	0	5	3	5	6	1	0	0	0	20
Atoll (completely land-ringed)	0	0	0	0	0	0	0	0	0	1	1
Reef island (with water body)	1	0	0	1	0	3	0	0	0	0	4
Reef island (without water body)	5	0	0	2	3	2	4	0	0	0	11
Total	57	1	10	29	17	34	14	1	15	7	128

Source: Holthus et al. (1993).

a. Excludes high islets and barrier reef islets in Pohnpei lagoon.

b. Only includes principal high islets in Chuuk island lagoon.

c. Excludes 504 individual Rock Islands.

d. Includes all Yap Proper islands.

e. Includes Faurulep Atoll (type unknown).

- coastal ecosystems, including coast type, coastline type, reef perimeter, shore area, reef islet area, and shorelines;

- marine ecosystems, including fringing, barrier, and perimeter reef areas (e.g., outer reef slope, reef top, and lagoon reef slope), and lagoon areas; and

- terrestrial ecosystems, including native vegetation, nonnative vegetation, and non-marine wetlands.

The existing *Island Directory* database (Dahl 1991) provided the framework for compiling and presenting the detailed ecosystem inventory information, and a Pacific Ecosystem Database Record for each island was compiled to record the results of the preliminary inventory (Holthus et al. 1993).

The inventory of marine ecosystems focused on intertidal and nearshore marine areas and was primarily based on the interpretation of maps and air photos, as there were few field survey reports available on the marine ecosystems of most of the islands investigated. However, the few marine field survey reports that were available did provide "ground truth" information for the inventory of marine ecosystems on other islands. For a few islands, detailed, comprehensive information was available in coastal, shoreline, or coral reef atlases and survey reports or in island-specific summaries in the *Coral Reefs of the World* (UNEP/IUCN 1988). Overall, the level of information available for each island was highly variable.

Interpretation and Application of Preliminary Ecosystem Inventory Results

Recognizing that conservation action is predominately undertaken at the country level, the preliminary inventory results were analyzed to indicate which islands within each country are more important for conservation (Table 13.3). Islands on which many kinds of ecosystems were identified were considered to have a "rich" diversity of ecosystems for

Table 13.3 Number of islands with important ecosystems and number of islands with "low" levels of information adequacy/availability

	FSM	Marshall Islands	Palau	Guam	Northern Mariana Islands	American Samoa
Islands rich in terrestrial and nonmarine wetland ecosystems	14	14	4	3	3	4
Islands with rare terrestrial and nonmarine wetland ecosystems	20	15	3	4	4	6
Islands rich in shore, marine, and marine wetland ecosystems	27	20	13	7	7	3
Islands with rare shore, marine, and marine wetland ecosystems	20	8	6	5	5	6
Islands "low" in information availability/adequacy	42	10	*	11	11	2

Source: Holthus et al. (1993).

*Information base is adequate due to a recent "rapid ecological assessment" field inventory, but this information has not yet been fully evaluated using the SPECS.

that country. Islands supporting ecosystems that had been identified on only one or a few islands in a country were considered to contain "rare" ecosystems.

Relative "richness" or "rareness" of ecosystems may provide an initial basis for determining the relative importance and interest of an island for conservation action or further investigations, although the determinations are likely to be biased toward those islands for which a more thorough and detailed dataset was available. The results of the preliminary inventory revealed that there are generally more islands rich in shore, marine, and marine wetland ecosystems than there are islands "rich" in terrestrial and nonmarine wetland ecosystems (Table 13.3). However, the number of islands in each country with "rare" shore, marine, and marine wetland ecosystems was variable in comparison to terrestrial ecosystems.

The adequacy of the information base used in this preliminary inventory was analyzed to indicate the validity of the results and to determine the need for further efforts of ecosystem inventory. Those islands ranked "low" in information availability or adequacy were considered as a high priority for additional inventory efforts (Table 13.3).

Interpretation and Application of Ecosystem Classification and Inventory Information

The inventory of marine ecosystems using a regionally accepted classification system can identify representative, rare, and unique ecosystems at a variety of scales. The interpretation and application of inventory results may be on an island, country, regional, or, if the classification system interfaces with a global ecosystem classification, global scale.

At present, the classification has only been used to conduct a preliminary inventory of the ecosystems of U.S. Pacific island areas. The classification is also being considered for inclusion in ReefBase as the detailed framework for classifying coral reefs. The marine ecosystem classification system developed for the tropical island Pacific may represent the most practical and concerted effort to analyze and classify the marine environment of most of the world's small tropical islands. The classification may thus be applicable or adaptable, with some revision, to other small tropical coastal and island areas (e.g., Caribbean, Indian Ocean).

The results of an inventory of marine ecosystems employing a hierarchical classification system will allow coastal and marine sites to be selected for a variety of purposes on a number of these scales. For example, for a given island, the inventory would provide the basis to select sites for the long-term marine ecosystems monitoring for local anthropogenic impacts. On a broader scale, the inventory could be used to select sites for the long-term monitoring of impacts associated with global change as part of an international program. An inventory of ecosystems can also provide valuable information indicating the likely impacts of climate change and a basis for planning response measures. In particular, the ecosystem inventory can aid in determining the potential vulnerability of different shoreline areas to sea-level rise and thereby aid in setting priorities for planning (Holthus et al. 1992).

Table 13.4 Proposed criteria for selecting ecosystem types and sites for conservation

1. Biological diversity	4. Ecological importance
a. Richness of habitats	5. Naturalness
b. Richness of taxa	6. Threats to the ecosystem
c. Richness of native taxa	7. Ecosystem fragility and resilience
d. Richness/fidelity of endemic taxa	8. Size
e. Richness of rare or threatened taxa	9. Economic importance
2. Ecosystems as critical habitat	10. Social and cultural importance
a. Critical habitat: Life stages or processes	11. Scientific importance
b. Refugia	12. Conservation feasibility (manageability)
3. Biogeographic importance	
a. Rare ecosystems	
b. Unique ecosystem aspects	
c. Representative ecosystems	
d. Biogeographic scale	

Source: Holthus et al. (1993).

At the level of an individual island, an inventory of marine ecosystems would provide a valuable information base for many aspects of development planning, such as environmental impact assessment and oil spill contingency planning. The distribution of marine ecosystems is also an essential component of the information base needed for ICZM, particularly for developing land and water use zoning and identifying candidate protected areas. The classification system potentially allows protected areas to be selected as representative, rare, or unique on an island, national, regional, or global scale. Identified as part of the preliminary inventory of ecosystems in U.S.-affiliated islands were criteria that may be used to evaluate the selection of ecosystem types and sites for conservation (Table 13.4).

CONCLUSIONS AND RECOMMENDATIONS

The inventorying of marine ecosystems using a regional classification system is an important tool to obtain marine science information in a timely and cost-effective manner. When used in concert with other sources of information, ecosystem inventory can provide a substantial amount of information for ICZM in support of the sustainable development of small islands and developing tropical coastal areas. However, the marine environments in most of these areas remain largely unknown, while the pressures and impacts of development and resource use continue to increase and expand. There is much that can be done to expand the application of the approaches described here to further support the development and implementation of ICZM in small islands of the tropical Pacific and elsewhere.

Further Application of the Marine Ecosystem Classification System

Work in several areas, including those listed here, needs to be continued to develop and apply the Marine Ecosystem Classification System (MECS) to the islands of the tropical Pacific.

- Finalize the marine component of the classification system through additional efforts to develop the detailed ecological unit level of classification; complete and publish a final working version of the marine classification to make it available to scientists and resource managers.

- Develop a handbook or field manual version of the MECS, complete with diagrams and photographs of ecosystem types and multi-lingua glossaries, which could be used for training, education, and in the field, particularly by in-country environment and natural resource agency officers.

- Conduct a preliminary inventory of marine ecosystems (using literature, maps, and air photos) for the rest of the island Pacific region as a basis for determining conservation priorities in these areas and at a regional level.

- Undertake more detailed field inventories using the MECS for those islands and areas determined to be a high priority for conservation (e.g., inventory existing marine protected areas to determine which ecosystems are already under some form of protection [ostensibly] at a national and regional scale).

- Further develop and refine the application of the ecosystem inventory and classification as a means to indicate the vulnerability of various shoreline types to sea-level rise (e.g., determine the relative vulnerability of the various combinations of atoll reef and shoreline geomorphological characteristics).

Application of MECS to Other Areas and Issues

The potential application of the MECS to tropical coastal areas should be explored. Furthermore, the interface of the system with global marine ecosystem classifications should be developed to allow information generated by its application in the Pacific to contribute to global programs. Several suggestions are offered, as follows.

- Hold a workshop on marine ecosystem classification for the Indo-Pacific to evaluate and adapt the classification system developed for the Pacific to the marine environments of the broader Indo-Pacific region.

- Test the applicability of the marine classification system developed for the tropical island Pacific by conducting pilot preliminary inventories of a range of coastal areas in the tropical Indo-Pacific.

- Hold a workshop on marine ecosystem classification in the Caribbean and tropical Atlantic, and conduct pilot preliminary inventories of islands in those areas to develop a global tropical MECS.

- Clarify the interface of the MECS with global classifications of marine environments used by agencies with a global mandate for the marine environment (e.g., Unesco Intergovernmental Oceanographic Commission; UNEP Ocean and Coastal Areas/ Program Activity Center; Global Environment Facility).

- Use the MECS to aid in systematically identifying sites in the island Pacific region for global projects (e.g., the long-term global monitoring of marine environments for the effects of climate change).

Acknowledgments—The initial development of the marine ecosystem classification emanated from a workshop held at the East-West Center in 1990. The authors thank all the participants in that workshop for contributing to development of the classification. We also acknowledge the assistance of Michel Ricard and Helmut Schumacher for preparing the French and German versions, respectively, of the glossaries.

APPENDIX 1

Marine Ecosystems of the Tropical Insular Pacific (Version of October 1994)

A. <u>MARINE REALM</u>
    A.A. OCEAN ECOSYSTEMS
        A.A.A.   OCEAN—OPEN (PELAGIC) ECOSYSTEMS
            <u>**1.**</u> <u>**Divergence (Oceanic Upwelling Zone)**</u>
                1.1 **Surface Microlayer**
                1.2 **Epipelagic (0–200 m)**
                      1.2.1 Epipelagic—Shallow
                              1.2.1.1 Waters—Surface
                              1.2.1.2 Waters—Mixed
                              1.2.1.3 Waters—Thermocline
                      1.2.2 Epipelagic—Deep
                          [Light Level modifiers: Photic, Aphotic]
                1.3 **Mesopelagic (200–1000 m)**
                1.4 **Bathypelagic (1000–4000 m)**
                1.5 **Abyssopelagic (4000–7000 m)**
                1.6 **Hadalpelagic (>7000 m)**
             <u>**2.**</u> <u>**Convergence (Oceanic Downwelling Zone)**</u>
                (as in <u>1</u>)
            <u>**3.**</u> <u>**Boundary Current**</u>
                (as in <u>1</u>)
            <u>**4.**</u> <u>**Water Mass—Open Ocean**</u>
                (as in <u>1</u>)
         A.A.B.   OCEAN—NEARSHORE (NERITIC) ECOSYSTEMS
            <u>**1.**</u> <u>**Eddy**</u>
                1.1 **Surface Microlayer**
                1.2 **Epipelagic**
                (as in A.A.A.1.2)
            <u>**2.**</u> <u>**Upwelling Zone—Coastal**</u>
                (as in <u>1</u>)
            <u>**3.**</u> <u>**Water Mass—Nearshore**</u>
                (as in <u>1</u>)
            <u>**4.**</u> <u>**Convergence**</u>
                (as in <u>1</u>)
    A.B. BENTHIC ECOSYSTEMS
        A.B.A.   CONTINENTAL SHELF (NONOCEANIC) ECOSYSTEMS
            <u>**1.**</u> <u>**Earthform—High Island**</u>
                [Geology modifiers: Continental; Volcanic; Limestone; Mixed Geology]
                1.1 **Coast—Noncoral Reef**
                      1.1.1 Area—Shore
                              1.1.1.1   Coastline—Undifferentiated (scale: 10 km)
                                    1.1.1.1.1 Shoreline—Sediment
                                      [Sediment Type modifiers: Calcareous; Terrigenous]
                                      1.1.1.1.1.1   Beach—Boulder/Cobble

1.1.1.1.1.2   Beach—Sand/Gravel
1.1.1.1.1.3   Beachrock
1.1.1.1.1.4   Boulder/Cobble Field
1.1.1.1.1.5   Sand/Gravel Flats
1.1.1.1.1.6   Mud/Silt Flats
1.1.1.1.1.7   Bar and Spit
1.1.1.1.1.8   Mangrove
1.1.1.1.2 Shoreline—Solid Substrate
1.1.1.1.2.1   Cliff—High (ht >10 m)
1.1.1.1.2.2   Cliff—Medium (ht 2–10 m)
1.1.1.1.2.3   Cliff—Low (ht <2 m)
1.1.1.1.2.4   Stack
1.1.1.1.2.5   Talus
1.1.1.1.2.6   Bench/Ramp—Marine
1.1.1.1.2.7   Notch/Cave—Marine
1.1.1.1.3 Shoreline—Artificial
1.1.1.1.3.1   Seawall/Revetment/Bulkhead
1.1.1.1.3.2   Landfill/Causeway/Groin
1.1.1.1.3.3   Fishpond/Fishtrap/Shipwreck
1.1.1.2   Coastline—Cove (scale: 10 km)
(as in 1.1.1.1)
1.1.1.3   Coastline—Bay (scale: 10 km)
(as in 1.1.1.1)
{Salinity modifiers: Marine; Estuarine}
1.1.1.4   Coastline—Coastal Lagoon/Lake/Pond (scale: 10 km)
(as in 1.1.1.1)
{Connectedness modifiers:
Subtidal Lagoon/Subtidal Connection;
Subtidal Lagoon/Intertidal Connection;
Intertidal Lagoon/Intertidal Connection (Barachois)}
{Salinity modifiers: Marine; Estuarine}
1.1.1.5   Coastline—Peninsula (scale: 10 km)
(as in 1.1.1.1)
1.1.1.6   Coastline—Irregular/Discontinuous/Islets
(as in 1.1.1.1)
1.1.2   Area—Nearshore Bottom
{Steepness/Slope Gradient modifiers}
1.1.2.1   High Islet—Ocean
(as in 1.1.1.1)
1.2 Coast—Fringing Reef
1.2.1   Area—Shore
(as in 1.1.1)
1.2.2   Area—Fringing Reef
{Exposure modifiers: Windward; Leeward}
1.2.2.1   Reef Top
{Reef Top Width modifier}
1.2.2.1.1 Reef Top Surface Features
1.2.2.1.1.1   Reef Pavement
1.2.2.1.1.2   Sand/Rubble/Rock Flats
1.2.2.1.1.3   Mud/Silt Flats
1.2.2.1.1.4   Sand/Gravel Flats
1.2.2.1.1.5   Boulder/Cobble Field
1.2.2.1.1.6   Rubble/Boulder Tract
1.2.2.1.1.7   Coral Bed/Microatolls
1.2.2.1.1.8   Algal Bed
1.2.2.1.1.9   Seagrass Bed

1.2.2.1.1.10	Algal Ridge	
1.2.2.1.1.11	Surge Channel	

1.2.2.1.2 Reef Top Subtidal Features

1.2.2.1.2.1	Hoa (Inter-Islet Channel)	
1.2.2.1.2.2	Moat and Depression	
1.2.2.1.2.3	Reef Pool (depth <5 m)	
1.2.2.1.2.4	Reef Hole (depth >5 m)	
1.2.2.1.2.5	Incomplete Reef Top	
1.2.2.1.2.6	Dredge Pit/Quarry/Channel/Basin	

1.2.2.1.3 Reef Top Supratidal Features

1.2.2.1.3.1	Storm Block	
1.2.2.1.3.2	Gravel/Boulder Ridge	
1.2.2.1.3.3	Beachrock	
1.2.2.1.3.4	Conglomerate/Reef Limestone Platform	
1.2.2.1.3.5	Aeolianite	
1.2.2.1.3.6	Coral/Algal Dam and Spillway	
1.2.2.1.3.7	Mangrove	
1.2.2.1.3.8	Fishpond/Fishtrap/Shipwreck	

1.2.2.1.4 Passes/Reef Top Openings
{No. Passes modifier} {Depth/Width modifier}
{Amount of Perimeter modifier}
{% of Reef Perimeter modifier}

1.2.2.1.4.1	Pass—Shallow (depth <10 m; width <2 km)	
1.2.2.1.4.2	Pass—Deep (depth >10 m; width <2 km)	
1.2.2.1.4.3	Reef Top Opening—Shallow (depth <10 m; width >2 km)	
1.2.2.1.4.4	Reef Top Opening—Deep (depth >10 m; width >2 km)	
1.2.2.1.4.5	Pass—False	
1.2.2.1.4.6	Channel—Fringing Reef	
1.2.2.1.4.7	Channel—Artificial	

1.2.2.2 Reef Islets
{Reef Islet Size modifier}
{Linear Ocean Extent modifier}
{% of Reef Perimeter modifier}
{No. of Reef Islets modifier}
{Water Body modifiers: Barachois; Anchialine Pond }

1.2.2.2.1 Shoreline—Sediment
{Orientation modifiers: Outer/Ocean; Inner/Lagoon/Shore}
(as in 1.1.1.1.1)

1.2.2.2.2 Shoreline—Solid Substrate
{Orientation modifiers: Outer/Ocean; Inner/Lagoon/Shore}
(as in 1.1.1.1.2)

1.2.2.2.3 Shoreline—Artificial
(as in 1.1.1.1.3)

1.2.2.3 Reef Slope—Outer
{Steepness/Slope Gradient modifiers}
{Substrate modifiers: Calcareous; Terrigenous; Volcanic}

1.2.2.3.1 Reef Slope Features (Nonterrace)

1.2.2.3.1.1	Spur and Groove	
1.2.2.3.1.2	Tunnel (Room and Pillar)	
1.2.2.3.1.3	Buttress and Valley	
1.2.2.3.1.4	Reef Edge Scarp	
1.2.2.3.1.5	Slope—Coral	
1.2.2.3.1.6	Slope—Coral/Sediment	
1.2.2.3.1.7	Slope—Solid Substrate	
1.2.2.3.1.8	Slope—Sand	
1.2.2.3.1.9	Slope—Sand/Rubble/Rock	

1.2.2.3.1.10   Slope—Boulder/Block
1.2.2.3.1.11   Submarine Cliff
1.2.2.3.1.12   Submarine Wall
1.2.2.3.1.13   Submarine Notch/Cave
1.2.2.3.2 Reef Slope Terrace/Submarine Platform Features
{Terrace/Submarine Platform Width modifier}
{Surface modifiers: with Furrows}
1.2.2.3.2.1   Terrace/Submarine Platform—Coral
1.2.2.3.2.2   Terrace/Submarine Platform—Coral/Sediment
1.2.2.3.2.3   Terrace/Submarine Platform—Solid Substrate
1.2.2.3.2.4   Terrace/Submarine Platform—Sand
1.2.2.3.2.5   Terrace/Submarine Platform—Sand with Coral Mounds
1.2.2.3.2.6   Terrace/Submarine Platform—Sand/Rubble/Rock
1.2.2.3.2.7   Terrace/Submarine Platform—Boulder/Block
1.2.3 Area—Nearshore Bottom
(as in 1.1.2)

1.3 Coast—Barrier Reef
1.3.1 Area—Shore
(as in 1.1.1)
1.3.2 Area—Lagoon Fringing Reef
(as in 1.2.2)
1.3.3 Area—Lagoon
{Lagoon Size modifier} {Lagoon Depth modifier}
{Lagoon Area modifiers: Sub-Lagoon(s); Perched Lagoon}
{No. of Patch Reefs/Pinnacles modifier}
1.3.3.1   Reef Top—Patch Reef/Pinnacle
(as in 1.2.2.1)
1.3.3.2   Reef Top—Reticulate Reef
(as in 1.2.2.1)
1.3.3.3   Reef Islets—Patch Reefs
(as in 1.2.2.2)
1.3.3.4   Reef Slope—Patch Reef/Pinnacle
(as in 1.2.2.3)
1.3.3.5   Reef Slope—Reticulate Reef Slope
1.3.3.6   Lagoon Floor
1.3.3.6.1 Lagoon Floor—Shallow Lagoon (<10 m deep)
1.3.3.6.2 Lagoon Floor—Algal Mound
1.3.3.6.3 Lagoon Floor—Deep
1.3.3.7   High Islet—Lagoon
(as in 1.1.1.1, 1.2.2.1–1.2.2.3)
1.3.4 Area—Barrier Reef
{Exposure modifiers: Windward; Leeward}
1.3.4.1   Reef Top
(as in 1.2.2.1)
1.3.4.2   Reef Islets
(as in 1.2.2.2)
1.3.4.3   Reef Slope—Outer
(as in 1.2.2.3)
1.3.4.4   Reef Slope—Lagoon
(as in 1.2.2.3)
1.3.5 Area—Nearshore Bottom
(as in 1.1.2)

2. Earthform—Atoll/Table Reef/Low Island (height <10 m)
(as in A.B.B.2)
3. Earthform—Submerged
3.1 Reef/Shoal—Nearshore

         3.2 Reef/Shoal/Bank—Mid-Shelf
         3.3 Reef/Shoal/Bank—Outer Shelf
         3.4 Plain—Nearshore
         3.5 Plain—Offshore
         3.6 Canyon
         3.7 Continental Slope

A.B.B.   OCEANIC (NONCONTINENTAL) ECOSYSTEMS

    1. Earthform—High Island
       {High Island modifier: Almost-Atoll}

       1.1 Coast—Noncoral Reef
           1.1.1 Area—Shore
               (as in A.B.A.1.1.1)
           1.1.2 Area—Nearshore Bottom
               (as in A.B.A.1.1.2)
           1.1.3 Area—Deep Bottom
               1.1.3.1   Bathyal (200–4000 m)
               1.1.3.2   Abyssal (4000–7000 m)
               1.1.3.3   Hadal (>7000 m)

       1.2 Coast—Fringing Reef
           1.2.1 Area—Shore
               (as in A.B.A.1.1.1)
           1.2.2 Area—Fringing Reef
               (as in A.B.A.1.2.2)
           1.2.3 Area—Nearshore Bottom
               (as in A.B.A.1.1.2)
           1.2.4 Area—Deep Bottom
               (as in A.B.B.1.1.3)

       1.3 Coast—Barrier Reef
           1.3.1 Area—Shore
               (as in A.B.A.1.1.1)
           1.3.2 Area—Lagoon Fringing Reef
               (as in A.B.A.1.2.2)
           1.3.3 Area—Lagoon
               (as in A.B.A.1.3.3)
           1.3.4 Area—Barrier Reef
               (as in A.B.A.1.3.4)
           1.3.5 Area—Nearshore Bottom
               (as in A.B.A.1.1.2)
           1.3.6 Area—Deep Bottom
               (as in A.B.B.1.1.3)

    2. Earthform—Atoll/Table Reef/Low Island (height <10 m)
       2.1 Atoll—Many Deep Passes/Open
          {Atoll Perimeter Length modifier}
           2.1.1 Area—Lagoon
               (as in A.B.A.1.3.3; excluding 1.3.3.7)
           2.1.2 Area—Perimeter Reef
               {Reef Islet Type modifier: Type 1; Type 2; Type 3; Type 4}
               (as in A.B.A.1.3.4)
           2.1.3 Area—Nearshore Bottom
               (as in A.B.A.1.1.2; excluding 1.1.2.1)
           2.1.4 Area—Deep Bottom
               (as in A.B.B.1.1.3)

       2.2 Atoll—Few/One Deep Pass(es) (pass depth >5 m)
          (as in A.B.B.2.1)

2.3 Atoll—No Deep Pass
(as in A.B.B.2.1)

2.4 Atoll—Completely Land-ringed
(as in A.B.B.2.1)

2.5 Table Reef—Reef Islet with Water Body
(as in A.B.B.2.1; excluding 2.1.1)

2.6 Table Reef—Reef Islet without Water Body
(as in A.B.B.2.1; excluding 2.1.1)

2.7 Table Reef—No Reef Islet
(as in A.B.B.2.1; excluding 2.1.1)

3.   Earthform—Submerged

3.1 Submerged Atoll-Reef (upper surface depth <20 m)

3.1.1 Near Surface (<200 m)

3.1.2 Bathyal (200–4000 m)

3.1.3 Abyssal (4000–7000 m)

3.1.4 Hadal (>7000 m)

3.2 Submerged Table-Reef (depth <20 m)
(as in 3.1)

3.3 Shoal (depth <20 m)
(as in 3.1)

3.4 Bank (depth 20–200 m)
(as in 3.1)

3.5 Seamount (depth >200 m)

3.6 Guyot (depth >200 m)

3.7 Ridge

3.8 Plain—Abyssal

3.9 Trench

3.10 Fracture

3.11 Volcano

3.12 Geothermal Vent

Diagrams That Illustrate Marine Features Within the Upper Levels of the Ecosystem Classification

Level 1 Choices

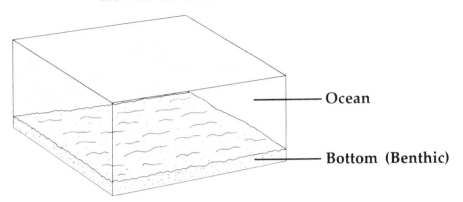

Ocean

Bottom (Benthic)

Level 2 Choices For Ocean

I. Open Ocean (Pelagic)
II. Nearshore Ocean (Neritic)

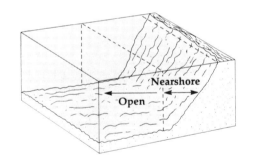

Level 2 Choices For Bottom

I. Continental Shelf (Non-Oceanic)
II. Non-Continental (Oceanic)

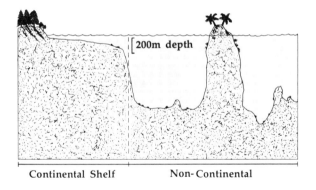

Level 3 Choices For Open Ocean (Pelagic)

1. Divergences
2. Convergences
3. Boundary Currents
4. Water Masses

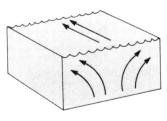

Divergence

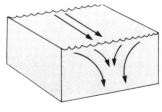

Convergence

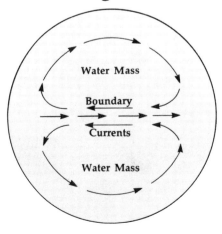

Level 3 Choices For Nearshore Ocean (Neritic)

1. Eddies
2. Coastal Upwelling Zone
3. Coastal Ocean
4. Convergences

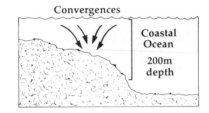

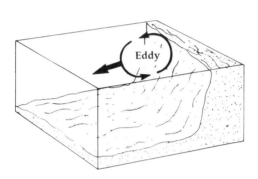

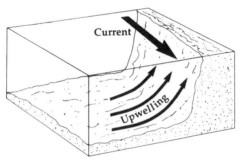

Level 4 Choices For Nearshore Ocean (Neritic)

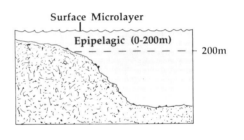

Level 4 Choices For Open Ocean and Non-Continental

Ocean
1.1. Surface Microlayer
1.2. Epipelagic (0-200m)
1.3. Mesopelagic (200-1000m)
1.4. Bathypelagic (1000-4000m)
1.5. Abyssopelagic (4000-7000m)
1.6. Hadalpelagic (>7000m)

Bottom
Near Surface (<200m)

Bathyal (200-4000m)

Abyssal (4000-7000m)
Hadal (>7000m)

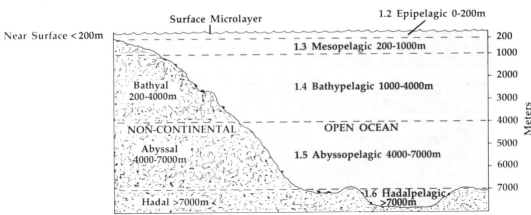

Level 5 and Level 6 Choices For Open Ocean Epipelagic

Level 5   Level 6

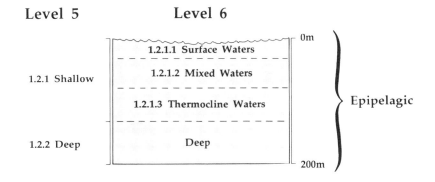

1.2.1 Shallow	1.2.1.1 Surface Waters	0m
	1.2.1.2 Mixed Waters	
	1.2.1.3 Thermocline Waters	
1.2.2 Deep	Deep	200m

Epipelagic

Level 3 Choices For Continental Shelf and Non-Continental

1. High Island Earthform (>10m height)
2. Low Island (Non-Oceanic) or Reef Island (Oceanic)
3. Submerged Earthform (non-land associated and below sea level)

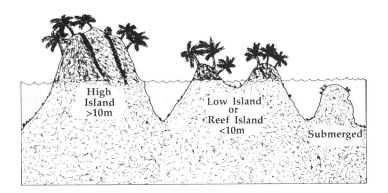

Level 4 Choices For Continental Shelf, Non-Continental High Island Earthform and Continental Shelf Low Island Earthform

1.1. Non-coral Reef Coast
1.2. Fringing Reef Coast
1.3. Barrier Reef Coast

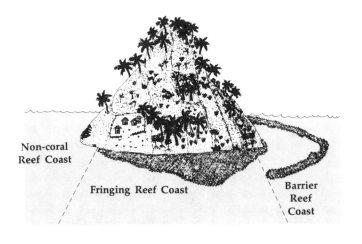

Level 4 Choices For Continental Shelf Submerged Earthform

3.1. Nearshore Reef/Shoal (including fringing reef)
3.2. Mid-shelf Reef/Shoal/Bank
3.3. Outer Shelf Reef/Shoal/Bank (including outer barrier reef)
3.4. Nearshore Plain
3.5. Offshore Plain
3.6. Submarine Canyon
3.7. Continental Slope

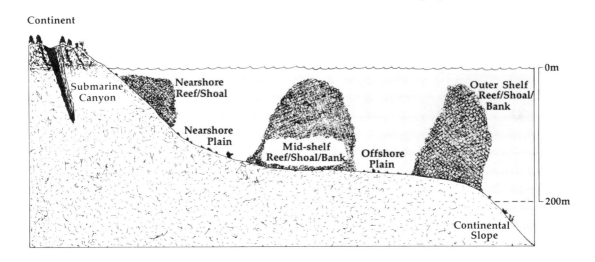

Level 4 Choices For Non-Continental Submerged Earthform

3.1. Atoll-reef (crest at <20m depth)   3.7. Ridge
3.2. Table Reef   3.8. Plain
3.3. Shoal (<20m depth)   3.9. Trench
3.4. Bank (20-200m depth)   3.10. Fracture
3.5. Seamount (>200m depth)   3.11. Submarine Volcano
3.6. Guyot (>200m depth)   3.12. Geothermal Vent

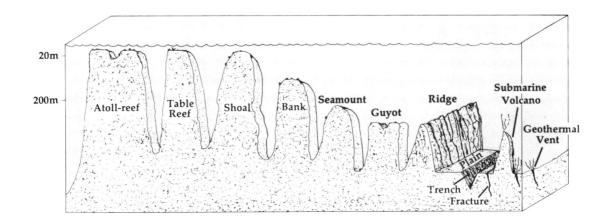

Level 4 Choices For Non-Continental Reef Island Earthform

2.1. Atoll with many deep passes/partially enclosed

2.2. Atoll with few/one deep pass(es), (>5m depth)

2.3. Atoll with no deep pass

2.4. Atoll completely land ringed

2.5. Reef island with water body

2.6. Reef island without water body

Examples of Each Reef Island Earthform

Johnston Atoll — 2.1 (169°32'W, 16°44'N; Akau I, Hikina I, Lagoon, Sand I, Johnston I; Nautical Miles 0–1)

Namorik Atoll — 2.3 (168°07'30"E, 5°37'30"N; Matamat Island, Lagoon, Namorik Island; Nautical Miles 0–1)

Lib Island — 2.5 (167°23'E, 8°19'N; Nautical Miles 0–.5)

Bokaak Atoll — 2.2 (169°E, 14°N; Lagoon, North I, Pass, Bylla I, Taongi I, South I; Nautical Miles 0–2)

Swains Island — 2.4 (171°05'W, 11°03'S; Nautical Miles 0–1)

Jemo Island — 2.6 (169°32'30"E, 10°05'00"N; Nautical Miles 0–1)

Level 5 Choices For Continental Shelf High or Low Island Earthform

- **Shore Area**

- **Fringing Reef Area**

- **Nearshore Bottom Area**

- **Lagoon Fringing Reef Area**

- **Lagoon Area**

- **Barrier Reef Area**

Level 5 Choices For Non-Continental High or Reef Island Earthform

- Shore Area

- Reef Area

- Lagoon Area

- Deep Bottom Area

- Fringing Reef Area

- Barrier Reef Area

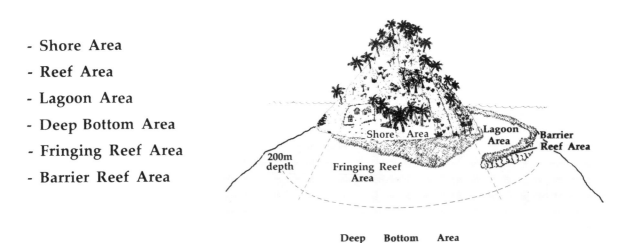

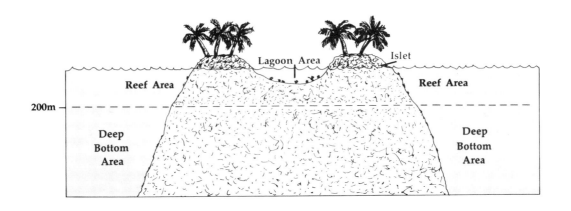

Level 6 Choices For Shore Areas

Birds Eye View

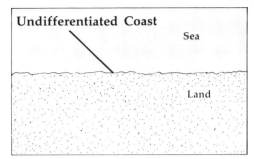

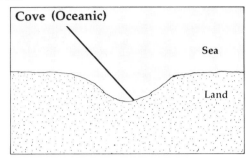

Birds Eye View With Depth Contours

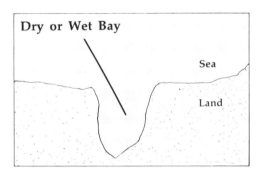

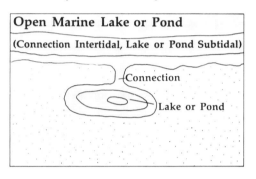

Birds Eye View With Depth Contours

Birds Eye View With Depth Contours

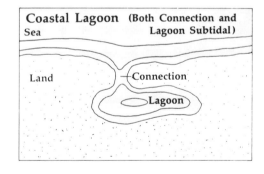

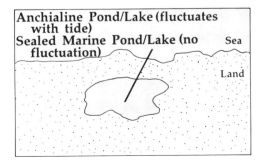

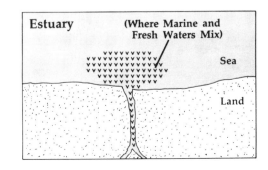

Level 7 Choices For Shore Area

- Boulder/Cobble Beach
- Sand/Gravel Beach
- Silt/Clay Sediment Shore
- Flat Shore - Boulder/Cobble
- Sand Flat
- Mud Flat
- Ramp
- Marine Bench

- High Cliff >10m
- Medium Cliff 2-10m
- Low Cliff <2m
- Marine Cave and Notch
- Talus Slope
- Beachrock
- Artificial Shore
- Fishpond
- Boulder/Cobble Field (Oceanic)

Cross Sections

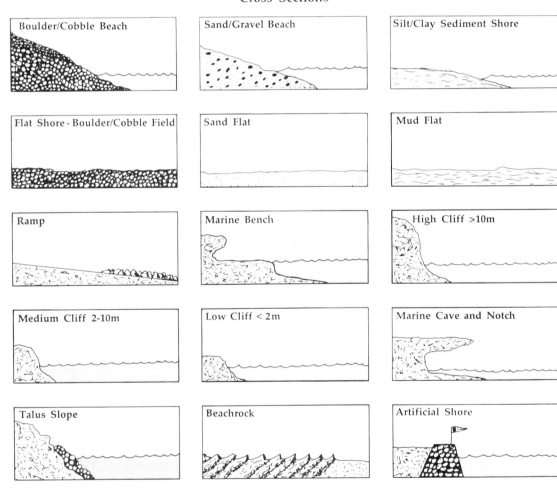

Birds Eye View

Level 7 Choices For Coral, Fringing, or Barrier Reef Area

- Reef Flat
- Algal Ridge
- Surge Channel
- Hoa
- Moat and Depression
- Rubble Platform
- Reef Conglomerate
- Storm Blocks
- Boulder Ramparts
- Sand and Gravel Sheets on Reef Flat
- Sand Cay
- Motu (<10m)
- High Islet (>10m)
- Reef Pool (≤5m)
- Reef Hole (>5m)
- Artificial Structure
- Lagoon Sand Bar and Spit

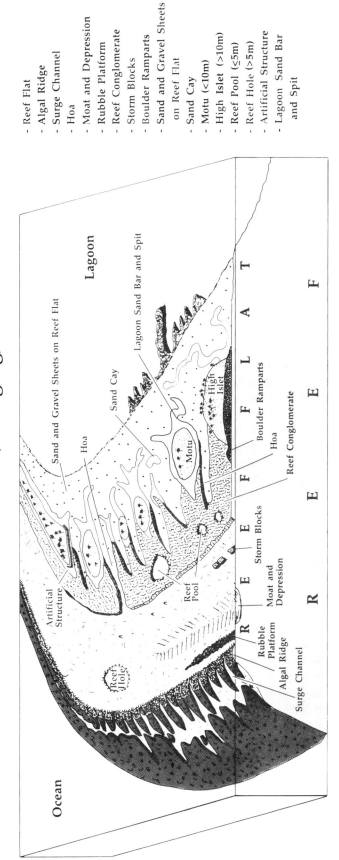

- Beachrock
- Aeolianite
- Coral/Algal Dam and Spillway
- Gravel Ridge

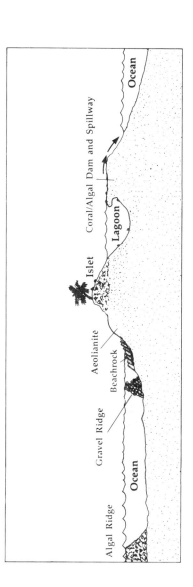

Level 7 Choices For Outer Slope

- Terrace
- Groove and Spur
- Buttress and Valley
- Tunnel (Room and pillar)
- Furrowed Platform
- Cliff or Vertical Slope
- Notch/Cave
- Detrital Boulders
  on Slope and
  Terrace

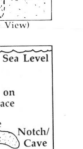

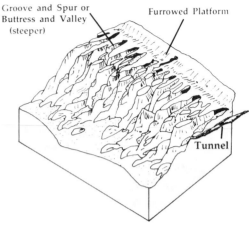

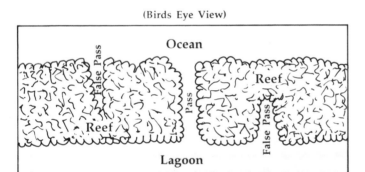

Atoll Or Barrier Reef Passes

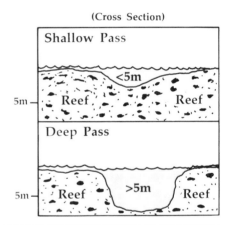

Lagoon Area

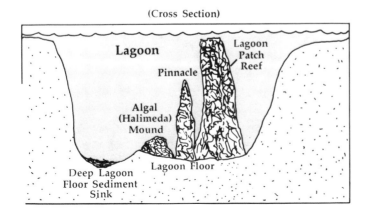

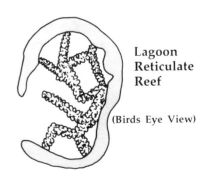

APPENDIX 3

Glossary of Marine Ecosystems of the Tropical Island Pacific

(Version of October 1994)

Note: Terms within definitions, which are in **bold**, are defined elsewhere in the glossary. Terms in *[bracketed italics]* are modifiers of ecosystem classes.

ABYSSAL: Deep Bottom Area or portion of **Submerged Earthform** between the depths of 4,000 and 7,000 m

ABYSSOPELAGIC: marine water layer between 4,000 and 7,000 m deep; below **Bathypelagic** Layer; ocean layer lacking light

AEOLIANITE: cemented dune sands and other wind sorted sediments; may be found on **Shoreline** or as **Reef Top Supratidal Feature**

ALGAL BED: Reef Top Surface Feature dominated by algae cover, usually brown algae (e.g., *Sargassum, Turbinaria*)

ALGAL RIDGE: asymmetric wave-resistant ridge of crustose coralline algae; characteristic **Reef Top Surface Feature** of the seaward margin of **Windward Reefs**

[ALMOST ATOLL]: modifier of **High Island Earthform**; an **Atoll** with one or more volcanic peaks emergent above the Reef Area and/or (usually) the **Lagoon**; essentially an **Atoll** with small remnants of the **High Island** present

[ANCHIALINE]: modifier of **Coastal Lagoon/Lake/Pond Coastline** and **Reef Islets**; marine or brackish water bodies which lack surface connection to the sea, usually located near the coast in permeable substrates and which, by the presence of salt and tidal fluctuations, show subsurface hydrologic connections to the sea

[APHOTIC]: light level modifier of **Deep Epipelagic Ocean Ecosystems**; areas never reached by natural light

AREA: portion of the **Coast** extending offshore

AREA—BARRIER REEF: portion of **Coast** consisting of offshore reef structure; i.e., **Reef Area** separated from **Shore Area** or **Lagoon Fringing Reef Area** by a **Lagoon Area**

AREA—DEEP BOTTOM: submerged portion of **Coast** from 200-m depth to deep ocean floor at the base of the **Earthform**

AREA—FRINGING REEF: portion of **Coast** consisting of **Reef** structure attached to shore, sometimes with narrow, shallow **Lagoon** feature; extends from lower limit of the **Shore Area** (i.e., low-tide line) to the bottom of the **Outer Reef Slope**

AREA—LAGOON: portion of **Barrier Reef Coast** between **Shore Area** or **Fringing Reef Area** and **Barrier Reef Area**; portion of **Atoll** surrounded by **Perimeter Reef Area**

AREA—LAGOON FRINGING REEF: portion of **Barrier Reef Coast** consisting of **Fringing Reef** structure

AREA—NEARSHORE BOTTOM: portion of **Coast** between the lower limit of the **Shore Area** (i.e., low-tide line) and a depth of 200 m; shallow marine substrate

AREA—PERIMETER REEF: portion of **Atoll** consisting of **Reef** structure which surrounds **Lagoon Area**; often cut through **Passes**; usually with **Reef Islet Areas** on **Reef Top**

AREA—SHORE: portion of **Coast** consisting of the marine **Benthic Ecosystems** between the terrestrial vegetation line (upper boundary) and the low-tide line (lower boundary)

ATOLL EARTHFORM: Earthform consisting of a ring-like **Perimeter Reef Area,** often with **Reef Islets;** encloses a **Lagoon Area**

ATOLL—COMPLETELY LAND RINGED: Atoll with a **Lagoon Area** which is completely land-locked from the open ocean by a continuous string of **Reef Islets;** may be occasionally breached by the open sea during storms

ATOLL—FEW/ONE DEEP PASS(ES): Atoll with **Lagoon Area** completely enclosed by a shallow **Reef** except for one or two **Passes** >10 m deep which bisect the **Perimeter Reef Area**

ATOLL—MANY DEEP PASSES/OPEN: Atoll with an open or partially enclosed **Lagoon Area;** characterized by 3 or more **Passes** >10 m deep

ATOLL—NO DEEP PASS: Atoll with shallow **Perimeter Reef Area** which completely encircles the **Lagoon Area;** lacking **Passes** >10 m in depth

BANK: Submerged Earthform with its crest at a depth of 20–200 m

[BARACHOIS]: see **Intertidal Lagoon with Intertidal Connection**

BAR AND SPIT: low accumulations of **Sand** forming intertidal or subtidal extensions of **Reef Islets;** often closing lagoon ends of **Hoas** and/or extending into **Lagoons**

BARRIER REEF AREA: see **Area—Barrier Reef**

BATHYAL: Deep Bottom Area or portion of **Submerged Earthform** between depths of 200 and 4,000 m

BATHYPELAGIC: intermediate layer of the ocean between depths of 1,000 and 4,000 m; between the **Mesopelagic** layer above and the **Abyssopelagic** layer below

BEACH: a sloped **Sediment Shoreline;** composed of **Sand, Gravel, Cobble,** or **Boulder-**sized sediments; sometimes with **Beachrock**

BEACH—BOULDER/COBBLE: sloped **Sediment Shoreline** composed mainly of **Cobbles** and/or **Boulders**

BEACH—SAND/GRAVEL: sloped **Sediment Shoreline** composed primarily of **Sand** and/or **Gravel;** may have **Beachrock**

BEACHROCK: sloped **Sediment Shoreline** consisting of exposed formations of cemented beach materials, usually **Sand;** formerly covered by unconsolidated materials in the intertidal zone; forms laterally extensive lines of rock dipping seaward at the slope of the **Beach;** may occur on **Reef Top** away from current **Shoreline** where it forms a **Reef Top Supratidal Feature**

BENCH/RAMP—MARINE: Solid Substrate Shoreline consisting of horizontal terrace, or terrace with low angle sloped toward ocean or **Lagoon**

BENTHIC ECOSYSTEMS: portion of the **Marine Realm** composed of, or dominated by, the bottom substrate and its organisms

BLOCK: a rock larger than 1 m in its longest dimension

BOULDER: a rock 250 mm–1 m in its longest diameter, usually rounded; larger than a **Cobble** and smaller than a **Block**

BOULDER/BLOCK SLOPE: Reef Slope consisting of a heterogeneous, unsorted mixture of **Boulder-** and **Block-**sized materials

BOULDER/COBBLE FIELD: a low angle **Sediment Shoreline** or shallow **Reef Flat Surface Feature** dominated by **Boulder-** and/or **Cobble-**sized materials; may include **Rubble**

BOUNDARY CURRENT: large-scale water stream in the upper ocean; separates **Water Masses**; driven by a combination of wind, temperature, geostrophic, or coriolis effects

BUTTRESS AND VALLEY: **Reef Slope Feature (Nonterrace)** consisting of large scale nearly vertical analogue of the **Spur and Groove** on deep and/or steep reef slopes; often found on **Leeward Reef Slope**

[CALCAREOUS]: modifier of **Sediment Shorelines** and **Reef Slope** substrate; light-colored (white, tan) sand, composed of coral, shell, foraminifera, *Halimeda* algae, and/or echinoderm fragments

CANYON: **Submerged Earthform** consisting of an incised large-scale submarine feature on a high angle slope; normally associated with **Continental Shelf**

CHANNEL—ARTIFICIAL: a man-made passage through a **Barrier Reef**, **Perimeter Reef**, or **Fringing Reef**; generally narrower and shallower than a natural passage

CHANNEL—FRINGING REEF: a natural passage transecting the edge of a **Fringing Reef** for a substantial distance through the **Reef Top** toward the **Shoreline**

CLIFF: high angle **Solid Substrate Shoreline**; includes Low Cliffs (height <2 m), Medium Cliffs (height 2–10 m), and High Cliffs (height >10 m)

COAST: the nearshore benthic marine environment; extends from the vegetation line to the deep ocean floor; a **Coast** section normally ranges from 10s to 100s of km in linear extent along the marine/terrestrial interface; **Coast** may be either: **Coast—Barrier Reef, Coast—Fringing Reef,** or **Coast—Noncoral Reef**; each Coast type contains different **Area** types

COAST—BARRIER REEF: **Coast** with **Barrier Reef Area** and **Lagoon Area**

COAST—FRINGING REEF: **Coast** dominated by a **Fringing Reef Area**

COAST—NONCORAL REEF: **Coast** without **Coral Reefs**; **Coral** communities and minor **Reef** structures may be present, but lacking flats, definable slopes, or significant thickness of **a Reef**

COASTLINE: subunits of **Shore Area**; **Coastline** is defined as either **Undifferentiated, Cove, Bay, Coastal Lagoon/Lake/Pond, Peninsula,** or **Irregular/Discontinuous/Islets**; consists of linear portion of **Coast** extending about 10 km along marine/terrestrial interface

COASTLINE—BAY: major indentation of **Shore Area** on otherwise **Undifferentiated Coastline** with at least partially protected conditions; the opening is generally less wide than the length of the embayment; may be primarily **Marine** (lacking permanent surface water discharge and major groundwater discharge) or **Estuarine** (having permanent or significant seasonal fresh surface water or groundwater inflow)

COASTLINE—COASTAL LAGOON/LAKE/POND: **Coastal Lagoon**: **Shore Area** indentation with narrow opening; shallower and smaller than a **Bay**; water body and opening may both be subtidal, have an intertidal connection with subtidal water body or both be intertidal (i.e., **Barachois**); **Coastal Lake**: enclosed haline water body >2 m deep; may be sealed or with some regular surface and/or subsurface marine water interaction; **Coastal Pond**: **Coastal Lake** 2 m or less in depth

COASTLINE—COVE: **Shore Area** with medium- or small-scale indentation of an otherwise **Undifferentiated Coastline**; smaller than a bay; ratio of mouth to depth of indention less than one

COASTLINE—IRREGULAR/DISCONTINUOUS/ISLETS: **Shore Area** consisting of many small indentations and headlands, disconnected areas, or small islets

COASTLINE—PENINSULA: major extension of **Shore Area** on otherwise **Undifferentiated Coastline**; length of extension generally longer than width

COASTLINE—UNDIFFERENTIATED: generally straight **Shore Area**, lacking any significant indentations or extensions; may be slightly irregular and contain occasional minor extensions and minor indentations, other than **Bays** and **Peninsulas**

COBBLE: rock fragment 50–250 mm in its longest diameter; usually rounded by weathering; larger than **Gravel**, smaller than a **Boulder**

CONGLOMERATE/REEF LIMESTONE PLATFORM: cemented sedimentary rock composed mainly of **Gravel**-sized materials and **Rubble** or reef rock; usually forms a flat-topped **Reef Top Supratidal Feature** <1 m high, although can be higher; often occurs associated with **Type 4 Reef Islets** or as an isolated structure on the **Reef Top**

[CONTINENTAL]: geology modifier of **High Island Earthform**; island lying on continental plate, possessing previous connections to large continental land masses; geological composition variable, but generally including andesitic, intrusive igneous rocks (e.g., granites) and metamorphic rocks

CONTINENTAL SHELF: extensions of the continents gently sloping seaward from the coast to the continental margins; submarine portion of continents between depths of 0 and 200 m; ocean bottom on the continental slopes and slopes of large islands composed of continental rocks

CONTINENTAL SHELF ECOSYSTEMS: nonoceanic **Benthic Ecosystems**; the benthic portion of the **Marine Realm** on the **Continental Shelf**

CONTINENTAL SLOPE: deeper, steeper extensions of the continents beyond the **Continental Shelf** and margin; submarine portion of continents at depths >200 m

CONVERGENCE: downward movement of surface ocean waters **in Open Ocean Ecosystem** (= Oceanic Downwelling Zone)

CONVERGENCE—COASTAL: downward movement of surface ocean waters in **Nearshore Ocean Ecosystem** (= Coastal Downwelling Zone)

CORAL: rock formation consisting of the calcareous skeletons of certain coelenterates (especially stony corals), coralline algae, mollusks, certain sponges (sclerosponges), and/or other organisms

CORAL BED/MICROATOLLS: **Reef Top Surface Feature** dominated by live **Coral** cover in the form of beds of dense **Coral** (usually short branching **Coral**), scattered **Coral** of mixed kinds, or scattered low mounds and **Microatolls**

CORAL/ALGAL DAM AND SPILLWAY: **Reef Top Supratidal Feature** composed of algal and **Coral**-rimmed platforms holding back pools or **Sub-Lagoons** at higher elevation (usually <0.5 m); forms interconnecting drainage systems through which the water drains over the **Reef Top**

DIVERGENCE: upward movement of deeper waters in **Open Ocean Ecosystems** attributed to the divergent movement of surface waters away from an alignment (= Oceanic Upwelling Zone)

DEEP BOTTOM AREA: see **Area—Deep Bottom**

DREDGE PIT/QUARRY/CHANNEL/BASIN: artificial **Reef Top Subtidal Feature**; consisting of a reef aggregate extraction site (Dredge Pit); a solid reef substrate extraction site (Quarry); a linear extraction, usually for boat passage (Channel); or a large deepened area, usually for ship movements (Basin)

EARTHFORM: a geomorphological feature of the earth's crust

EARTHFORM—ATOLL/TABLE REEF/LOW ISLAND: a marine **Benthic Ecosystem Earthform** capped with an **Atoll** or **Table Reef**; may be referred to as a **Low Island** when a **Reef Islet** is present

EARTHFORM—HIGH ISLAND: a marine **Benthic Ecosystem Earthform** which forms an island with an elevation more than 10 m above high tide

EARTHFORM—SUBMERGED: a marine **Benthic Ecosystem Earthform** which is completely submarine; the upper surface is at least >20 m below sea level

EDDY: a mesoscale circulation feature in the ocean formed in island wakes or from spinoffs or meanders in ocean currents

EPIPELAGIC: the uppermost (normally **Photic**) layer of the ocean between the ocean surface and the **Thermocline**; the upper mixed layer of the ocean normally between depths of 0 and 200 m

EPIPELAGIC—DEEP: lower **Epipelagic** layer; below the **Thermocline**

EPIPELAGIC—SHALLOW: upper **Epipelagic** layer; above the **Thermocline**

{ESTUARINE}: salinity modifier of **Bay Coastlines**; for **Coastlines** where marine and freshwaters meet and mix; waters often brackish (i.e., mixohaline, with salt content 0.5–30 ‰)

FISHPOND/FISHTRAP/SHIPWRECK: **Artificial Shoreline** or **Reef Top Supratidal Feature** consisting of a walled or fenced enclosure built in shallow marine waters (Fishpond), usually connected to the **Shoreline**; permanent or semipermanent structure to channel fish through a narrow opening into a small enclosure (Fishtrap), usually near, but not connected to, the **Shoreline**; or the substantial remains of a vessel in an intertidal area (Shipwreck)

FRACTURE: **Submerged Earthform** consisting of a large-scale elongated crack on the deep ocean floor; a fault line or zone attributed to differential movements of the ocean crust

FRINGING REEF AREA: see **Area—Fringing Reef**

{FURROWS}: modifier of **Reef Slope Terrace/Submarine Platform**; shallow linear channels

GEOTHERMAL VENT: **Submerged Earthform** consisting of a vent of hot, mineral-rich waters on the ocean floor; generally located on or near spreading oceanic **Ridges** or on the continental margins of subduction **Trenches**

GRAVEL: angular, unsorted, unconsolidated sediment with a diameter of 2–50 mm, i.e., larger than **Sand** and smaller than **Cobbles**

GRAVEL/BOULDER RIDGE: deposit of **Gravel** to **Boulder**-size **Reef** materials which form a **Reef Top Supratidal Feature**; often dominated by poorly sorted and/or loosely cemented **Rubble**; often close to the seaward edge of **Reefs**, on the seaward **Reef Top**, or fronting the seaward **Shoreline** of **Reef Islets**

GUYOT: **Submerged Earthform** at depths >200 m consisting of a flat-topped **Seamount** with a cap of the carbonate remains of a drowned **Atoll**

HADAL: the deepest **Deep Bottom Area** or portion of **Submerged Earthform**, at depths >7,000 m

HADALPELAGIC: deepest layer of ocean waters >4,000 m deep; below the **Abyssopelagic** layer

HIGH ISLAND EARTHFORM: **Earthform** which forms an island with an elevation more than 10 m above high tide

HIGH ISLET—LAGOON: small isolated land mass more than 10 m above high tide in **Lagoon** of **High Island Earthform**

HIGH ISLET—OCEAN: small isolated land mass more than 10 m above high tide located on **Nearshore Bottom Area** of **High Island Earthform**; larger and farther offshore than a **Stack**

HOA (Inter-Islet Channel): Polynesian term for **Reef Top Subtidal Feature** consisting of a shallow channel or inlet extending into or between **Reef Islets**

INCOMPLETE REEF TOP: **Reef Top Subtidal Feature** consisting of a portion of **Reef** upper surface that is partly submerged and/or broken up into a series of small, shallow **Reef** sections, instead of a solid, continuous **Reef Top**

[INNER/LAGOON/SHORE]: modifier of the **Shorelines** of **Reef Islets**; lagoon or nonocean facing side of **Reef Islet**

[INTERTIDAL LAGOON with INTERTIDAL CONNECTION]: modifier of **Coastal Lagoon** indicating level of permanence and connectedness; intertidal lagoon with intertidal connection means that presence of coastal lagoon and interchange with larger marine water body is tide dependent; also known as **Barachois**

LAGOON AREA: see **Area—Lagoon**

LAGOON FLOOR: portion of **Lagoon Area** making up bottom surface of a **Lagoon**; excludes **Reef Slope**, **Reef Top**, and the **Reef Islets** of **Patch Reefs**, **Pinnacles**, and **Reticulate Reef**

LAGOON FLOOR—ALGAL MOUNDS: portion of **Lagoon Floor** dominated by growth of *Halimeda* algae, with mounds created by the algae deposits

LAGOON FLOOR—SHALLOW: portion of **Lagoon Floor** that is <10 m deep

LAGOON FLOOR—DEEP: portion of **Lagoon Floor** that is >10 m deep; sections may function as a sediment sink

LAGOON FRINGING REEF AREA: see **Area—Lagoon Fringing Reef**

LANDFILL/CAUSEWAY/GROIN: **Artificial Shoreline** consisting of solid extension and filling of shoreline into marine environment (Landfill); linear extension connecting islets (Causeway), which may have a road; or solid extension perpendicular for shoreline (Groin), which may serve as a dock or wharf

[LEEWARD]: modifier of **Reef Area** exposure indicating the side opposite to direction of prevailing winds

[LIMESTONE]: geology modifier of **High Island Earthform**; geological composition of generally ancient **Reef** and **Reef** materials

LOW ISLAND EARTHFORM: an **Atoll Earthform** or **Table Reef Earthform** with **Reef Islets**; an island with an elevation less than 10 m above high tide, often composed only of carbonate rock and/or sediments

MANGROVE: **Sediment Shoreline** dominated by mangrove species adapted to or dependent on brackish conditions; may be present away from **Shoreline** as **Reef Top Supratidal Feature**

[MARINE]: salinity modifier of **Bay Coastlines** for **Coastlines** where waters are seawater (i.e., euhaline, with salt content 30–40 ‰)

MARINE REALM: portion of the earth's biosphere, geosphere, and waters generally seaward of mean high water and/or vegetation line, including embayments, lagoons, and other water bodies with measurable salinity and surface water connection to the ocean

MESOPELAGIC: a middle layer of the ocean between the **Epipelagic** and **Bathypelagic** layers; aphytal zone of the ocean between depths of 200 and 1,000 m

MICROATOLL: **Coral** colony in shallow water which is dead on the upper surface due to low tide, and continues to grow outward on the perimeter of the colony; usually 1–3 m in diameter

[MIXED GEOLOGY]: geology modifier of **High Island Earthform**; geological composition of several types; a combination of **Continental, Volcanic,** and/or **Limestone**

MOAT AND DEPRESSION: linear or rounded depression on a **Reef Top**; **Reef Top Subtidal Feature** characteristically immediately inshore of **Algal Ridge**

MUD/SILT FLAT: low angle-sloped, low-energy **Sediment Shoreline**; primarily composed of fine grain sediment, i.e., finer than **Sand**

NEAR SURFACE: portion of **Submerged Earthform** <200 m deep

NEARSHORE BOTTOM AREA: see **Area—Nearshore Bottom**

NERITIC: **Nearshore Ocean Ecosystems**; i.e., those associated with the coasts because the waters are overlying **Continental Shelves** and/or the waters are <200 m deep in areas of coastal submarine slopes

NOTCH/CAVE—MARINE: sea-level indentation into high-angle **Solid Substrate Shoreline**; indentation of 1–3 m is a notch; indentation of >3 m is a cave; both formations usually at base of **Cliff**, often associated with **Marine Bench/Ramp**

OCEAN ECOSYSTEMS: portion of the **Marine Realm** composed of, or dominated by, marine waters and their organisms

OCEAN—CONTINENTAL: portion of ocean and marine floor overlying continental crust (e.g., andesite, granite); shallower than the ocean overlying oceanic crust (e.g., basalt); includes **Continental Shelves, Continental Slopes,** and overlying waters

OCEAN ECOSYSTEMS: portion of the **Marine Realm** consisting of water bodies and their organisms; i.e., not including the **Benthic Ecosystems**

OCEAN ECOSYSTEMS—NEARSHORE—see **Neritic**

OCEAN ECOSYSTEMS—OPEN—see **Pelagic**

OCEANIC ECOSYSTEMS—noncontinental **Benthic Ecosystems**; portion of the benthic **Marine Realm** not associated with the continents; ocean bottom beyond the continental slopes and slopes of large islands composed of continental rocks; portion of the earth's crust composed primarily of basalt

[OUTER/OCEAN]: modifier of the **Shorelines** of **Reef Islets**; ocean or nonlagoon facing side of **Reef Islet**

PASSES/REEF TOP OPENINGS: narrow passage or wide opening transecting the **Reef Top** of a **Barrier Reef Area** or **Perimeter Reef Area** or indenting the **Reef Top** of a **Fringing Reef Area** or **Lagoon Fringing Reef Area**

PASS—DEEP: natural passage through the **Reef Top** of a **Barrier Reef** or **Perimeter Reef**; with a floor depth of >10 m; width <2 km

PASS—FALSE—dead-end passage; a natural passage incompletely transecting a **Barrier Reef** or **Perimeter Reef**; usually <10 m deep and blocked by reef growth or sediment deposition

PASS—SHALLOW: a natural passage transecting the **Reef Top** of a **Barrier Reef** or **Perimeter Reef**; floor depth of <10 m; width <2 km

PATCH REEF/PINNACLE: **Reef** rising from the **Lagoon Floor** to near the water surface; **Patch Reefs** are larger than **Pinnacles**, reach close to the sea surface, and are flat topped with a substantial **Reef Top**; the upper surface of **Pinnacles** are usually not flat topped and are limited in area, usually being pointed, irregular, or conical; pinnacles often do not reach the **Lagoon** surface and lack a substantial **Reef Top**

PELAGIC: offshore **Ocean Ecosystems**; i.e., those not associated with the coasts because the waters are not overlying **Continental Shelves** and/or the waters are >200 m deep in areas of coastal submarine slopes

[PERCHED LAGOON]: modifier of **Lagoon Area**; **Sub-Lagoon** which is slightly raised above level of main **Lagoon**; usually with **Coral/Algal Dams and Spillways**

PERIMETER REEF AREA: see **Area—Perimeter Reef**

[PHOTIC]: light level modifier of **Deep Epipelagic Layer**; zone where light is sufficient for photosynthesis; in oceanic waters above approximately 200 m in depth

PINNACLE: see **Patch Reef/Pinnacle**

PLAIN—ABYSSAL: flat **Submerged Earthform** expanse of the deep marine floor seaward of the **Continental Slope**; ocean floor at depths greater than 200 m

PLAIN—INSHORE: flat **Submerged Earthform** expanse of the marine floor on the inshore portion of the **Continental Shelf**, at depths of 200 m or less and usually covered with sediments

PLAIN—OFFSHORE: flat **Submerged Earthform** expanse of the marine floor on the outer portion of the **Continental Shelf**, at depths of 200 m or less and usually covered with sediments

REEF: a largely consolidated wave-resistant feature; the upper surface is within 0–20 m of the ocean surface; in the tropics the upper portion is typically composed mainly of living and nonliving remains of **Coral** and coralline algae

REEF AREA: the reef portion of the **Coast** beyond the lower limit of the **Shore Area** to a maximum depth of 200 m; can be **Fringing Reef Area**, **Lagoon Fringing Reef Area**, **Barrier Reef Area**, or **Perimeter Reef Area**

REEF AREA—BARRIER: see **Area—Barrier Reef**

REEF AREA—FRINGING: see **Area—Fringing Reef**

REEF EDGE SCARP: 1–3 m vertical or high angle drop off from the upper edge of the **Reef Top** to upper **Terrace** surface; usually on the **Lagoon Reef Slope**

REEF HOLE: **Reef Top Subtidal Feature** of variable dimensions; always contains marine water; usually >5 m and <60 m deep

REEF ISLET: land area on **Reef Top** of **High Island Earthform** or on **Reef Top** of **Atoll/Table Reef/Low Island Earthform**; usually composed only of carbonate rock and/or sediments, with an elevation less than 10 m above high tide; usually have geological stability and support substantial vegetation; with presence of **Reef Islet Area**, an **Atoll Earthform** or **Table Reef Earthform** is often referred to as a **Low Island Earthform**

REEF ISLET—PATCH REEF: land area on **Reef Top** of **Patch Reef**

[REEF ISLET—TYPE 1]: modifier for types of **Reef Islets** on **Atolls**; usually singular isolated islets composed mostly of **Sand**, with **Beachrock** common; usually found near **Passes** or on **Patch Reefs**

[REEF ISLET—TYPE 2]: modifier for types of **Reef Islets** on **Atolls**; large, wide islets at convex bends in **Perimeter Reef**; **Gravel** may be common on **Outer/Ocean Shoreline**

{REEF ISLET—TYPE 3}: modifier for types of **Reef Islets** on **Atolls**; narrow, elongate islets which are usually curved or sinuous; usually on narrow, straight **Perimeter Reef** sections

{REEF ISLET—TYPE 4}: modifier for types of **Reef Islets** on **Atolls**; complex, irregular-shaped islets developed around **Gravel/Boulder Ridges** and **Conglomerate/Reef Limestone Platforms**; usually on wider **Perimeter Reef** sections

REEF PAVEMENT: **Reef Top Surface Feature** composed of solid horizontal substrate; little or no accumulation of sediment

REEF POOL: **Reef Top Subtidal Feature** consisting of depression of variable dimensions which always contains marine water; maximum depth 5 m

REEF SLOPE—LAGOON: portion of **Reef Area** which descends from **Reef Top** to **Lagoon Floor**; may include **Terraces**

REEF SLOPE—OUTER: portion of **Fringing Reef Area** or **Barrier Reef Area** which descends from **Reef Top** to **Nearshore Bottom Area**; may include **Terraces**

REEF SLOPE—PATCH REEF/PINNACLE: sloped portion of **Patch Reefs/Pinnacles** which descend from **Reef Top** to **Lagoon**; may include **Terraces**

REEF SLOPE—RETICULATE REEF: sloped portion of **Reticulate Reefs** which descend from **Reef Top** to **Lagoon Floor**; may include **Terraces**

REEF SLOPE WITH TERRACE: type of **Reef Slope** with substantial low angle or horizontal section >10 m wide

REEF SLOPE WITHOUT TERRACE: type of **Reef Slope** without substantial low angle or horizontal section; may have small sections <10 m wide that are low angle or horizontal

REEF TOP: horizontal portion of **Fringing Reef Area**, **Lagoon Fringing Reef Area**, or **Barrier Reef Area**; usually shallow (0–3 m depths), may have some deeper or permanently exposed features; composed of **Surface Features**, **Subtidal Features**, and **Supratidal Features**; terminates at beginning of **Outer Reef Slope** and **Lagoon Reef Slope** (on **Barrier Reef** or **Perimeter Reef**)

REEF TOP—PATCH REEF/PINNACLE: horizontal portion of **Patch Reef** or **Pinnacle**; usually shallow (0–3 m depths), may have some deeper or permanently exposed features; composed of **Surface Features**, **Subtidal Features**, and **Supratidal Features**; terminates at beginning of **Patch Reef Slope** or **Pinnacle Slope**

REEF TOP—RETICULATE REEF: horizontal portion of **Reticulate Reef**; usually shallow (0–3 m depths), may have some deeper or permanently exposed features; composed of **Surface Features**, **Subtidal Features**, and **Supratidal Features**; terminates at beginning of **Reticulate Reef Slope**

REEF TOP OPENING—DEEP: extensive deep **Reef Top Opening**; predominately >10 m deep; opening extends >2 km along length of **Barrier Reef** or **Perimeter Reef**

REEF TOP OPENING—SHALLOW: extensive shallow **Reef Top Opening**; predominately <10 m deep; opening extends >2 km along length of **Barrier Reef** or **Perimeter Reef**

REEF TOP SUBTIDAL FEATURES: submerged features of the otherwise shallow, horizontal surface of the **Reef Top**; includes the following features: **Hoa**, **Moat and Depression**, **Reef Pool**, **Reef Hole**, **Submerged/Incomplete Reef Top**, and artificial subtidal features (**Dredge Pit/Quarry/Channel/Basin**)

REEF TOP SURFACE FEATURES: formations on the generally flat surface of the **Reef Top**; lacking significant **Subtidal Features** or **Supratidal Features**

REEF TOP SUPRATIDAL FEATURES: features permanently exposed above high tide on the otherwise shallow, horizontal surface of the **Reef Top**; includes the following features: **Storm Block, Gravel/Boulder Ridge or Rampart, Beachrock, Conglomerate/Reef Limestone Platform, Aeolianite, Coral/Algal Dam and Spillway, Mangrove, Fishpond/Fishtrap/Shipwreck**

REEF/SHOAL/BANK—INSHORE: **Submerged Earthform** rising from the inner **Continental Shelf** to within 20 m of the ocean surface

REEF/SHOAL/BANK—MID-SHELF: **Submerged Earthform** rising from the mid-**Continental Shelf** to within 20 m of the ocean surface

REEF/SHOAL/BANK—OUTER SHELF: **Submerged Earthform** rising from outer **Continental Shelf** to within 20 m of the ocean surface

REEF WALL: submerged vertical or near vertical portion of **Reef Slope** with >5 m in vertical extent and >10 m along linear **Reef Slope**

RETICULATE REEF: linear reefs arranged in polygonal patterns, usually in **Lagoon Areas** of **Atolls**

RIDGE: elevated and elongated **Submerged Earthform** feature on the deep ocean floor

RUBBLE: poorly sorted, irregular-shaped fragments of **Coral** and **Reef** rock; normal size ranges from **Sand** to **Cobble**

RUBBLE/BOULDER TRACT: **Reef Top Surface Feature** consisting mainly of **Rubble** and **Boulders** deposited in a linear arrangement; often aligned perpendicular to the **Outer Reef Slope**

SAND: rock particles larger than 0.05 mm and smaller than or equal to 2 mm in diameter; smaller than **Gravel**

SAND/GRAVEL FLATS: low angle-sloped, low-energy **Sediment Shoreline** or **Reef Top Surface Feature**; primarily composed of **Sand** or **Gravel** size sediment

SAND/RUBBLE/ROCK FLATS: **Reef Top Surface Feature** consisting of a heterogeneous, unsorted mixture of **Sand, Rubble,** and rock reef materials

SAND SLOPE: **Reef Slope** primarily composed of **Sand** or **Gravel**-size sediment

SEAGRASS BED: **Reef Top Surface Feature** dominated by seagrass cover

SEAMOUNT: **Submerged Earthform** which is a submarine mountain with its crest at a depth of 200 m or more

SEAWALL/REVETMENT/BULKHEAD: linear **Artificial Shoreline** consisting of solid vertical wall (Seawall); sloped boulder-like material (Revetment); or back-filled sheeting (Bulkhead)

SHOAL: a **Submerged Earthform** composed of loose materials (sediments), within <20 m of the ocean surface but not uncovered at low tide

SHORE AREA: see **Area—Shore**

SHORELINE: the narrow linear boundary between aquatic and terrestrial environments on the **Coastline**; types of shoreline are **Sediment Shoreline, Solid Substrate Shoreline,** or **Artificial Shoreline**

SHORELINE—ARTIFICIAL: human-made shoreline structures or conditions; includes linear structures (**Seawall/Revetment/Bulkhead**), perpendicular structures and shore extensions (**Landfill/Causeway/Groin**), marine structures attached to shore (**Fishpond/Fishtrap/Shipwreck**) or extraction sites (**Dredge Pit/Quarry/Channel/Basin**)

SHORELINE—SEDIMENT: **Shoreline** consisting of predominately unconsolidated materials; includes **Beaches** (high angle shorelines) of large (**Boulder/Cobble**), small (**Sand/Gravel**),or recently cemented (**Beachrock**) sediments; low angle shorelines of large (**Boulder/Cobble Field**), small (**Sandflat**), or fine (**Mud/Silt Flat**) sediments; **Bar and Spit** extensions of **Shorelines**; and **Sediment Shorelines** dominated by marine vegetation (**Seagrass Bed** or **Mangrove**)

SHORELINE—SOLID SUBSTRATE: **Shoreline** consisting of consolidated, cemented materials; includes vertical Shorelines (**Cliffs**) of various heights (High, Medium, Low), **Stacks**, **Talus**, **Ramps**, **Marine Benches**, and **Marine Notches/Caves**

SLOPE—BOULDER/BLOCK: **Reef Slope** consisting primarily of **Boulder-** and **Block-**size deposits

SLOPE—CORAL: **Reef Slope** consisting primarily of live **Coral** cover

SLOPE—CORAL/SEDIMENT: **Reef Slope** consisting of a mix of live **Coral** cover and sediment substrate

SLOPE—SAND: **Reef Slope** dominated by **Sand** and **Gravel-**size sediments

SLOPE—SAND/RUBBLE/ROCK: **Reef Slope** consisting of a heterogeneous, unsorted mixture of **Sand, Rubble,** and rock **Reef** materials

SLOPE—SOLID SUBSTRATE: **Reef Slope** dominated by nonliving solid substrate

SPUR AND GROOVE: **Reef Slope Terrace Feature** consisting of **Sand** or rock-floored linear depressions (grooves) separated by spurs formed by **Coral** growth; found on upper **Reef Slope** with **Terrace** aligned at right angles to the **Reef** front; usually found on **Windward Reefs**; more gently sloping and with smaller topographic relief than **Buttress and Valleys**

STACK: small steep-sided remnant sections of **Solid Substrate Shoreline,** separated from primary **Shoreline** by a short distance (usually <100 m); usually associated with **Cliffs**

STORM BLOCK: large block on the **Reef Top; Reef Top Supratidal Feature** usually 0.5–2 m in diameter, sometimes larger; either individual coral heads or large detached fragments of **Reef** framework deposited by storm surge, waves, or tsunamis

[SUB-LAGOON]: modifier of **Lagoon Area**; portion of **Lagoon** which forms smaller lagoon enclosed by **Reef** development

SUBMARINE CLIFF: submerged vertical or near-vertical portion of **Reef Slope** of at least 2 m vertical extent; when >5 m in vertical extent and >10 m along linear **Reef Slope,** this kind of formation is a **Reef Wall**

SUBMARINE NOTCH/CAVE: permanently submerged indentation on **Reef Slope**; indentation of 1–3 m is a **Notch**; indentation of >3 m is a **Cave**

SUBMERGED ATOLL-REEF: a **Submerged Earthform** with the form of an **Atoll**; the upper surface is more than 20 m below sea level

SUBMERGED TABLE-REEF: a **Submerged Earthform** with the form of a **Table Reef**; the upper surface is more than 20 m below sea level

[SUBTIDAL LAGOON with SUBTIDAL CONNECTION]: modifier of **Coastal Lagoon** indicating level of permanence and connectedness; **Subtidal Lagoon** with subtidal connection persists, and connected to larger marine water body, through all stages of tide

[SUBTIDAL LAGOON with INTERTIDAL CONNECTION]: modifier of **Coastal Lagoon** indicating level of permanence and connectedness; subtidal lagoon is present through all stages of tide; intertidal connection means that interchange with larger marine water body is tide dependent

SURFACE MICROLAYER: the very thin surface layer of ocean waters where neuston is concentrated

SURGE CHANNEL: Reef Top Surface Feature consisting of shallow channels that intersect **Algal Ridge;** usually oriented at right angles to the **Reef** edge and often extend onto **Reef Top** a short distance

TABLE REEF EARTHFORM: Earthform consisting of a flat-topped **Reef** structure; may have **Reef Islet,** with or without a water body on the islet

TABLE REEF—REEF ISLET WITH WATER BODY: Table Reef with **Reef Islet;** with a water body which is completely, or mostly, enclosed by the islet land mass

TABLE REEF—REEF ISLET WITHOUT WATER BODY: Table Reef with **Reef Islet;** without a permanent water body

TABLE REEF—NO REEF ISLET: Table Reef without **Reef Islet,** only a shallow **Reef** platform

TALUS: Blocks and **Boulder** materials accumulated at base of high angle **Solid Substrate Shoreline;** often associated with **Cliffs** and/or **Marine Bench/Ramps**

TERRACE/SUBMARINE PLATFORM: low angle or horizontal shelf of >10 m width on **Reef Slopes;** when >50 m wide this kind of formation is a **Submarine Reef Platform;** both features may have **Furrows**

TERRACE—BOULDER/BLOCK: Terrace/Submarine Platform consisting primarily of **Boulder-** and **Block-**size deposits

TERRACE—CORAL: Terrace/Submarine Platform consisting primarily of live **Coral** cover

TERRACE—CORAL/SEDIMENT: Terrace/Submarine Platform consisting of a mix of live **Coral** cover and sediment substrate

TERRACE—SAND: Terrace/Submarine Platform dominated by **Sand** and **Gravel-**size sediments

TERRACE—SAND WITH CORAL MOUNDS: Sand Terrace/Submarine Platform with substantial numbers of large massive **Corals** or knolls of **Reef** rock and **Coral** growth; mounds are usually 1–10 m in diameter

TERRACE—SAND/RUBBLE/ROCK: Terrace/Submarine Platform consisting of a heterogeneous, unsorted mixture of **Sand, Rubble,** and rock **Reef** materials

TERRACE—SOLID SUBSTRATE: Terrace/Submarine Platform dominated by nonliving solid substrate

[TERRIGENOUS]: modifier of **Sediment Shorelines** and **Reef Slope** substrate; dark-colored sediment (black, green) composed of particles of basalt, olivine, etc.; light-colored sediment particles of andesite, rhyolite, etc., and/or weathered crystals of granitic minerals (e.g., quartz, mica) from continental areas

THERMOCLINE: transition layer between the mixed layer and deeper layer zone of the ocean where temperature changes rapidly over a small depth range; transition layer in any temperature-based stratification in any water body; inhibits mixing between shallow (mixed) and deep layers

TRENCH: an elongated **Submerged Earthform** depression on the deepest margin of the ocean floor, generally at depths >7,000 m; typically associated with subduction zones along the boundaries between oceanic and continental plates

TUNNEL (ROOM AND PILLAR): Reef Slope Terrace Feature; an elaboration of **Spur and Groove** in which the spur is roofed over by **Coral** and coralline algal growth-forming **Reef** galleries

UPWELLING ZONE—COASTAL: nearshore upward movement of deeper nearshore ocean waters attributed to prevailing offshore wind along a coast and/or often associated with coastal **Boundary Currents**

[VOLCANIC]: geology modifier of **High Island Earthform**; island lying on oceanic plate without previous connection to continental landmasses; geological composition generally basalt

VOLCANO: **Submerged Earthform** equivalent to terrestrial volcano

WATER MASS–OPEN OCEAN: relatively stable masses of pelagic water with similar temperature, salinity, and specific gravity characteristics separated from other water masses by density stratification or **Boundary Currents**; large-scale features of the **Open Ocean** which are not a part of **Boundary Currents**

WATER MASS–NEARSHORE: relatively stable masses of water with similar temperature, salinity, and specific gravity characteristics separated from other water masses by density stratification or **Boundary Currents**; large-scale features of the **Nearshore Ocean**

WATERS—MIXED: **Shallow Epipelagic** ocean waters not part of **Surface Waters** or **Thermocline Waters**

WATERS—SURFACE: **Shallow Epipelagic** ocean waters mixed by shallow water waves, wind stress, and tides

WATERS—THERMOCLINE: **Shallow Epipelagic** ocean waters within the **Thermocline** layer

[WINDWARD]: modifier of **Reef Area** exposure indicating the side facing the direction of prevailing winds

REFERENCES

Brodie, J., C. Arnould, L. Eldredge, L. Hammond, P. Holthus, D. Mowbray, and P. Tortell. 1990. Review of the state of the marine environment: South Pacific action plan regional report. UNEP Regional Seas Reports and Studies No. 127. UNEP.

Connell, J. 1984. Islands under pressure: Population growth and urbanization in the South Pacific. *Ambio* 13(5–6): 306–12.

Dahl, A. L. 1980. Regional ecosystems survey of the South Pacific area. SPC Technical Paper No. 179. South Pacific Commission.

———. 1984. Oceania's most pressing environmental concerns. *Ambio* 13(5–6): 296–301.

———. 1988. *Protected areas of Oceania.* IUCN.

———. 1991. *Island directory.* UNEP Regional Seas Directories and Bibliographies No. 35. UNEP.

Fosberg, F. R., and S. H. Pearsall III. 1993. Classification of non-marine ecosystems. *Atoll Research Bulletin* 389.

Holthus, P. F. In press. Coastal and marine environments of Pacific islands: Ecosystem classification, ecological assessment, and traditional knowledge for coastal management. In *Small island oceanography,* ed. G. A. Maul. American Geophysical Union Coastal and Estuarine Series.

Holthus, P., P. Brennan, S. Gon, L. Honigman, and J. Maragos. 1993. Preliminary classification and inventory of ecosystems of U.S.-affiliated islands of the tropical Pacific. Report prepared for U.S. Fish and Wildlife Service and The Nature Conservancy Pacific Program.

Holthus, P., M. Crawford, C. Makroro, E. Nakasaki, and S. Sullivan. 1992. Case study on vulnerability to sea level rise: Majuro Atoll,

Marshall Islands. SPREP Reports and Studies Series No. 60. South Pacific Regional Environment Programme.

Holthus, P. F., and J. E. Maragos. 1992. Marine biological diversity and marine protected areas in the South Pacific region: Status and prospects. Prepared for the IVth World Parks Congress and GEF/World Bank Workshop on Marine Biodiversity Conservation, February, Caracas, Venezuela.

Maragos, J. E. 1992. A marine ecosystem classification system for the South Pacific region. In *Coastal resources and systems of the Pacific basin: Investigations and steps toward protective management,* 253–99. UNEP Regional Seas Reports and Studies No. 147. UNEP.

Pearsall, S., J. Maragos, R. Thaman, E. Guinther, D. Polhemus, P. Holthus, and P. Thomas. 1992. An ecosystem classification and criteria for biodiversity conservation in the tropical insular Pacific. The Nature Conservancy, Pacific Region, Honolulu.

Polhemus, D. A., J. Maciolek, and J. Ford. 1992. An ecosystem classification system of inland waters for the tropical Pacific islands. *Micronesica* 25(2): 155–73.

Stoddart, D. R. 1992. Biogeography of the tropical Pacific. *Pacific Science* 46(2): 276–93.

Thomas, P., P. Holthus, and I. Reti. 1992. Pacific regional review of national parks and protected areas. Chapter 11, *Regional reviews.* Proceedings of the IVth World Parks Congress and GEF/World Bank Workshop on Marine Biodiversity Conservation, February, Caracas, Venezuela.

Thomas, W. L., Jr. 1963. The variety of physical environments among Pacific islands. In *Man's place in the island ecosystem,* ed. F. R. Fosberg, 7–37. Honolulu: Bishop Museum Press.

UNEP/IUCN. 1988. *Coral reefs of the world.* Vol. 3, *Central and Western Pacific.* UNEP Regional Seas Directories and Bibliographies. IUCN, Gland, Switzerland, and Cambridge, U.K./UNEP, Nairobi, Kenya.

Information Management Systems

UNEP-IOC(UNESCO)-ASPEI-IUCN Global Task Team on the Implications of Climate Change on Coral Reefs

Charles Birkeland

ABSTRACT

The objectives of the species systematics and the information management sections of this workshop have different criteria; the former requires taxonomic certification, the latter requires broad-scale time-series assessments. Because it is impractical to achieve both criteria simultaneously, it is most important to code all data with indicators of levels of taxonomic certification in order to allow the data to be retrieved automatically according to level of certifications needed for the question at hand. Coral-reef ecosystems may have already been changed substantially by human activities before the advent of scuba so we need to compile valid data from the past, especially data from surveys taken over three or four decades ago. Resurveying previous sites is of far greater priority than selecting and establishing new sites to monitor for baseline data. The size distribution (taken as a proxy for age distribution) of corals, fishes, molluscs, and some other animals is more informative of the status of the condition of the reef community and more predictive of future changes than are surface cover or community composition. A factorial or paired-comparisons design for monitoring would be the most powerful for factoring out random variation and determining which changes are anthropogenic, which are natural, and which are interactions between the two. Standardization of survey methods is necessary for testing hypotheses and for comparative studies, but standardization of methods should be done with caution when trying to find answers to open questions or explore for latent resources. At least ten handbooks have been produced with similar standard methodologies for coral-reef surveys using techniques that do not require expensive equipment or sophisticated taxonomic expertise.

It is now more urgent to begin fieldwork than to have more workshops to come to agreement on standardized methods that have already been developed. The UNEP-IOC-ASPEI-IUCN Global Task Team recommends the basic structure of data management to be a central, dedicated database, connected by e-mail to a network of databases within each of the participating countries. The Task Team recommended that sponsoring agencies (e.g., UNEP, IOC[UNESCO], WMO, IUCN, ASPEI) fund the central, dedicated database facility. The trained personnel of the central database facility must develop data storage and checking protocol, prepare the training methods for data collection and entry, ensure that all data are verified upon submission, report back immediately to the scientist who submitted the data whenever errors are found, prepare regular summaries of the contents of the database and regional comparisons, and prepare reports on the status of reefs throughout the world for international agencies. Although entries must be checked for accuracy at the central database facility, the corrections must be made by the scientists who originally submitted the data.

INTRODUCTION

The objectives of participants in this workshop fell into two categories that had different requirements for data management. The first set of questions concerned biodiversity per se—the **status** of our knowledge of biodiversity and biogeography in the tropical Pacific at the taxonomic, systematic, habitat, and ecosystem levels. For these questions it is critical that species identifications are accurate and certified. If errors in identification without voucher specimens are accepted into the database, it is quite likely that the species could be falsely and permanently recorded as present at some place and time in the past, then interpreted as becoming locally extinct as when the area is resurveyed.

The focus of the UNEP-IOC-ASPEI-IUCN Global Task Team is on **change** in biodiversity that might be occurring currently. The former set of objectives on status requires large-scale, long-term programs that produce taxonomically certified data sets. The latter programs on changes are concerned with rapidly determining whether substantial changes are occurring now, and the taxonomic specificity is not so critical. For example, it is considered important to know if a significantly greater portion of the coral-reef substrata is covered by algae and significantly less is covered by living coral than was the case a few years ago, even if the identification of the algae and corals was not at the level of species.

In order to gain access to data with appropriate levels of taxonomic certification, there must be an indicator for each observation, datum, or record that categorizes the record as, perhaps, 1 = field observation, 2 = identified by a qualified taxonomist, 3 = voucher specimen number for museum collection. The data can then be automatically sorted and only those records at the appropriate level of taxonomic certification, depending on the nature of the question being investigated, are retrieved.

CHANGES IN BIODIVERSITY—PAST AND FUTURE

The questions concerning biodiversity were addressed by the UNEP-IOC-ASPEI-IUCN Global Task Team on the Implications of Climate Change on Coral Reefs: To what extent is change in biodiversity occurring now? What proportions of the variance in diversity through time are caused by natural cycles, by human activities, and by their interactions? Has biodiversity undergone substantial changes prior to the invention of scuba and prior to our organized studies of tropical coastal regions, but recently enough to have been under the influence of human activities? Is what we accept as "normal" (brought about mainly by factors not under human influence) actually "abnormal" (brought about mainly by human activities)? What changes are predicted? Resurveying sites for which adequate data were recorded several decades ago is a means of obtaining insight into the past. Size distributions of populations is a method of predicting future trends at the location.

Resurveying Previous Coral-Reef Study Sites

In order to determine whether the coral-reef ecosystems already have been changed substantially by human activities before the advent of scuba, we need to compile valid data from the past, especially data from over three or four decades ago. We must assess changes that already have occurred or that are currently occurring. Unless we can determine whether the starting point was "normal," the changes observed by monitoring will be difficult to interpret. For example, *Diadema antillarum* may have been extraordinarily abundant in the western tropical Atlantic prior to its mass mortality in 1983, and herbivorous fishes may be extraordinarily few in number because of overfishing during the past centuries. This hypothesis is suggested by the inability of the coral-reef ecosystem to recover to dominance of coral over algae for over a decade since the dieoff of *D. antillarum* (Hughes 1994).

Although scuba was invented in the early 1940s, and not widely used for coral-reef research for the next decade, a number of coral-reef surveys have been conducted in decades before this time. Alexander Agassiz mapped the distribution of major corals in the Dry Tortugas in 1888. A resurvey in 1976 demonstrated the nearly complete disappearance of the previously dominant coral *Acropora palmata* from the area and three episodes of mass mortality of *A. cervicornis* (Davis 1982). Likewise, reefs in American Samoa and in Palau have been surveyed before World War II.

A number of programs of coral-reef monitoring and surveying have been active in recent decades. The ASEAN-Australia Project on Living Coastal Resources has been compiling data on coral reefs in the Southeast Asian region since 1984. Broadscale surveys of living and dead coral cover have been made annually on many reefs over great distances on the Great Barrier Reef since 1985 by the Australian Institute of Marine Science. Thirty transects around American Samoa have been surveyed on several occasions since 1979 (Birkeland et al. 1987, 1994). Some marine reserves (e.g., Fagatele Bay in American Samoa [Birkeland et al. 1987, 1994] and Ngerukewid Islands Wildlife Preserve in Palau [Birkeland and Manner 1989]) have permanent transects established. It is important to make use of baseline data that already exist before starting new monitoring projects.

Some of these previous surveys have not been resurveyed recently, but provide the opportunity to answer important questions about changes in biodiversity over the past several decades. Biological surveys were undertaken on Pohnpei decades ago, some about half a century ago. Have there been changes? It is not surprising if populations of species that are targeted for food have been declining in recent years, but are populations of nontargeted, small, cryptic species also declining in abundance or changing in other ways?

These nontargeted species on remote islands such as Pohnpei should serve as a broader index of the general condition of the less developed, relatively undisturbed world, especially on Pacific islands. For example, *Conus* is a diverse genus of coral-reef gastropod not targeted by humans for harvest on Pohnpei. In August 1956, Dr. Alan J. Kohn of the University of Washington surveyed gastropods of the genus *Conus* around Pohnpei, at varying distances from human habitation. He collected about 180 specimens from about twenty species and took stomach contents. From the stomach contents, he derived data on polychaete species (from the perspective of sampling via *Conus*). Alan Kohn is available to resurvey his study sites.

A substantial trade has developed on Pohnpei for *Trochus niloticus* shells for use as mother-of-pearl material for manufacturing buttons. It would be informative to determine whether some of the sites that Kohn previously surveyed are now or have been extensively gleaned for shells to be used in the curio and handicraft trade. A comparison of changes in molluscan community structure between gleaned and ungleaned sites would be valuable.

An example of an almost-certain rapid extinction of species on tropical Pacific islands that is occurring but not being documented is the extinction of landsnails resulting from introduced predators. Sixty-six species of landsnails have been recorded on Pohnpei, of which 40 (61 percent) are endemic. In 1936, Dr. Y. Kondo surveyed sixty-six sites around Pohnpei and provided a detailed account of the sampling sites and of the distributions and abundances of the various species. In the 1980s, a triclad flatworm *Platydemus manokwari* from Papua New Guinea, a predator of landsnails, was released (or escaped) on Pohnpei. This species of flatworm also appeared in Guam in 1979 and has already caused the extinction of six endemic species of landsnails; in fact, these six species were last seen about five years ago. *P. manokwari* has also become established on Palau, Rota, and Oahu.

P. manokwari has apparently spread throughout Pohnpei, having been found all the way up to the peak of Mount Ngihneni, at 2,256-ft elevation (Barry D. Smith, pers. com.). All the endemic landsnails of Pohnpei have been officially nominated for the *IUCN Invertebrate Red Data List* for endangered species. A resampling of the exact sites surveyed in 1936 by Dr. Kondo would allow a paired comparison of the landsnail populations between 1936 and 1994, about ten years after the invasion of the predator. Dr. Kondo also surveyed numerous sites in Palau (Peleliu, Angaur, rock islands, Koror, Babeldaob), Chuuk, and Kosrae.

Several meetings have been sponsored by international agencies to discuss the best locations to set up study sites to monitor global changes. In regards to coral reefs, the changes already have been under way for decades. Previously surveyed sites should have priority over starting over with new sites. With new sites, we have even less insight into what is "normal" or "pristine."

Size Distributions to Identify Trends

At a recent UNESCO/Coastal Marine program conference, coral-reef scientists agreed that community-scale parameters such as surface cover by living coral tissue, filamentous algae, and so forth are not a good measure of the "health" of a reef by itself. A reef with 20 percent cover by living coral tissue might be a "healthy and stable" reef community. Extensive cover of live coral, but dominated by large colonies with no recruitment, could indicate that large and aged colonies were still surviving but the reef was dying by attrition (i.e., there was no recruitment). Conversely, a situation in which there are no large corals, but the juveniles are numerous, could indicate that there was a very destructive event recently, but the coral community might recover. The "healthiest" reef would have a fairly even age distribution.

The Sanctuaries and Reserves Division of NOAA, U.S. Department of Commerce, has sponsored the monitoring of marine life in Fagatele Bay National Marine Sanctuary, American Samoa, in 1979, 1982, 1985, and 1988. A total of thirty permanent transects have been established on all sides of Tutuila, including those in the Fagatele Bay National Marine Sanctuary (Birkeland et al. 1987, 1994). The recovery of coral communities from the *Acanthaster* attack in 1978 and 1979, the normal fluctuations of reefs not attacked, and the decrease in the health of the coral community near the Rainmaker Hotel have all been monitored. The total living coral cover has not changed much near the Rainmaker, but through time the larger colonies of *Porites* have become the only things there; there has been no recruitment, no small colonies, and the other species have been dropping out. We would not have seen anything if we relied on surface-cover measurements alone. Size-distribution data have given us much more added insight into what is going on. This decrease in coral community health near the Rainmaker has been occurring as the other communities around Tutuila have been recovering nicely from the outbreak of *Acanthaster* in 1978.

One must be careful with interpreting size distributions of corals because they may fragment or separate into smaller distinct colonies. Smaller colonies can be older than larger colonies (Hughes and Jackson 1980). Nevertheless, this problem can be largely overcome through experience in examining the dead coral between the patches of living tissue and the structure of attachment at the base of reattached branches (see Figures 39 A, B in Birkeland and Lucas 1990).

PAIRED-COMPARISONS MONITORING: NATURAL AND ANTHROPOGENIC CHANGES

The PACICOMP (Pacific Coastal Marine Productivity) Project has advocated focusing on establishing long-term coral-reef monitoring programs in which reefs under the influence of human activities are monitored in comparison with reefs relatively unexploited. This factorial or paired-comparisons design for monitoring would be the most powerful for factoring out random variation and determining which changes are anthropogenic, which are natural, and which are interactions. Pairs of sites have been surveyed since the 1970s in Samoa, Palau, and Guam, and these have been producing interesting insights that could not have been seen in short-term studies. These sets of paired comparisons (pristine vs.

human impact) are widely separated along a steep gradient in species diversity from west (Palau) to east (Samoa).

For SPREP, IUCN, UNEP, and the government of Palau, twelve permanent transects were established in the Ngerukewid (seventy islands) Marine Reserve in Palau (Birkeland and Manner 1989). Five transects off the southern tip of Malakal in Palau, which were established January 1976 for the government of Palau before a sewer disposal was placed at the site (Birkeland et al. 1976, 1993), were resurveyed in 1993. These long-term studies of both reserves and sewer outfall sites have provided some surprising insights that would not have been possible with one-time surveys, no matter how quantitative or detailed.

On Guam, permanent transects in Tumon Bay (where hundreds of thousands of tourists visit each year) were established in 1977. These tourists have been wading and jet-skiing all around these corals, sea cucumbers, fishes, etc.; yet the changes have been remarkably little (Amesbury et al. 1993). In contrast to this area of human activity, the United States Air Force has commissioned a permanent monitoring program in a marine reserve on their guarded property.

A compilation of data from previous surveys is of prime interest, especially from sites across a range of distances from concentrations of humans. Special priority should be given to data from surveys made several decades ago.

MONITORING CORAL REEFS FOR GLOBAL CHANGE

The establishment of a cooperative global program of monitoring the state of coral-reef systems has been the goal of international agencies for the past decade. The state-of-the-art techniques for studying coral reefs prior to 1981 was compiled by Stoddart and Johannes (1978).

It was then recognized that if a large number of sites around the tropics were to be assessed and/or monitored, there had to be a standard methodology of coral-reef survey using techniques that did not require expensive equipment or sophisticated taxonomic expertise. Arthur Lyons Dahl produced the first such handbook in 1981. Since that time, Dahl's handbook has been developed into at least nine more recent versions (see Appendix, this paper). The most extensive version has been developed by the ASEAN-Australia Project on Living Coastal Resources and used for the past decade. The UNEP-IOC-ASPEI-IUCN Global Task Team recommends this basic handbook of field survey techniques be used for the international cooperative coral-reef assessment program.

It should be noted, however, that the basic design of Dahl's handbook has not changed through the fourteen years and among the numerous handbooks. Workshops are often sponsored by international agencies to bring scientists together to establish standardized methods for low-cost monitoring. The meetings have been redundant because we have basically agreed on Dahl's handbook from the beginning. The handbook has just grown larger, culminating in English et al. (1994). It is now more urgent to begin our work than to have more workshops to agree on standardized methods.

DANGERS OF STANDARDIZATION

Standardized techniques are necessary for testing hypotheses and comparing data among monitoring sites. However, they are restrictive in discovering answers to refractory questions. Standardized techniques constrain our questions and can block discovery of new concepts. For example, plankton nets defined plankton for many years and framed the question, "How can so many suspension-feeders make a living on so few plankton drifting in over coral reefs?" The invention of traps for demersal plankton demonstrated that the question was not relevant to coral reefs because most zooplankton are resident; they do not just drift in. Residency was excluded as an alternative hypothesis because plankton was defined by standardized technique rather than by biological considerations. We should be aware that standardization is necessary for testing and for comparative hypotheses, but should be done with caution when trying to find answers to open questions or explore for latent resources.

DATABASE MANAGEMENT

A major concern of the design of monitoring systems is the easy accessibility of the information. If a cooperative international program is established and operating successfully, then a vast amount of information can be rapidly accumulated at a number of sites around the world. Both the UNEP-IOC-ASPEI-IUCN Global Task Team (meetings at Guam in June 1992 and at Miami in June 1993) and the participants in the UNEP-UNESCO-IOC-WMO-IUCN-SPREP-SCOR-CARICOMP-PACICOMP workshop (held at the Seventh International Coral Reef Symposium at Guam in June 1992) concluded that it is critical to have an efficient protocol of data storage and retrieval designed before the monitoring system is established and put into operation. Otherwise, we would have files full of data that are never used. Both computer facilities and trained personnel must be established before the monitoring program begins.

The UNEP-IOC-ASPEI-IUCN Global Task Team recommends the basic structure to be a central, dedicated database, connected by e-mail to a network of databases within each of the participating countries. The Task Team recommended that sponsoring agencies (e.g., UNEP, IOC[UNESCO], WMO, IUCN, ASPEI) fund the central, dedicated database facility. The Australian Institute of Marine Science (AIMS) was suggested as a location.

The trained personnel of the central database facility must develop data storage and checking protocol, prepare the training methods for data collection and entry, ensure that all data are verified upon submission, report back immediately to the scientist who submitted the data whenever errors are found, prepare regular summaries of the contents of the database and regional comparisons, and prepare status reports of the reefs throughout the world for international agencies.

As with the field methods for the monitoring program, the UNEP-IOC-ASPEI-IUCN Global Task Team recommended using the database structure for regional and in-country information management systems developed by the ASEAN-Australia Project on Living Coastal Resources as a model. Appendix I, "The Database," from the UNEP and IUCN handbooks on *Monitoring Coral Reefs for Global Change* describes the protocol of raw data sheets, database modules referred to as "tables" for sample identification, ambient (physical environmental) data, surveys (e.g., transects and manta tows taxa), and taxa. Each of the tables is subdivided into columns (fields) and records (rows).

An extensive list of code names for coral species and for geographic regions is available in both the UNEP and the IUCN handbooks, as well as the ASEAN-Australia manual by English et al. (1994). The coral-reef monitoring handbooks are standardized down to detail such as the data entry codes.

The UNEP-IOC-ASPEI-IUCN Global Task Team recommended that all participating scientists must be formally trained on the protocol of data collection, encoding, entry, and preparation of summaries. A formal training is expected to impress upon the participating scientists the necessity of being organized within the protocol so that information is easily accessible when it is time to retrieve it.

CONCLUSIONS OF THE UNEP-IOC-ASPEI-IUCN GLOBAL TASK TEAM

The UNEP, IOC, ASPEI, and IUCN are interested in the potential effects of global climate change on coral reefs because human populations associated with coral reefs, especially those on atolls, are especially vulnerable to sea-level rise. Furthermore, coral reefs may be the clearest and earliest harbingers of climate change. In the past, biogenic reef systems have broken down approximately a million years before the final extinctions that marked the boundaries of geological periods. "Thus reefs appear to be reliable advance indicators of mass extinctions phases, and probably better markers for global environmental crises than biodiversity shifts or terminations of lineages" (Copper 1994).

The activities of the UNEP-IOC-ASPEI-IUCN Global Task Team have been published in four products:

1. A report by Clive R. Wilkinson and Robert W. Buddemeier on *Global Climate Change and Coral Reefs: Implications for People and Reefs* (124 pages),

2. A brief glossy pamphlet on *Reefs at Risk: A Programme of Action* (a 24-page extended abstract or executive summary),

3. A field manual of monitoring methods for *Monitoring Coral Reefs for Global Change* published by UNEP and IUCN, and

4. A training session for techniques of assessing and monitoring coral-reef resources in the Cook Islands with participants from Fiji, Papua New Guinea, Solomon Islands, and Cook Islands.

The Task Team concurred with conclusions of the report by Wilkinson and Buddemeier—specifically that climate change by itself was unlikely to eliminate coral reefs and may even be advantageous to reefs in some areas. However, climate change will create hardships for people dependent on reefs because of changes in reef structure and diversity. For example, the corals on atolls may grow faster if sea level rises, but the production of sand will not keep up and the islets or *motus* on which people live will disappear or become smaller.

The pressures of human activities are a far greater immediate threat to coral reefs than is climate change. The rates of change in coral reef diversity, community structure (dominance by corals shift to dominance by algae), and resource availability resulting from increasing rates of exploitation of resources, coastal ecosystem modification, nutrient loading, pollution, and human population growth make the next decade or two a critical time in implementing effective and wise management programs for coral reefs.

APPENDIX

Coral-Reef Survey Handbooks

These handbooks are for a global cooperative coral-reef monitoring program. The techniques do not need expensive equipment and taxonomic expertise.

1981   Dahl, A. L. 1981. *Coral reef monitoring handbook.* South Pacific Commission Publications Bureau, Noumea, New Caledonia.

1984   Kenchington, R. A., and B. Hudson, eds. *Coral reef management handbook.* UNESCO Regional Office for Science and Technology for Southeast Asia.

1984   UNESCO. *Comparing coral reef survey methods.* Report of a UNESCO/UNEP Workshop, Phuket Marine Biological Centre, Thailand. UNESCO Report in Marine Science No. 21.

1986   Dartnall, A. J., and M. Jones. *A manual of survey methods for living resources in coastal areas.* Australian Institute of Marine Science.

1990   Coyer, J., and J. Witman. *The underwater catalog: A guide to methods in underwater research.* Shoals Marine Laboratory, Ithaca, New York.

1991   CARICOMP (Caribbean Coastal Marine Productivity). *Manual of methods for mapping and monitoring of physical and biological parameters in the coastal zone of the Caribbean.* Florida Institute of Oceanography.

1993   IUCN Publications. *Monitoring coral reefs for global change.* A review of interagency efforts compiled by John C. Pernetta. IUCN Publications Unit, Cambridge, England.

1993   UNEP/AIMS. *Monitoring reefs for global change.* Reference Methods for Marine Pollution Studies No. 61. United Nations Environment Programme.

1994   English, S., C. Wilkinson, and V. Solev. *Survey manual for tropical marine resources.* ASEAN-Australia Marine Science Project—Living Marine Resources. Australian Institute of Marine Science, Townsville.

1994   Rogers, C. S., G. Garrison, R. Grober, Z.-M. Hillis, and M. A. Franke. *Coral reef monitoring manual for the Caribbean and western Atlantic.* National Park Service, Virgin Islands National Park.

REFERENCES

Amesbury, S. S., R. T. Tsuda, R. H. Randall, A. M. Kerr, and B. Smith. 1993. Biological communities in Tumon Bay, 1977–1991. University of Guam Marine Laboratory Technical Report No. 99.

Birkeland, C., S. S. Amesbury, and R. H. Randall. 1993. Koror wastewater facility: Biological survey. Unpublished report for Parsons Overseas Company.

Birkeland, C., and J. S. Lucas. 1990. Acanthaster planci: *Major management problem of coral reefs.* Boca Raton, Fla.: CRC Press.

Birkeland, C., and H. Manner, eds. 1989. Resource survey of Ngerukewid Islands Wildlife Preserve, Republic of Palau. A report to the Government of Palau, the South Pacific Regional Environment Programme (SPREP), World Wildlife Fund, and the International Union for the Conservation of Nature. SPREP, Noumea, New Caledonia.

Birkeland, C., R. H. Randall, and S. S. Amesbury. 1994. Coral and reef-fish assessment of the Fagatele Bay National Marine Sanctuary. NOAA Technical Memorandum, NOS MEMD.

Birkeland, C., R. H. Randall, R. C. Wass, B. D. Smith, and S. Wilkins. 1987. Biological resource assessment of the Fagatele Bay National Marine Sanctuary. NOAA Technical Memorandum, NOS MEMD 3.

Birkeland, C., R. T. Tsuda, R. H. Randall, S. S. Amesbury, and F. Cushing. 1976. Limited current and underwater biological surveys of a proposed sewer outfall site on Malakal Island, Palau. University of Guam Marine Laboratory Technical Report No. 25.

Buddemeier, R. W., and D. G. Fautin. 1993. Coral bleaching as an adaptive mechanism. *BioScience* 43:320–26.

Copper, P. 1994. Ancient reef ecosystem expansion and collapse. *Coral Reefs* 13:3–11.

Davis, G. E. 1982. A century of natural change in coral distribution in the Dry Tortugas: A comparison of reef maps from 1881 and 1976. *Bull. Mar. Sci.* 32:608–23.

English, S., C. Wilkinson, and V. Solev, eds. 1994. *Survey manual for tropical marine resources.* ASEAN-Australia Marine Science Project—Living Marine Resources. Australian Institute of Marine Science, Townsville.

Hughes, T. P. 1994. Catastrophes, phase shifts, and large-scale degradation of a Caribbean coral reef. *Science* 265:1547–51.

Hughes, T. P., and J. B. C. Jackson. 1980. Do corals lie about their age? Some demographic consequences of partial mortality, fission, and fusion. *Science* 209:713–15.

Stoddart, D. R., and R. E. Johannes. 1978. *Coral reefs: Research methods.* Paris: UNESCO.

CHAPTER 15

The World Conservation Monitoring Centre: Handling Global Biodiversity Data

Mark D. Spalding

ABSTRACT

The data handling and management activities of the World Conservation Monitoring Centre (WCMC) are reviewed to illustrate the wider issues of biodiversity data management at the international level. WCMC compiles and disseminates global information on biological resources, their conservation, and sustainable use. These data include information relating to threatened and endemic species, protected areas, and habitats. They are stored, managed, and maintained on purpose-built databases using a variety of software packages, where they can be accessed by a wide range of users.

This paper gives a short overview of three of the main databases used at WCMC: the Species Database, the Protected Areas Database, and the Biodiversity Map Library (BML). Information relating to these databases is now available on-line over computer networks. None of these are specifically marine/coastal, but all hold considerable volumes of information relating to marine/coastal species, sites, and habitats. Mention is made of the considerable links already made between these databases and "external systems": both information management systems and organizations that act as data sources. Particular focus is given to the tropical coastal sensitivity mapping work being undertaken by WCMC, with its links to other regional and global initiatives.

The database systems in operation at WCMC mirror the much wider array of existing global, regional, and national databases covering various aspects of the marine and coastal resources of Oceania. Prior to any proposal for a new database system for this region a critical assessment should be made, both of the "user community" needs for such a database and of the applicability of existing systems. At any level, new databases may be able to draw design features and even base-line data from existing systems, thereby saving time and money.

One clear need that arose from this conference was for the better management of taxonomic data for marine and coastal species in the region. Such data management could be achieved through the development of a new database that would combine elements of raw data and metadata in its structures. Care should be taken in any such development to make full use of both existing taxonomic database structures and of data already held in these and other databases.

INTRODUCTION

One of the key products of a workshop such as this should be the development of a more coordinated and efficient approach to biodiversity data management for the tropical island Pacific region. This should include increased networking and data exchange, thereby reducing costs and preventing repetition in developing national biodiversity data management systems while providing a wider resource base for government officials, scientists, and managers throughout the region.

The more specific development of a "new" marine/coastal database for Oceania should be considered in the light of these comments (i.e., as a system that would facilitate data exchange and the wider use of existing data by scientists and decision makers, and one that would also provide encouragement for the gathering of new data). Such a database could be a single system, based in one organization but freely available to the wider community, or a more loosely linked series of existing regional databases with a number of specialist "nodes" throughout the region. This paper gives an overview of some of the data management activities at WCMC, which will give concrete examples of some of the issues involved in all stages from networking to database design, to data-sharing.

The WCMC in Cambridge, U.K., was established in 1988 as a company limited by guarantee with charitable status. Prior to this date the Centre existed in a similar form, with similar objectives as the IUCN Conservation Monitoring Centre. WCMC is managed as a joint venture among the three partners in the *World Conservation Strategy* and its successor *Caring for the Earth*: IUCN (the World Conservation Union), UNEP (United Nations Environment Programme), and WWF (World Wide Fund for Nature). Its mission is to provide information on the status, security, management, and utilization of the world's biological diversity to support conservation and sustainable development.

Information is the key resource behind most of WCMC's activities, and the Centre holds substantial information resources covering plant and animal species of conservation concern; important natural habitats and areas of special biological richness; national parks and protected areas; wildlife utilization, and the volume and impact of the international trade in wildlife; financial investments in biodiversity conservation; and a large conservation bibliography, including both published and unpublished literature.

WCMC's information service is utilized by many users ranging from governments, development agencies, non-governmental organizations, and multinational corporations to individual scientists, journalists, and conservationists. The specific requirements of these users range from standard data printouts to data for emergency response (e.g., for major oil spills), to the commissioning of major reports and books.

Handling these quantities of information can be divided into a number of tasks:

- gathering information on species, habitats, and sites through an extensive network of contacts;

- managing this information so that it can be accessed easily;

- disseminating this information as widely as possible so that it can be applied by the conservation and development organizations; and

- promoting the establishment of information networks to improve the flow and exchange of data, including the development of data centers in developing countries.

Most of the remainder of this paper deals specifically with the data management tools being developed and used at WCMC, and on aspects of data sharing between organizations. A more detailed description is also given of particular work relating to global coastal sensitivity mapping. Further information concerning these databases, including some very large datasets, is now available on-line over the international computer networks (notably through a World Wide Web Server on the Internet: http://www.wcmc.org.uk).

BIODIVERSITY DATABASES

A number of linked databases are managed by WCMC, and others are also held for which WCMC is a subsidiary user (see Box 1: Data Sources and Data Sharing). There follows a short overview of three of the main databases managed at WCMC.

Species Database

Since the late 1970s, WCMC has gathered and provided information on animals and plants of conservation concern. This work began with the preparation and/or support of global Red Data Books for different species groups and later Red Lists for animals (e.g., IUCN 1994b). WCMC stores computerized plant data (for more than 81,000 taxa) in a series of database files managed by *BG-BASE* (see following paragraphs). Animal data are held on a FoxPro database for over 27,000 animal taxa, with subsidiary datasets covering globally threatened taxa (6,000 species), single-country endemic species (12,000 species), and further datasets relating to species covered under global and regional conventions and protocols (e.g., the Convention on International Trade in Endangered Species of Wild Flora and Fauna [CITES] and the Specially Protected Areas and Wildlife [SPAW] Protocol of the Cartegena Convention). Marine and coastal species information is held in all of these databases, as well as data on the important numbers of threatened and endemic species frequently associated with island floras and faunas. Work is under way to strengthen and consolidate the animal databases onto *BG-BASE* to create a single system for species data management.

BOX 1: **DATA SOURCES AND DATA SHARING**

The data held in computer files and databases frequently represent only the "tip of the iceberg" in terms of the data held at WCMC as these data have been drawn from a very wide range of references, including a considerable amount of "grey" literature difficult to obtain elsewhere. More important than the files, however, are the networks of collaborating organizations and individuals who supply this information to WCMC. Increasingly, many of these contacts have highly sophisticated computer networks and databases of their own, and many of the data that come into WCMC are in digital form and may be sent via computer networks.

When large datasets are exchanged with other organizations, problems of ownership and copyright may arise. The principle adhered to by WCMC of free access to data is an essential measure in such a data network. There is no question of "selling" data, which are given and exchanged freely, and any charges that are made by WCMC to clients are simply to recover costs of data handling. Often the development of memoranda of understanding between collaborating organizations helps to clarify the ownership and data management responsibilities. It is helpful to consider a number of different ways in which data from other organizations can be shared, and how such data are handled and used. A sequence of examples are given, taking specific cases from WCMC's own experience.

Sole data holding:

Where WCMC acts as the key data holder for an external organization, usually under a detailed agreement or memorandum of understanding. For example, WCMC holds and manages the global datasets on protected areas for the IUCN CNPPA and also manages data on wildlife trade under contract to the CITES Secretariat.

Collaborative data holding and enhancement:

Where the user holds a near-identical copy of data being developed by another organization, but performs a specific agreed role in enhancing these data. E.g., WCMC is preparing or supporting the mapping activities of BirdLife International and the ReefBase database.

Subsidiary data holding:

Where the user holds a copy of a dataset but does not have a primary role in developing or enhancing data. E.g., WCMC's holding of FishBase and Digital Chart of the World.

Summary data holding:

Where the user holds certain summary data from another organization, but not entire datasets. E.g., WCMC's holding of The Nature Conservancy's plants data, waterfowl counts, and site-details derived from the International Waterfowl and Wetlands Research Bureau, and digital habitat and protected areas maps from multiple sources around the world.

Catalogue holding:

An extreme form of summary data holding where the user holds little other than references to key aspects of another dataset. WCMC is the biodiversity node for the CEISIN (Consortium of International Earth Science Information Network) Information Cooperative and will soon hold a major biodiversity metadataset describing the location and content of biodiversity databases globally.

These relationships are all two-way and many organizations, including those listed above as providers of data, use other parts of WCMC's data in one or more of these ways also. It should further be borne in mind that, although data-enhancement is only mentioned under one of the points above, it clearly has a role in all cases where data are amalgamated with other datasets.

BG-BASE is a variable-length field relational database application built using *Advanced Revelation*. *BG-BASE* was originally designed to manage information on the collections maintained by botanic gardens and arboreta. Since its inception in 1985, it has been installed in fifty-two institutions around the world, including botanic gardens, arboreta, zoos, universities, horticultural societies, and conservation organizations. Handling an extremely broad range of functions, it has grown to be a large application, with some 3,000 data fields spread across nearly 150 files (see Figure 15.1). WCMC uses only a small portion of these fields and files, but remains compatible with all users of *BG-BASE* worldwide because the underlying structure and data dictionary are identical. Currently, WCMC maintains the largest number of names, distributions, and data sources of any *BG-BASE* users (or of any other plant conservation organization anywhere).

The NAMES file is the central area for taxonomic and nomenclatural data. In it are fields for, among other things, each component of the scientific name; the source of the name; synonymy; IUCN Red Data Book conservation category at the world level; endemism code; distribution completeness flag; as well as flags for presence in the international trade and in cultivation. In addition, this file allows many-to-many linkages between scientific names and their synonyms and their common (vernacular) names.

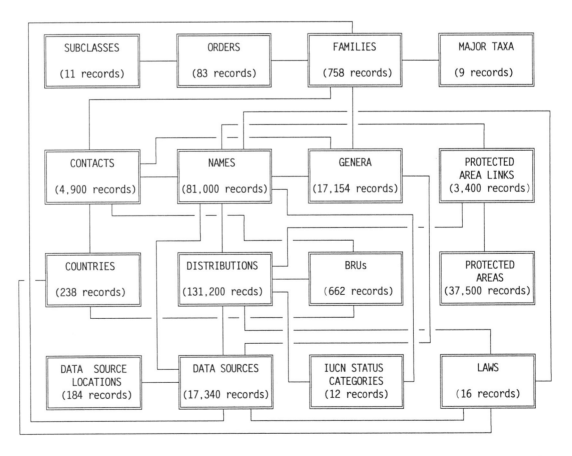

Figure 15.1 A simplified structure of *BG-BASE* showing some of the main files used by WCMC and their linkages

The DISTRIBUTIONS file links a taxon record to a Biological Recording Unit (BRU) record. These divide the world into 622 geopolitical units in which major political boundaries and oceanic island groups are recognized. Within the distributions file are fields for free-text geographic qualifier, doubtful occurrence; conservation status; and data sources for both the distribution and the conservation status information, number of individuals/populations left, habitat, elevation, threat(s), and legal status.

Records in the DATA SOURCES file are linked to many other files. A data source record contains fields for type of reference; date of publication; author(s); title; subtitle; source; publisher; language; location; key words; citation notes; abstract; relevance; and linkages to COUNTRIES, BRUs (Biological Recording Units), FAMILIES, GENERA, and NAMES. This file was used to produce the *World Plant Conservation Bibliography* (WCMC/RBG 1990).

Information is retrieved either through programmed reports or through direct database queries. Following the main body of each printout is a listing of the data sources used to create the report and a statistical count of Red Data Book categories.

The strengths of the *BG-BASE* system are clear, and it will obviously prove a powerful and workable system for holding animal data. From the marine viewpoint, these will include not only the standard data on threatened marine species (notably turtles, sirenians, cetaceans, and pinnipeds) but also wider datasets, such as the full species listings and distribution data for species in the Caribbean protected under the SPAW Protocol and similar listings for all species listed under CITES (including, for example, all scleractinian and black corals). External digital datasets, from which WCMC would also like to draw information for use in its species database, would undoubtedly include FishBase being prepared in the Philippines, CoralBase from the Australian Institute of Marine Science (AIMS), and the Coral Reef Fishes Specialist Group reef fish distribution work (which has had partial support from WCMC).

Protected Areas Database

WCMC works closely with the IUCN Commission on National Parks and Protected Areas (CNPPA) and national or provincial protected area agencies in maintaining a global database of protected areas. It is further recognized by CNPPA as being the key holder of a global dataset of marine protected areas that are flagged within the full dataset. Statistical data on over 37,000 sites are held, covering both nationally designated sites and sites of recognized international importance (e.g., natural World Heritage natural sites, biosphere reserves, and Ramsar wetlands). This Protected Areas Database is regularly used to produce the *United Nations List of National Parks and Protected Areas* (IUCN 1994a). It also links to maps showing the location of many of these sites and to text documents describing some of them. The boundaries of many protected areas are held in digital form on the Biodiversity Map Library (see next section), linked to the Protected Areas Database through common identifiers.

The data are held on a database built on a FoxPro application developed using the FoxPro

2.0 project manager, screen builder, report generator, and general purpose libraries. Key data for nationally designated sites include the following:

Site name
Designation
Area covered
Year of establishment
IUCN management category
Latitude/longitude
Altitude
Administrative responsibility
Location maps
Basic habitat information (marine habitats only)
Biogeographic information (after Udvardy 1975)
Relationship to other sites
Detailed text description of key sites (20% of sites)
Wider geographic location (region, country, Biological Recording Unit)

Further data are linked at levels above the site data, including:

- National statistics (country area, population, economic indicators)
- Definition of each designation used
- Management agencies
- Legislation
- Involvement in international conventions/programs
- Text description of protected area systems

Further structures have been incorporated into the database for ongoing and proposed enhancement. Data are being entered describing agencies responsible for protected areas and their budgets and staffing. It is further intended to expand and incorporate key contacts and bibliographic information. Future projects will add data covering threats to sites and management effectiveness, and will strengthen the links to the species data.

Biodiversity Map Library

WCMC began a detailed mapping program in 1989 using an ARC/INFO geographic information system (GIS) on a project to map the world's tropical forests. Since that time considerable datasets have been builtup including:

- Potential vegetation (global and regional)
- Tropical forests
- Global bathymetry and coastlines
- Internationally important wetlands
- Protected areas
- Coastal ecosystems (coral reefs and mangrove forests)
- Critical conservation sites lacking legal protection
- Species distribution data

Although many of the datasets are terrestrially based, all have relevance to marine and coastal environments. There is a particular drive to greatly expand and enhance the marine and coastal elements.

The complexity and nature of the data builtup over the years led to the requirement for a powerful GIS package. Hence, it was decided that version 6.1.1 of ARC/INFO be used, to be run on SUN SPARC workstations (this was upgraded to version 7.0 in April 1995). Further to this, the need was seen to develop a graphical user interface to allow rapid access to these data. This has led to the development, since 1991, of the Biodiversity Map Library (BML), which is designed to integrate much of the mapped and nonmapped information managed by WCMC—on species, ecosystems, and protected areas—into a common geographical format.

Through the BML it is also possible to link to the other databases to pull up related data on, for example, protected areas, critical wetlands, or turtle-nesting beaches. In addition to material prepared at the Centre, numerous datasets have been provided from external sources prepared globally, regionally, or nationally. The working scale of the different datasets varies enormously within the BML. Much of the data were originally gathered at scales of around 1:1,000,000, but much higher resolution has been achieved for a number of countries, particularly smaller nations and small islands, with highest resolutions being 1:20,000 or 1:50,000 for some countries.

The BML has been developed in the Librarian module of ARC/INFO, a module that has been designed to handle large datasets in a simple, yet structured, way. The BML can be split into two separate functional units: a data management component and a graphical user interface (GUI). The GUI has been written in Arc Macro Language as a user-friendly front-end. It provides most of the simpler GIS tools and the ability to construct graphical output from the large datasets quickly and easily. ARC/INFO is a well-established GIS with a large user-base; it has good facilities for exporting to other spatial data standards, while the internal export format for ARC/INFO has itself become a *de facto* standard. This is clearly important for data dissemination, which forms a key part of WCMC's work.

In addition to being a user-friendly front-end, the BML now provides the medium for managing most of the large GIS datasets collected by WCMC over the last few years. The software also provides tools to maintain the structure of a database, ensuring consistent data definition and projection specifications.

TROPICAL COASTAL SENSITIVITY MAPPING— A CASE STUDY

The Biodiversity Map Library offers a clear illustration of a system that links data of many different types from many different sources, both internal and external. WCMC is undertaking two major initiatives to improve its coastal digital map holdings—the first for coral reefs and the second for mangrove forests.

These projects are both collaborative—the former with the International Center for Living Aquatic Resources Management (ICLARM) in the Philippines, the latter with the International Society for Mangrove Ecosystems. The coral reef maps will be linked to a database being prepared at ICLARM to be known as ReefBase; the mangrove maps will be published as an atlas. In both cases WCMC will be responsible for preparing the maps, but it will rely heavily on international collaboration.

There are a number of stages to any projects of this nature. Initially it is important to establish a set of data standards, considering the user needs, the costs (both time and money), and the data availability. It is then important to test these standards and establish a process for data collection and handling using a number of case studies. Once these have been used to sort out major difficulties with the process, there follows a major drive to gather and process all the available information. Depending on the final output, subsequent stages will involve mail-outs and checking by regional experts; filling in gaps from other sources; further summarization or standardization; and maintaining, updating, or amalgamating to other existing datasets. End-products are prepared in whatever form and circulated to all collaborators.

The coral reef mapping project has completed the data-standards and the case-study phases and has now reached the stage of a major data-gathering exercise. Coastal zone mapping varies enormously in quality and resolution from place to place. Thus, this project is expected to handle data in quality ranging from low-resolution sketch maps to high-resolution, processed, remotely sensed images. Due to time and money constraints, it will not be possible to undertake any direct interpretation of remotely sensed data. The working scale of most maps will be 1:250,000, with more detailed scales (1:50,000 or 1:100,000) for specific sites and small nations. Where digital datasets do not exist, WCMC is actively seeking to obtain the best available maps in hard copy, which it will then digitize. For those countries with little or no information at the national scale relating to marine resources, this project could thus be seen to provide maps, which could be incorporated into national databases. For countries with existing coastal resource maps at the national level, either on hard copy or in digital format, such data will prove very valuable for the project. Nondigitized maps will be digitized and made available to suppliers of the data. Digital maps will be incorporated directly where these are available. High-resolution maps will be summarized as necessary. All the data will be fully acknowledged and referenced.

The mangrove mapping project is at an earlier stage, although it should be possible to reduce a number of the early stages, using the experiences gained through the reef-mapping work. Although the coral reef and mangrove-mapping projects can be perceived as separate projects, they will be closely linked. When combined with the other data held at WCMC, through the BML it is clear that they will provide much of the data required for fairly detailed coastal sensitivity mapping, including the following features:

- Coastline (differentiated into shoreline type if available)
- Mangrove
- Emergent reef crest/reef ridges
- Shallow reef areas or reef flats (as polygons where available)
- Bathymetric data (20, 50, and/or 100 m wherever available)

- Seagrass
- Turtle-feeding and -nesting sites
- Protected areas
- Species distribution data (country level only)
- Reef fish distribution
- Seabird colonies

The current status of mapping work in these projects is constantly changing, and for this reason an overview cannot be provided without being almost immediately out-of-date. At the beginning of 1995, high-resolution work was begun or completed for six countries in the tropical insular Pacific region. WCMC hopes to greatly increase this figure by focusing particularly on this region in the 1995–96 period. The scales for these maps vary from 1:20,000 to 1:100,000. Digital maps have been independently prepared for other nations in the region—work is under way to identify and contact the holders of these data.

CONCLUSIONS AND RECOMMENDATIONS

One of the principal themes underpinning the Convention on Biological Diversity (UNEP 1992) is the need to empower countries to sustainably manage their own renewable resources. Efficient biodiversity data management is a crucial component of this process. WCMC has considerable experience in developing and using large biodiversity databases. It has learned a great deal over that time relating to all of the processes involved from hardware and software needs to data gathering to preparing outputs. The expertise gained in this process has enabled it to become actively involved in supporting the development of national and regional databases to monitor biodiversity and conservation at different scales.

As the development of computer networks and databases relating to biodiversity information increases, seemingly at an exponential rate, it brings with it the potential for increased efficiency in data handling and in utilizing large quantities of accurate information to support the conservation and sustainable development of the world's biological resources. It is, however, very important to consider these large and increasing datasets and their associated guidelines in designing and developing new systems. Not only do they increase the data availability in digital formats, but they provide a very useful base or source of information for database design with considerable cost-cutting potential. By contrast, ignoring existing databases would lead to considerable inefficiencies in later aspects of data-sharing and transfer. There may also be cases and all scales from local to global where new databases per se are not required. In these situations, it may be more efficient to consider adaption or amalgamation of existing systems or simple linkage through metadatabase structures and live linkages over systems such as the Internet.

In many ways the database systems in operation at WCMC mirror the much wider array of existing global, regional, and national databases that already cover various aspects of the marine and coastal resources of Oceania. Prior to any proposal for a new database system a critical assessment should be made, both of the "user community" needs for such a database, and of the applicability and adaptability of *existing* systems to meet these needs. At the very least, new databases should be able to draw design features and even base-line data from existing systems, thereby saving time and money.

The papers presented in the current work provide an appraisal of existing data relating to the marine environment in Oceania (particularly taxonomic data) and of existing databases (notably regional and global, but with limited representation of the growing number of national level databases). One clear need that arises from these presentations is the better management of taxonomic data for marine and coastal species in the region. None of the database systems described provides for managing taxonomic data for all marine and coastal species in the region. The development of a new database should enable wider dissemination of data, currently concentrated in the hands of relatively few individuals. It should further provide for the wider use of these data and encourage the gathering of new data. Taxonomic data form the basic building blocks of all studies of biodiversity and all programs to protect and sustainably use that biodiversity. A new database could use raw data from, and establish metadatabase links to, other systems such as those previously described and in other papers described in this volume. Further considerations of this database and proposals for its structure were considered during the working group sessions of this meeting and are outlined in Chapter 1 and Appendix B of this volume.

Acknowledgments—Much of what is contained in this document is an amalgamation of other work compiled at WCMC, taken from countless unpublished documents and conversations. Many thanks must, however, go to the following: Harriet Gillett and Kerry Walter (BGBase information), Chris Magin ("animals" data), James Paine (protected areas), Jonathan Rhind and Ian Barnes (GIS), and Ian Crain, Jeremy Harrison, and Richard Luxmoore for general advice.

REFERENCES

IUCN. 1994a. *1993 United Nations list of national parks and protected areas.* Prepared by the World Conservation Monitoring Centre and IUCN Commission on National Parks and Protected Areas. IUCN, Gland, Switzerland, and Cambridge, U.K.

———. 1994b. *1994 IUCN red list of threatened animals.* Compiled by the World Conservation Monitoring Centre in association with the IUCN Species Survival Commission and BirdLife International. IUCN, Gland, Switzerland, and Cambridge, U.K.

Udvardy, M. D. F. 1975. A classification of the biogeographic provinces of the world. IUCN Occasional Paper No. 18, 1–48. IUCN, Gland, Switzerland.

UNEP. 1992. *Convention on biological diversity.* Environmental Law and Institutions Programme Activity Centre, United Nations Environment Programme.

WCMC/RBG. 1990. *World plant conservation bibliography.* World Conservation Monitoring Centre, Cambridge, U.K., and Royal Botanic Gardens, Kew, U.K.

CHAPTER 16

The United States Coral Reef Initiative

Michael P. Crosby and James E. Maragos

ABSTRACT

Evidence supports the consensus that coastal ecosystems are not only experiencing changes but are also being stressed and destroyed through increasing anthropogenic perturbations. Economic demands and heightened coastal population growth have imposed unforeseen strains on our nation's coral reef ecosystems including resource exploitation, loss of biodiversity, marine-based and coastal nonpoint source pollution. As an initial effort to counter this trend, the United States is developing an inter-agency Coral Reef Initiative (CRI) to create the base for a combined domestic and international effort aimed at the conservation and effective management of coral reefs and related ecosystems (mangroves and seagrass beds). The U.S. CRI will build on existing federal, state, territorial, commonwealth, and local partnerships through communication with relevant stakeholders at all levels. Mechanisms for ongoing consultation among stakeholders have been initiated and are being expanded to take into account local needs, priorities, and opportunities in developing the U.S. CRI. Working in partnership at the federal, state, territory, and commonwealth level, agencies are seeking to integrate their operational management activities in an ecosystem-wide approach, increase monitoring, conduct assessments to provide better information for decision-makers, provide education and outreach to increase public understanding, and undertake a more pro-active effort in understanding and main-taining biodiversity of coral reef ecosystems. Efforts to achieve holistic management must consider not only the fish and the coral reef resources but also the ecological, social, economic, and political aspects that involve all stakeholders. A key compo-nent of such a strategy would be promotion of healthy ecosystems by ensuring that economic development in the United States is managed in ways that maintain biodiversity and long-term productivity for sustained use of our coral reef ecosystems.

The primary objective of the CRI is to foster innovative cross-disciplinary approaches to sustainable management and conservation of coral reef biodiversity and ecosystems through the development of cooperative relationships among the various stakeholders. Perhaps the most important element within the U.S. CRI is support for community involvement in developing and implementing local and regional CRI suited to those community needs and situations. For the U.S. and international CRI to succeed in effectively conserving and managing coral reef ecosystems for long-term sustainable use, programs should essentially be based on local community involvement. Scientists, government managers, nongovernmental organizations, and other interested residents are now actively planning local coral reef initiatives throughout the American Flag Pacific Island. As of June 1995, the Hawaii State Coastal Zone Management Program has taken the leadership in organizing support and initiating Hawaiian CRI activities by the end of summer 1995 with substantial support from the East-West Center, the Pacific Basin Development Council, the U.S. Fish and Wildlife Service, the NOAA National Marine Fisheries Service, the University of Hawaii (UH) Sea Grant Program, other UH departments, and other state agencies, especially the Department of Health and the Department of Land and Natural Resources. Efforts are also being made to enlist the active support of county/local governments, other federal and state agencies, Hawaiian groups, nature conservation organizations, educational groups, business organizations, and ocean recreational organizations.

INTRODUCTION

Coral reefs are one of the most biologically diverse and productive natural ecosystems in the United States. They are vital to the ecological sustainability and to the economies of coastal regions. Coral reef ecosystems, and associated mangroves and seagrass beds, provide sheltered waters and high productivity that have long attracted human settlement. Hence, coral reef ecosystems support not only tremendous biodiversity but are also the basis of significant economic and cultural activities through fisheries and tourism, and often form the centerpieces for subsistence and spiritual linkages for the human population in the United States and its territories.

Evidence supports the consensus that coastal ecosystems are not only experiencing changes but are also being stressed and destroyed through increasing anthropogenic perturbations. Economic demands and heightened coastal population growth have imposed unforeseen strains on our nation's coral reef ecosystems including resource exploitation, loss of biodiversity, marine-based and coastal nonpoint source pollution. This consensus reaches beyond the U.N. Agenda 21, the final document of the 1992 Earth Summit, which identifies coral reefs as ecosystems of great biodiversity and production to be accorded high priority for protection, and calls for an integrated management approach for their protection and sustainable use. The call by scientific, management, and environmental advocacy communities for action toward a coordinated strategy for coral reef research and management has grown dramatically in recent years (Grigg 1994; Crosby et al. 1995). Effective management of coral reef ecosystems addresses national and global environmental concerns such as marine biodiversity, land-based sources of marine pollution, sustainable development, and climate change. Some sources estimate that 10 percent of all reefs have been degraded beyond recovery, and another 20–30 percent are in peril over the next 10–20 years. As an initial effort to counter this trend, the United States is developing an interagency Coral Reef Initiative to create the base for a combined domestic and international effort aimed at the conservation and effective management of coral reefs and related ecosystems (mangroves and seagrass beds).

Natural resource managers, scientists, industry, environmental advocacy groups, and the general public are beginning to share a vision of the United States and the world in which societal and economic decisions will be strongly coupled with an increasingly comprehensive understanding of the environment. A principal component of achieving this goal is the implementation of ecosystem approaches to environmental management and natural resource development. Efforts to achieve holistic management must consider not only the fish and the coral reef resources but also the ecological, social, economic, and political aspects that involve all stakeholders. A key component of such a strategy would be promotion of healthy ecosystems by ensuring that economic development in the United States is managed in ways that maintain biodiversity and long-term productivity for sustained use of our coral reef ecosystems.

The United States, including the states, territories, and commonwealths within or associated with it, has much to contribute to addressing the crisis of coral reef degradation. It has, at the federal level and in partnership with the states, territories, and commonwealths, jurisdiction over coral reefs in the Pacific, Atlantic, Gulf of Mexico, and the Caribbean; extensive experience in integrated coastal and coral reef management; worldwide and world-renowned research capability; and existing partnerships with many nations and international organizations. Working in partnership at the federal, state, territory, and commonwealth level, agencies are seeking to integrate their operational management activities in an ecosystem-wide approach, increase monitoring, conduct assessments to provide better information for decision-makers, provide education and outreach to increase public understanding, and undertake a more pro-active effort in understanding and maintaining biodiversity of coral reef ecosystems.

The U.S. Coral Reef Initiative (CRI) will build on existing federal, state, territorial, commonwealth, and local partnerships through communication with relevant stakeholders at all levels. Mechanisms for ongoing consultation among stakeholders have been initiated and are being expanded to take into account local needs, priorities, and opportunities in developing the U.S. CRI. Important among these mechanisms are meetings to consult with agency staff from each of the affected states, commonwealths, and territories, together with nongovernmental organizations (NGOs) and other interested parties.

These forums provide for developing specific proposals to strengthen domestic partnerships among federal, state, territory, and commonwealth agencies and stakeholders, as well as for input on the planning and funding of the CRI's international components. The domestic implementation of the CRI will make full use of the various existing institutional and jurisdictional arrangements for management and conservation of these coral reef ecosystems.

VISIONS AND STRATEGIES OF THE U.S. CORAL REEF INITIATIVE

The long-term vision for the U.S. CRI is to build a comprehensive national program for the conservation and effective management of coral reef ecosystems, including mangroves and seagrass beds, combining new and existing activities and programs. Unifying principles for the CRI include the concept of sustainable development and the need to conserve coastal and marine ecosystems, in general, and coral reef ecosystems, in particular, as globally significant reservoirs of marine diversity. This program is intended to reverse the current trend toward degradation of these valuable ecosystems, both domestically and internationally. It will seek to create partnerships to address the problems facing these ecosystems and increase the capacity of states, territories, and commonwealths for management and sustainable use of marine biodiversity.

The primary objective of the CRI is to foster innovative cross-disciplinary approaches to sustainable management and conservation of coral reef biodiversity and ecosystems through the development of cooperative relationships among the various stakeholders. The U.S. CRI will be based on partnerships developed among federal agencies, states, territories, and commonwealths, and other interested and affected parties, including NGOs, industry, the scientific community, the public, and other countries, and international and regional organizations. Such partnerships will recognize the legal authority and responsibility of the states, territories, and commonwealths to manage coral reefs within their jurisdictions. In order to utilize resources as efficiently and effectively as possible while avoiding duplication of already existing efforts, it is essential that collection, synthesis, and exchange of information be conducted in a well-coordinated manner. The CRI will also endeavor to ensure that all critical ecosystem components (genetic, community, landscape/seascape) and linkages are taken into account and that pertinent research, assessment, monitoring, and management programs are comprehensively integrated.

A critical variable in achieving long-term conservation and sustainable use of coral reef ecosystems and their associated biodiversity in the United States and around the world will be support for "capacity-building." This involves strengthening technical and human resources for effective management of coral reefs, through cooperative education, training, and infrastructure development. In employing these strategies, the CRI will undertake the installation and use of new monitoring technologies needed for the conservation and sustainable use of these ecosystems. These activities include research, mapping and resource assessment, monitoring, and management for sustainable use (including protection and integrated coastal zone management and environmental damage assessment and restoration).

THE CORAL REEF INITIATIVE AS A TOOL FOR THE CONSERVATION AND SUSTAINABLE USE OF MARINE BIODIVERSITY RESOURCES

The CRI complements ongoing cross-cutting U.S. domestic management and science initiatives including the Ecosystems Management and Sustainable Use Initiatives and the National Council on Science and Technology's Committee on Environment and Natural Resources. In particular, the CRI will provide significant benefits to the United States by:

- Strengthening existing capabilities and responsibilities of coastal and marine ecosystem management and conservation of marine biodiversity in U.S. states, commonwealths, and territories with coral reef resources (e.g., Florida, Texas, Puerto Rico, the U.S. Virgin Islands, Hawaii, Guam, American Samoa, and the Northern Mariana Islands);

- Strengthening domestic capabilities and programs to conserve and sustainably manage U.S. coastal and marine biodiversity through sharing experience and information with other countries on common problems;

- Integrating the private sector into efforts to provide solutions to environmental problems and expand markets for U.S. environmental technologies, management strategies and techniques; and

- Demonstrating U.S. commitment to sustainable development and biodiversity conservation in fragile tropical marine ecosystems.

Nonpoint source pollution from on-island development, sedimentation, nutrient enrichment, contaminated storm water, and marine debris significantly affects water quality, reef health, and ultimately marine biodiversity. Soil runoff is a frequent source of sediment content in coastal waters, which is considered one of the most damaging impacts on the coral reefs. The adverse effect of siltation on the coral is compounded by the effects that the agricultural herbicides and fertilizers have on coastal water quality. The United States has long advocated international action to address the problem of land-based marine pollution, which is a major cause of degradation of coral reefs and loss of biodiversity.

Coral reefs may experience significant impacts from global climate change. Reefs in many parts of the world have been undergoing local bleaching that is potentially related to above-normal local sea-surface temperatures in many reef areas. While short-term temperature increases have resulted in bleaching related to the effects of El Niño, surface warming as a result of climate change could have similar impacts. Thus changes in the health of coral reefs could be an indicator of long-term global climate change. The monitoring program proposed under the CRI would be an important feature of the U.S. efforts to predict climate change and sea-level rise, and their effects on biodiversity.

The measurement and maintenance of biodiversity of coral reef ecosystems is now perceived as an extremely important topic. The United States signed the Convention on Biological Diversity on 4 June 1993 and the convention is currently undergoing ratification. The CRI is expected to be an important element of the U.S. position at the first meeting of the Parties and to contribute to the 1995 review by the Commission on Sustainable Development of the biodiversity chapter of Agenda 21. Chapter 17 of Agenda 21 calls for the sustainable use and conservation of marine living resources under national jurisdiction. It identifies coral reefs, mangroves, and seagrass beds as marine ecosystems of high biodiversity and production and recommends that they be accorded high priority for identification and protection. Chapter 17 also calls for an integrated management approach for the protection and sustainable use of coastal resources. Management of coral reefs and associated ecosystems must be integrated with land-use planning and national development priorities if it is to be effective in promoting both sustainable use and conservation of coral reef biodiversity. Therefore, integrated coastal zone management is a critical element of the CRI.

Capacity-Building: Partnerships for Effective Management

The U.S. CRI supports the aforementioned objectives of Agenda 21 and recognizes that effective management of coral reefs should be integrated with land use and urban development planning. Nonpoint source pollution has been cited as a major source of damage to coral reefs. The reefs in the Florida Keys are subject to significant stresses from development in southern Florida. Other reefs in Hawaii and in the U.S. territories and commonwealths in the Caribbean and the Pacific are being increasingly stressed as these areas are developed. The most important sources of stress include nutrient enrichment from sewage and agriculture, overfishing, and sedimentation from deforestation, agriculture, vessel traffic, and coastal runoff. Tourist and navigational damage, urban pollution, harvesting of nonrenewable reef resources, and destructive fishing have further aggravated reefs already stressed by natural disturbances and disease.

A significant effort of the U.S. CRI will be to assist state, territory, and commonwealth coastal zone management programs at the local/regional levels in completing and implementing plans for nonpoint source pollution control. Specific projects could focus on addressing agricultural and urban nonpoint sources of pollution in watersheds with drainage that affects coral reefs, and monitoring the effectiveness of management plans implemented to mitigate against such impacts. The U.S. CRI envisages a full partnership with states, territories, and commonwealths who are generally the U.S. managers of coral reef ecosystems. Particular attention will also be given to partnerships and the stimulation and enhancement of community-level efforts aimed at the management of coral reef ecosystems for sustainable use. Expanded international coastal zone management activities, for which the U.S. domestic experience provides a strong foundation, are a critical element of the CRI.

Although a few developed countries have coral reef resources (the United States, France, Australia, and Japan, among others), the majority of these resources are in developing countries. Most of these countries are aware of their rich and rapidly deteriorating coral reef resources. Many need to strengthen their scientific and managerial expertise to adequately assess, monitor, and manage their resources. Capacity-building, focusing particularly on the concept of integrated coastal zone management, is an important part of an effort to reverse the deterioration of coral reefs globally.

Capacity-building also recognizes that physical oceanographic boundaries seldom correspond to territorial borders. The health of one country's marine environment, including water quality and availability of larvae to replenish harvested or otherwise damaged fish stock, frequently depends on the actions of neighboring countries. For example, the Florida spiny lobster spends 7–9 months in a planktonic stage in the Caribbean, Gulf Stream, and the Gulf of Mexico; and southeast Atlantic fishery stocks are interdependent with the health of stocks in the Caribbean due to adult migration and larval transport. Thus the effectiveness of management efforts in U.S. waters, particularly in the Caribbean, partially depends on similar efforts by other nations.

The overall objective of the capacity-building program as part of the CRI is to improve the management of coral reefs and associated coastal ecosystems by:

- Developing education and outreach capabilities to raise awareness levels of local communities as to the importance of protecting, conserving, and using coral reef ecosystems in ways that will promote sustainable use and functional viability of these ecosystems in perpetuity. More classroom and community education is needed to increase public knowledge and understanding about reef ecosystems and sources of degradation.

- Facilitating the development of private/public partnerships to develop educational and marketing programs for tourists and standard operating procedures for tour operators.

- Transferring knowledge (data, information, management skills) to developing countries via training programs to improve their ability to assess coral reef ecosystem health and manage those reef resources.

- Involving developing countries in global monitoring networks to facilitate the exchange of physical, biological, and socioeconomic data with global databases.

- Facilitating the development of necessary legislative frameworks, implementation, and enforcement capabilities.

This effort will have both global and regional components. Globally, the effort will focus on building networks for management and coordinating efforts with ongoing activities of international organizations. Regional efforts will focus initially on the Caribbean and the Pacific. A regional effort in the Caribbean region could focus on shared fisheries stocks and pollution threats, building on strong existing intergovernmental and nongovernmental cooperation, and relatively well-developed infrastructure. A focus on the Pacific region is also important because of the significance of coral reef ecosystems to the ecology, culture, and economies of the State of Hawaii, the territories of Guam and American Samoa, the Commonwealth of the Northern Mariana Islands, and the U.S. Freely Associated States.

Coral Reef Ecosystems: Monitoring and Assessment

The U.S. CRI will foster the development of a national and global monitoring program that will establish an internationally coordinated monitoring network involving scientists and managers to facilitate the exchange of physical, chemical, biological, and socioeconomic data. This proposed network is an important component of the Coastal Zone nodule of the Global Ocean Observing System. Data such as biological diversity, community structure and productivity, toxic and nutrient inputs, disease, bleaching events, sea level, water temperature, and salinity must be available on a timely, global scale to interpret correctly the causes of coral reef degradation. This coral network would have appropriate links to existing systems of ecosystem monitoring networks. Initial contributions to this network would include:

- Establishment of an International Coral Reef Research and Monitoring Office, under the auspices of the Intergovernmental Oceanographic Commission, the United Nations Environment Programme, the International Union for Conservation of Nature and Natural Resources, with contributions from the National Oceanic and Atmospheric Administration (NOAA), the United Nations Development Programme, and others.

- Establishment of a U.S. Coral Reef Monitoring and Assessment Program, building on existing activities of NOAA, the National Biological Survey (NBS), the National Park Service, the National Science Foundation (NSF), and other federal agencies, and the states, territories, and commonwealths. Sites should be located in American Samoa, Commonwealth of the Northern Mariana Islands, Florida, Guam, Hawaii, Puerto Rico, Texas, and the U.S. Virgin Islands.

- Establishment of a Pacific Regional Coral Reef Assessment and Monitoring Program (PACICOMP) to complement the existing Caribbean Program (CARICOMP). Initial activities should be developed through partnerships among federal agencies, state, territorial, and commonwealth governments, countries such as Australia, Japan, France, and independent Pacific island governments including the U.S. Freely Associated States, NGOs, academia, and other stakeholders.

Coral Reef Ecosystems: Strategic Research

While still in the developmental stage, the CRI Research Program plans to focus on coordinating current NSF, NOAA, and NBS activities with partnerships among the states, territories, and commonwealths of the United States and other nations directly concerned with coral reefs. Components of this research program may include:

- Paleoclimate/reef history, to derive information on the history of global climate change to predict temporal changes in global climate and responses of corals to future change.

- Coral bleaching, to develop an understanding of the reactions of reefs to global climate changes (increases in sea-surface temperature, increased ultraviolet penetration, and sea-level rise) and to discover the mechanisms of the "bleaching" phenomenon at the organismal and cellular level.

- Ecosystems function, including research on the biology and ecology of "keystone" organisms in order to understand the overall dynamics of ecosystems function.

- Natural products (biodiversity), including research on chemically mediated ecological interactions of reef organisms to provide compounds that can be tested for medicinal and other purposes.

- Eutrophication, to provide information on the effects of nonpoint source inputs of nutrients from upland sources on coral reef biodiversity and functional ecology.

- Sustainable use, including research on the importance of conserving coral reef bio-diversity and integrated management as the basis for long-term sustainability of fisheries.

Tropical Pacific Scientific Initiative

During the early, formative stages of the U.S. Coral Reef Initiative, coral reef scientists, extension agents, and educators in the tropical Pacific worked together to form a network and consortium to support the U.S. CRI. In March 1994, eleven institutions in Hawaii and Guam drafted a preliminary science-based proposal covering the U.S. sovereign tropical Pacific areas (Hawaii, Guam, American Samoa, Northern Mariana Islands, and the isolated U.S. island possessions of Johnston, Wake, Midway, Palmyra, Kingman, Jarvis, Howland, Baker) and the Freely Associated States of the Marshall Islands, Federated States of Micronesia, and Palau. The draft proposal was also circulated among forty of the known government agencies, NGOs, academic institutions, and research laboratories within this region, as well as to important institutions outside the region with interest or responsibilities in coral reefs in the tropical insular Pacific. More than thirty responses were received, with many desiring to join the consortium. The proposal also prompted the governments of Hawaii, Guam, American Samoa, and the Northern Mariana Islands to become more involved in the U.S. CRI, especially in developing management strategies for coral reefs within their jurisdictions. Subsequent planning led to the convening of the December 1994 American Flag Pacific Island (AFPI) CRI Management Planning meeting described below.

The March 1994 scientific proposal described the scope and status of coral reefs within the region and covered the following components: information management, networking, assessment, monitoring, research, marine protected areas, fisheries management, sustainable development, integrated coastal zone management, and public education and training. The proposal will be revised and submitted to various sources for funding and implementation. First, however, the details of government-based coral reef management initiatives will be integrated into the proposal when the specific details and priorities have been developed. Second, more institutions from around the tropical Pacific region will be added to the consortium. Third, the proposal will probably have several parts, including the U.S. flag area initiative, the international/Pacific initiative, the coral reef monitoring initiative (already under way), and the U.S. Pacific and French Pacific initiative (already in the planning stages).

An essential requirement for any coral reef initiative in the Pacific will be for coral reef managers, researchers, and educators to work together as equal partners in furthering the conservation and stewardship of coral reefs. Equally important will be the need for capacity-building to improve management, monitoring, research, and education regarding coral reefs among the island governments and academic/research institutions of the tropical Pacific. Close partnerships are now being forged through the collective efforts of the universities, museums, and research institutions of the region working in association with additional international organizations (e.g., SPREP, East-West Center, Pacific Science Association, The Nature Conservancy, World Conservation Monitoring Centre). The prospect looks optimistic for meaningful cooperation and action among the teams now being forged within the region. The approach being used may serve as a role model for similar initiatives in other coral reef regions.

THINK GLOBALLY, ACT LOCALLY

Perhaps the most important element within the U.S. CRI is support for community involvement in developing and implementing local and regional CRI suited to those community needs and situations. For the U.S. and international CRI to succeed in effectively conserving and managing coral reef ecosystems for long-term sustainable use, programs should essentially be based on local community involvement.

A significant step toward achieving this goal occurred when the American Flag Pacific Island (AFPI) CRI Management Planning Meeting convened 5–7 December 1994 in Honolulu, Hawaii, on behalf of the governments of American Samoa, Commonwealth of the Northern Mariana Islands, Guam, and Hawaii. The participants sought to develop a draft plan for the AFPI CRI that involves all stakeholders, addresses the unique problems facing the AFPI in managing coral reef ecosystems, and identifies possible contributions of AFPI governments and institutions to the international and U.S. CRI.

The governors' appointed points of contact for the CRI (see Appendix) agreed that each of the AFPI governments should initiate a local CRI as a basis for involvement in regional, national, and international CRI activities. Each jurisdiction established a working group consisting of agencies responsible for coastal zone management, natural resources, and water quality, and representatives of research institutions and NGOs.

Resource management must maintain the health of the coral reef ecosystems while providing for the economic needs of the local people. To encourage ecologically sustainable use of the coral reef ecosystems, the CRI Management Program in the American Flag Pacific Island will:

- Increase public support for perpetuating coral reef ecosystems and instilling stewardship for future generations.

- Maintain and enhance the high quality of coral reef ecosystems and ensure that their quality and use are sustainable for future generations.

- Build effective public–private-sector partnerships among regional governments and organizations, educational and research institutions, and NGOs to plan and manage land- and water-use activities that affect coral reef ecosystems.

A similar CRI Management Planning Meeting for the U.S. Virgin Islands, Puerto Rico, Florida, and Texas was held in May 1995 in St. Croix, U.S. Virgin islands. The final report from that meeting will be published in the fall of 1995 and is expected to concur with many of the conclusions of the AFPI meeting.

The long-term vision for the CRI may be viewed as a global effort to conserve and restore coral reef ecosystems at local levels for the use and enjoyment of future generations. The U.S. CRI will seek to build on existing activities and programs through partnerships and, where appropriate, develop new activities. The CRI will strive to reverse the trend toward degradation of these valuable ecosystems by increasing the capacity of the states, territories, and commonwealths for management and sustainable use of coral reef ecosystems. A strong commitment to the U.S. CRI will improve domestic programs that will serve as models for managing coral reef ecosystems as part of sustainable development strategies having global applications. The CRI will also provide linkages between domestic and international programs so that U.S. and foreign coral reef management strategies will have a basis in sound science and provide for sustainable economies, communities, and environmental health, and develop new partnerships among all stakeholders.

THE HAWAII CORAL REEF INITIATIVE: AN EXAMPLE OF "ACTING LOCALLY"

Scientists, government managers, NGOs, and other interested residents are now actively planning local coral reef initiatives throughout the AFPI. As of June 1995, the Hawaii State Coastal Zone Management Program has taken the leadership in organizing support and initiating Hawaiian CRI activities by the end of summer 1995 with substantial support from the East-West Center, the Pacific Basin Development Council, the U.S. Fish and Wildlife Service, the NOAA National Marine Fisheries Service, the University of Hawaii (UH) Sea Grant Program, other UH departments, and other state agencies, especially the Department of Health and the Department of Land and Natural Resources. Efforts are also being made to enlist the active support of county/local governments, other federal and state agencies, Hawaiian groups, nature conservation organizations, educational groups, business organizations, and ocean recreational organizations. The Hawaii Coral Reef Initiative (HCRI) envisions three major steps at this time: (1) statewide assessment of the status of coral reefs, (2) initiation of a long-range monitoring and research program for Hawaiian coral reefs, and (3) implementation of other management strategies to protect and promote sustainable use and benefits afforded by coral reefs.

At this time (June 1995), the HCRI organizers have agreed to use methodologies developed during the 1992 Hawaii Environmental Risk Ranking (HERR) Project to accomplish the statewide assessment of coral reefs (Carpenter et al. 1992). The ecological component of the HERR Project involved the development of a methodology to assess the risks of human stressors or impacts on both marine and terrestrial ecosystems in Hawaii. The 1992 effort involved classifying Hawaii's ecosystems into about twenty categories; defining about twelve kinds of human stressors that affect ecosystems in Hawaii; mapping specific sites for each ecosystem category; developing criteria on how to rate the value of each site in terms of biodiversity, recreational, cultural, and economic values; and assessing the risks of the stressors to each ecosystem site. The methodology also involved the development of a geographic information system (AeGIS) to map, analyze, and summarize the information from the risk rankings. The 1992 HERR study completed a preliminary ecological risk assessment of the native ecosystems of Molokai Island and about 100 additional sites elsewhere in the State of Hawaii.

In 1995, the HERR Project resumed to accommodate risk assessment of several hundred ecosystem sites throughout Hawaii, including about 100 sites for all of the major Hawaiian Islands. Workshops on each of the main islands are planned in the late summer of

1995 and will involve scientists, residents, and government officials on these islands. The workshops will seek public input regarding risk assessment of additional ecosystem sites. Risk assessment of these sites will be accomplished during subsequent workshop sessions.

The HCRI will attempt to take advantage of these workshops to accomplish a complete risk assessment of all coral reef sites in the State of Hawaii, essentially using a refined HERR methodology. Subsequent workshop sessions are now being organized and planned specifically for risk assessment of the coral reefs following the risk assessment of other ecosystem sites for each island. The results of the combined HERR and HCRI workshops will lead to a thorough assessment of coral reefs in Hawaii and the causes for their degradation. The methodology will allow mapping and presentation of results that can be understood by government leaders, NGOs, and scientists. These results will lead to priorities on specific ecosystem (including coral reef) sites that need protection, monitoring, restoration, and other management interventions. The assessments will also lead to priorities on reducing the impacts of human stressors on ecosystems, including planning and regulatory approaches (i.e., nonpoint source pollution control, coastal zone management, protected areas designation, and educational/public awareness programs). Because both the HERR and HCRI methodologies being used in Hawaii are flexible, they can readily be modified for use in other island groups in the insular tropical Pacific and other regions.

APPENDIX

Governors' Points of Contact for the Coral Reef Initiative

American Samoa
Mr. Lelei Peau, Manager
American Samoa Coastal Management Program
American Samoa Economic Development and Planning Office
American Samoa Government
Pago Pago, American Samoa 96799
Phone: 684-633-5155
Fax: 684-633-4195

Commonwealth of the Northern Mariana Islands
Mr. Nicholas P. Strauss
Special Assistant for Policy and Research
Commonwealth of the Northern Mariana Islands
Saipan, MP 96950
Phone: 670-322-5091/2/3
Fax: 670-322-5096

Florida
Ms. Paula Allen, Senior Analyst
Environmental Community and Economic Policy Development Unit
Room 1501, The Capitol
Tallahassee, Florida 32399-0001
Phone: 904-488-5551
Fax: 904-922-6200

Guam
Mr. Michael J. Cruz
Acting Director, Bureau of Planning
Government of Guam
P.O. Box 2950
Agana, Guam 96910
Phone: 671-472-4201/3
Fax: 671-477-1812

Hawaii
Mr. Douglas S. Y. Tom, Chief
Coastal Zone Management Program
Office of State Planning
State of Hawaii
P.O. Box 3540
Honolulu, Hawaii 96811-3540
Phone: 808-587-2875
Fax: 808-587-2899

Puerto Rico
Mr. Pedro Gelabert, Secretary
Department of Natural & Environmental Resources
P.O. Box 5887
San Juan, Puerto Rico 00906
Phone: 809-724-8774
Fax: 809-723-4255

Texas
Dr. Larry McKinney
Director, Resource Protection
Texas Parks and Wildlife Department
4200 Smith School Road
Austin, Texas 78711
Phone: 512-389-4636
Fax: 512-389-4394

Virgin Islands

Mrs. Beulah Smith, Manager and Acting Commissioner

Coastal Zone Management Project

Department of Planning and Natural Resources

Nisky Plaza, Suite 231

St. Thomas, VI 00802

Phone: 809-774-3320

Fax: 809-775-5706

U.S. CRI Co-Chairs

Dr. Michael P. Crosby

National Research Coordinator

Ocean and Coastal Resource Management

NOAA, SSMC-4, Room 11536

1305 East West Highway

Silver Spring, Maryland 20910

Phone: 301-713-3155

Fax: 301-713-4012

Ms. Nancy Fanning, Chief

Territorial Liaison

DOI, Territorial and International Affairs

1849 C Street, NW, MS 4328

Washington, D.C. 20240

Phone: 202-208-6816

Fax: 202-501-7759

REFERENCES

Carpenter, R. A., W. Mitter, N. Convard, S. Edgerton, J. Maragos, and K. Smith. 1992. Environmental risks to Hawaii's public health and ecosystems. A report of the Hawaii Environmental Risk Ranking Study. Prepared by the East-West Center, Honolulu, for the Department of Health, State of Hawaii.

Crosby, M. P., S. F. Drake, C. M. Eakin, N. B. Fanning, A. Paterson, P. R. Taylor, and J. Wilson.

1995. The United States Coral Reef Initiative: An overview of the first steps. *Coral Reefs* 14(1): 1–3.

Grigg, R. W. 1994. The International Coral Reef Initiative: conservation and effective management of marine resources. *Coral Reefs* 13(4): 197–98.

CHAPTER 17

ReefBase: An International Database on Coral Reefs

John W. McManus, Benjamin M. Vallejo Jr.,
Lambert A. B. Meñez, and Grace U. Coronado

ABSTRACT

ReefBase, which is currently being prepared, is an international database on the coral reefs of the world. It was initiated in response to the demand for summary information on coral reefs globally. The system will include maps designed specifically for it, as well as other maps and imagery as available. The basic unit of information will be the reef by year. Most data are acquired from reports and publications, and the system evolves according to data trends and questions to be answered. ReefBase will include interactive linkages to other databases including FishBase and CoralBase. Major spin-offs of the project include improved reef terminology; determinations of reef health; estimates of reef contributions to food, income, and atmospheric processes; and standardization of reef survey and monitoring methods. The current phase will end with a release of a preliminary CD-ROM version by 1996. Future expanded phases are planned. Databases of this type must be adaptive to trends in available data as well as to an evolving set of questions for which the database should provide answers. We recommend that such databases involve a centralization of data entry efforts and the occasional use of trial queries to ensure that data demands are being met efficiently.

BACKGROUND

The rise of global environmental consciousness in the late 1980s led to increasingly frequent inquiries from the media about the status of the coral reefs of the world. By this time, many useful estimates were available concerning the state of the world's rain forests, such as the coverage of virgin and secondary rain forest by country, the loss in area per day, and the percentage lost in the last decade. However, no comparable figures were available for coral reefs. Part of this difference can be attributed to the comparable ease with which rain forests in most areas can be inventoried with satellite sensors. Indeed, while the technology exists to build satellite sensors with features that could greatly improve our ability to analyze coral reefs (McManus 1989), the focus of most satellites deployed to date has been on land vegetation or low-resolution oceanographic analyses. However, there have been many surveys of coral reefs around the world. Most of these studies were considered too "descriptive" to be suitable for process-oriented ecological publications. The existing reports were generally locally published or unpublished, and difficult to obtain. A first-order understanding of the usage and status of coral reefs could be obtained by summarizing such literature.

The three-volume *Coral Reefs Of the World* (UNEP/UNDP 1988) was the first systematic attempt to summarize this literature. These volumes continue today to be an important starting point for research on coral reefs in any given locality. The summary tables in the beginning of this work are particularly useful. One can learn, for instance, that approximately forty countries or island states had reported blastfishing, and fifteen had reported poison-fishing as problems affecting their reefs. However, this summary aspect of the work was limited in scope. It soon became apparent that a new effort must be initiated that would focus primarily on summarizing data across reefs and over time, and present the information in a flexible, user-friendly computer database format.

The concept for the database evolved from 1988 to 1991, at which time it was strongly influenced by the maturing FishBase system of ICLARM. In 1992, it was one of the primary foci of an invitational workshop held in Australia, "The Management of Coral Reef Resource Systems" (Munro and Munro 1994). A preliminary design for the structure of the database was presented at that time (Froese 1994). A proposal for a two-year start-up project was approved by the European Commission, and the project began in October 1993.

It was realized early on that global summaries of the utility of and threats to coral reefs would be highly dependent on studies of the geomorphologies, locations, and areal extents of reefs. Such studies had been initiated on a global scale with the work of Darwin ([1842] 1962) and his successors (e.g., Dana 1872; Davis 1928). The most commonly cited estimate of the areal extent of the world's reefs is that of Smith (1978), who used bathymetric charts to estimate reef area. However, his figure of 600,000 km² for the reefs of the world was six times higher than that of de Vooys (1979), but *less* than that estimated by John Munro (1983; in press) for the Caribbean region alone. It was thus obvious that a concerted effort would be necessary in mapping the major reef systems of the world and determining total reef areas by country and region. Thus, the World Conservation Monitoring Centre was tasked to digitize existing reef maps, while concurrent efforts were directed toward making reef area estimates based on bathymetric and bottom classification data. The latter effort is expected to be substantially facilitated by ongoing work at the National Center for Atmospheric Research in Colorado, USA, toward modeling the potential distribution of the world's coral reefs based on environmental parameters (Kleypas 1994).

A preliminary design for the database formed a basis for discussions for a second international workshop held in Cambridge, U.K., in January 1994. In the same month, the ReefBase effort was presented at the initial workshop of the U.S. Coral Reef Initiative, later to evolve into the International Coral Reef Initiative. Seminars were also presented at the offices of the National Oceanic and Atmospheric Administration (NOAA) and the U.S. Agency for International Development (USAID). A few days later, discussions were held with social scientists at the University of Rhode Island concerning ways that useful data on the human populations affecting coral reefs could be incorporated. This led to the initiation several months later of a separate, complementary project on "Rapid Assessment of Management Parameters (RAMP)" (discussed below).

The suggestions from the Cambridge and University of Rhode Island workshops led to changes in the ReefBase data fields. The revised table input forms were sent out for review by selected workshop participants by May 1994. Several further revisions were made based on the comments received. The project was presented in poster sessions in conferences in the Philippines and Germany. Oral presentations were made during the United Nations Conference on the Sustainable Development of Small Island States in Belize, 25 April to 6 May 1994; the Sixth Pacific Congress on Marine Science and Technology (PACON 94) in Australia, 4–8 July 1994; and the Second European Regional Meeting of the International Society for Reef Studies (ISRS) in Luxembourg, 6–9 September 1994. The ISRS meeting included a workshop entitled "ReefBase: Prioritizing Global Information Needs," during which the concepts and progress of the database were discussed in the light of international reef monitoring efforts. The ReefBase effort was discussed in other workshops in the same meeting, concerning "Monitoring and Management Implications" and "Volunteer Programmes in Marine Research and Reef Management." The database was also discussed in the ISRS forum, "Year of the Reef," as a core activity of the proposed 1996 International Year of the Reef. Suggestions were accumulated from these discussions and, wherever possible, were incorporated into the database design.

More than a hundred inquiries were generated from the reprinting of a one-page project description by at least fifteen international newsletters, magazines, and journals. In many cases, our response to these inquiries led to the accumulation of reports and papers from specific countries from which data could be extracted for the database. A substantial number of these inquiries included a request for suggestions as to what data should be gathered on future expeditions. This was a clear indication that ReefBase could be an important catalyst for standardizing reef survey methods.

MAJOR QUESTIONS

The current ReefBase effort is directed toward answering a variety of questions, including:

- How much coral reef area is there at what depths?
- How are coral reefs distributed globally, relative to patterns of oceanic current, biodiversity, etc.?
- How much of each major reef zone exists and where globally?
- What is the average bottom cover (by broad category) of each zone in each region?
- How has the bottom cover changed over time?
- How much harvest production comes from reefs?
- What species are harvested and with what type of equipment in each region?
- What are the nonextractive uses of reefs (types of tourism, etc.)?

- What constitutes a "healthy" reef?
- How "healthy" are reefs now, and how has this changed over time?
- What are the types and extent of perturbations and stresses to coral reefs?
- What management systems govern the reefs (traditional and governmental)?
- What is the distribution of coral reef protected areas?
- What social and economic aspects of reef-user populations are indicative of likely future change?
- What data gaps exist that require priority research in the future?

EXPECTED PRODUCT AND HARDWARE REQUIREMENTS

ReefBase will be released in a preliminary version on a CD-ROM (along with FishBase and other ICLARM products) in early 1996. The software will require an IBM-compatible computer running the Windows operating system, with a VGA monitor, eight megabytes RAM, a mouse, and a CD-ROM drive. The CD-ROM will be distributed at production cost.

STRUCTURE

The current design of ReefBase includes tables linked to the reef-year unit, and those that are based on other key fields (Figure 17.1). Each reef is given a code based on its latitude and longitude to the whole degree, and then a sequential number is assigned within that

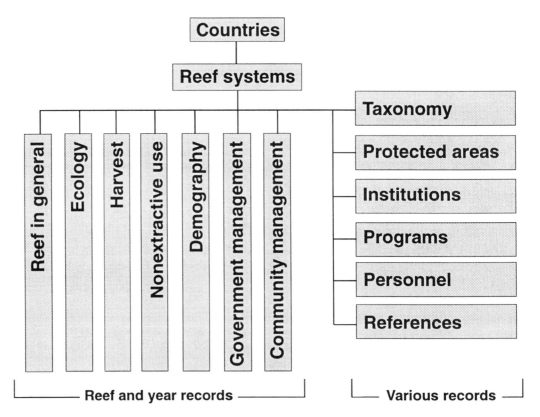

Figure 17.1 Generalized database structure for ReefBase. The actual database has more than fifty tables and continues to grow as new data trends and anticipated uses are identified.

block. The low-resolution system is necessary because of inaccuracies in describing locations of reefs. For very large reef systems, the approximate center of planar area is used. The major reef-year tables include ecology, nonextractive use, harvest, oceanography, aquaculture, human stresses, natural stresses, socioeconomics, government management, and traditional management. Other tables include information on countries, protected areas, institutions involved in reef research and management, key collaborating individuals and experts, references, and a taxonomic table into which data on harvested, stressed, or dominant species are maintained. Some other tables are being considered for inclusion. The data fields are modified continuously to accommodate new information and data needs. Occasional testing is carried out, including the running of preliminary queries to ensure that data requirements are being met efficiently.

The system is programed in Microsoft Access. The first-release version will be in the Access Developers Kit compiled version, consisting of data queries and output forms with an extensive help screen system.

LINKAGES

ReefBase was designed from its earliest conceptual stages to be highly compatible with FishBase, the international database on fish (Froese and Palomares, this volume). The two development teams interact regularly, and the resulting user-interfaces will be as similar as possible. The two systems will incorporate hot links to facilitate jumping between systems, and data from both systems will be displayable simultaneously on the same "Mapper" geographic information system mapping subsystem. Additionally, efforts are under way to make ReefBase as interactive as possible with CoralBase (Navin and Veron, this volume).

MAPPING ACTIVITY

The major effort in producing maps for inclusion in ReefBase rests with the World Conservation Monitoring Centre (WCMC) in Cambridge, U.K. These maps are produced primarily by digitizing existing maps and rescaling to a limited number of standard scales. Additionally, other maps of specific coral reef systems will be digitized as bit images for display, along with imagery from satellites and aerial photographs whenever official permission can be obtained to do so.

PROBLEMS AND STRATEGIES

The greatest difficulty in the development of ReefBase involves the definition of the basic record unit, the "coral reef." Entities referred to in the literature as coral reefs vary in scale from thousands of kilometers to a few meters. Some coral reefs are defined by their substrates—others by the benthic communities the substrates support. Many entities considered to be coral reefs in some papers are merely components of larger coral reefs identified in other papers. No matter how reefs are defined, the problem remains that substantial amounts of coral exist in benthic communities which defy inclusion (Goreau 1969; McManus 1988).

The problem is currently handled by creating a separate record for each entity identified as a coral reef in the literature, and then classifying the entities on a hierarchical scale as to whether they are country reefs, reef systems, reef subsystems, individual reefs, or subreef units. This will permit data retrieval at selected levels and extrapolations from lower to higher units where appropriate. The disadvantages include the fact that workers who input the data must be well-versed in coral reef science and they must be centrally located so that similar classifications are made by each worker.

Similar problems arise concerning the definition of a coral reef zone. In some areas, reef flats tend to be dominated by seagrass. Thus, for many purposes, it makes little sense to group reef flats together with reef slopes when determining the coral cover of reef areas. However, many reports do not clearly specify the zone investigated or use terminology and definitions that make reasonable summaries by zone extremely difficult. Our current strategy here involves defining a set of broadly applicable reef zones into which reported data can be classified and yet keeping track of original terminology for future reference. Hierarchical zonations are being developed as well, including "reef," "reef portion," and "reef zone," where a "portion" includes several "zones."

A third difficulty involves data accessibility. Many researchers are concerned that the data they have gathered will be used for future publications without credit to the source. One strategy involves emphasizing the use of reports and papers as source materials. The source of each item of information is identified by a reference number, and a click on the reference number reveals the entire citation. People who spend time gathering reports for ReefBase are also acknowledged. Data downloaded from other databases (with permission) are similarly noted. The danger that the data will be used secondarily without proper citation differs little from that involved if the information were published in papers or reports, particularly with the advent of optical character reading systems that can digitize tables from printed reports. The exchange of scientific information has always depended on a degree of adherence to protocol, and it is primarily a matter now of overcoming anxiety emanating from the "newness" of databases as exchange media.

The question of reef health is both an objective of ReefBase and a target of descriptive data input. The process of developing the concept of reef health, from information such as average coral cover by zone, depends on the data chosen for input, and the data chosen depends on the definition of health. Thus, the two are expected to interact over time, with the data fields undergoing modifications as improved concepts of reef health are evolved.

The identification of data concerning social and economic aspects of reef users is problematic because of the vast array of potential indicator data to choose from. It is widely realized that information about people impacting a coral reef, such as net income, work ethics, and educational attainment, is important to assessing the future of coral reefs. However, no single set of variables have been shown to be particularly useful in such assessments. Thus, a complementary project has been initiated with the University of Rhode Island to develop a "report card" of appropriate variables. This project, "Rapid Assessment of Management Parameters (RAMP)," is expected to produce guidelines prior to the completion of the first release version of ReefBase, and appropriate data will be input for selected reefs.

The identification of data gaps is both a product of the program and a sign of hindrance to its ultimate success. Coral reef literature is extremely patchy, with most of the descriptive work having originated from a few, unrepresentative areas of the world and from reefs that, on a world scale, are primarily anomalous. The body of literature prior to 1960 focused primarily on atolls and other large oceanic surface reefs that, although important for biogeographic and other reasons, represent a negligible amount of total reef area in the world. The recent emphasis on the Great Barrier Reef in Australia and the fringing reefs of Southeast Asia and the Caribbean has brought concepts of reef processes somewhat closer to an appropriate global view. However, for every reef that breaks the surface, there are many more that do not. Corals grow on hundreds of thousands of subsurface reefs on tropical shelf areas of the world. They grow on rocky shores with no distinct geomorphology and in clumps scattered over the shelves themselves. Recently, we have resurrected evidence gathered by Shepard, Emery, and others from hundreds of charts prior to 1970 that very large reefs may exist on the Southeast Asian shelves, virtually unknown to science today (Shepard, Emery, and Gould 1949; Emery 1969).

Another source of data gap arises from the uneven distribution of trained coral reef researchers throughout the developing world. Most of the coral reefs of the world are found in developing country seas. There is an immediate need to develop standardized survey methods of broad applicability and to institute training and technical assistance programs to gather information on reefs in the little-known areas of the tropics (i.e., most of the tropical marine world). Efforts are under way to implement appropriate programs regionally, and ICLARM is seeking funds to organize a collaborative effort on a global scale.

SPINOFFS: BY-PRODUCTS OF THE REEFBASE EFFORT

Because of the novelty of the database, the ReefBase development effort is expected to lead to a number of useful advances. The need to categorize reef information has led to intensified work on standardizing reef terminology (Holthus and Maragos, this volume). Investigations into the distributions and extent of coral reefs will lead to a better understanding of the role of coral reefs in the economies and life support systems of the earth, including as sources of food, medicine and livelihood, and as components of global CO_2 cycling. This work will also lead to better regional and national assessments of available marine resources, including biodiversity, food, tourist potential, and other resources. The networking involved in the development of the database has led to better coordination among research programs around the world. For example, researchers on reef fish in Brazil learned about a major reef fish database program in the Caribbean through contacts with ReefBase. Finally, a large number of requests have been received for advice on what data to gather on future reef surveys. Thus, ReefBase is leading to better standardization of reef survey and monitoring efforts.

FUTURE VERSIONS

The current funding is for two years, ending in October 1995. This "Phase I" covers the costs of one project leader, two research assistants, a programmer, and a secretary at ICLARM, as well as two individuals at the WCMC for the mapping component. Because the literature being covered in this phase is extensive, only a sparse and selective sampling will be completed in the two-year period. We will propose a second phase, which will allow for adjustments to the system based on feedback from the Phase I release, as well as a far more thorough coverage of available data. In the meantime, several major advances in technology are anticipated to have made large volumes of digital imagery data available from satellite, aerial, and underwater sources. Thus, a third phase is planned, which will concentrate on accommodating these data and coordinating efforts to conduct a more meaningful inventory of the coralline areas of the globe.

RECOMMENDATIONS FOR AN ENVIRONMENTAL OCEANIA DATABASE

Management of the ecosystems of Oceania would be greatly facilitated by the development of regional databases. A possible biodiversity database has been discussed elsewhere (see Froese and Palomares, this volume) and involves determining distributions of a wide range of species based on museum records and other data sources. However, regional management would be further enhanced with the development of an environmental database for Oceania. Such a database would focus on the ecosystems as units and permit inquiries regarding their ecological and management status.

Environmental databases tend to fall somewhere on a continuum between two types: (1) those intended for a specific sampling program and (2) those intended to accommodate existing and future data sources. The data fields for Type I databases can be primarily determined before implementation of the database. Data entry can often be broadly distributed geographically provided that clear definitions of data requirements are available in a standard database manual.

Type II databases must be adaptive. Data inputs will be determined by a combination of trends in data availability as discovered and an evolving set of questions for which the database is expected to provide answers. The Oceania database would probably be primarily the latter. Thus, the database should involve a preliminary design, as well as a system of periodical reviews and adjustments based on perceived data trends and information needs. In general, data entry should be centralized to ensure uniformity of interpretation and compatibility as the system is upgraded. The latter can be greatly facilitated through the use of a computer network, so that only one version of the database is active at any time. Some distribution of effort will be possible, as people in the field can input data into simple subsets of the database or computer spreadsheets. However, the incorporation of these data sets into the central database must be closely supervised, as there will often have been changes in both the database program and the concepts of data requirements while the field data were being collected.

There must be a continuous effort to ensure that the data being input can specifically answer the questions the database is designed to answer. One method for facilitating this is to periodically query the database as it develops. For example, it may become apparent that a question such as "How many species of hardwood-dwelling insects are found on the average island?" might be less useful to resource managers than "How many species of hardwood-dwelling insects are found on the average island with hardwood trees?" At this point, it is important to ensure that islands with hardwood trees are easily distinguished (based on query logic) from those without. This might require the addition of a new data field or the conversion of memo-field data to categorical data. Such conversions are best made in the early stage. This also highlights the need to archive all original sources for later reference as the database evolves.

On the other hand, it is important that strict controls be made on the use of the database prior to planned release dates. The process of "debugging" the database will be lengthy and must be done in a conscious, deliberate effort in planned stages. Following this, the database will still need corrections, but the complexity will make it impossible for the designers alone to identify all-important errors. Thus, a preliminary release of the database must be made to a selected group of potential users and very close contact maintained with them so that they are kept abreast of changes and do not release the database to other users. When the final database of a particular version is released, all test versions must be destroyed. Otherwise, conflicting levels of confidence in the system will arise out of confusion between official and unofficial release versions. This has particular significance if the "release" is to be made on-line. The test versions must have very strict access requirements. If possible, users of the first few versions, even after testing has been completed, should be listed with contact addresses so that they can be updated about errors and necessary revisions. This kind of control is simpler with in-house CD-ROM production, but may be possible in a carefully designed on-line system. On-line systems are not desirable when the primary target users do not have ready access to the network system, and this is certainly the case with many Pacific islands currently. Thus, a release as a CD-ROM may be an important option to consider, at least in the early stages.

Finally, there may be too much concern initially about what the database "looks like." The "face" of the database for input purposes should be designed specifically for those responsible for input of data. If such people have multiple levels of experience, it is possible to have two or more sets of electronic input "forms" addressing the same tables. The actual arrangement of data on the storage tables should be governed by consideration of computational, programming, and debugging efficiency, and database integrity. With modern database software, this need not have much to do with the arrangement of data fields on the input forms. The most important "face" of the database will be that designed to access the data for user queries. Here again, the database can have different "faces" for different groups of users, and these "faces" need have little to do with the actual arrangement of data on the storage tables. These "user faces" should be designed in the later evolutionary stages of the database, within a given release cycle. Unfortunately, there will be continuous pressure for outsiders to "see" the database in its early stages. It must be made clear that looking at the data entry forms will be a very poor way to judge the appearance of the database upon release to a user, other than perhaps as an early indicator of the kinds of data that are entered into the database. The "look" of a database upon release will indeed be extremely important in determining its usefulness, and a concerted effort should be made into enhancing user-friendliness through simplifying screen designs and the prolific use of explanatory help screens and similar devices.

Acknowledgments—ReefBase owes its conception to a large number of people, especially to John Munro, Daniel Pauly, Rainer Froese, and Susan Wells—all of whose persistent efforts combined to get the project started. This project is funded by the European Commission. This is ICLARM contribution number 1120.

REFERENCES

Dana, J. D. 1872. *Corals and coral islands.* New York: Dodd and Mead.

Darwin, C. [1842] 1962. *The structure and distribution of coral reefs.* Reprint, Berkeley: University of California Press.

Davis, W. M. 1928. *The coral reef problem.* Special Publication No. 9. American Geographical Society, New York.

De Vooys, C. G. N. 1979. Primary production in aquatic environments. In *The global carbon cycle,* ed. B. Bolin, E. T. Degens, S. Kempe, and P. Ketner, 259–92. SCOPE Report 13. Chichester: Wiley.

Emery, K. O. 1969. Distribution pattern of sediments on the continental shelves of western Indonesia. *Tech. Bull.* (ECAFE) 2:79–82.

Froese, R. 1994. ReefBase: A global database of coral reef systems and their resources. In *The management of coral reef resource systems,* ed. J. L. Munro and P. E. Munro, 52–54. ICLARM Conference Proceedings 44, Manila, Philippines.

Goreau, T. F. 1969. Post-Pleistocene urban renewal in coral reefs. *Micronesica* 5:323–26.

Kleypas, J. A. 1994. Modeling present-day coral reef distribution: Stepping into the data void, p. 40. Abstracts, Joint Scientific Conference on Science, Management, and Sustainability of Marine Habitats in the 21st Century. James Cook University, Jamestown, Australia.

McManus, J. W. 1988. Coral reefs of the ASEAN region. *Ambio* 17(3): 189–93.

———. 1989. Earth Observing System and coral reef fisheries. *Coastal Zone* 89(5): 4936–49.

Munro, J. L., ed. 1983. Reprint. Coral reef fish and fisheries of the Caribbean Sea. Chapter 1, *Caribbean coral reef fishery resources.* ICLARM Studies and Reviews 7, Manila, Philippines. Original edition, Research Report, Zoology Department, University of West Indies 3(1), 1973.

Munro, J. L. In press. The scope of coral reef fisheries. Chapter 1, *Reef fisheries,* ed. N. Polunin and C. Roberts. Fish and Fisheries Series. London: Chapman and Hall.

Munro, J. L., and P. E. Munro, eds. 1994. *The management of coral reef resource systems.* ICLARM Conference Proceedings 44, Manila, Philippines.

Shepard, F. P., K. O. Emery, and H. R. Gould. 1949. *Distribution of sediments on east Asiatic continental shelf* (Charts 1–3, Figures 1–26). Occasional Paper 9, Allan Hancock Foundation Publications. Los Angeles: University of Southern California Press.

Smith, S. V. 1978. Coral reef area and the contributions of reefs to processes and resources of the world's oceans. *Nature* (London) 273:225–26.

UNEP/IUCN. 1988. *Coral reefs of the world.* 3 vols. United Nations Environment Programme and International Union for the Conservation of Nature and Natural Resources, Cambridge, U.K.

CHAPTER 18

CoralBase: A Taxonomic and Biogeographic Information System for Scleractinian Corals

Kim F. Navin and J. E. N. Veron

ABSTRACT

CoralBase is a distributable Scleractinian coral information system of taxonomic and biogeographic data being developed at the Australian Institute of Marine Science. While concentrating on the Indo-West Pacific Region, it contains a database of approximately 700 coral species from 350 locations worldwide, a large image collection, and a 4,000-reference bibliography. Utilities to plot distributions, carry out common analyses, draw graphs, and compile reports are also included. CoralBase, with its ease of use, comprehensive set of utilities, and frequent updates, is intended to provide a standard reference for all researchers and students working on coral reefs.

INTRODUCTION

CoralBase, an initiative of the Australian Institute of Marine Science, is being established to disseminate current information on Scleractinian coral taxonomy and biogeography to the widest possible range of users. The primary goal of CoralBase is to present data in a form that is easily accessible to users whose level of expertise range from students to reef researchers. This goal will be achieved by using powerful relational database techniques via a user-friendly interface. The secondary goal, but of no less importance, is to ensure that CoralBase becomes a truly worldwide facility by encouraging the coral research community to submit data for inclusion in the system being developed.

BACKGROUND

The study of corals has received added impetus in the last twenty-five years from three unrelated but coincidentally providential factors. The first is a growing political awareness of the vital importance of the role of coral reefs in both developed and developing countries. The second is that the technology needed to study corals in detail has been provided, with the advent of SCUBA, while, thirdly, advances in information technology have allowed researchers to undertake highly complex analyses of this increased data load. This situation, which has resulted in more funding being made available for reef research and the establishment of management proposals for coral reefs, has in turn highlighted the problems associated with accessing and efficiently disseminating existing data.

In the field of Scleractinian taxonomy and biogeography, studies have been carried out in most major regions of the world, which have resulted in extensive data sets being accumulated by researchers and research institutions. The problem is collating and disseminating these data. In this field as in any other, although much of these data are published, a large proportion are not. To compound this problem, data published in both journals and books may be rendered obsolete soon after publication in the light of more recent findings. Unpublished data very often reside in reference collections, reports, annotations to previous publications, or, more recently, in "in-house" databases for years to decades before publication. These data are not easily accessible to users outside the originating institution.

Advances in knowledge are usually incremental, and it is not always feasible to communicate them to the research community as quickly as one would like. This often results in a lag of years between discovery and publication. Whereas in the past it could be argued that this delay was acceptable because there was not the urgency that exists today to protect reefs from steadily increasing human impacts, this is no longer the case. With data-handling technologies that existed before 1980, there was very little that could be done about the situation. Recent advances in computer technology have made it possible to distribute data almost instantly to all corners of the world, and given the current situation with regards to coral reefs, every effort should be made to do so.

In many countries today, access to accurate taxonomic information is crucial to allow researchers to conduct pre-emptive research on which to base management decisions that directly affect the survival of reefs. Now that the political will to protect and manage reefs is growing, the ability to access previously conducted research results by reef managers and policy makers is at least as important as conducting the research in the first place.

It is very often the case, especially in developing countries, that those most in need of access to reliable data are the very people who lack the resources to tap into the relevant mainstream literature needed to undertake this critical research. Most countries also lack the expertise needed to undertake research in specific areas, and the solution is often to import experts in that field. While sometimes this may be the only feasible solution, it is of far greater long-term benefit to the country concerned if a pool of local expertise can be developed. The underlying philosophy of CoralBase is to provide not only background data but to attempt to transfer some of the expertise needed to develop this local knowledge base.

BACKGROUND

CoralBase, as it will be initially distributed, is based on research conducted over the past twenty years by Veron and colleagues. The database is centered around a "species-in-site" data set that includes approximately 700 coral species from 350 locations worldwide. Systematic data are available at the species, genus, and family levels. For each species the family, genus, species, and authority are recorded. Site data include site name, latitude and longitude, ReefBase reef codes, and, in the case of East Australia, the Great Barrier Reef Marine Park Authority's gazetteer reef code. Sites are grouped into zones based on species composition. Zones are further grouped into regions based broadly on biogeographic divisions. Data may be queried or plotted at any of these levels. A unique identification number for each species, genus, family, site, zone, and region provides the key for fast database searches.

The central tables are linked to a 4,000-reference bibliography and an image base containing several thousand images. Taxonomic data for eastern Australian species from the AIMS Monograph Series *Scleractinia of Eastern Australia* are directly accessible from within the program (Veron 1993; Veron and Pichon 1976, 1980, 1982; Veron, Pichon, and Wijsman-Best 1977; Veron and Wallace 1984).

Comprehensive query capabilities are available throughout the database. By selecting a "key species" the user can, for example, quickly find its synonyms and the sites where it has been recorded, view a list of available images of the species, or plot its distribution.

Digital versions of two recently published books, *Hermatypic Corals of Japan* and *A Biogeographic Database of Hermatypic Corals* (Veron 1992, 1993), will be included with, and accessible from, CoralBase.

CORAL IDENTIFICATION

As a means of passing on to other researchers the "in-house" expertise in coral identification, a software module called CoralKey is being written. CoralKey will lead the user through an identification of coral species using both textual and pictorial prompts without the need to know a large vocabulary of taxonomic terms. It is being designed to allow users without previous knowledge of Scleractinian taxonomy to identify corals to species level starting from a general description of gross morphology. More competent users can specify a "start-point" for the identification at the family-, genus-, or species-group level. Fuzzy or missing data can be catered for while ecological and biogeographic information can also be used to assist identification. CoralKey, which is based on expert system technology, will also keep track of the steps taken during the identification sequence and calculate the probability of a correct identification.

PLOTTING DISTRIBUTIONS

CoralMap, a simple raster-based plotting facility, allows coral distributions to be plotted at the worldwide, regional, or zonal level. Taxa can be plotted at species, genus, or family level, and each plotted site can be interrogated for species lists by clicking on the plotted point. Because data are plotted directly from the database, using latitude and longitude values, distribution plots can be generated using any plotting program or GIS capable of reading data files.

IMAGE DISPLAY

The coral image base is available by direct linkage to the database or by independent browsing. CoralBase includes an image display module (CoralImg) capable of displaying still images, video, and animation, as well as being able to play recorded sound files on machines fitted with Creative Labs SoundBlaster™ compatible sound cards. CoralImg can display any digital image in windows bitmap (.BMP) or compressed JPEG format, including maps and aerial photographs.

THE UTILITIES

CoralBase contains a number of utilities that will give it maximum usefulness in remote or underresourced locations. Experience has shown that not everyone has access to, or the need for, high-powered, expensive, analytical software packages; therefore, some basic statistical and pattern analysis functions are being included within CoralBase. Included also are graph- and report-generation facilities, data management tools, and a pop-up notes utility. A data export facility is provided to output data for use in other analytical programs.

USING AND UPDATING CORALBASE

Rather than just being a reference work, CoralBase will be a fully functional database, and new records may be added to the database as required. New data will be handled seamlessly with the distributed data set for all database operations, but users will not be able to change or delete original records as this would compromise the integrity of the data set. A facility is included to extract added data. We hope that users will be prepared to regularly forward data sets, or corrections to the original data set, for inclusion, with appropriate credit, in future updates of CoralBase.

A number of standard queries will be included in the QuickSearch facility (e.g., species lists for a given site) while more complex queries can be generated using the Custom-Query module. The results of queries can be viewed on screen, stored to a file, or output as a report. As with queries the report facility contains a number of standard report forms, but the user may generate customized reports.

Throughout the database, the ability exists to create flagged lists of, for example, species, genera, sites, and references that can be stored and recalled for later use. Flagged lists allow multiple use of the system or enable one user to research several independent projects simultaneously. Users are able, for example, to store preferred system settings, defaults, and queries for use in future sessions.

To allow multiple users to work with CoralBase effectively, a user identification and password system is used when a CoralBase session is initiated. While anyone is able to bypass this security to use the system, only by entering an identification and a password will users be able to recall information stored from their previous sessions.

DISTRIBUTION AND HARDWARE REQUIREMENTS

In keeping with the aim to make CoralBase as widely available as possible, CoralBase will be released on CD-ROM to run on PC under Microsoft Windows™ V3.1+. A 386DX-33 PC fitted with a CD-ROM drive and with 8Mb RAM is the minimum configuration required. SVGA 800*600 screen resolution has been used to allow for extensive use of images.

THE CORALBASE ENVIRONMENT

The Microsoft Windows™ environment was chosen as the platform for developing the first version of CoralBase because it provides a user-friendly environment and has the widest acceptance in the small computer market. Because CoralBase includes a suite of utilities, it was not feasible to program entirely within a proprietary database product; thus, while the database files are in Microsoft Access™ V2.0 native format, the interface and utilities are being written in Microsoft Visual Basic™ V3.0. While Microsoft Access™ is not the optimum database from a performance standpoint, it is fast becoming a de facto standard for PC and is therefore the logical choice to allow maximum integration with other applications in the future.

The original CoralBase concept called for development of a number of software modules that could be run independently from the computer desktop or in conjunction with each other as required. This approach has proved to have a number of disadvantages, particularly in data integration, while providing no specific advantages. The current version consists of a fully integrated suite of modules that are run from a central module called CoralBar. Each module can be used independently within CoralBase. Only the CoralNotes utility can be started separately from the desktop.

THE FUTURE

AIMS has undertaken to support future updates of CoralBase, which may be funded by a subscription fee similar to any scientific journal. It is envisaged that updates, to be added to the database using a utility supplied with the system, will be available via FTP while the updated CD-ROM will be available periodically to subscribers.

It is conceivable that in the future CoralBase may be ported to the Apple Macintosh platform and almost certainly will be ported to the new Microsoft "Windows 95" operating system when it arrives in 1995. There are no plans to produce any mainframe versions of the product, although very little extra programming would be needed to produce a fully networkable PC version.

EXTENDING "AIMS-BASE" SOFTWARE TO OTHER TAXONOMIC GROUPS

The "AIMS-Base" software is generic enough to be used with any taxonomic group with little modification and has been written with that future purpose in mind. Because of its modular construction, it is not necessary to have all the data types to use the database. If, for instance, there was no image collection or synonymy data the other modules could be used without them, and they could be implemented at a later date if required. Conversely if a new data type was to be added (e.g., genetic or morphometric data), it would not be difficult to add new database files and software modules within the current system to handle them. The utilities modules are data-set independent and so can be used with any data set.

INTEGRATION AND COMPATIBILITY

While CoralBase will be an independent product, we are aiming to make it as compatible as possible with other database initiatives that are emerging. We expect that the CoralBase will be able to run from within the ICLARM ReefBase and vice versa. As stated previously, the site data in CoralBase will contain the reef codes to be used in ReefBase. A standardization of site codes in all developing database projects is the obvious way to ensure compatibility between projects even when being developed on different platforms.

CONCLUSION

CoralBase aims to overcome many of the problems associated with printed publications by providing a distributable, updateable, and easy-to-use method of accessing the latest information on Scleractinian corals from all over the world. With cooperation from researchers worldwide and frequent updates, CoralBase will become the "standard" reference on Scleractinian corals and will expand to cover all areas of the globe.

REFERENCES

Veron, J. E. N. 1993. Reprint. *Corals of Australia and the Indo-Pacific*. Honolulu: University of Hawaii Press. Original edition, Queensland: Australian Institute of Marine Science.

———. 1992. *Hermatypic corals of Japan*. Australian Institute of Marine Science Monograph Series No. 8, Queensland.

———. 1993. *A biogeographic database of hermatypic corals of the Central Indo-Pacific and genera of the world*. Australian Institute of Marine Science Monograph Series No. 9, Queensland.

Veron, J. E. N., and M. Pichon. 1976. Scleractinia *of Eastern Australia. I. Families Thamnasteriidae, Astrocoeniidae, Pocilloporidae*. Australian Institute of Marine Science Monograph Series No. 1, 1–86, Queensland.

———. 1980. Scleractinia *of Eastern Australia. III. Families Agariciidae, Siderastreidae,* Fungiidae, Oculinidae, Merulinidae, Mussidae, Pectiniidae, Caryophylliidae, Dendrophyliidae. Australian Institute of Marine Science Monograph Series No. 4, 1–422, Queensland.

———. 1982. Scleractinia *of Eastern Australia. IV. Family Poritidae*. Australian Institute of Marine Science Monograph Series No. 5, 1–159, Queensland.

Veron, J. E. N., M. Pichon, and M. Wijsman-Best. 1977. Scleractinia *of Eastern Australia. II. Families Faviidae, Trachyphylliidae*. Australian Institute of Marine Science Monograph Series No. 3, 1–233, Queensland.

Veron, J. E. N., and C. Wallace. 1984. Scleractinia *of Eastern Australia. V. Family Acroporidae*. Australian Institute of Marine Science Monograph Series No. 6, 1–485, Queensland.

CHAPTER 19

The Role of Taxonomic Databases, with Special Emphasis on Fishes

William N. Eschmeyer

ABSTRACT

Managing information about animals and plants depends on accurate taxonomic databases. Understanding taxonomy and the role of taxonomists is important for all aspects of biological information management. Very large databases for fishes have been built at the California Academy of Sciences with funding from the National Science Foundation. The building of these databases is explained. The value and utilization of such databases are addressed. Fields in each database, search entry programs, and output programs are partially presented in an appendix to this paper.

INTRODUCTION

The building of accurate taxonomic databases is an essential step in addressing issues of conservation, management, resource planning, and other societal matters involving man and the other species. Effective information management depends on such databases.

RESULTS

The Role of Taxonomy

The primary way basic information about animals and plants is organized and stored is by taxonomic categories (typically species). Since many members of the workshop are not taxonomists, I will begin with a brief discussion of taxonomy, using slides from my own research. It is important to understand (1) why good taxonomic databases are essential for studying biodiversity, (2) what taxonomy entails, (3) why a hierarchical classification is useful, and (4) why classifications and names change, thereby making it more difficult to accumulate and keep track of information for many purposes from management to inventories, to species entering into commerce.

Taxonomists do two important tasks: name organisms and make classifications. The system of hierarchical classification and a two-word system for naming species began with Linnaeus in 1758. The system was codified in 1842 (Strickland et al. 1843), and it became the system used by all zoologists worldwide from 1843 to the present. The two-word name for species consists of a generic name and a specific name. A genus may contain more than one species, and species are placed together in a genus based on perceived genetic affinity (as determined mostly by morphological differences and similarities). Taxonomists discover or describe species (1) by assembling specimens through fieldwork and/or by borrowing from museum collections, (2) by studying variation, (3) by grouping the specimens into species categories, (4) by comparing these with previously described species, (5) then naming the new species following specific rules (International Code of Zoological Nomenclature 1985), and (6) by publishing the information in scientific journals. Monographs contain thorough treatments of all the species in a larger group, such as a family, and monographs represent the latest summary of information for that group.

A second thing taxonomists do is make classifications. Classifications are useful because they contain information about relationships. For example, when a chemical suitable for pharmaceutical products is found in one species, biochemists can quickly learn from classifications the close relatives (e.g., other species in the same genus) that might contain similar or better chemicals. All species in the same genus should share many behavioral, biochemical, ecological, and biological properties because they are closely related evolutionarily. The effect of pollution on a species at Palau should be similar to the effect on a close relative occurring in Hawaii. Those in the same family (next category up) similarly share many but fewer features. Classifications thereby have predictive value. Since the late 1960s, most taxonomists have used "cladistic" methods of forming classifications, basing them on shared advanced (new) features. This approach results in cladograms or trees that reflect ancestry as well as relatedness of individual taxa.

The changing nature of classifications and scientific names (because of changing ideas of relationships and because of technical [nomenclatural] rules) makes it almost impossible to know under which specific, generic, or even family names one will find pertinent information in the prior literature or in specimen collections. For example, in 1989 both the genus name and specific name of the Rainbow trout were changed (see Smith and Stearley 1989). Thousands of publications cite *Salmo gairdneri*; now, we call it *Oncorhynchus mykiss*. The genus name was changed from *Salmo* to *Oncorhynchus* partly based on fossil evidence because the Pacific trout was thought to be more closely related to the Pacific salmon than to the Atlantic salmon (the name carrier or type for *Salmo*). Pacific trout and salmon are now classified in *Oncorhynchus*. The species name *gairdneri* was changed to *mykiss* when it was thought that *mykiss* from Kamchatka was the same as *gairdneri*; since *mykiss* was described first, that name had priority for use over *gairdneri*.

Another major activity (or duty) of taxonomists is to make "synonymies" that summarize prior accumulated knowledge about each species. Unfortunately, scientific names change for several reasons, which make inventory especially difficult since information about a single species may be found under several scientific names. Names change because:

1. One species may have been described more than once (such as from different geographical areas, from different sexes, from atypical specimens, or from a lack of knowledge of earlier descriptions). As these "duplicates" are discovered, the first-

described name is selected as the valid name, often resulting in a name change, as in the Rainbow trout.

2. Scientists may differ on what species to include in a particular genus, or species are moved to different genera based on perceived relatedness. This results in the first half (generic) of the name changing; sometimes the ending of the specific name also changes since, if it is an adjective, it must agree in gender with the genus.

3. Sometimes names are changed for technical reasons.

Another problem is that scientific names are frequently misspelled in scientific publications, in collection records for museums holdings, and by abstracting services. Often a name is misspelled because the spelling as originally presented was not verified by subsequent workers. Although there are current arguments about how to incorporate fossils into classifications, and especially how to treat them in higher taxa, the present system probably will continue for many years. Numbering taxa has not worked either. Often common names are more stable than scientific names, and they can be very useful in some groups.

Clearly, information management priorities must deal effectively with taxonomy. Building inventories, for example, involves accurately tracking information. However, this is complicated by a number of factors such as inadequate knowledge of the fauna and changes in names and synonymies. It involves having accurate, up-to-date taxonomic databases.

Taxonomic Databases for Fishes

The building of accurate taxonomic databases is time consuming and requires excellent library facilities and a fair amount of money. There are not many shortcuts.

Extensive databases for all fish genera, described species, and pertinent taxonomic literature have been built at the California Academy of Sciences. In 1984, we began building a database of fish genera, and shortly thereafter received a grant from the National Science Foundation to assist in this work. The strategy used for building the genera of fish databases was to enter information from the most complete sources first and then to edit. For this phase we used Jordan's *Genera of Fishes* (published in the 1920s and containing about 50 percent of the genera); Fowler's *Fishes of the World* (partially published in the Journal of the Taiwan Museum, with much of the remaining parts in manuscript form); cards of genera, described since Jordan, that were maintained at the University of Michigan; the Zoological Record; and 3" × 5" cards prepared by Henry W. Fowler (to about 1905). At this point, we had more than 10,000 generic group names and the corresponding literature.

The database contained misspellings, and there were many problems including where and when a genus was published and its original spelling. We then began editing and proofing. I examined nearly all the original descriptions from their original source (first publication) for content and for the reference citation. This phase involved solving nomenclatural problems as well (and an "interpretation" of the Code was prepared). Using current literature sources, the status (valid or a synonym of another genus) was established. The genera were allocated to a recent classification. This work was published in 1990 and consisted of three main parts: the 10,000 genera arranged alphabetically with their associated information and a bibliography of about 4,500 references. We continued to update the genera database, and in early 1994 it and the associated appendices were placed on the Internet Gopher System of the California Academy of Sciences. It was also adopted as an authority standard by the Zoological Record.

While the genera database was being readied for publication, we began preparing the database for species. We used the same basic strategy (few initial sources) including (1) the Zoological Record for 1864–present, (2) Fowler's 3" × 5" cards—1858 to about 1905, and (3) Fowler's *Fishes of the World* to the mid-1950s. This consists of about 55,000 records and now a total of about 20,000 references. The rough database was completed in 1991. A three-year continuation grant was received from the National Science Foundation in April 1992 to complete the species database to (1) improve information in records, (2) tie all records to the original publication, (3) proof information against original descriptions (verify spelling, authorship, reference date, pages and figure, type specimens, type localities), (4) record information on location of types (from type catalogs and museum visits), (5) document status by using current literature (major journals and recent monographs over the last ten years or so), (6) ready the database for publication (about 3,000 pages, similar sections as used in the "Genera" [Eschmeyer 1990]), and (7) make available electronically, as a CD-ROM, and as a dictionary for collection databases in fishes. We expect to complete the project in about twenty-four months.

The approximate cost of building the fish databases, including the most recent grant, will be about $490,000, which is the equivalent of about 14 person-years, including my time.

DISCUSSION

Obviously, once an authoritative database is available, other databases can be associated with it. The databases are of tremendous value to individual taxonomists. Much time is saved in locating nominal species in a taxon, working with date problems and nomenclatural problems, locating type specimens, finding literature, and preparing bibliographies. Databases will make taxonomists much more efficient.

There are now several examples of how these databases can be useful in achieving the goals of this workshop.

1. The first is a project called NEODAT (Neotropical Database Project), which is available on Internet (ca. 200,000 records from perhaps twenty institutions). The NEODAT project provided computer equipment where needed to museums housing South American freshwater fishes. Other work involved preparing a mapping system (FISHMAP) and gathering information on river basins for publication in a drainage dictionary. We assisted them by downloading records for all families with species in the freshwaters of South America (about 10,000 records) to get them started.

2. A second is the MUSE system of collection management developed by Julian Humphries of Cornell University. The genera database was provided to Cornell in a classifi-

cation, which they made into a spell-check authority file for all fish collections using MUSE. They are anxious to have access to the species database for the same purpose. Interestingly, the MUSE project has put on Internet the computerized holdings of several museums that have North American fishes. This database runs together with a mapping program, such that a mere click on a distribution dot will call up the collection record responsible for that dot and provide access to catalog numbers, full locality information, and related data. Museums typically have many bottles under "old" names or under misspellings, and an authority file for species should make museum data more useful.

3. FISHBASE (Froese and Palomares, this volume) is another project that received information from the "Genera" (Eschmeyer 1990) database.

4. I recently received information on MAPFISH, which covers North American freshwater fishes (ca. 930 species) and appears to offer similar features as does FISHBASE, including dot distribution maps, color illustrations of specimens, some taxonomic information, and biological information (e.g., habitats, spawning times).

Accurate taxonomic databases will be essential for many aspects of studying and managing the information on the aquatic biota of the tropical island Pacific region. Building them is something that taxonomists and museums are ideally suited to undertake.

Acknowledgments—Programming was done by Barbara Weitbrecht (mostly genera) and David Boughton. Others who assisted in authorship of parts are Reeve M. Bailey, Carl J. Ferraris, Jr., Mysi Hoang, William Smith-Vaniz, and Barbara Weitbrecht. Data entry assistance was provided by Frances Bertetta, Pamela Donegan, Jon Fong, Melissa Gibbs, Lezlie Skjeetz, and Geraldine Stockfleth. Many others helped in many ways. Funding for this research was supported by the National Science Foundation grants NSF 8416085, 8801702, and 9108603.

APPENDIX

Slides were shown at the workshop, detailing the databases maintained, fields in each database, search entry programs, and output programs. Examples of such data are partially presented here.

Databases: Species, genera, families, fossil species, fossil genera, classification, references, journals, museums, place names. The fossil databases have the same structure as the recent ones, plus fields for geologic time period and parts, if not a whole specimen.

1. Species database. **Fields**: Genus, species, level (sp., subsp., var.), author, date, reference number, journal acronym, journal or book descriptor, citation (e.g., volume), page, plates. Fields to handle classification. Fields to handle synonymies. Availability of name. Large field for information on type catalogs, nomenclatural items, and current status with citations to author, date, reference number that document status. Fields for types and type locality. "Notes to us" field, and a "done" field. Common names field is not used at present. Plus we use tracking fields, such as vol./p. in Zoological Record. **Links to other databases**: References, place names, museums, genera, classification.

Seek programs for entry: Species by author, original genus, species + author, species + genus, family number, journal acronym, journal name, reference number, current genus, append new record. **Output programs**: Alphabetical by species, by reference (short version), by reference (long version), first author, author (complete), original genus, journal, locality, by museum (primary types), museum (all types), by museum on cards, family/subfamily, family/subfamily (short version).

2. Genera database. **Fields**: Generic name, subgenus of, author, date, page, gender, classification, type designation, fields involved with status, remarks and memo field for nomenclatural items and status references, notes, done, reference number. Plus tracking fields for sources. **Links**: Classification, references (+ journal). **Seek programs for entry**: Genera by genus + author, author, by classification, reference number, append new record. **Output programs**: Genera by reference (short version), reference (long version), name, author, family/subfamily.

3. Families database. **Fields**: Family name, original level (e.g., fam. or subfam.), author, year, journal, classification fields, technical nomenclatural fields (such as type genus, author, date), fields for status, notes. **Links**: Classification, references. **Seek programs for entry**: By author, original level, original name, reference number, type genus, zoological record number/p. **Output programs**: Families by reference (long and short version), original name, author, journal, type genus.

4. Classification database. **Fields**: Name of taxon, common name, index name (to make index), age (fossil, recent or both), current status, classification fields, "tracking" fields. **Links**: Species, genera. **Seek programs for entry**: In phylogenetic order, by higher taxon name, update classification. **Output program**: Printout of classification.

5. References database. **Fields**: First author, initials, second author, initials, other authors, year, month and day, title, journal acronym, journal name, pages, plates, remarks to be published, notes, done, fields to limit to one project (e.g., only references dealing with genera). **Links**: Species, genera, families. **Seek programs for entry**: By author(s) + (optional) year, journal name, journal acronym, reference number, append record. **Output programs**: References by first author, by any author, by journal.

6. Journals database. **Fields**: Acronym, standard abbreviation, full name, library call number, notes, Biosis journal acronym. **Links**: Species. **Seek programs for entry**: By acronym, by journal name, append record. **Output programs**: Journals by acronym, by abbreviated title.

7. Museums database. **Fields**: Museum acronym, full name of museum or institution, notes. **Links**: Species. **Seek programs for entry**: By acronym, by museum or institution name, append record. **Output programs**: Directory of museums, acronyms.

8. Place names database. **Fields**: Place name (string, specific to general), a usable name (yes or no), notes. **Links**: Species. **Seek programs for entry**: By place name, append new record. **Output program**: Geographic dictionary.

Other programs: Generation of bibliographies based on selected references (several versions). Use of "filters" can produce a variety of outputs, such as all species with a type locality in Brazil.

REFERENCES

Eschmeyer, W. N., ed. 1990. *Catalog of the genera of recent fishes.* California Academy of Sciences.

International Code of Zoological Nomenclature. 1985. *The international trust for zoological nomenclature.* London.

Smith, G. R., and R. F. Stearley. 1989. *Fisheries* (Bethesda) 14(1): 4.

Strickland, H. E., et al. 1843. Report of a committee appointed "to consider of the rules by which the nomenclature of zoology may be established on a uniform and permanent basis." *Brit. Assoc. Adv. Sci.* (rept. 12th meeting), 1842:105–21.

CHAPTER 20

FishBase as Part of an Oceania Biodiversity Information System

Rainer Froese and Maria Lourdes D. Palomares

ABSTRACT

FishBase is an electronic encyclopedia on fish being developed at the International Center for Living Aquatic Resources Management with the cooperation of the Food and Agriculture Organization of the United Nations and funded by the European Commission. To date, FishBase contains key information for about 12,000 fish species globally, including about 2,800 species of the Oceania region. FishBase is available on CD-ROM for personal computers with Windows 3.1 or later. A design for an Oceania Biodiversity Information System is presented, and the integration of FishBase into such a system is discussed.

INTRODUCTION

FishBase, a large biological database on fish, is being developed in a joint project between the International Center for Living Aquatic Resources Management (ICLARM) and the Food and Agriculture Organization of the United Nations (FAO) funded by the European Commission. FishBase contains key information on fish populations such as nomenclature, ecology, population dynamics, aquaculture, genetics, physiology, and occurrence of fishes (Froese 1990, 1993; Palomares et al. 1991; Pauly and Froese 1991a, 1991b; Froese, Palomares, and Pauly 1992; Palomares and Pauly 1992a, 1992b; Pauly, Palomares, and Froese 1993; Froese and Pauly 1994b). It was conceived as a "tool" to help fisheries researchers and managers to better understand and manage their natural resources.

FishBase was designed to answer, for example, the following questions for the fishes of a given country:

- What is the current scientific name and classification?
- What is the international (FAO, American Fisheries Society, FishBase) common name?
- What is the global commercial importance?
- What key information (life history, population dynamics) is available?
- What actually are these key parameters?
- What are the research gaps?
- What species are (or can be) used for aquaculture?

- What species are introduced?
- What species are of interest to sports fishers?
- What species are of interest to the ornamental trade?
- What species are threatened?
- What species are endemic?
- What species are dangerous to humans?

FishBase also offers a structure for national information such as occurrence, importance, use, or restrictions in a given country. Such information can only be obtained from national researchers and managers, and the FishBase Project has started collaborative efforts with Hawaii, Taiwan, and Mexico to identify the most efficient procedure for such data exchange.

MATERIALS AND METHODS

FishBase contains more than 1,000 data fields organized in 50 tables, totaling half a million records. More than 200 procedures access this information and create a variety of outputs. FishBase has been moved recently from a DOS (DataEase) to a Windows (Microsoft Access) database software, an undertaking that was much more difficult than we had anticipated. For a decent performance, FishBase now requires at least a 486 PC with at least 8 megabytes of memory. FishBase can be run from the CD-ROM on which it is distributed; however, it will be slow. There are options to install the data on the hard disk (faster; 120 megabytes required) and access the pictures on the CD-ROM, or to install everything on the hard disk (fastest; 430 megabytes required). FishBase can also be installed on a local area network (LAN) so that several persons can use it at the same time.

FishBase makes use of published literature (e.g., journal articles, technical reports, theses) and recent revisions of fish species or families such as those produced by FAO. Key information on any of the topics mentioned earlier is extracted and entered by a group of research assistants (all with graduate degrees in marine biology) based at the ICLARM headquarters in Manila. Collaborators worldwide participate in data entry through data collection forms. Some collaborators have developed similar electronic databases on specific topics (e.g., Eschmeyer, this volume) and have opted to make their databases available by incorporation and distribution through FishBase. Contributions are given proper citations and acknowledgments in FishBase: every datum is attached to the reference where it stems from, and every record bears a "stamp" identifying who entered, contributed, modified, or validated the information.

The quality of the information in the database is monitored in-house through several loops of verification by the research assistants who check each other's work and also through collaborating scientists who offer to check the information in FishBase for their different fields of expertise.

The Project has purchased a CD-ROM recorder with software (~US$8,000) to produce CD-ROMs in-house, allowing for continuous creation of updates. The production time for one CD-ROM takes about one hour, and a blank CD-ROM costs about US$20.

RESULTS AND DISCUSSION

As of October 1994, the FishBase team has been able to incorporate more than 12,000 species extracted from more than 7,000 references, representing half of the estimated 25,000 species of fish in the world (Figure 20.1). Of these, over 2,800 are from marine waters in Oceania, extracted from more than 2,000 references.

In addition, FishBase has become the repository of several outstanding collections—the first four of which have been compiled in-house:

- the largest collection of population dynamics data (growth parameters, natural mortality, length-weight relationships, maximum ages and sizes) entered in more than 4,000 records covering over 1,000 species;

- the largest collection of ecological data (prey, 4,400; predators, 800; diet composition, 700; and food consumption, 150);

- the largest available collection of electrophoretic data entered in 3,600 records for about 50 populations/strains;

- the largest collection of common names (by country and language) entered in about 50,000 records for more than 8,000 species;

- the genera of recent fishes (from Eschmeyer 1990; see also Eschmeyer, this volume);

- the largest collection of data on fish metabolism (from Thurston and Gehrke 1991) entered in 7,400 records for about 300 species;

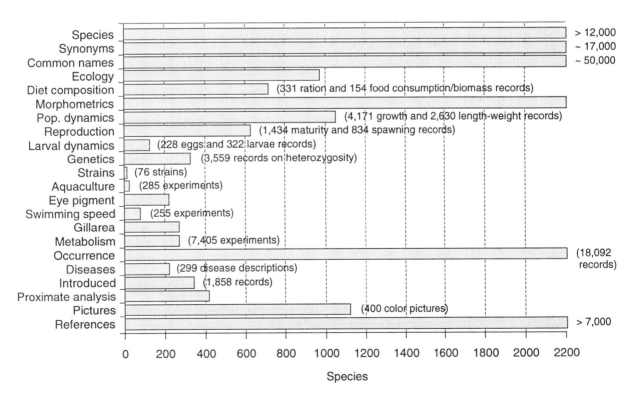

Figure 20.1 Content of FishBase as of October 1994 in number of species covered and/or number of records

- the largest collection of data on larval dynamics (from Houde and Zastrow 1992) covering about 100 species;

- the largest collection of data on introduced fishes (mainly from Welcomme 1988) entered in 1,815 records covering about 360 species; and

- about 1,000 line drawings or color pictures of adults, larvae, eggs, or diseases.

The WINMAP software, a low-level geographic information system that forms part of the FishBase package, can be used to display a variety of maps based on the occurrence and abundance records in the database. This feature is specially useful in fish biodiversity research. Maps showing the global occurrence of members of an order, family, subfamily, genus, or a species can be created.

When representative levels of occurrence and abundance records are reached, it will be possible to use FishBase to determine a theoretical status of threat for certain species or species groups by combining trends in distribution and abundance with a set of biological characters such as environmental tolerance, fecundity, longevity, and age at first maturity.

The CD-ROM version of FishBase is distributed at nominal costs to fisheries institutions worldwide. Special emphasis will be given to developing countries, some of which will be supported in purchasing the necessary hardware and receiving training on how to use FishBase and related analytical tools.

The success of FishBase will depend on its acceptance (= use) by researchers, managers, teachers, and students. It also depends on collaboration of experts in the various fields to ensure quality of the information. Collaborators are entitled to a free copy of FishBase. Others can order the FishBase CD-ROM and manual from ICLARM for US$95 (including air mail).

RECOMMENDATIONS FOR A MARINE/COASTAL DATABASE SYSTEM FOR OCEANIA

Before a database system can be designed, one has to be clear about the purpose it should serve. The objectives mentioned in the background information for the workshop call for a "data management system ... [to assist in a] ... long-range program to conserve biodiversity" and to "summarize information on the systematics and taxonomy of nearshore marine and coastal species in Oceania." We quote the abstract from Froese and Pauly (1994a):

> In order to improve our understanding of aquatic biodiversity, it is suggested to assemble in a single database the huge amount of existing data on the occurrence of aquatic species in space and time. Such data are available in museum collections, research vessel surveys, tagging studies, the scientific literature, and a variety of other sources, often in digitized form. The database would be distributed on CD-ROM with annual upgrades. It would preserve data which might otherwise be lost; it would provide baseline data on biodiversity from historic data sets; in combination with data derived from existing biological, oceanographic, and meteorological databases it would allow for analyses of biodiversity which are currently not possible; and it would guide the ongoing efforts towards collection of data that are most useful for analytical models. We suggest to establish a network of institutions that hold relevant data and are willing to share it.

The database structure suggested by Froese and Pauly (1994a) is presented here again, in more detail and with some modifications. The database for aquatic biodiversity should consist of the following tables (Figure 20.2):

> The <u>Specimens</u> table is the center of the database. It contains one record per specimen encountered and is similar to existing collection databases. It contains the following fields: genus, species, taxonomic group, catalog code, record type (museum, survey, literature, tagging), collector, station code, number of specimens, length, weight, sex, live stage, abundance, . . . , remarks, contributor code.
>
> Taxonomic group, genus, and species fields of the <u>Specimens</u> table are linked to taxonomic dictionaries such as Eschmeyer's <u>Genera</u> and forthcoming <u>Species</u> databases for fishes or CoralBase for corals (see Eschmeyer, and Navin and Veron, respectively, this volume). The dictionaries ensure the validity and proper spelling of the scientific name and provide the classification into higher taxa. The same fields will also provide the link to biological databases such as FishBase.

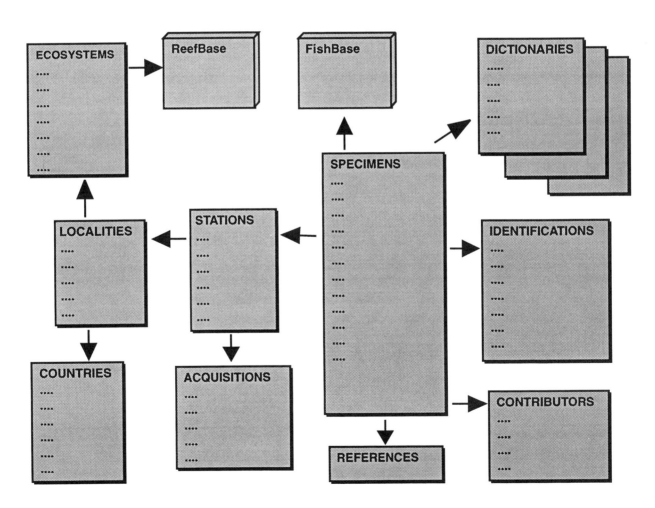

Figure 20.2 Structure suggested for the Oceania Biodiversity Information System

Through the taxonomic group and catalog code, the <u>Specimens</u> table is linked to the <u>Identifications</u> tables that are specific for each taxonomic group and contain the following fields: taxonomic group, catalog code, identifier, date, identified as, length, head length, . . . (other group-specific morphometric counts or measurements), remarks, reference. These tables keep track of the taxonomic work and verify identifications where no specimen is preserved (e.g., during surveys).

The station code links the <u>Specimens</u> table to the <u>Stations</u> table containing the following fields: station code, station name, locality code, acquisition code, date, time, latitude, longitude, depth, altitude, temperature, salinity, gear, remarks, reference. The <u>Stations</u> table, which is similar to those used in trawl surveys, facilitates data entry because station data are entered only once for all specimens encountered.

The locality code links the <u>Stations</u> table to the <u>Localities</u> table, which is a gazetteer of aquatic localities and contains the following fields: locality code, locality name, locality type (river, lake, lagoon, reef, seamount), locality description, ecosystem code, country code, latitudinal range, longitudinal range, other names used, remarks. This table should provide links to ecosystem databases such as ReefBase.

The acquisition code links the <u>Stations</u> table to the <u>Acquisitions</u> table, which contains the following fields: acquisition code, expedition/survey name, date from, date to, collector, donor, reference, remarks. This table is similar to the acquisition file maintained in museums. It facilitates data entry in that this information need not be repeatedly entered for each station of a survey.

The country code links the <u>Localities</u> table to the <u>Countries</u> table, which contains the following fields: country code, country name, area, shelf area, coastline, population, coastal population, languages, . . . , reference. This table standardizes country names and provides additional information that might be useful in comparative studies. This table already exists in ReefBase and FishBase.

The ecosystem code links the <u>Localities</u> table to the <u>Ecosystems</u> table, which contains the following fields: ecosystem code, ecosystem name, ecosystem type, latitudinal range, longitudinal range, surface area, drainage area, depth range, average depth, salinity range, surface temperature range, average surface temperature, . . . , remarks, reference.

The ecosystem classification should closely follow the one described by Holthus and Maragos, this volume. For reefs, this table can draw on information contained in ReefBase (McManus et al., this volume).

The contributor code in the <u>Specimens</u> table establishes the link to the <u>Contributors</u> table, which identifies the collaborating individuals and institutions and contains the following fields: contributor code, name, institution, address, e-mail, remarks.

A <u>References</u> table will be linked to the references mentioned in the other tables and will contain the full citation.

This structure, consisting of more than ten tables with many code fields, will be largely hidden from the user who will just click on buttons and select options from choice lists to access the desired information.

As Froese and Pauly (1994a) have pointed out, "A species name, a date, and a locality do not seem to be much of a base for sophisticated analytical models. However, one has to realize that these three bits of data each represent a vast amount of information: the species name actually provides us with all the biological information on an organism; the locality leads to all the ecological information available for a site; and the date provides us with information about seasonal and historical environmental conditions." As a result of subsequent discussions, the source (i.e., a reference or the name of the provider of the information) is added as the fourth bit of core biodiversity data. The source gives an indication of the reliability of the information, leads to other reports and species identified by the collector, and gives feedback on the experience of the collector.

Thus, the proposed system will be able to answer, for example, the following questions:

- What species occur in a certain country or ecosystem?
- Where and when has a certain sensitive species last been collected?
- What surveys have been made in a certain ecosystem or country?
- What species have been collected during a certain survey?
- What are all the ecosystems in which a certain species occurs?
- What biological characters are shared between species of a certain ecosystem?
- What is the percentage of indicator species in a given area compared to other areas?

A crucial point of the suggested system is the quality of the identifications. It will be necessary to determine expert centers for each species group that will classify the probability of correctness (probably correct, doubtful, probably misidentified) for each new record. There will probably be a need to include experts outside of Oceania.

Froese and Pauly (1994a) suggested, "Start this exercise with a [international] project which, if successful, will be turned into a permanent activity of an appropriate international body, similar to the institutionalized gathering of meteorological, oceanographic, and recently also coral reef data."

Acknowledgments—We thank the members of the FishBase team—Daniel Pauly, Susan Luna, Crispina Binohlan, Armi Torres, Liza Agustin, Maria Teresa Cruz, Pascualita Sa-a, Roberto Cada, Rodolfo Reyes, Jr., Emily Capuli, Rachel Atanacio, and Portia Bonilla—for their dedicated work. We also thank our collaborators throughout the world who believed in our vision.

Note: This is ICLARM Contribution No. 1104.

REFERENCES

Eschmeyer, W. N. 1990. *Catalogue of the genera of recent fishes.* California Academy of Sciences, San Francisco.

Froese, R. 1990. FishBase: An information system to support fisheries and aquaculture research. *Fishbyte* 8(3): 21–24.

————. 1993. Study group report on FishBase. ICES C. M. 1993/L:6.

Froese, R., M. L. D. Palomares, and D. Pauly. 1992. Draft user's manual of FishBase. ICLARM, Manila, Philippines.

Froese, R., and D. Pauly. 1994a. A strategy and a structure for a database on aquatic biodiversity. In *Data sources in Asian-Oceanic countries*, ed. J.-L. Wu, Y. Hu, and E. F. Westrum, Jr., 209–20. Committee on Data for Science and Technology, Ann Arbor, Michigan.

————, eds. 1994b. *FishBase user's manual.* ICLARM Software No. 7, Manila, Philippines.

Houde, E. D., and C. E. Zastrow. 1992. Ecosystem- and taxa-specific dynamic and energetic properties of fish larvae assemblages. *Bull. Mar. Sci.* 53(2): 290–335.

Palomares, M. L. D., J. Moreau, P. Reyes-Marchant, R. Froese, and D. Pauly. 1991. FishBase: Une base de données sur les poissons. *Fishbyte* 9(2): 58–61.

Palomares, M. L. D., and D. Pauly. 1992a. FishBase as a worldwide computerized repository of ethnoichthyology or indigenous knowledge on fishes. Paper presented at the International Symposium on Indigenous Knowledge and Sustainable Development, 20–26 September, International Institute for Rural Reconstruction, Silang, Cavite, Philippines.

————. 1992b. Use of FishBase for assembling information on the fishes of Cambodia, Laos, and Vietnam. Paper presented at the 3d Asian Fisheries Forum, 26–30 October, World Trade Center, Singapore.

Pauly, D., and R. Froese. 1991a. FishBase: Assembling information on fish. *Naga, ICLARM Q.* 14(4): 10–11.

————. 1991b. The FishBase project, or how scattered information on fish can be assembled and made useful for research and development. *EC Fisheries Cooperation Bulletin*, December.

Pauly, D., M. L. D. Palomares, and R. Froese. 1993. Some prose on a database of indigenous knowledge on fish. *Indigenous Knowledge and Development Monitor* 1(1): 26–27.

Thurston, R. V., and P. C. Gehrke. 1991. Respiratory oxygen requirements of fishes: Description of OXYREF, a data file based on test results in the published literature. In *Fish physiology, toxicology, and water quality management. Proceedings of an International Symposium, Sacramento, California, USA, 18–19 September 1990*, ed. R. C. Russo and R. V. Thurston, 95–108. U.S. Environmental Protection Agency EPA/600/R-93/157.

Welcomme, R. L. 1988. International introductions of inland aquatic species. FAO Fish. Tech. Paper 294, Rome.

CHAPTER 21

Hawaii Biological Survey: A Model for the Pacific Region

Allen Allison, Scott E. Miller, and Gordon M. Nishida

ABSTRACT

Bishop Museum, which houses the world's largest natural history collections from Hawaii—nearly four million specimens, was designated by the Hawaii State Legislature in 1992 as the Hawaii Biological Survey. Under this arrangement the museum has implemented a six-point plan to fully develop its information resources to support a wide range of applications in Hawaii, including research and resource management. We describe this plan, review its initial accomplishments, and argue that museum collections are important information resources for the Pacific region and could provide the basis for a Pacific-wide biological survey.

INTRODUCTION

The Bernice Pauahi Bishop Museum was established in 1889 by Charles Reed Bishop, noted businessman and philanthropist of nineteenth century Hawaii, as an independent organization dedicated to collecting, preserving, studying, and disseminating knowledge of the cultural and natural history of Hawaii and the Pacific. The museum now has about twenty-three million items in its collections, including four million specimens of plants and animals from Hawaii, and is among the five largest natural history museums in the United States. Early on, the museum established strong programs to study and document the plants and animals of Hawaii, and that effort has made Bishop Museum the world's largest single source of information on the Hawaiian biota. Virtually all definitive published treatments and manuals of terrestrial and aquatic Hawaiian organisms, beginning with *Fauna Hawaiiensis* in 1890 and including a number of current marine projects such as *Reef and Shore Fauna of Hawaii* and *Indo-Pacific Fishes*, have been produced by the museum or in close collaboration with the museum, with heavy reliance on information from the museum's collections.

In recent years, increasing awareness by conservation organizations and the general public of the potential value of biodiversity and the accelerated decline in species richness has sparked a renewed interest in better understanding our biological heritage.

In 1992, the Hawaii State Legislature designated Bishop Museum as the Hawaii Biological

Survey (HBS) and charged it with the task of developing a comprehensive inventory of the terrestrial, freshwater and marine plants and animals of the state. The museum subsequently developed a comprehensive plan to accomplish this task. This involves, for each major group of plants and animals: (1) developing a computerized database of the literature; (2) preparing a checklist based on the literature, collections, and consultation with experts; (3) databasing museum collections (localities are geocoded to facilitate GIS analysis and presentation); (4) databasing information from other collections or from other organizations conducting biological surveys (or establishing computer linkage to this information); (5) filling gaps in information through additional fieldwork and research; and (6) making this information widely available. In actual practice many of these activities take place simultaneously or out of sequence, and some are undertaken by, or in cooperation with, partner agencies such as the Nature Conservancy's Hawaii Heritage Program, Center for Plant Conservation, University of Hawaii, state agencies, National Biological Service, and other federal agencies.

The museum is now extending the approach used in the HBS to other parts of the Pacific. This includes the development of a national biological survey for Papua New Guinea and involvement in the development of a biodiversity center for Indonesia.

We see the role of HBS in Hawaii as providing a service to the scientific and local communities as an information clearing center. The activities of HBS focus on the gathering, processing, synthesis, and distribution of information related to the biological resources of the Hawaiian Islands. Information from the collections are critical to provide authority files, data points for distribution maps, additional ecological information, and provide a historical perspective. As completeness is required to improve functionality, HBS may also play a role in centralizing and facilitating the distribution of information from many partner organizations. Adequate levels of technology are already available (e.g., fast computers, sufficient information-storage capabilities, broad avenues of distribution such as Internet) to provide potential users with a wealth of up-to-date information. The strategy outlined herein streamlines the process and ensures the rapid development of products that can be of immediate use, while continuing development of longer-term projects, and refining all products continuously.

In this paper we discuss the overall accomplishments of HBS to mid-1995, with an emphasis on the marine sciences. Substantial funding to support this work has been provided by the John D. and Catherine T. MacArthur Foundation, the National Science Foundation, National Biological Service, and the U.S. Fish and Wildlife Service.

RESULTS

The overall size and strengths of the museum's collections of Hawaiian and Pacific organisms are shown in Table 21.1. Literature databases and authority files have been created for most of these groups, and the museum is now focusing its efforts on databasing these collections (Figure 21.1). A significant portion of the marine collections have been databased (Table 21.2).

Table 21.1. Strengths of natural history collections regarding Hawaii at Bishop Museum

Group	Total specimens	Hawaiian specimens	%
Plants (including algae)	500,000	175,000	35
Marine invertebrates	500,000	250,000	50
Mollusks	6,000,000	3,000,000	50
Insects and mites	13,500,000	500,000	4
Fish	130,000	15,000	12
Terrestrial vertebrates	85,000	20,000	24
Total	20,715,000	3,960,000	19

Notes: The comprehensive collections of Bishop Museum are a core resource for the Hawaii Biological Survey. This chart indicates the relative sizes of the Hawaiian collections, plus related material from the Pacific region and elsewhere that provides the context for understanding the Hawaiian biota.

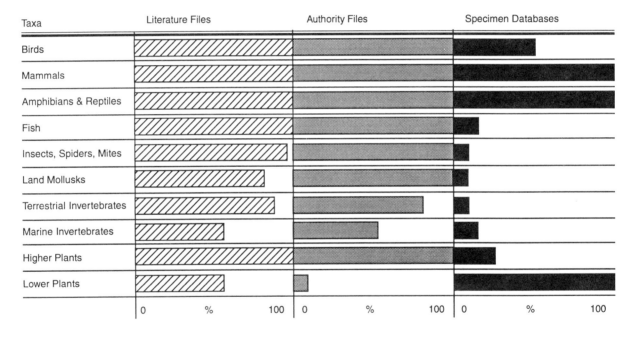

Figure 21.1 Bishop Museum electronic databases and other information services (% completion as of June 1995). The figure shows current status of primary databases of the Hawaii Biological Survey: comprehensive bibliographic databases of what is already known in the literature; authority files for taxonomic names; and specimen data from Bishop Museum collection (see Table 21.1). As these core databases are completed, information from other sources is interfaced or integrated.

Table 21.2 Bishop Museum marine collections (lots)

Group	Total lots	Hawaiian lots	% databased
Algae	55,000	25,000	100
Invertebrates	30,000	20,000	2
Mollusks	68,000	34,000	0
Fish	35,000	5,000	10
Mammals	60	30	100
Birds	1,500	900	100
Total	189,560	84,930	32

Note: The marine collections at Bishop Museum are maintained as a series of discrete lots, a significant percentage of which is databased.

In terms of specific achievements of HBS, since 1992 Bishop Museum and its collaborators have:

- Published a summary list of the 8,552 species of terrestrial arthropods in Hawaii (G. Nishida 1994).

- Produced a catalog of 779 native Hawaiian land snail species (Cowie, Evenhuis, and Christensen 1995).

- Continued progress on the book series *Reef and Shore Fauna of Hawaii* with Vol. 6 (Echinodermata, Chaetognatha, Hemichordata, and Chordata [excluding vertebrates]) scheduled for publication in 1996 (Eldredge et al., in prep.).

- Continued progress on the publication of a marine algae manual (Abbott and Magruder, in prep.).

- Produced three issues of *Indo-Pacific Fishes* (Randall and McCosker 1992; Gon 1993; Hoese and Larson 1994).

- Signed a memorandum of agreement with the Smithsonian Institution to create a database of all Hawaiian plant specimens housed at both institutions. Data entry is virtually complete at this time.

- Continued significant upgrades of the quality of care and organization of Hawaiian biological collections, especially insects, plants, and land snails, with funding from the National Science Foundation.

- Developed a new publication series, *Records of the Hawaii Biological Survey*, to document changes in understanding of the Hawaiian biota. The first in the series was published in March 1995 and included a paper by Eldredge and Miller (1995) documenting the overall species richness of the Hawaiian biota (Evenhuis and Miller 1995a, 1995b).

- Museum staff and resident research associates in the natural sciences produced 200 books, book chapters, and journal articles since 1992.

- Produced and distributed a bibliographic database of Hawaiian plant literature in cooperation with the Center for Plant Conservation and the U.S. Fish and Wildlife Service.

- Provided access to databases and other information resources via the Internet through World Wide Web (http://www.bishop.hawaii.org) and Gopher (bishop.bishop.hawaii.org) servers.

- Uploaded collections data into FishGopher (gopher://muse.bio.cornell.edu:70//11/fishgopher), a project coordinated at Cornell University.

Information on species diversity in the Hawaiian marine environment is provided in Table 21.3. This table illustrates one type of information that can be made available through a survey program such as HBS.

Table 21.3 Species diversity in the Hawaiian marine environment

Taxon	Total	Endemic	NIS[a]
PROTOCISTA	1539+	3?	7
Algae	470	?	5
Protozoa	1042+	2?	0
Fungi	27	?	2
PLANTAE	9	1	0
ANIMALIA	4861+	1128+	83+
Placozoa	1	0	0
Porifera	84	24	4+
Cnidaria	339	74	8
Ctenophora	10	0	1
Platyhelminthes	560	391+	0
Nemertinea	26	5?	0
Gnathostomulida	8	4	0
Rotifera	3	0	0
Nematoda	65+	23	0
Priapulida	1	0	0
Mollusca	787	190	6
Annelida	281+	80+	7?
Pogonophora	2	2	0
Echiura	6	0	0
Sipuncula	14	0	0
Arthropoda	915	42	22
Pycnogonida	12	?	0
Phoronida	3	0	0
Bryozoa	150+	?	?
Entoprocta	55+	?	?
Brachiopoda	3	0	0
Chaetognatha	8	1	0
Echinodermata	278+	150	0
Hemichordata	4	1	0
Chordata	1246+	141+	35?
Total	6406+	1132+	90?+

Notes: For details and references, see Eldredge and Miller (1995). Most of the taxa currently listed as endemic will probably eventually prove to be more widely distributed. Nonindigenous species (NIS) are those that do not naturally occur in Hawaiian waters (also known as alien, exotic, adventive, or introduced species).

a. NIS = Nonindigenous species

DISCUSSION AND CONCLUSION

The Hawaii Biological Survey represents the first successful attempt to create an information management system for the entire biota of any area in the world approaching the diversity of Hawaii. The success that HBS has achieved serves to illustrate the importance of museum collections to document biodiversity and provide information that is crucial for environmental planning and natural resource management (see Allison 1991).

A biological survey provides the framework to bring together widely dispersed or inaccessible information resources to support a wide range of applications. This approach is crucial to the Pacific region with its widely dispersed island groups and many political divisions. There are large collections from the Pacific region in museums from throughout the world (probably in excess of seventy million specimens), representing an enormous information resource. The approach that we have used to capture and synthesize information on the Hawaiian biota to guide research and resource management priorities can, with appropriate international support, be adopted Pacific-wide to develop regional priorities.

Inasmuch as museums throughout the world are actively databasing collections (databases for U.S. museums now include more than eleven million specimens; estimate based on updated figures from Cooley, Harrington, and Lawrence 1993), we need to develop a mechanism to link this information with other biologically important data (e.g., marine surveys and individual research projects) in order to assist resource managers in the challenging task of managing the enormous diversity of the Pacific region (e.g., Miller 1994). More than two million specimen records are already available through various Internet sources (Miller 1993).

The many nations of the Pacific region have achieved considerable success in commerce, health, education, and other activities through cooperation. Much of this has taken place under the guidance and structure of the United Nations, South Pacific Commission, South Pacific Forum, and other regional organizations. The South Pacific Regional Environment Programme (SPREP) has made good progress toward the task of identifying regional environmental priorities.

However, museums have not taken much of an active role in these programs. We hope that this will change. Information from museum collections can be captured into databases for about 10 percent of the cost of collecting the same information from the field (Armstrong 1992), and additionally provides an invaluable historical perspective. The development of FishBase and ReefBase, which incorporate some museum information into their databases, together with ever improving technology to access information directly via the Internet, gives museums the potential to provide a broad range of information products to document and manage biodiversity (Miller 1994). We believe that the development of a Pacific Biological Survey would have considerable merit by unifying all our efforts, and during the next few years we will be working toward this goal.

Acknowledgments—We thank Jim Maragos and Lu Eldredge for organizing the symposium and for helpful comments on the manuscript, Jack Randall for providing information on Indo-Pacific fishes, and Walter Appleby and Clyde Imada for information on Hawaiian plants. We are also grateful to the John D. and Catherine T. MacArthur Foundation for supporting the early development of the Hawaii Biological Survey, the National Science Foundation for support of collections facilities, and the National Biological Service and the U.S. Fish and Wildlife Service for support of data entry projects.

REFERENCES

Allison, A. 1991. The role of museums and zoos in conserving biological diversity in Papua New Guinea. In *Conservation and environment in Papua New Guinea: Establishing research priorities,* ed. M. Pearl, B. Beehler, A. Allison, and M. Taylor, 59–63. Wildlife Conservation International.

Armstrong, J. 1992. The funding base for Australia's biological collections. *Austr. Biol.* 5(1): 80–88.

Cooley, G. P., M. B. Harrington, and L. M. Lawrence. 1993. Analysis and recommendations for scientific computing and collection information management for free-standing museums of natural history and botanical gardens. 2 vols. MITRE Corporation, McLean, Virginia.

Cowie, R. H., N. E. Evenhuis, and C. C. Christensen. 1995. *Catalog of the native land and freshwater molluscs of the Hawaiian Islands.* Leiden: Backhuys Publishers.

Eldredge, L. G., and S. E. Miller. 1995. How many species are there in Hawaii? *Bishop Museum Occasional Papers* 41:3–18.

Evenhuis, N. L., and S. E. Miller. 1995a. Records of the Hawaii Biological Survey for 1994. Part 1: Articles. *Bishop Museum Occasional Papers* 41:1–80.

————. 1995b. Records of the Hawaii Biological Survey for 1994. Part 2: Notes. *Bishop Museum Occasional Papers* 42:1–68.

Gon, O. 1993. Revision of the cardinalfish genus *Cheilodipterus* (Perciformes: Apogonidae), with description of five new species. *Indo-Pacific Fishes* 22:1–59.

Hoese, D. F., and H. K. Larson. 1994. Revision of the Indo-Pacific gobiid fish genus *Valenciennea,* with descriptions of seven new species. *Indo-Pacific Fishes* 23:1–71.

Miller, S. E. 1993. Biological collections databases available on Internet. *Pacific Science Information Bulletin* 45(3–4): 14–15.

————. 1994. Development of world identification services: Networking. In *The identification and characterization of pest organisms,* ed. D. L. Hawksworth, 69–80. Wallingford: CAB International.

Nishida, G. M., ed. 1994. Hawaiian terrestrial arthropod checklist. 2d ed. *Bishop Museum Technical Report* No. 4. Bishop Museum, Honolulu.

Randall, J. E., and J. E. McCosker. 1992. Revision of the fish genus *Luzonichthyes* (Perciformes: Serranidae: Anthiinae), with descriptions of two new species. *Indo-Pacific Fishes* 21:1–21.

CHAPTER 22

The Hawaii Natural Heritage Program: Natural Diversity Database

Samuel M. Gon III and Roy Kam

MISSION	The mission of the Hawaii Natural Heritage Program (HINHP) is "to compile and maintain a set of databases and geographic information systems (GIS) of the location and status of rare and endangered species and ecosystems of Hawaii. This information is used to synthesize, interpret, and distribute accurate, current, and comprehensive information to a wide set of appropriate users for the purpose of making a positive impact on biodiversity protection." This paper discusses the scope, structure, database, and mapping conventions, strengths and limitations of the HINHP system and its links with other efforts, and ends with some recommendations for establishing a resource-tracking database.
SCOPE	In pursuing its mission, HINHP tracks rare, threatened, and endangered taxa and ecosystems of the Hawaiian archipelago. This set of taxa and ecosystems is referred to as "rare elements" in biodiversity shorthand, and currently amounts to over 1,000 different native taxa and types of ecosystems (Appendix 1). The historic base for location and status information on these rare elements is broad but corresponds to the history of scientific discovery and biological exploration in the islands, from 1778, shortly after James Cook's arrival, to the present. Database and GIS updates are required continuously to keep abreast of ongoing biological discoveries and changes in the status of rare taxa and ecosystems. Over 10,000 rare element location records are mapped statewide. The geographic scope includes the main Hawaiian Islands, offshore islets, and the northwestern Hawaiian Islands. A manual mapping convention uses standard USGS (U.S. Geological Survey) 1:24000 quadrangles, but GIS-generated mapping is available in a wide range of scales.
ELEMENT OCCURRENCE DATABASE	The basic database for rare taxon/ecosystem tracking is called the *Element Occurrence Database*, with its individual records referred to as *Element Occurrence Records* (EORs). The EOR is a summary of all the available information for a single element at a single location or occurrence. The record is produced by combining information from museum collections, published and unpublished reports, communications from knowledgeable individuals, and field surveys. A sample EOR and field descriptions are provided in Appendix 2. The most relevant fields of the EOR describe location, precision, federal and

state agency endangerment status, global and local status and rarity ranking index, issues related to management needs and threats to the occurrence, and landownership and tenure. They also provide dates of first and most current observations, a site-specific status report, and cross-references to source citations for information in the record.

MANUAL MAPPING CONVENTIONS

Manual maps of HINHP use standard USGS 7.5-minute topographic quadrangles as a base, upon which numbered mapped symbols are affixed. The conventions described here are summarized in Appendix 3. Map symbols are segregated by color, corresponding to four rare element types: plants (yellow), vertebrates (orange), invertebrates (purple), and ecosystems/natural communities (green). Shapes of map symbols indicate relative precision of the source information extracted and used to map the location: circles indicate the most precise location information and represent confirmation of location to at least a map second (1/360 degree, roughly 0.16-mile radius); triangles indicate location within minute precision (roughly a 0.75-mile radius), while squares indicate the most general precision (approximately quad size, with a radius of about 5 miles). Each map symbol is uniquely numbered in sequence per quadrangle, connecting mapped symbols to database records for that quadrangle. Unique database codes assigned to each map symbol identify the type of element and specific taxon and are arranged according to an international Heritage database network convention.

GEOGRAPHIC INFORMATION SYSTEM (GIS) CONVENTIONS

The HINHP GIS uses a UNIX workstation running ArcINFO GIS software. The workstation is networked to a Pentium PC running ArcCAD, AutoCAD, and ArcVIEW in a menu-integrated system. A D-scale plotter provides for large-format map output, while a 90 × 120 centimeter digitizing table is used for full USGS quad scale input. A 24-bit color flatbed scanner allows for image input (opaque or transparency sources) to attach as attributes for mapped locations. A wide variety of symbols and colors are available to distinguish mapped information (see Figure 22.1 for the GIS system set-up).

STRENGTHS OF THE SYSTEM

One of the main strengths of the mapping convention of HINHP is that databases are arranged primarily by location. Location-wise queries of the database and mapping system allow landowners, managers, and a wide variety of scientific and other users to address geographically bound information rapidly. Taxon or ecosystem-specific summaries are also available, indicating rapidly the global range for a particular element. As the single, most comprehensive source for rare element location and status information, HINHP provides assessments rapidly, especially in requests for summaries of the biological significance of a particular area. HINHP anticipates agency-protective action by tracking taxa considered rare by the scientific community despite lack of state/federal listing status. Thus, when agencies decide to take formal action in the listing process, HINHP has generally already compiled much of the salient information needed to facilitate the task. A highly useful feature of each location record is that the information source or

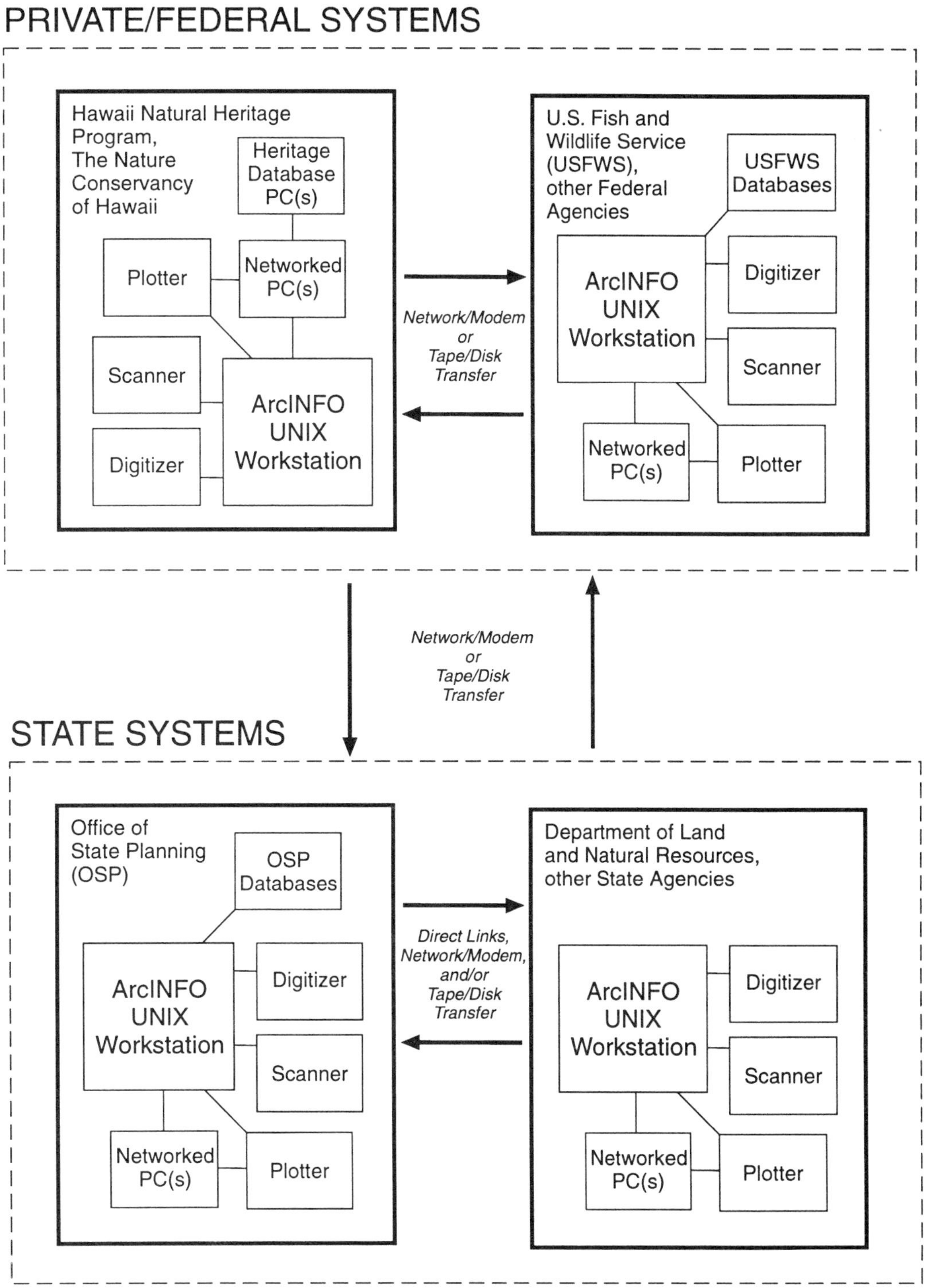

PRIVATE/FEDERAL SYSTEMS

Hawaii Natural Heritage Program, The Nature Conservancy of Hawaii

Heritage Database PC(s)

Plotter

Networked PC(s)

Scanner

Digitizer

ArcINFO UNIX Workstation

Network/Modem or Tape/Disk Transfer

U.S. Fish and Wildlife Service (USFWS), other Federal Agencies

USFWS Databases

ArcINFO UNIX Workstation

Digitizer

Scanner

Networked PC(s)

Plotter

Network/Modem or Tape/Disk Transfer

STATE SYSTEMS

Office of State Planning (OSP)

OSP Databases

ArcINFO UNIX Workstation

Digitizer

Scanner

Networked PC(s)

Plotter

Direct Links, Network/Modem, and/or Tape/Disk Transfer

Department of Land and Natural Resources, other State Agencies

ArcINFO UNIX Workstation

Digitizer

Scanner

Networked PC(s)

Plotter

Figure 22.1 System design for Hawaii Natural Heritage Program database/GIS in a multiparticipant GIS

sources used to map and summarize the information accompany each record. HINHP is widely available to users and restricts data only to the extent needed to protect particularly sensitive taxa or ecosystems. For example, endangered land-snail locations are generalized in public records, while detailed location information is held for qualified researchers and managers. A listing of major users of the HINHP database and a report summary is provided in Appendix 4.

LIMITATIONS OF THE SYSTEM

HINHP database updates are labor-intensive and involve quality control that requires review by staff biologists. The sheer bulk of taxa and location records that need to be updated dictates that the 1,000+ elements already tracked require updating every two years. To obtain maximum use, the information in HINHP records requires a basic level of expertise/familiarity with rare species and ecosystems. Thus, agencies with a mandate for working with rare taxa have been the most effective users of HINHP records. Another limitation is external to HINHP—much of the information relies on third-party data (published reports, field notes, and observations by individuals) that sometimes require confirmation. An important limitation of HINHP-mapped data is that only known locations can be reported. Empty spots on the HINHP maps may truly be devoid of rare elements but may be relatively unexplored and contain highly significant, undocumented biological resources. Presuming that an empty spot on an HINHP map is lacking significant resources is one obvious way that HINHP information can be misused or abused if its caveats and limitations are ignored. The standard caveat statement for HINHP products and reports is as follows:

Special Notice to Developers and Consultants

The quantity and quality of data collected by the Heritage Program are dependent on the research and observations of many individuals and organizations. In most cases, this information is not the result of comprehensive or site-specific field surveys and it has not been confirmed by the Heritage staff. Many natural areas in Hawaii have never been thoroughly surveyed, and new species of plants and animals are still being discovered. For these reasons, the Heritage Program cannot provide a definitive statement on the presence, absence or condition of biological elements in any part of Hawaii. Heritage reports summarize the existing information known to the program at the time of the request regarding the biological elements or location in question. These reports should never be regarded as final statements on the elements or areas being considered, nor should they be substituted for on-site surveys required for environmental assessments. Furthermore, Heritage reports do not represent or imply a position or policy taken by The Nature Conservancy of Hawaii on any related matter. If information from a Hawaii Heritage Program report is distributed in any way, the above statement must accompany that information.

Finally, a major limitation in the HINHP coverage for Hawaii is the complete lack of a marine component and incomplete coverage of an emerging set of rare invertebrate taxa.

LINKAGE AND COMPATIBILITY

The HINHP GIS is compatible with those in use or being developed by the state (Office of State Planning) and the federal government (U.S. Fish and Wildlife Service, Department of Defense). The database system used (Advanced Revelation) allows for import/export of data with dBASE and other emerging database standards. The UNIX-PC network allows data to be shared in both workstation and PC environments. Modem links to exchange data have been accomplished with other GIS systems and are envisioned as a method to link HINHP data with both state and federal colleagues.

RECOMMENDATIONS

On establishing a database and mapping system to track resources, there are several specific recommendations. Involve an advisory committee of pertinent individuals, agencies, and organizations to help guide the development of the databases and mapping systems and set information priorities. Explore the scope of other related database and mapping efforts and strive for complementarity. Continuously assess the true cost of database maintenance beyond start-up (a worthwhile database requires a long-term investment that is not unsubstantial!). Continuously assess the users of the database and work to service both their short- and long-term needs.

APPENDIX 1

List of Rare Taxa and Types of Ecosystems

<u>Natural Communities</u>

Acacia koa Lowland Mesic Forest
Acacia koa/Metrosideros polymorpha Lowland Mesic Forest
Acacia koa/Metrosideros polymorpha Lowland Wet Forest
Acacia koa/Metrosideros polymorpha Mixed Montane Mesic Forest
Acacia koa/Metrosideros polymorpha Montane Wet Forest
Acacia koa/Metrosideros polymorpha/Sapindus saponaria Montane Mesic Forest
Acacia koa/Santalum paniculatum Montane Mesic Forest
Acacia koa/Sophora chrysophylla Montane Dry Forest
Acacia koaia Lowland Dry Forest

Bidens menziesii Lowland Dry Shrubland
Bidens sp. Lowland Dry Shrubland
Bulboschoenus sp./*Schoenoplectus* spp./*Cyperus laevigatus* Coastal Wet Grassland

Canthium odoratum Lowland Dry Shrubland
Canthium odoratum Mixed Lowland Dry Shrubland
Carex alligata Montane Wet Grassland
Chamaesyce celastroides Montane Dry Forest
Chamaesyce olowaluana Montane Dry Forest
Chenopodium oahuense Mixed Coastal Dry Shrubland
Chenopodium oahuense Subalpine Dry Shrubland

Deschampsia nubigena Subalpine Mesic Grassland
Dicranopteris sp. Lowland Wet Shrubland
Diospyros sandwicensis Lowland Dry Forest
Diospyros sandwicensis Lowland Mesic Forest
Diospyros sandwicensis/Colubrina oppositifolia Lowland Dry Forest

Diospyros sandwicensis/*Metrosideros polymorpha* Lowland Mesic Forest
Dodonaea spp./*Styphelia tameiameiae* Lowland Mesic Shrubland
Dodonaea viscosa Lowland Dry Shrubland

East Maui Diverse Montane Mesic Forest
Eragrostis atropioides Subalpine Dry Grassland
Eragrostis variabilis Coastal Dry Grassland
Eragrostis variabilis Lowland Mesic Grassland
Erythrina sandwicensis Lowland Dry Forest

Fimbristylis sp. Coastal Dry Grassland

Gossypium tomentosum Coastal Dry Shrubland
Gyroxiphium sandwicense/*Dubautia menziesii* Alpine Dry Shrubland

Hawaiian Alpine Aeolian Desert
Hawaiian Alpine Lake
Hawaiian Coastal Hypersaline Lake
Hawaiian Coastal Mesic Boulder Community
Hawaiian Continuous Perennial Stream
Hawaiian Estuary
Hawaiian Fumarole Communities
Hawaiian Intermittent Stream
Hawaiian Mixed Shrub Coastal Dry Cliff
Hawaiian Montane Lake
Hawaiian Springs & Seep
Heliotropium anomalum Coastal Dry Shrubland
Heteropogon contortus Lowland Dry Grassland
High Salinity Lava Anchialine Pool
High Salinity Lava Tube Anchialine Pool
High Salinity Limestone Anchialine Pool

Kauai Diverse Lowland Mesic Forest

Lanai Diverse Lowland Mesic Forest
Lepturus repens Coastal Dry Grassland
Lipochaeta integrifolia Coastal Dry Shrubland
Lipochaeta sp. Coastal Mesic Shrubland
Low Salinity Lava Anchialine Pool
Low Salinity Lava Tube Anchialine Pool
Low Salinity Limestone Anchialine Pool
Lowland Brackish Lake
Lowland Freshwater Lake

Marsilea villosa Lowland Dry Herbland
Meioneta gagnei/*Hawaiioscia parvituberculata* Lowland Dry Cave
Metrosideros polymorpha Lowland Dry Forest
Metrosideros polymorpha Lowland Mesic Forest
Metrosideros polymorpha Lowland Mesic Shrubland
Metrosideros polymorpha Mixed Montane Bog
Metrosideros polymorpha Mixed Montane Dry Forest
Metrosideros polymorpha Montane Mesic Forest
Metrosideros polymorpha Montane Wet Shrubland
Metrosideros polymorpha Subalpine Dry Forest
Metrosideros polymorpha/*Cheirodendron* sp. Lowland Wet Forest
Metrosideros polymorpha/*Cheirodendron* spp. Montane Wet Forest

METROSIDEROS POLYMORPHA/CIBOTIUM SPP. LOWLAND WET FOREST
METROSIDEROS POLYMORPHA/CIBOTIUM SPP. MONTANE WET FOREST
METROSIDEROS POLYMORPHA/DICRANOPTERIS SPP. LOWLAND MESIC FOREST
METROSIDEROS POLYMORPHA/DICRANOPTERIS SPP. LOWLAND WET SHRUBLAND
METROSIDEROS POLYMORPHA/DICRANOPTERIS SPP. MONTANE WET FOREST
METROSIDEROS POLYMORPHA/DODONAEA ERIOCARPA MONTANE DRY SHRUBLAND
METROSIDEROS POLYMORPHA/FREYCINETIA ARBOREA LOWLAND WET FOREST
*METROSIDEROS POLYMORPHA/*MIXED SHRUB MONTANE WET FOREST
METROSIDEROS POLYMORPHA/RHYNCHOSPORA SP./*DICRANOPTERIS* LOWLAND WET BOG
MIXED FERN/SHRUB MONTANE WET CLIFFS
MIXED SEDGE AND GRASS MONTANE BOG
MYOPORUM SANDWICENSE COASTAL DRY SHRUBLAND

NAMA SANDWICENSIS COASTAL DRY HERBLAND
NESTEGIS SANDWICENSIS LOWLAND MESIC FOREST
NESTEGIS SANDWICENSIS MONTANE MESIC FOREST
NESTEGIS SANDWICENSIS/DIOSPYROS SANDWICENSIS LOWLAND DRY FOREST

OAHU DIVERSE LOWLAND MESIC FOREST
OLIARUS LORETTAI COASTAL DRY CAVE
OLIARUS PRIOLA/CACENOMOBIUS HOWARTHI LOWLAND WET CAVE
OSTEOMELES ANTHYLLIDIFOLIA LOWLAND MESIC SHRUBLAND

PANDANUS TECTORIUS COASTAL MESIC FOREST
PIONEER VEGETATION ON LAVA FLOWS
PIPTURUS SPP. LOWLAND WET SHRUBLAND
PISONIA SPP./*CHARPENTIERA* SPP. LOWLAND MESIC FOREST
PLEOMELE AUWAHIENSIS LOWLAND MESIC FOREST
PRITCHARDIA HILLEBRANDII COASTAL MESIC FOREST
PRITCHARDIA KAALAE LOWLAND MESIC FOREST
PRITCHARDIA MARTII LOWLAND WET FOREST
PRITCHARDIA REMOTA COASTAL MESIC FOREST

RACOMITRIUM LANUGINOSUM MONTANE BOG

RUBUS HAWAIIENSIS MONTANE WET SHRUBLAND

SANTALUM ELLIPTICUM COASTAL DRY SHRUBLAND
SAPINDUS OAHUENSIS LOWLAND DRY FOREST
SCAEVOLA CORIACEA MIXED COASTAL DRY SHRUBLAND
SCAEVOLA SERICEA COASTAL DRY SHRUBLAND
SESBANIA TOMENTOSA LOWLAND DRY SHRUBLAND
SESUVIUM PORTULACASTRUM COASTAL DRY HERBLAND
SIDA FALLAX COASTAL DRY SHRUBLAND
SOPHORA CHRYSOPHYLLA SUBALPINE DRY FOREST
SOPHORA CHRYSOPHYLLA/MYOPORUM SANDWICENSE SUBALPINE DRY FOREST
SPOROBOLUS VIRGINICUS COASTAL DRY GRASSLAND
STYPHELIA TAMEIAMEIAE SUBALPINE DRY SHRUBLAND

TETRAMOLOPIUM ROCKII COASTAL DRY SHRUBLAND
TRIBULUS CISTOIDES COASTAL DRY SHRUBLAND

UNCHARACTERIZED ANCHIALINE POOL

WIKSTROEMIA SPP./*DODONAEA* SPP./*OSTEOMELES ANTHYLLIDIFOLIA* LOWLAND DRY SHRUBLAND

Plants

Abutilon eremitopetalum
A. menziesii
A. sandwicense
Acacia koaia
Acaena exigua
Achyranthes atollensis
A. mutica
A. splendens var. *rotundata*
A. splendens var. *splendens*
Adenophorus periens
Alectryon macrococcus var. *auwahiensis*
A. macrococcus var. *macrococcus*
Alsinidendron lychnoides
A. obovatum
A. trinerve
A. viscosum
Amaranthus brownii
Argyroxiphium caliginis
A. kauense
A. sandwicense ssp. *macrocephalum*
A. sandwicense ssp. *sandwicense*
A. virescens
Artemisia mauiensis
Asplenium fragile var. *insulare*
A. hobdyi
A. leucostegioides
A. schizophyllum
A. sp. 3
A. sp. 4
A. sp. 5

Bidens campylotheca ssp. *campylotheca*
B. campylotheca ssp. *pentamera*
B. campylotheca ssp. *waihoiensis*
B. conjuncta
B. forbesii ssp. *kahiliensis*
B. micrantha ssp. *ctenophylla*
B. micrantha ssp. *kalealaha*
B. molokaiensis
B. populifolia
B. sandvicensis ssp. *confusa*
B. wiebkei
Bobea sandwicensis
B. timonioides
Bonamia menziesii
Botrychium subbifoliatum
Brighamia insignis
B. rockii

Caesalpinia kavaiensis
Calamagrostis expansa
C. hillebrandii
Canavalia molokaiensis
C. napaliensis

C. pubescens
Capparis sandwichiana
Carex wahuensis ssp. *herbstii*
Cenchrus agrimonioides var. *agrimonioides*
C. agrimonioides var. *laysanensis*
Centaurium sebaeoides
Chamaesyce arnottiana
C. atrococca
C. celastroides var. *kaenana*
C. celastroides var. *laehiensis*
C. celastroides var. *stokesii*
C. celastroides var. *tomentella*
C. deppeana
C. halemanui
C. herbstii
C. kuwaleana
C. olowaluana
C. remyi var. *hanaleiensis*
C. remyi var. *kauaiensis*
C. remyi var. *remyi*
C. skottsbergii var. *skottsbergii*
C. skottsbergii var. *vaccinioides*
C. sp. 1
C. sparsiflora
Charpentiera densiflora
Cheirodendron dominii
Cleome spinosa ssp. 1
Clermontia arborescens ssp. *arborescens*
C. calophylla
C. drepanomorpha
C. lindseyana
C. multiflora
C. oblongifolia ssp. *brevipes*
C. oblongifolia ssp. *mauiensis*
C. peleana
C. peleana ssp. *peleana*
C. peleana ssp. *singuliflora*
C. pyrularia
C. samuelii ssp. *hanaensis*
C. samuelii ssp. *samuelii*
C. tuberculata
C. waimeae
Colubrina oppositifolia
Ctenitis squamigera
Cyanea arborea
C. asarifolia
C. asplenifolia
C. comata
C. copelandii ssp. *copelandii*
C. copelandii ssp. *haleakalaensis*
C. crispa
C. dunbarii
C. eleelaensis
C. giffardii
C. glabra
C. grimesiana ssp. *cylindrocalyx*

C. grimesiana ssp. grimesiana
C. grimesiana ssp. obatae
C. hamatiflora ssp. carlsonii
C. horrida
C. humboldtiana
C. kolekoleensis
C. koolauensis
C. kunthiana
C. lanceolata ssp. calycina
C. leptostegia
C. linearifolia
C. lobata
C. longiflora
C. longissima
C. macrostegia ssp. gibsonii
C. mannii
C. marksii
C. mceldowneyi
C. membranacea
C. obtusa
C. parvifolia
C. pinnatifida
C. platyphylla
C. pohaku
C. procera
C. profuga
C. purpurellifolia
C. pycnocarpa
C. quercifolia
C. recta
C. remyi
C. scabra
C. shipmanii
C. solanacea
C. solenocalyx
C. sp. 1
C. st.-johnii
C. stictophylla
C. superba ssp. regina
C. superba ssp. superba
C. tritomantha
C. truncata
C. undulata
Cyperus trachysanthos
Cyrtandra biserrata
C. crenata
C. cyaneoides
C. dentata
C. filipes
C. giffardii
C. gracilis
C. halawensis
C. hematos
C. kamooloaensis
C. kaulantha
C. kealiae

C. kohalae
C. limahuliensis
C. lydgatei
C. macrocalyx
C. munroi
C. oenobarba
C. olona
C. oxybapha
C. pickeringii
C. polyantha
C. pruinosa
C. sandwicensis
C. sessilis
C. subumbellata
C. tintinnabula
C. viridiflora
C. waiolani
Cystopteris douglasii

Delissea fallax
D. laciniata
D. lauliiana
D. parviflora
D. rhytidosperma
D. rivularis
D. sinuata ssp. lanaiensis
D. sinuata ssp. sinuata
D. subcordata
D. undulata ssp. kauaiensis
D. undulata ssp. niihauensis
D. undulata ssp. undulata
Diellia erecta
D. falcata
D. laciniata
D. mannii
D. unisora
Diplazium molokaiense
Dissochondrus biflorus
Doodia lyonii
Dryopteris sp. 1
D. sp. 2
D. sp. 3
Dubautia arborea
D. herbstobatae
D. imbricata ssp. acronaea
D. imbricata ssp. imbricata
D. knudsenii ssp. filiformis
D. knudsenii ssp. knudsenii
D. knudsenii ssp. nagatae
D. laevigata
D. latifolia
D. microcephala
D. pauciflorula
D. plantaginea ssp. humilis
D. sherffiana
D. waialealeae

Eragrostis deflexa
E. fosbergii
E. hosakai
E. mauiensis
Eugenia koolauensis
Euphorbia haeleeleana
Eurya sandwicensis
Exocarpos gaudichaudii
E. luteolus

Festuca hawaiiensis
Fimbristylis hawaiiensis
Flueggea neowawraea

Gahnia lanaiensis
Gardenia brighamii
G. mannii
G. remyi
Geranium arboreum
G. hanaense
G. humile
G. kauaiense
G. multiflorum
Gnaphalium sandwicensium var. molokaiense
Gouania hillebrandii
G. meyenii
G. vitifolia

Haplostachys bryanii
H. haplostachya
H. linearifolia
H. munroi
H. truncata
Hedyotis cookiana
H. coriacea
H. degeneri var. coprosmifolia
H. degeneri var. degeneri
H. elatior
H. fluviatilis
H. foliosa
H. formosa
H. littoralis
H. mannii
H. parvula
H. schlechtendahliana var. remyi
H. st.-johnii
H. tryblium
Hesperocnide sandwicensis
Hesperomannia arborescens
H. arbuscula
H. lydgatei
Hibiscadelphus bombycinus
H. crucibracteatus
H. distans
H. giffardianus
H. hualalaiensis

H. sp. 1
H. wilderianus
Hibiscus arnottianus ssp. immaculatus
H. brackenridgei ssp. brackenridgei
H. brackenridgei ssp. mokuleianus
H. clayi
H. kokio ssp. kokio
H. kokio ssp. saintjohnianus
H. waimeae ssp. hannerae
Huperzia mannii
H. sulcinervia

Ischaemum byrone
Isodendrion hosakae
I. laurifolium
I. longifolium
I. pyrifolium
Isoetes sp. 3

Joinvillea ascendens ssp. ascendens

Kokia cookei
K. drynarioides
K. kauaiensis
K. lanceolata
Korthalsella degeneri

Labordia cyrtandrae
L. helleri
L. kaalae
L. lydgatei
L. pumila
L. tinifolia var. lanaiensis
L. tinifolia var. wahiawaensis
L. triflora
Lagenifera erici
L. helenae
L. maviensis
Lepidium arbuscula
L. bidentatum var. o-waihiense
L. bidentatum var. remyi
L. serra
Lindsaea repens var. macraeana
Lipochaeta bryanii
L. degeneri
L. fauriei
L. kamolensis
L. lobata var. leptophylla
L. lobata var. lobata
L. micrantha var. exigua
L. micrantha var. micrantha
L. ovata
L. perdita
L. remyi
L. tenuifolia
L. tenuis

L. venosa
L. waimeaensis
Lobelia dunbarii ssp. *dunbarii*
L. dunbarii ssp. *paniculata*
L. gaudichaudii ssp. *koolauensis*
L. hypoleuca
L. monostachya
L. niihauensis
L. oahuensis
L. remyi
L. yuccoides
Lycopodium nutans
Lysimachia daphnoides
L. filifolia
L. forbesii
L. glutinosa
L. kalalauensis
L. lydgatei
L. maxima
L. venosa

Mariscus fauriei
M. kunthianus
M. pennatiformis ssp. *bryanii*
M. pennatiformis ssp. *pennatiformis*
M. rockii
Marsilea villosa
Melicope adscendens
M. balloui
M. christophersenii
M. cinerea
M. cruciata
M. degeneri
M. haleakalae
M. haupuensis
M. hawaiensis
M. knudsenii
M. lydgatei
M. macropus
M. makahae
M. mucronulata
M. munroi
M. nealae
M. obovata
M. orbicularis
M. ovalis
M. pallida
M. paniculata
M. puberula
M. quadrangularis
M. reflexa
M. saint-johnii
M. sandwicensis
M. sp. 1
M. sp. 2
M. waialealae

M. wailauensis
M. zahlbruckneri
Munroidendron racemosum
Myrsine fosbergii
M. juddii
M. knudsenii
M. linearifolia
M. mezii
M. petiolata
M. vaccinioides

Neraudia angulata var. *angulata*
N. angulata var. *dentata*
N. kauaiensis
N. melastomifolia
N. ovata
N. sericea
Nesoluma polynesicum
Nothocestrum breviflorum
N. latifolium
N. peltatum
Nototrichium humile

Ochrosia haleakalae
O. kauaiensis
O. kilaueaensis
Ophioglossum concinnum

Panicum beecheyi
P. fauriei var. *carteri*
P. lineale
P. niihauense
Peperomia degeneri
P. rockii
P. subpetiolata
Peucedanum sandwicense
Phyllostegia bracteata
P. brevidens
P. floribunda
P. glabra var. *lanaiensis*
P. helleri
P. hillebrandii
P. hirsuta
P. imminuta
P. kaalaensis
P. knudsenii
P. mannii
P. mollis
P. parviflora var. 1
P. parviflora var. *glabriuscula*
P. parviflora var. *parviflora*
P. racemosa
P. rockii
P. stachyoides
P. variabilis
P. velutina

P. vestita
P. waimeae
P. warshaueri
P. wawrana
Pisonia wagneriana
Pittosporum napaliense
Plantago hawaiensis
P. princeps var. *anomala*
P. princeps var. *laxiflora*
P. princeps var. *longibracteata*
P. princeps var. *princeps*
Platanthera holochila
Platydesma cornuta var. *cornuta*
P. cornuta var. *decurrens*
P. remyi
P. rostrata
Pleomele forbesii
P. hawaiiensis
Poa mannii
P. sandvicensis
P. siphonoglossa
Portulaca molokiniensis
P. sclerocarpa
P. sp. 1
P. villosa
Pritchardia affinis
P. aylmer-robinsonii
P. glabrata
P. hardyi
P. kaalae
P. lanigera
P. lowreyana
P. munroi
P. napaliensis
P. remota
P. schattaueri
P. viscosa
Psychotria grandiflora
P. hexandra var. *hosakana*
P. hexandra var. *oahuensis*
P. hexandra var. *rockii*
P. hobdyi
Pteralyxia kauaiensis
P. macrocarpa
Pteris lydgatei

Ranunculus hawaiensis
R. mauiensis
Remya kauaiensis
R. mauiensis
R. montgomeryi
Rubus macraei

Sanicula kauaiensis
S. mariversa
S. purpurea

S. sandwicensis
Santalum freycinetianum var. *lanaiense*
Scaevola coriacea
S. kilaueae
S. sp. 1
S. sp. 2
Schiedea adamantis
S. amplexicaulis
S. apokremnos
S. diffusa
S. globosa
S. haleakalensis
S. helleri
S. hookeri
S. implexa
S. kaalae
S. kealiae
S. ligustrina
S. lydgatei
S. mannii
S. membranacea
S. menziesii
S. nuttallii
S. pubescens var. *pubescens*
S. pubescens var. *purpurascens*
S. salicaria
S. spergulina var. *leiopoda*
S. spergulina var. *spergulina*
S. stellarioides
S. verticillata
Sesbania tomentosa
Sicyos alba
S. cucumerinus
S. erostratus
S. hillebrandii
S. macrophyllus
S. sp. 1
S. waimanaloensis
Silene alexandri
S. cryptopetala
S. degeneri
S. hawaiiensis
S. lanceolata
S. perlmanii
Solanum incompletum
S. nelsonii
S. sandwicense
Spermolepis hawaiiensis
Stenogyne angustifolia
S. bifida
S. campanulata
S. cinerea
S. cranwelliae
S. haliakalae
S. kanehoana
S. kealiae

S. macrantha
S. microphylla
S. oxygona
S. scrophularioides
S. viridis

Tetramolopium arenarium var. *arenarium*
T. arenarium var. *confertum*
T. capillare
T. consanguineum ssp. *consanguineum*
T. consanguineum var. *kauense*
T. consanguineum var. *leptophyllum*
T. conyzoides
T. filiforme var. *filiforme*
T. filiforme var. *polyphyllum*
T. humile var. *sublaeve*
T. lepidotum ssp. *arbusculum*
T. lepidotum ssp. *lepidotum*
T. remyi
T. rockii var. *calcisabulorum*
T. rockii var. *rockii*
T. sylvae
T. tenerrimum
Tetraplasandra gymnocarpa
Thelypteris boydiae
Torulinium odoratum ssp. *auriculatum*
Trematolobelia grandifolia
T. singularis
Trisetum inaequale

Urera kaalae

Vicia menziesii
Vigna o-wahuensis
Viola chamissoniana ssp. *chamissoniana*
V. helenae
V. kauaensis var. *wahiawaensis*
V. lanaiensis
V. oahuensis

Wikstroemia bicornuta
W. hanalei
W. skottsbergiana
W. villosa
Wilkesia hobdyi

Xylosma crenatum

Zanthoxylum dipetalum var. *tomentosum*
Z. hawaiiense
Z. oahuense

Invertebrates

Achatinella abbreviata
A. apexfulva
A. bellula
A. bulimoides
A. byronii
A. cestus
A. concavospira
A. curta
A. decipiens
A. dimorpha
A. elegans
A. fulgens
A. fuscobasis
A. juddii
A. juncea
A. leucorraphe
A. lila
A. lorata
A. mustelina
A. phaeozona
A. pulcherrima
A. pupukanioe
A. rosea
A. sowerbyana
A. stewartii
A. swiftii
A. taeniolata
A. turgida
A. valida
A. viridans
A. vittata
A. vulpina
Adelocosa anops
Agrotis crinigera
A. fasciata
A. kerri
A. laysanensis
A. procellaris
Amastra aemulator
A. affinis
A. agglutinans
A. albocincta
A. albolabris
A. amicta
A. anthonii
A. assimilis
A. aurostoma
A. badia
A. baldwiniana
A. biplicata
A. borcherdingii
A. breviata
A. conica
A. conifera

A. cookei
A. cornea
A. crassilabrum
A. cyclostoma
A. cylindrica
A. davisiana
A. decorticata
A. delicata
A. durandi
A. dwightii
A. elegantula
A. elephantina
A. elliptica
A. elongata
A. eos
A. erecta
A. farcimen
A. flavescens
A. forbesi
A. fragilis
A. fraterna
A. frosti
A. globosa
A. goniops
A. gouveii
A. grayana
A. gulickana
A. hawaiiensis
A. humilis
A. hutchinsonii
A. implicata
A. inflata
A. inopinata
A. intermedia
A. irwiniana
A. johnsoni
A. juddii
A. kalamaulensis
A. kauaiensis
A. kaunakakaiensis
A. knudseni
A. laeva
A. lahainana
A. lineolata
A. luctuosa
A. magna
A. makawaoensis
A. malleata
A. mastersi
A. melanosis
A. metamorpha
A. micans
A. mirabilis
A. modesta
A. modicella
A. moesta

A. montagui
A. montana
A. montivaga
A. mucronata
A. nana
A. nannodes
A. neglecta
A. nigra
A. nubifera
A. nubigena
A. nubilosa
A. nucleola
A. nucula
A. obesa
A. oswaldi
A. peasei
A. petricola
A. pilsbryi
A. porcus
A. porphyrea
A. porphyrostoma
A. praeopima
A. problematica
A. pullata
A. pusilla
A. reticulata
A. ricei
A. rubens
A. rubida
A. rubristoma
A. rugulosa
A. seminigra
A. seminuda
A. sericea
A. sola
A. solida
A. soror
A. spaldingi
A. sphaerica
A. spicula
A. spirizona
A. subcrassilabris
A. subobscura
A. subrostrata
A. subsoror
A. sykesi
A. tenuilabris
A. tenuispira
A. textilis
A. thaanumi
A. transversalis
A. tricincta
A. tristis
A. turritella
A. ultima
A. umbilicata

A. undata
A. uniplicata
A. variegata
A. violacea
A. viriosa
A. whitei
Apterocychus honoluluensis
Armsia petasus
Atyoida bisulcata
Auriculella aff. *castanea* n. sp. 1
A. aff. *perpusilla* n. sp. 1
A. ambusta
A. amoena
A. auricula
A. brunnea
A. canalifera
A. castanea
A. cerea
A. crassula
A. diaphana
A. expansa
A. flavida
A. kuesteri
A. lanaiensis
A. malleata
A. minuta
A. montana
A. newcombi
A. olivacea
A. perpusilla
A. perversa
A. petitiana
A. pulchra
A. serrula
A. straminea
A. tantalus
A. tenella
A. tenuis
A. turritella
A. uniplicata
A. westerlundiana

Banza nihoa
Bryania bipunctata

Caconemobius howarthi
C. schauinslandi
C. varius
Campsicnemus mirabilis
Carelia bicolor
C. kalalauensis
C. tenebrosa
C. turricula
Cavaticovelia aaa
Chersodromia hawaiiensis
Conocephaloides Remotus

Cookeconcha sp. 1
Cyrtopeltis phyllostegiae

Deinocossonus nesiotes
Deinomimesa hawaiiensis
D. punae
Drosophila lanaiensis

Ectemnius curtipes
E. fulvicrus
E. giffardi
E. haleakalae
E. nesocrabo bidecoratus
Eidoleon perjurus
Empicoris pulchrus
Endodonta sp. 1
Eopenthes spp.
Eupelmus nihoaensis

Fletcherana ioxantha

Gulickia alexandri

Hedylepta anastrepta
H. anastreptoides
H. asaphombra
H. epicentra
H. euryprora
H. fullawayi
H. giffardi
H. iridias
H. laysanensis
H. meyricki
H. monogona
H. musicola
H. pritchardii
H. telegrapha
Heliothis confusa
H. minuta
Heteramphus filicum
Heterocrossa viridis
Holcobius pikoensis
Hypena laysanensis
H. newelli
H. plagiota
H. senicula

Ithamar annectans
I. hawaiiensi
Itodacnus sp. 1
I. sp. 2

Kalania hawaiiensis
K. sp. 1

Laminella alexandri

L. aspera
L. bulbosa
L. citrina
L. concinna
L. gravida
L. kuhnsi
L. picta
L. remyi
L. sanguinea
L. straminea
L. tetrao
L. venusta
Leptachatina lepida
L. sp. 1
L. sp. 2
L. sp. 3
L. sp. 4
L. sp. 5
L. sp. 7
L. sp. 8
Leptogryllus deceptor
Lyropupa perlonga
L. sp. 1

Machiloides heteropus
M. perkinsi
Macrobrachium grandimanus
Manduca blackburni
Margaronia cyanomichla
M. exaula
Megalagrion adytum
M. amaurodytum fallax
M. amaurodytum peles
M. amaurodytum waianaenum
M. jugorum
M. leptodemus
M. molokaiense
M. nesiotes
M. nigrohamatum
M. nigrolineatum
M. oahuensis
M. oceanicum
M. pacificum
M. xanthomelas
Metabetaeus lohena
Metrarga obscura

Neritina granosa
Neseis alternatus
N. haleakalae
Nesidiolestes ana
N. insularis
N. roberti
N. selium
Nesocryptias villosa
Nesomimesa kauaiensis

N. sciopteryx
Nesoprosopis andrenoides
N. angustula
N. anomala
N. anthricina
N. assimulans
N. blackburni
N. caeruleipennis
N. chlorosticata
N. comes
N. coniceps
N. connectens
N. crabronoides
N. difficilis
N. dimidiata
N. erythrodemas
N. facilis
N. filicum
N. finitima
N. flavifrons
N. flavipes
N. fuscipennis
N. fuscipennis obscuripes
N. haleakalae
N. hilaris
N. hirsutula
N. homeochroma
N. hostilis
N. hula
N. insiginis
N. kauaiensis
N. koae
N. kona
N. laeta
N. laticeps
N. longiceps
N. mauiensis
N. melanothrix
N. mutata
N. neglecta
N. nivalis
N. obscurata
N. ombrias
N. pele
N. perkinsiana
N. perspicua
N. psammobia
N. pubescens
N. rubrocaudatus
N. rugulosa
N. satellus
N. setosifrons
N. simplex
N. specularis
N. sphecodoides
N. unica

N. vicina
N. volatilis
Nesorestias filicicola
Nesosydne acuta
N. bridwelli
N. cyrtandrae
N. cyrtandricola
N. kuschei
N. leahi
N. longipes
N. sulcata
Nesothauma haleakalae
Nesotocus giffordi
N. kauaiensis
N. munroi
Newcombia canaliculata
N. carinella
N. cinnamomea
N. cumingi
N. perkinsi
N. pfeifferi
N. philippiana
N. plicata
N. sulcata
Nysius frigatensis
N. fullawayi
N. neckerensis
N. nihoae
N. suffusus

Oceanides bryani
O. perkinsi
O. rugosiceps
Odynerus niihauensis
O. soror
Oeobia dryadopa
Oliarus consimilis
O. discrepans
O. lanaiensis
O. lihue
O. myoporicola
O. priola
Oodemas sp. 1
O. sp. 2
O. sp. 3
O. sp. 4
Orobophana baldwini
O. meineckei
O. uberta

Partulina anceyana
P. carnicolor
P. confusa
P. crassa
P. crocea
P. dolei

P. dubia
P. dwightii
P. fusoidea
P. germana
P. gouldii
P. grisea
P. horneri
P. induta
P. kaaeana
P. lemmoni
P. marmorata
P. mighelsiana
P. mucida
P. mutabilis
P. nattii
P. nivea
P. perdix
P. physa
P. plumbea
P. porcellana
P. proxima
P. radiata
P. redfieldii
P. rufa
P. semicarinata
P. splendida
P. subpolita
P. tappaniana
P. terebra
P. tessellata
P. thaanumiana
P. theodorei
P. thwingii
P. ustulata
P. variabilis
P. virgulata
P. winniei
Pentarthrum blackburni
P. obscurum
Perdicella carinella
P. fulgurans
P. helena
P. kuhnsi
P. mauiensis
P. ornata
P. zebra
P. zebrina
Petrochroa neckerensis
Plagithymysus spp.
Pleuropoma hawaiiensis
P. kauaiensis
P. nonouensis
P. oahuensis
P. rotelloidea
P. sandwichiensis
P. sulculosa

Proterhinus spp.
Pseudopsectra cookeorum
P. lobipennis
P. swezeyi
P. usingeri
Pterodiscus alatus
P. cookei
P. discus
P. heliciformis
P. rex
P. thaanumi
P. wesleyi

Rhyncogonus spp.

Sclerodermus nihoaensis
Scotorythra megalophylla
S. nesiotes
S. paratactis
Siacella smithi
Spelaeorchestia koloana
Spheterista oheoheana
S. pterotropiana
S. reynoldsiana
Stenotrupis pritchardiae
Strymon bazochii

Thaumatogryllus cavicola
T. variegatus
Tinostoma smaragditis
Tritocleis microphylla

Vaga blackburni

Zorotypus swezeyi

Vertebrates

Acrocephalus familiaris kingi
Anas laysanensis
A. wyvilliana
Asio flammeus sandwichensis
Awaous stamineus

Branta sandvicensis
Buteo solitarius

Chasiempis sandwichensis ibidis
Chelonia mydas
Corvus hawaiiensis

Dermochelys coriacea

Eretmochelys imbricata

Fulica alai

Gallinula chloropus sandvicensis
Gymnothorax hilonis

Hemignathus lucidus affinis
H. lucidus hanapepe
H. munroi
H. procerus
H. virens wilsoni
Himantopus mexicanus knudseni
Himatione sanguinea sanguinea

Lasiurus cinereus semotus
Lentipes concolor
Lepidochelys olivacea
Loxioides bailleui
Loxops coccineus coccineus
L. coccineus ochraceus
L. coccineus wolstenholmi

Melamprosops phaeosoma
Moho bishopi
M. braccatus
Monachus schauinslandi
Myadestes lanaiensis rutha
M. myadestinus
M. palmeri

Oreomystis mana

Palmeria dolei
Paroreomyza flammea
P. maculata
Pseudonestor xanthophrys
Psittirostra psittacea
Pterodroma phaeopygia sandwichensis
Puffinus newelli

Sicyopterus stimpsoni

Telespyza cantans
T. ultima

Vestiaria coccinea (Lanai only)
V. coccinea (Molokai only)
V. coccinea (Oahu only)

APPENDIX 2   **Sample Element Occurrence Record and Field Definitions**

Sample EOR:

05/20/91

HAWAII HERITAGE PROGRAM
ELEMENT OCCURRENCE RECORD

------- ---------- ------

EOCODE: PDFAB3M090.001   BMCODE: FAB

NAME: SESBANIA TOMENTOSA

COMNAME: 'OHAI

QUADNAME: KAENA   PRECISION: SC
MAPREF: 42   TENTEN: 1,4   ISLANDCODE: OA   EORANK: B
EORANKCOMM: SEVERE HABITAT DEGRADATION
SURVEYDATE: 1987-12-17   FIRSTOBS: 1915   LASTOBS: 1987-12-17
GRANK: G2   SRANK: S2.1   IDENT: Y

DIRECTIONS: /KAENA PT, NR SEA LEVEL TO 100 FT/ KAENA PT, 100 YRD E OF
            LIGHTHOUSE, TRAN 2 STA 4.0, 4.1, 4.2 & ABOVE RD NR WELL-KNOWN LG
            ROCK (F87HHP01); ABOVE & BELOW MAJOR RD, (CONT)
GENDESC: SAND DUNES W/ ROCK OUTCROPS W/ SCAEVOLA SERICEA MIXED COASTAL
            DRY SHRUBLAND (CTASSC00B0.001), W/ SCAEVOLA, HELIOTROPIUM,
            JACQUEMONTIA, MYOPORUM, PROSOPIS, (CONT)
ELEV: 25   SIZE: 0

EODATA: CA 30 PLANTS NR LIGHTHOUSE W/ FL, FR & 3 PLANTS ON RD BANK NR LG
            ROCK "SOULS LEAP" (F87HHP01); ABOVE RD: 3 PLANTS SEEN IN 1976
            (U77NAG01); BELOW RD: 36 PLANTS COUNTED IN 1987 (P87PER01);
            PLANTS PROSTRATE, MOST SPEC FERTILE

COMMENTS: REFERRED TO S. ORICOLA INED. BY W. CHAR (U83CHA01)

TMK1: 6-09-02-009   TMK2:   TMK3:

PROTCOMM: THREATS FROM ORVS
MGMTCOMM:

BESTSOURCE: HERITAGE STAFF. 1987. FIELD FORMS FOR KAENA NAR SURVEY,
            DECEMBER 1987. UNPUBL.
SOURCECODE: F87HHP01 U87SOU01 U83CHA01 S68PEABM S68VANBM S67HERBM
            S68WEBBM S58???BM S60KINBM S40STJBM
UPDATE: 88-10-27 MMB

    DIRECTIONS: DETAILED MAPS AVAILABLE (U87SOU01, U77NAG01, F87HHP01)

    GENDESC: LEUCAENA FLATS & ON ROAD BANK, DRY AREAS (F87HHP01); ACACIA
    FARNESIANA, W/ HERB LAYER OF CHLORIS INFLATA, HETEROPOGON CONTORTUS,
    SETARIA VERTICILLATA & TRICACHNE INSULARIS FOUND IN EXPOSED AREAS
    (U77NAG01)

SOURCECODE: S36FOSBM S31DEGBM S15FORBM S76CARUH S60LAMUH S64LONUH S72YAMUH S67HERUH S77NICUH S68NAGLY U77NAG01 S30RUSBM S50HATBM S63STEBM P87PER01

EOR Field Descriptions:

EOCODE (Element Occurrence Code)
Unique 14-character identifier code for each EOR. The first 10 characters preceding the decimal point identify the element. The three-digit number following the decimal point identifies the occurrence of this element. For example, PDFAB3M090.001 is the first occurrence of *Sesbania tomentosa* entered in the Heritage database.

BMCODE (Bishop Museum Code)
Standard three-letter code used by Bishop Museum's Herbarium Pacificum and other herbaria worldwide to identify plant families.

NAME
Scientific name of element.

COMNAME (Common Name)
Hawaiian or English common name.

QUADNAME (Quadrangle Name)
Name of USGS 7.5-minute quadrangle(s) where EO is mapped.

PRECISION
Precision of EO as mapped on quadrangle map.

SC = Specific with exact location confirmed
S   = Specific—EO reported within a 0.33-mile radius of mapped symbol
M   = Medium—EO reported within a 1.5-mile radius of mapped symbol
G   = General—EO reported within approximately 5-mile radius of mapped symbol
U   = Unmappable—inadequate information to map EO
N   = Not mapped—primarily cultivated plants and vague locations for which more specific occurrences are already mapped

See also Key to Map Symbols (Appendix 3, this paper).

MAPREF (Map Reference Number)
Map reference number indicates location of occurrence on Heritage maps.

TENTEN (Ten, Ten)
Coordinates from a 10 × 10 grid system superimposed on each quadrangle. The first number indicates the horizontal position (with 1 at far left; 10 at far right). The second number indicates the vertical position (top = 1, bottom = 10).

ISLANDCODE (Island Code)
A two-letter code indicating an island location (see Codes Used in Heritage Reports).

EORANK (Element Occurrence Rank)
Rank based on quality, condition, viability, and defensibility of the occurrence.

A   = Excellent
B   = Good
C   = Marginal
D   = Poor
X   = Destroyed (no longer exists at location)
Blank = Unknown

EORANKCOMM (Element Occurrence Rank Comments)
    Provides information to explain the EORANK.

SURVEYDATE
    Date of field survey used to assign EORANK. May be same as LASTOBS date.

FIRSTOBS (Year First Observed)
    Year EO was first reported from this site. (NOTE: Same as LASTOBS if EO is based on only
    one report. Field is left blank if all historic data are not included [e.g., for snails].)

LASTOBS (Date Last Observed)
    Date element was last observed extant at this site; not necessarily date site was last visited.

GRANK (Global Element Rank)
    Indicates the priority assigned by The Nature Conservancy for data collection and protection
    (for definitions, see section on Global Ranking Form in Heritage Reports).

SRANK (State Element Rank)
    Indicates state subdivisions of the global ranks (see section on Global Ranking Form in
    Heritage Reports).

IDENT (Identification)
    Indicates whether the identification has been checked and is believed to be correct.

        Blank   = Unknown whether identification correct or not
          Y   = Yes, identification confirmed and believed to be correct
          P   = Identification evaluated as possible or probable
          N   = No, identification determined to be wrong
          ?   = Whether identification is correct or not is confusing or disputed

DIRECTIONS
    Place name for location of EO and description of how to get there. If there are many reports of
    apparently one occurrence, all directions are listed to document the range of elevations and
    place names where the element has been recorded.

GENDESC (General Site Description)
    General site description of the place where EO occurs, including associated species.

ELEV (Elevation)
    Elevation of the precision symbol and map reference number on the Heritage map, as mapped
    (in feet).

SIZE (Size)
    Size of the area that contains the EO (in acres).

EODATA (Element Occurrence Data)
    Details about element at site such as reproductive status, numbers, size, condition, distur-
    bance, and viability.

COMMENTS
    Any information not covered by other fields.

TMK1–TMK3 (Tax Map Key)
    Tax map codes for up to three principal land parcels that contain the EO, in standard format
    (zone-section-plat-parcel). If the exact location or parcels cannot be determined, the appropri-
    ate plats or sections are indicated (e.g., 6-09-02-000 indicates the EO is found in Zone 6,

Section 9, Plat 2, but the specific parcels cannot be determined from available data). NOTE: Available only for select elements.

PROTCOMM (Protection Comments)
Comments on new or additional legal protection needed.

MGMTCOMM (Management Comments)
Comments on new or additional management needed.

BESTSOURCE (Best Source)
Best source of information on this EO, usually most recent.

SOURCECODE (Source Code)
Codes for all sources containing information on this EO. (See Codes and Abbreviations in Heritage Reports for a more detailed explanation.)

UPDATE
Date and initials of Heritage staff member who last updated this EOR.

APPENDIX 3   **Hawaii Natural Heritage Program Map Symbol Conventions**

HERITAGE QUAD MAP

Duplicates of the Heritage quad maps (U.S. Geological Survey topographic quadrangle maps—7.5-minute series) are very useful for illustrating the distribution and location of rare elements in an area of interest. Duplicate maps are available in three formats:

- Photocopy of USGS topographic quad maps
  (manually produced duplicates of Heritage maps)
  Scale   1:24,000
  Color   Yes

- GIS-produced USGS topographic quad map overlays
  Scale   Any level
  Color   Yes

- Partial photocopies of USGS topographic quad maps
  Scale   1:18,000 to 1:48,000
  Color   No

<u>Key to Map Symbols</u>

TYPE OF ELEMENT
     GREEN   =   Natural Community
     YELLOW   =   Plant
     ORANGE   =   Vertebrate
     PURPLE   =   Invertebrate
     BLUE   =   Special Managed Areas

PRECISION SYMBOL
     O   =   Specific location (locality mapped within 0.33-mile radius).
     △   =   Medium precision (locality mapped within approximately 0.75-mile radius), typically based on an older record containing only place name information.
     □   =   General locality (locality mapped within 5-mile radius, approximately quad size), typically based on an older record containing only *ahupua'a* or regional information.

APPENDIX 4   **Major Users and Reports of the Hawaii Natural Heritage Program**

<u>Past Users</u>
Amfac/JMB Hawaii Inc.
Bishop Estate
CH2M Hill
County of Hawaii
Department of the Army
Department of the Navy
The Estate of James Campbell
Hawaii Army National Guard
Hawaii Coastal Zone Management Program
Hawaii State Department of Business and Economic Development
Hawaii State Department of Hawaiian Home Lands
Hawaii State Department of Land and Natural Resources
      Division of Forestry and Wildlife
      Division of State Parks
Hawaii Stream Assessment
Kahoʻolawe Island Conveyance Commission
Kamehameha Schools/Bishop Estate
Maui Land and Pineapple Company, Inc.
National Park Service
Natural Area Reserves System
The Nature Conservancy
Outdoor Recreation and Historic Sites
State of Hawaii Office of State Planning
United States Fish and Wildlife Service, Pacific Division
University of Hawaii Environmental Center
Vandenberg Air Force Base

Hawaii Natural Heritage Program Reports (1987–March 1995)

1987

Biological Database and Reconnaissance Survey of Nanakuli. Prepared for Department of Hawaiian Home Lands. 37 pp.

Biological Database and Reconnaissance Survey of Waimanalo. Prepared for Department of Hawaiian Home Lands. 52 pp.

Biological Database of Rare Species and Anchialine Pond Types of the State of Hawaii. Prepared for the County of Hawaii. 52 pp.

Biological Overview of Hawaii's Natural Area Reserves System. Prepared for Department of Land and Natural Resources. 40 pp.

Biological Overview of Hawaii's Natural Area Reserves System. Appendices. Prepared for Department of Land and Natural Resources. 40 pp.

Biological Survey of Hawaiiloa Ridge Trail. Prepared for Department of Land and Natural Resources, Division of Forestry and Wildlife. 40 pp.

Biological Survey of Kaala Trail. Prepared for Department of Land and Natural Resources, Division of Forestry and Wildlife. 50 pp.

Biological Survey of Kuaokala Road Easement. Prepared for Department of Land and Natural Resources, Division of Forestry and Wildlife. 31 pp.

Biological Survey of Kuliouou Mauka Trail. Prepared for Department of Land and Natural Resources, Division of Forestry and Wildlife. 22 pp.

Biological Survey of Nuuanu Trail. Prepared for Department of Land and Natural Resources, Division of Forestry and Wildlife. 28 pp.

Biological Survey of Puu Pia Trail. Prepared for Department of Land and Natural Resources, Division of Forestry and Wildlife. 31 pp.

Historical Record of Rare Species and Natural Communities in Honouliuli. Prepared for The Estate of James Campbell. 48 pp.

1988

Biological Database of Rare Coastal Plants and Animals in the State of Hawaii. Prepared for Department of Land and Natural Resources, Division of Forestry and Wildlife; Hawaii Coastal Zone Management Program; and Department of Business and Economic Development. 40 pp.

Natural Area Reserves System Inventory Field Manual. Prepared for Department of Land and Natural Resources, Natural Area Reserves System, Division of Forestry and Wildlife. 43 pp.

1989

Ahihi-Kinau Natural Area Reserves Resource Information. Prepared for Department of Land and Natural Resources, Natural Area Reserves System, Division of Forestry and Wildlife. 75 pp.

Hanawi Natural Area Reserves Resource Information. Prepared for Department of Land and Natural Resources, Natural Area Reserves System, Division of Forestry and Wildlife. 150 pp.

Hono O Na Pali Natural Area Reserves Resource Information. Prepared for Department of Land and Natural Resources, Natural Area Reserves System, Division of Forestry and Wildlife. 200 pp.

How to Read Heritage Database Reports. 28 pp.

Kaena Point Natural Area Reserves Resource Information. Prepared for Department of Land and Natural Resources, Natural Area Reserves System, Division of Forestry and Wildlife. 100 pp.

Kahaualea Natural Area Reserves Resource Information. Prepared for Department of Land and Natural Resources, Natural Area Reserves System, Division of Forestry and Wildlife. 200 pp.

Kipahoehoe Natural Area Reserves Resource Information. Prepared for Department of Land and Natural Resources, Natural Area Reserves System, Division of Forestry and Wildlife. 100 pp.

Kuia Natural Area Reserves Resource Information. Prepared for Department of Land and Natural Resources, Natural Area Reserves System, Division of Forestry and Wildlife. 150 pp.

Laupahoehoe Natural Area Reserves Resource Information. Prepared for Department of Land and Natural Resources, Natural Area Reserves System, Division of Forestry and Wildlife. 200 pp.

Manuka Natural Area Reserves Resource Information. Prepared for Department of Land and Natural Resources, Natural Area Reserves System, Division of Forestry and Wildlife. 150 pp.

Mt. Kaala Natural Area Reserves Resource Information. Prepared for Department of Land and Natural Resources, Natural Area Reserves System, Division of Forestry and Wildlife. 150 pp.

Native Biological Resources of Amfac Lands, West Maui, Hawaii. Prepared for Amfac Hawaii, Inc. 72 pp.

Natural Area Reserves Resource Information Reference Sources and Archive Diskettes Guide. Prepared for Department of Land and Natural Resources, Natural Area Reserves System, Division of Forestry and Wildlife. 75 pp.

Natural Area Reserves Resource Information System User's Manual. Prepared for Department of Land and Natural Resources, Natural Area Reserves System, Division of Forestry and Wildlife. 38 pp.

Olokui Natural Area Reserves Resource Information. Prepared for Department of Land and Natural Resources, Natural Area Reserves System, Division of Forestry and Wildlife. 50 pp.

Pahole Natural Area Reserves Resource Information. Prepared for Department of Land

and Natural Resources, Natural Area Reserves System, Division of Forestry and Wildlife. 75 pp.

Puu Alii Natural Area Reserves Resource Information. Prepared for Department of Land and Natural Resources, Natural Area Reserves System, Division of Forestry and Wildlife. 100 pp.

Puu Makaala Natural Area Reserves Resource Information. Prepared for Department of Land and Natural Resources, Natural Area Reserves System, Division of Forestry and Wildlife. 150 pp.

Puu O Umi Natural Area Reserves Resource Information. Prepared for Department of Land and Natural Resources, Natural Area Reserves System, Division of Forestry and Wildlife. 100 pp.

Rare Plants, Animals and Natural Communities Reported from Maui Land and Pineapple Company Lands, West Maui, Hawaii. Prepared for Maui Land and Pineapple Company, Inc. 70 pp.

State of Hawaii Natural Area Reserves System Biological Resources and Management Priorities. Appendices. Prepared for Department of Land and Natural Resources, Natural Area Reserves System, Division of Forestry and Wildlife. 50 pp.

State of Hawaii Natural Area Reserves System Biological Resources and Management Priorities. Summary Report. Prepared for Department of Land and Natural Resources, Natural Area Reserves System, Division of Forestry and Wildlife. 26 pp.

West Maui Natural Area Reserves Resource Information. Prepared for Department of Land and Natural Resources, Natural Area Reserves System, Division of Forestry and Wildlife. 250 pp.

1990

Biological Database and Reconnaissance Survey of Na Pali Coast State Park, Island of Kauai. Prepared for Department of Land and Natural Resources, Division of State Parks. 80 pp.

Biological Database and Reconnaissance Survey of Rare Species and Natural Communities in the Kahuku Study Area. Prepared for The Estate of James Campbell. 72 pp.

Biological Database and Reconnaissance Survey of the Department of Hawaiian Home Lands Kawaihae Parcel. Prepared for Department of Hawaiian Home Lands. 38 pp.

Biological Database and Reconnaissance Survey of the Department of Hawaiian Home Lands Kula Parcel. Prepared for Department of Hawaiian Home Lands. 47 pp.

Biological Database and Sources of Na Pali Coast State Park and Adjacent Areas. Prepared for Department of Land and Natural Resources, Division of State Parks, Outdoor Recreation and Historic Sites. 150 pp.

Biological Information on Distribution of Indicator Species in the North Kauai Streams. Prepared for Division of Aquatic Resources, Department of Land and Natural Resources, and University of Hawaii Environmental Center. 44 pp.

Distribution of Plants Selected for United States Fish and Wildlife Service Listing Proposals. Prepared for Department of Land and Natural Resources, Division of Forestry and Wildlife. 700 pp.

Hawaii Stream Assessment, Aquatic Resources. Prepared for the Hawaii Stream Assessment. 200 pp.

Interpretive Opportunities for Na Pali Coast State Park, Halelea District, Island of Kauai. Prepared for Department of Land and Natural Resources, Division of State Parks, Outdoor Recreation and Historic Sites. 15 pp.

Management Recommendations for Na Pali Coast State Park, Halelea District, Island of Kauai. Prepared for Department of Land and Natural Resources, Division of State Parks, Outdoor Recreation and Historic Sites. 19 pp.

Puu Kukui Watershed Management Area Management Plan. Prepared for Maui Land and Pineapple Company. 69 pp.

Puu Kukui Watershed Management Area Resource Information. Prepared for Maui Land and Pineapple Company. 100 pp.

1991

Biological Database and Reconnaissance Survey of the Department of Hawaiian Home Lands Hoolehua Parcel. Prepared for Department of Hawaiian Home Lands. 54 pp.

Biological Information on Rare Plants, Animals, and Natural Communities of the Humuula Saddle Proposed Line Area. Prepared for CH2M Hill. 60 pp.

Biological Reconnaissance of Rare Element Database for the Proposed Ka Iwi National Park, Island of Oahu. Prepared for National Park Service. 26 pp.

Biological Reconnaissance Survey of the Maunawili Trail Alignment. Prepared for Department of Land and Natural Resources, Division of Forestry and Wildlife. 33 pp.

Management Report, Amfac/JMB West Maui Watershed. Prepared for Amfac/JMB Hawaii Inc. 44 pp.

Maui Lava Tubes Preserve Resource Information. Prepared for The Nature Conservancy, Maui Projects Office. 50 pp.

Moomomi Preserve Resource Information. Prepared for The Nature Conservancy, Molokai Projects Office. 100 pp.

Pelekunu Preserve Resource Information. Prepared for The Nature Conservancy of Hawaii. 150 pp.

Potential Interpretive and Recreational Sites, Amfac/JMB West Maui Watershed and Adjacent Amfac/JMB Lands. Prepared for Amfac/JMB Hawaii Inc. 11 pp.

Proceedings of the Native Ecosystems and Rare Species Workshops. Prepared for State of Hawaii Office of State Planning. 182 pp.

Waikamoi Preserve Resource Information. Prepared for The Nature Conservancy of Hawaii. 150 pp.

1992

Biological Database and Reconnaissance Survey of Kahoʻolawe Island. Prepared for Kahoʻolawe Island Conveyance Commission. 141 pp.

Biological Database and Reconnaissance Survey of the Waianapanapa Area, Island of Maui. Prepared for Department of Land and Natural Resources, Division of State Parks. 55 pp.

Biological Database and Sources for the Waianapanapa Area, Maui. Prepared for Department of Land and Natural Resources, Division of State Parks. 15 pp.

Biological Inventory of Honomalino, Island of Hawaii. Prepared for Department of Land and Natural Resources, Division of Forestry and Wildlife. 42 pp.

Biological Inventory of Waioli Valley, Island of Kauai. Prepared for Department of Land and Natural Resources, Division of Forestry and Wildlife. 35 pp.

Botanical Inventory of Hanahanapuni Crater, Island of Kauai. Prepared for Department of Land and Natural Resources, Division of Forestry and Wildlife. 22 pp.

Botanical Survey of Selected Portions of the Puu Waa Waa Game Management and Lease Area, Island of Hawaii. Prepared for Department of Land and Natural Resources, Division of Forestry and Wildlife. 44 pp.

Database Report on Rare Natural Communities, Plants and Animals Reported from Kamehameha Schools/Bishop Estate Lands, Kona, Hawaii. Prepared for Kamehameha Schools/ Bishop Estate. 69 pp.

Database Report on Rare Natural Communities, Plants and Animals Reported from Keauhou, Kau, Hawaii. Prepared for Kamehameha Schools/Bishop Estate. 46 pp.

Hawaii Heritage Program, Natural Diversity Database Maps and General Information for Bishop Estate North and South Kona. Prepared for Bishop Estate. 140 pp.

1993

Biological Database and Field Survey of the Molokai Receiver Station, Molokai, Hawaii. Prepared for 30th Space Wing, Vandenberg Air Force Base. 25 pp.

Biological Database and Reconnaissance Survey of the Coastal Lands of the Kiholo Bay Area, Island of Hawaii. Prepared for Department of Land and Natural Resources, Division of State Parks. 110 pp.

Biological Database and Reconnaissance Survey of the Department of Hawaiian Home Lands Makuu Parcels, Island of Hawaii. Prepared for Department of Hawaiian Home Lands. 63 pp.

Biological Database and Sources for the Polihale Area, Kauai. Prepared for Department of Land and Natural Resources, Division of State Parks. 15 pp.

Biological Reconnaissance Survey of the Department of Hawaiian Home Lands Kamaoa-Puueo Parcel, Island of Hawaii. Prepared for Department of Hawaiian Home Lands. 91 pp.

Biological Reconnaissance Survey of the Kanaio Training Area, Maui, Hawaii. Prepared for Hawaii Army National Guard. 87 pp.

Biological Resources of the Keamuku to Kailua 138 kV Transmission Line Project Area. Prepared for CH2M Hill. 66 pp.

Botanical Database and Reconnaissance Survey of the Polihale Area, Kauai. Prepared for Department of Land and Natural Resources, Division of State Parks. 64 pp.

Hawaii Heritage Program Natural Diversity Database Summary Report, Map Keys and Maps of United States Navy Lands in Hawaii and the Pacific. Prepared for The Nature Conservancy. 30 pp.

Preliminary Classification of Ecosystems of United States Affiliated Islands of the Tropical Pacific. Prepared for United States Fish and Wildlife Service, Pacific Division, Department of Interior. 300 pp.

Vertebrate Inventory Surveys at the Multipurpose Range Complex Pohakuloa Training Area (PTA), Island of Hawaii. Prepared for Department of the Army. 120 pp.

1994

Biological Boundary Assessment for Waipio Parcel, Oahu, Hawaii. Prepared for United States Fish and Wildlife Service, Region 1, Department of the Interior. 42 pp.

Biological Database and Reconnaissance Survey of the Poamoho and Schofield-Waikane Trails Area, Island of Oahu. Prepared for Department of Land and Natural Resources. 80 pp.

Biological Database and Reconnaissance Survey of the Waimano Area, Island of Oahu. Prepared for Department of Land and Natural Resources. 80 pp.

Biological Inventory and Management Assessment for the Kahuku Training Area, Oahu, Hawaii. Prepared for 25th Infantry Division (Light) and U.S. Army Pacific Command, Hawaii. 150 pp.

Biological Inventory and Management Perspectives for the Naval Magazine, Headquarters, Lualualei, Oahu, Hawaii. Prepared for Department of the Navy. 138 pp.

Biological Inventory for the Makua Military Reservation, Oahu, Hawaii. Prepared for Department of the Army. 202 pp.

Biological Inventory of the Kawailoa Training Area, Oahu, Hawaii. Prepared for 25th Infantry Division (Light) and U.S. Army Pacific Command, Hawaii. 176 pp.

Biological Inventory of the Makaula Ooma Tract, Island of Hawaii. Prepared for Department of Land and Natural Resources. 50 pp.

Biological Inventory of the Schofield Barracks Military Reservation, Oahu, Hawaii. Prepared for Department of the Army. 213 pp.

Management Assessment for the Kawailoa Training Area, Oahu, Hawaii. Prepared for 25th Infantry Division (Light) and U.S. Army Pacific Command, Hawaii. 49 pp.

Management Assessment for the Makua Military Reservation, Oahu, Hawaii. Prepared for Department of the Army. 57 pp.

Management Assessment for the Schofield Barracks Military Reservation, Oahu, Hawaii. Prepared for Department of the Army. 52 pp.

1995

Biological Database and Botanical Field Survey of Kalalau Rim, Island of Kauai. Prepared for Department of Land and Natural Resources. 80 pp.

CHAPTER 23

Marine Biosystematic/Biodiversity Priorities: A Canadian Perspective

Don E. McAllister

ABSTRACT

Biological surveys, preservation of voucher specimens, and analysis on the geographic distribution using tools like geographic information systems (GISs) help pinpoint potential protected areas. Sound GIS analyses depend on geographically adequate sampling, correctly identified specimens, preferably verifiable by voucher collections in museums, and ongoing support for systematic research and natural history museums.

Useful GIS methods include analysis of all species or a subset in a taxonomic group, endemic species, and the variety of higher taxa in which lie kinds of diversity not expressed by simple species counts. Information from coral reef fishes was used to illustrate these approaches. The carrying out of one or more marine all-taxa biodiversity inventories (ATBIs) was proposed to measure the total diversity in one or more areas in the ocean.

The International Convention on Biological Diversity is one of the most pivotal conventions ever ratified. Its articles describe a number of roles in which biosystematics and natural history museums could play important parts. These include inventory, conservation, sustainable use, development of new bio-industries, and scientific cooperation. The preparation of country studies on biodiversity, a review of existing knowledge, programs, and policies, as well as gaps, offer museums an opportunity to assist in their own and other countries. Similarly, national biodiversity strategies and action plans offer both museums and biosystematists an opportunity to participate and to increase awareness of the societal roles that they can play. Country studies and national biodiversity strategies were exemplified from the Canadian experience.

Museums play a role in communicating knowledge about biodiversity and popularizing it. Attention should be drawn to these activities if museums are to obtain the resources needed to carry out research, complete national and global biodiversity inventories, and provide the knowledge needed for conservation and sustainable use. Environmental non-governmental organizations also have an important role to play in increasing public awareness, in communicating knowledge, and in undertaking projects in conservation and sustainable use. Partnerships between museums and environmental organizations should be considered in the changing role of museums in the modern world.

Thirteen recommendations are made on biosystematics/museum involvement in country studies, Biodiversity Convention negotiations, national biodiversity strategies and action plans, popularization of systematics, geographic analyses of species/taxa occurrence, development of all taxa biodiversity inventories, prioritizing funding for lesser-known taxonomic groups, use of databases, creation of posts for systematists, harmonizing taxonomic codes, use of higher taxa, development of a global computer catalog of all species, and supporting and cooperating with environmental organizations.

INTRODUCTION

This paper discusses global biosystematic and biodiversity priorities. The discussion is based on experience with a geographic information system (GIS) analysis of coral reef fishes of the world and work at the Canadian Museum of Nature's Canadian Centre for Biodiversity on the International Convention on Biodiversity, the Canadian Biodiversity Strategy, and *Canada's Biodiversity* (the Canada country study on biodiversity). These projects are then used as the basis for discussions on the needs of biosystematics and biodiversity. Recommendations are made based on these sources and on those made at the International Union for Biological Sciences Conference held in Paris, September 1994.

Biosystematics serves societal roles in science, conservation, sustainable use, and new bio-industries. These need information based on sound biosystematic research, on the distribution of species, and on our sister science, ecology. Biosystematic research, in some cases, and knowledge of distribution are hampered by inadequate collections and poor geographic sampling. Research is also held back by the declining number of taxonomists and graduating students in taxonomy, decreasing positions for taxonomists, and declining support for collections. Behind this declining support are the swings in support to emerging fields in biology. But also involved is the failure of biosystematists to communicate the nature and societal roles of their research to a broader public. The publication of research findings solely in refereed scientific journals has meant that the general public and decision makers are not aware of the significance of scientific work or its needs by biosystematists. Museums conceal some of their other key activities from the public behind the facade of displays. Biosystematists have also failed to develop priorities for research and training of students. For example, vertebrate taxonomy is relatively well resourced compared to the invertebrates, and the study of wild microorganisms is especially poorly funded.

GLOBAL CORAL REEF FISH GIS STUDY

Spot distribution maps for more than 800 species of coral reef fishes in seven families were prepared by Callum and Julie Roberts for the IUCN Species Survival Commission's Coral Reef Fish Specialist Group with the intent of locating areas rich in species or taxa worthy of proposal as protected areas. The data were entered into a QUIKMAP GIS and analyzed by Frederick Schueler. Species were recorded as present or absent in a global equal-area grid system; point positions were also retained. The numbers of species and taxa were computed for each grid "cell" and plotted on maps in color to show areas of high diversity. Species with small ranges, those with ranges less than four cells apart, were selected and plotted to find areas of endemism. Species counts by latitude and longitude were plotted to study geographic trends. Geographic ranges could be measured by the number of cells occupied. The study located a number of areas rich in species and in endemic species. (For more information, consult McAllister et al. 1994.) This work was carried out by Ocean Voice International for the IUCN Species Survival Commission's Coral Reef Fish Specialist Group. More species are being added to the database, and another database of human impacts on coral fishes and habitats will be created.

Findings included the following. Equal-area global grids provide a useful tool for discovering biodiversity hot spots and recognizing geographic trends. Species numbers increased toward the equator but dropped off at the equator, because of fewer coral reefs along the equatorial band. The Indo-Pacific had more species than the Atlantic, and there were various hot spots within the Indo-Pacific. The view that most coral reef fishes are unlikely to be threatened by human activities because they are geographically widespread was refuted—59 percent of the species ranges spanned less than 6,600 km and 33 percent spanned less than 2,200 km. Endemics comprised 18 percent of all species studied. Generic richness was correlated with species richness. The latter suggests that higher taxa can be used as surrogates to identify biodiversity hot spots. Identifications to higher taxa such as genera can be made more rapidly and could be used in RAP (rapid assessment process) surveys. Thus, taxonomy and classifications can be important in conservation.

Equal-area grid databases have a number of advantages. Unlike those for counties, states, countries, islands, bays, and seas, the cell areas within the grid are equal, making cell comparisons more comparable than those of other geographic units. It makes finding species/taxa hot spots or other patterns easier and simplifies measuring north-south and east-west clines. Since the cells are equal, quantification of data for systematic, biogeographic, or ecological hypothesis is easily accomplished. If a standard grid is accepted, then data between taxa and environment can be readily shared.

Equal-area grids provide one of several options for geographic analysis. For example, isopleth contour maps are also useful to portray changes in species richness, though less amenable to quantitative analyses. In some cases, for political or other scientific reasons, other political/unit or geographic polygon data (e.g., island or country data) for biodiversity are more appropriate. For studies close to the equator or in a confined geographic area, latitude-longitude cells will be sufficiently close to equal in area so that findings will not be contorted. Although geographic polygons or grids are useful for many purposes, point maps and databases should always be maintained in parallel or relationally linked to polygon/grid databases. Each point should be tied to museum catalog numbers or literature citations. This practice ensures that unusual patterns or records can be verified, and it may help in resolving finer scale patterns not shown on coarser grids of polygons. If databases like those of natural history museums, World Conservation Monitoring Centre, the International Center for Living Aquatic Resources Management (ReefBase, FishBase), and the Australian Institute of Marine Science (CoralBase) retain their information as points defined by latitude and longitude, then this raw information can be transferred to other geographic grids or polygons.

Inspection of the species density maps revealed two other findings. Species were often relatively denser in areas like Guam and the Ryukyu Archipelago that have been scientifically better studied. Conversely there were a number of areas that had fewer species than expected, probably reflecting a low level of sampling. This suggests the need for geographically broad, even global, inventories with standardized levels of sampling (e.g., an equal-area grid sampling pattern).

A few long-term, extra-heavily sampled locations, under an all-taxa biodiversity inventory (ATBI), would provide an anchor as to the thoroughness of the standardized sampling. Terrestrial ATBIs are discussed in Janzen and Hallwachs (1994). One or more marine ATBIs should be planned, and the Center for Marine Conservation and Ocean Voice International plan to develop a proposal for them. Coupled with RAP (rapid assessment process) surveys, the ATBIs would also provide an approximate index of what percent of the fauna would be detected during RAP surveys. RAPs are quick surveys, first used in the tropical ecosystems, designed to quickly measure the relative richness of a threatened landscape. That concept should be tested in aquatic and marine ecosystems. Other biodiversity survey tools are discussed in McAllister (1994).

A third finding is quite obvious to any taxonomist but not necessarily to those outside the field. The state of taxonomic knowledge is quite uneven. Thus, the butterflyfish family Chaeodontidae is quite well studied taxonomically, while study of the seahorse family Hippocampidae is in a state of disarray in several parts of the world. Poorly known taxa interfere with biodiversity studies, conservation of biodiversity, sustainable use of biological resources, and other applications.

BIODIVERSITY CONVENTION

The International Convention on Biological Diversity has been signed by more than 200 nations, has been ratified by more than 117 nations as of 5 April 1995, and came into force in December 1993. The first conference of the convention parties will be held in November 1995 in the Bahamas. The Biodiversity Convention has been styled "the mother of conventions" because of its broad significance covering conservation, funding mechanisms, sustainable use, sharing of benefits and technology, biotechnology, inventory, research, and education. The recommendations to carry out biological inventories and to conduct and share relevant research are among areas of interest to biosystematists.

During the convention negotiations, the parties decided to carry out a few country studies of biodiversity to provide information that would assist negotiators. These studies were found so useful that the guidelines for their preparation were revised, and they are being recommended as part of national strategy plans for countries applying for biodiversity aid to the Global Environment Facility of the World Bank. But there is no reason developed countries should not undertake country studies for their own ends, and some have. Country studies survey a country's biodiversity, its worth, identify knowledge gaps, what resources are expended in its study, management, and conservation, and what resources are needed to fully conserve biodiversity and sustainably use biological resources.

A taxonomic census is included as part of the country study inventorying process. The census estimates the number of scientifically described and yet to be described species for each major taxon at the phylum and class level. Those figures are useful in determining resources needed and priorities in biosystematic research. For Canada we estimated that there were about 140,000 nonviral species, only half of which were scientifically described. Canada is the first country to produce an estimate of the number of species of all taxa. We did it by consulting experts for individual taxa and by researching literature. A crude estimate of viral species meant that altogether there are about 280,000 species in Canada. Costa Rica, although much smaller than Canada, was estimated to have 510,000 species, with a much higher proportion of undescribed species. In Canada more phyla and classes were marine than terrestrial. Neglecting the insects, marine and aquatic organisms comprised half the species biodiversity of Canada.

The Canadian country study on biodiversity by Mosquin, Whiting, and McAllister is expected to be published by the museum in late 1995. Country studies on biodiversity are valuable in assembling facts for planning. I believe every country should carry one out and that museums and biosystematists should be major players in producing such studies.

NATIONAL BIODIVERSITY STRATEGIES

Canada has developed a draft national biodiversity strategy and released it for public discussion. The strategy should be officially completed, approved, and released in mid-1995. It was a two-year process led by the federal Department of the Environment's Biodiversity Convention Office. Four committees were involved: a steering committee, a federal committee involving all interested government departments, a federal-provincial-territorial committee, and the BCAG (Biodiversity Convention Advisory Group). The BCAG included representatives of governments, industry, NGOs, academe, and others. The Draft Canadian Biodiversity Strategy is a 60-page document, representing the results of eighteen months of discussions and compromises among interested parties.

The Canadian Museum of Nature participated in all four committees working on the strategy. A national biosystematics task force was under way during the national strategy negotiations. Production of task force recommendations was accelerated so they could be provided to the national strategy process. Several focus groups were developed during formulation of the strategy to advise the committees. One focus group dealt with taxonomic research, collections, and databases. Contributions from these sources assured that biosystematics was not forgotten in the strategy. The findings of *Systematics Agenda 2000* provided valuable reinforcement (Anonymous 1994).

BIODIVERSITY ACTION PLANS

In 1995 Canada will prepare biodiversity action plans based on the national strategy. Federal, provincial, and territorial governments will be involved. Industry, NGOs, academe, and other organizations and individual citizens are free to prepare biodiversity action plans, should they so choose.

COMMUNICATION AND POPULARIZATION OF BIOSYSTEMATICS/ BIODIVERSITY

Organizations such as the Canadian Centre for Biodiversity and environmental nongovernmental organizations play a major role in communicating and popularizing biosystematics and biodiversity.

Canadian Centre for Biodiversity, Canadian Museum of Nature

The Canadian Museum of Nature established the Canadian Centre for Biodiversity (CCB) in 1990. Among its goals are to communicate knowledge about and appreciation of the diversity of life on Earth; provide access to knowledge of biodiversity, biosystematics, and the environment; and increase awareness of the museum biological surveys, scientific databases, collections, research, publications, and displays. The CCB has only four part-time employees but is able to secund staff from other offices. A primary focus of the CCB was to make natural history museums and biosystematics better known and to increase appreciation of their relevance to modern society. We felt museum resources were weak in large part because the public and decision makers did not know about museums and their societal and environmental contributions..

As indicated earlier, the CCB was involved in negotiations for the International Convention on Biological Diversity and in the development of the Canadian Biodiversity Strategy. The CCB also acts as an interface to direct requests for information to the appropriate part of the museum. The CCB has been involved in developing relationships and agreements with the private sector. Recently CCB has launched an information service on Carleton University's Internet, which can be accessed by:

World Wide Web:
  URL to http://www.ncf.carleton.ca/freeport/....
    ...social services/eco/gorgs/cc-biodiv/menu

Or by Telnet to: freenet.carleton.ca
  Login as: guest
  Type: go ccb

Or by Direct Dial (long distance rates will apply for those outside the Ottawa area):
  2400 BAUD - (613) 564-3600, Login as: guest, Type: go ccb
  14400 BAUD - (613) 564-0808, Login as: guest, Type: go ccb

Guest users may read any of the material; however, only registered users of Freenet can write material in the "discussions" area at the moment. But if you e-mail material to Noel Alfonso at:

ah201@freenet.carleton.ca

he can post it for you. Your contributions are most welcome.

Global Biodiversity (available in French as *La biodiversité mondiale*) is the quarterly bulletin of the Canadian Centre for Biodiversity. It has similar goals to CCB. *Global Biodiversity* tries to communicate information on biosystematics and other aspects of biodiversity between biological disciplines and to a broad audience of citizens and decision makers.

A greater variety of tools is needed to help popularize systematics if it is to receive the understanding and resources it needs to carry out its global tasks.

Role of ENGOs exemplified by Ocean Voice International

Environmental non-governmental organizations (ENGOs) have an important role to play in increasing public awareness, communicating information, carrying out research and "on-the-ground" or underwater projects in biodiversity education, training, conservation, sustainable use, and other areas. They often enhance their capacities through partnerships and networking with other ENGOs. There are fewer marine than terrestrial ENGOs— examples include the Center for Marine Conservation, Washington, D.C.; Greenpeace (various national offices); Marine Conservation Society; Ross-on-Wye, Herefordshire, U.K.; and Ocean Voice International, Ottawa, Canada. The IUCN Species Survival Commission includes a number of specialist groups with marine foci including whales, coral fishes, and sea turtles.

Ocean Voice International was incorporated in 1987 as a non-profit charitable organization dedicated to conservation of marine diversity and ecosystems and equitable and sustainable harvesting of marine resources. It works through research, education, training, and field projects, on its own or in partnership with other ENGOs, governments, and individuals. Ocean Voice International publishes a quarterly bulletin, *Sea Wind*.

Ocean Voice's projects include an environmental study of the impacts of an airport being planned for Shiraho Coral Reef in southernmost Japan (with the IUCN Species Survival Commission); helping marine aquarium fish gatherers in the Philippines switch to using small fence nets instead of stunning fish with a solution of sodium cyanide; educating coastal communities with a manual about coral reefs, their ecological needs, what degrades them, and how to conserve them (McAllister 1993); publishing a secondary school educational manual, *Green School Biodiversity Booklet* (McAllister 1995); and analyzing by GIS coral reef fishes of the world, in partnership with the Species Survival Commission's Coral Reef Fish Specialist Group. Marine ENGOs play useful roles by complementing those undertaken by academic organizations and natural history museums. For more information, contact Carleton University's Freenet (described earlier) and type in <go ovi>. For a World Wide Web home page, consult URL:

http://www.conveyor.com/oceanvoice.html

RECOMMENDATIONS

1. Promote, prepare, and publish national country studies on biodiversity, as recommended by the Biodiversity Convention. Use UNEP guidelines, but go beyond them. Include a census of all the major taxonomic groups down to order or class (insects) with the number of known species and an estimate of the number yet to be described. Include an analysis of existing systematic resources and needs. Country studies provide data for establishing priorities and developing a national strategy. Stress marine and coastal issues.

2. Become involved in the Biodiversity Convention negotiations. Promote and participate in the development of national and state/province biodiversity strategies. Address in the strategies the needs of systematics and their relationship to societal goals. Follow strategies with action plans.

3. Popularize systematics with stories about your work. Inform the public of the value of systematics to conservation, sustainable use of natural resources, and other sciences. Prepare assured nonreactionary responses to points raised by conservationists, scientists, funders, and others who argue other priorities. Make partners of ecologists, conservationists, NGOs, etc. NEVER repeat the "old saws" about museums being dusty or ivory towers. Share your excitement about taxonomic research. Send out a press release at least every quarter or half a year. Work to overcome the bias in favor of terrestrial studies.

4. Develop plans for global, national, and state/provincial equal-area grid-based biological inventories, with deposition of collections in accessible institutions. Promote support for such studies in developing countries. Develop technology for computer capture of data in the field. Train parataxonomists and partner with NGOs, sharing data with them.

   Develop plans for at least one marine, aquatic, and terrestrial all-taxa biodiversity inventory (ATBI) in every country, plus an ATBI site network in areas beyond national boundaries—offshore areas are everyone's responsibility. Give high priority to developing the world's first marine ATBI.

5. Develop priorities for funding the taxonomic study of major taxa—more funding for poorly known groups.

6. Develop geographic, environmental, and conservation priorities for funding biosystematic research. Using museum computer data, find geographic and environmental gaps in collecting—marine organisms need special attention. Study first those biotic areas that are rich in taxa and threatened by human development. Certain taxonomic groups may require prioritizing for socioeconomic reasons.

7. Creatively solve the conflict between conservation organization and museum databases. Support the need for ecological studies and encourage them to suppport biosystematics.

8. Urge the creation of new posts for systematists in universities and museums, using the new priorities.

9. Support harmonization of the taxonomic codes of zoology, botany, and microorganisms.

10. Promote use of cladistic and taxonomic data, as well as species data, in biodiversity, ecological, and other inventories. Hot spots based on counts of higher taxa or clades may differ from those based on counts of species. An island with a peregrine falcon and a duck is more diverse genetically, behaviorally, ecologically, etc., than another island with two species of ducks.

11. Support the development of an on-line global computer catalog of all species of plants, animals, and microorganisms, either as a central bank or as a collection of databases on different nodes on Internet. Those databases could be national, regional, or environmental. Eschmeyer's fish genera database is already on Internet (see Eschmeyer, this volume).

12. Develop popular guides to marine organisms for education, recreation, and eco-tourism. Ensure the public knows that these are based on biological inventories, museum collections, and databases, and biosystematic research.

13. Encourage the support for and cooperation with non-governmental organizations to help create public awareness, broaden knowledge of scientific findings, and carry out projects in the conservation of biodiversity and the sustainable use of biological resources.

REFERENCES

Anonymous. 1994. *Systematics Agenda 2000: Charting the biosphere*. Technical Report of Systematics Agenda 2000. American Museum of Natural History, New York.

Janzen, D. H., and W. Hallwachs. 1994. All-taxa biodiversity inventory (ATBI) of terrestrial ecosystems. A generic protocol for preparing wildland biodiversity for non-damaging use. Biodiversity and Biological Collections Gopher and anonymous ftp archive at Harvard.

McAllister, D. E. 1993. *Save our coral reefs.* Ottawa: Ocean Voice International. (In English; Filipino language edition in press.)

———. 1994. Tools for conserving biodiversity: Inventories, biosystematics, museums, TAP, RAP, GAP, ETAP, and ATI. A species at a time is too slow. *Global Biodiversity* 4(2): 16–21.

———. 1995. *The Green School Biodiversity Booklet*. Ottawa: Ocean Voice International and Canadian Coalition for Biodiversity. (English and French editions available.)

McAllister, D. E., F. W. Schueler, C. M. Roberts, and J. P. Hawkins. 1994. Mapping and GIS analysis of the global distribution of coral reef fishes on an equal-area grid. In *Mapping the diversity of nature*, R. I. Miller, 155–75. London: Chapman & Hall.

Mosquin, T., P. Whiting, and D. E. McAllister. 1995. *Canada's biodiversity*. Ottawa: Canadian Centre for Biodiversity, Canadian Museum of Nature. In press.

Circular and Invitation for the
Workshop and Lists of Sessions
and Presentations

MARINE/COASTAL BIODIVERSITY
IN THE TROPICAL ISLAND PACIFIC REGION
I. SPECIES SYSTEMATICS AND INFORMATION MANAGEMENT PRIORITIES

Hawaii Imin International Conference Center at Jefferson Hall
East-West Center, Honolulu, Hawaii, 2–4 November 1994

Data management is an essential aspect of any long-range program to conserve biodiversity. For the region of Oceania (tropical island Pacific), systematics for marine/coastal species has not been summarized and pulled together to assist in information management. Furthermore there is no established database management system for the region, although several new initiatives have emerged. We are attempting to bring together experts from the region to discuss the status of information management needs for nearshore tropical island Pacific biodiversity conservation and to help design a system specifically tailored to the region. To pursue the objectives of this workshop, we are inviting you as an expert to participate (see attached instructions and list of experts). If you are unable to accept, please recommend someone who might or whether you would still be prepared to write a manuscript on your subject area that would be presented at the workshop on your behalf. In order to make the workshop effective, we need to have brief written status reviews prepared and distributed to all participants prior to the workshop. We intend to publish the results of the workshop as a proceedings volume and, in addition, pursue the fashioning of a book at a later date. We plan to hold the workshop on 2–4 November 1994 at the East-West Center, which is on campus at the University of Hawaii, Honolulu. Accommodations for out-of-town participants will be on campus at the East-West Center's Lincoln Hall. To be effective, we would like papers completed by mid-October.

This is the first of a two-part series of workshops on Marine/Coastal Biodiversity in the Tropical Island Pacific. The second of the series, II. Population, Development, and Conservation Priorities, has already been organized and is scheduled to be held during the following week on 7–9 November 1994. One goal of holding the workshops back-to-back is to allow participants to attend both workshops and to increase the level of participation and quality of interchange at the workshops. The flyer for both workshops is attached.

The first workshop participants will need to make their own arrangements if they intend to extend beyond dorm check-out time on 6 November 1994. Other major institutional participants of the first workshop on Species Systematics and Information Management Priorities include the University of Hawaii, the Pacific Science Association (PSA), and The Nature Conservancy. The East-West Center (EWC) and PSA are the sponsors of the second workshop on Population, Development, and Conservation Priorities with The Nature Conservancy and the University of Hawaii also serving as major institutional participants.

The successful completion of the workshops will be valuable to other ongoing or planned marine biodiversity conservation initiatives and will also provide the opportunity for closer collaboration among the participants in future biodiversity work in the tropical Pacific. We look forward to your accepting this invitation and your attending the first workshop where we hope to have stimulating interactions between scientists and information managers.

Schedule

- Finalize decisions on workshop participants—15 September.

- Submit expanded outline of paper now due.

- Submit draft of presentation paper by mid-October.

- Distribute papers to all participants by the end of October before travel to Honolulu.

- Discuss papers at workshop 2–4 November 1994.

- Submit comments and review other papers by mid-December 1994, following the workshop.

- Edit and return own papers to Honolulu for final editing and collation by the end of January 1995.

If you have agreed to participate in the first workshop, please indicate whether you would be willing to extend your stay to attend the second workshop (at your own expense). Please indicate if you need assistance in finding accommodations beyond the check-out time of 6 November 1994. Accommodations at the Waikikian Hotel for $45 per nite plus tax can be arranged through the Bishop Museum (Dr. Lu Eldredge).

FLYER FOR WORKSHOPS   **Background**

The region of Oceania, or the tropical Pacific island region encompassing Melanesia, Micronesia, and Polynesia, is the most poorly understood by marine scientists in the world. Stretching nearly 15,000 km in an east-west dimension and 5,000 km in a north-south dimension, Oceania covers more than 10,000 islands, 500 atolls, and literally millions of coral reefs, of which less than 1 percent have been visited and assessed by marine and coastal scientists. Oceania also serves as the last major wilderness region for coral reef ecosystems in the world. Hampered by geographic isolation and lack of infrastructure, the Pacific islands have been buffered from the socioeconomic upheavals in other developing countries in the tropics, and economic development has been slow. Yet, since the end of World War II, the region has experienced phenomenal human population growth, averaging 3 percent per year. Greater demand for manufactured goods and processed foods has spurred a cash economy, supplanting some of the subsistence-based economy and eroding traditionally based marine tenure and conservation systems for coastal and marine resources. Land-use planning, coastal zone management, protected areas management, and environmental impact assessment are conservation tools now widely established, but to date they have been only marginally embraced by Pacific island cultures.

At present Oceania is at a crossroads, with a strong tradition of living in harmony with nature but with a strong future likelihood of heavy development and exploitation of natural resources. In particular, marine fisheries, mariculture, tourism, agriculture, handicrafts, and other development all hold promise in the tropical Pacific. Infrastructure development is accelerating on many islands.

However, marine and coastal conservation and associated educational programs are lagging behind economic development. Documentation of the status of biodiversity of Oceania is also lagging. Excluding the United States, Australia, New Zealand, and Japan, the region supports only a handful of universities and other institutions of higher learning. All lack adequate funding and facilities. There are shortages of resident marine and coastal scientists to assist in the region.

The South Pacific Regional Environment Programme (SPREP) in Apia, Western Samoa, has emerged as the most important environmental organization in the region and has established a new biodiversity conservation initiative for eleven island countries with the support of substantial outside funding. Nevertheless, a lack of scientists and long-range strategies for most of the twenty-five governments in the region threaten to delay meaningful progress on conservation in the face of economic development, and marine and coastal resources are most at risk.

The two workshops to be held in Honolulu on Marine/Coastal Biodiversity in the Tropical Island Pacific Region in early November 1994 are planned to pull together leading scientists, island resource managers, and data/information managers to discuss the needs of biodiversity conservation in the island Pacific. There is now the need to develop a more thorough and well-considered action plan and priorities for the conservation of marine and coastal biodiversity in the region.

Species Systematics and Information Management

There is an urgent need to summarize information on the systematics and taxonomy of nearshore marine and coastal species in Oceania to assist in the development of database management systems. An ecosystem-based classification system has already been developed through the cooperation of SPREP, the support of USAID and USFWS, the leadership of The Nature Conservancy, and the participation of scientists from the Pacific Science Association, island governments, and universities/institutions of the region. The region also needs to establish a network of institutions and activities that assembles, analyzes, and uses important data and information on marine and coastal biodiversity. The overall goal would be to develop an action plan that assembles, organizes, and makes available biodiversity data and information to a variety of resource managers and scientists. A fundamental premise is that good information is the basis of good management. A variety of activities and programs throughout the region are in the process of developing and assembling databases. Now is the appropriate time for representatives of these initiatives to meet and discuss how to proceed in a more organized and cooperative manner that benefits all parties.

The Ocean Policy Institute of the Pacific Forum, Center for Strategic and International Studies, in collaboration with the East-West Center, the Pacific Science Association, the U.S. National Oceanic & Atmospheric Administration, the U.S. Environmental Protection Agency, the University of Hawaii, and other organizations are initiating the dialogue and associated analysis to plan a database for oceanic life, using the tropical Pacific region as an initial pilot program for the database system.

Objectives

The geographic information and database study would review and summarize information from the following perspectives: (1) the scientific or technical components, especially on species and ecosystems diversity, and (2) information management components, including database and geographic information systems. These components would then contribute to a societal needs assessment that would be largely the scope of the second workshop of the series (Population, Development, and Conservation Priorities). Following both workshops, we also predict that there will be a government agency and institutional analysis of marine/coastal biodiversity conservation needs.

The Region and Subject Area

The region includes all of the Pacific island countries and territories of the South Pacific Commission and SPREP, with the addition of Hawaii and the subtropical Pacific islands of Japan (Marcus, Bonins, Ryukyus), Chile (Easter and Sala-y-Gómez), Australia (Norfolk and Lord Howe), and New Zealand (Kermadecs). The subject area covers the shallow nearshore marine environment, the littoral zone, and narrowly defined immediate coastal zone (e.g., immediate coastal low islands or within 0.1 to 1.0 km from the shoreline).

Workplan

Day 1   Discuss systematics, taxonomy, and diversity of major species groups, and the established ecosystem classification

Day 2   Discuss ongoing or planned geographic information and database management systems for tropical nearshore marine and coastal biodiversity

Day 3   Formulate an action plan, indicating priorities, approaches, geographic area, and components for an integrated information management system for Oceania including linkages to other systems

Outputs

- A review of the status of the taxonomy and systematics of major or important organism groups likely to be found in the nearshore marine and coastal ecosystems of Oceania

- An overview of ongoing or planned geographic information and database management systems for nearshore marine and coastal ecosystems both within and outside the Oceania region

- An action plan for designing and assembling an Oceania geographic information and database management system and linkages to other systems

PRELIMINARY LIST OF PARTICIPANTS TO BE INVITED

Organizers

L. G. Eldredge, J. E. Maragos, M. N. A. Peterson

November 2

State-of-the-species systematics and taxonomy of principal nearshore marine organism groups. Status of the marine ecosystem classification for the tropical Pacific (L. G. Eldredge, Convener, fax 808-841-8968).

Keynote address:
　　Dr. Sylvia Earle, former Chief Scientist of NOAA

Marine Algae:
　　Dr. Isabella A. Abbott, Department of Botany, University of Hawaii, Honolulu, HI 96822 (808-956-3923)

Mangroves:
　　Dr. Joanna Ellison, Australian Institute of Marine Science, PMB No. 3 MC, Townsville, Queensland 4810, Australia (fax 61-77-725-852)

Seagrasses:
　　Dr. Robert Coles, Northern Fisheries Centre, Queensland Department of Primary Industries, P. O. Box 5396, Cairns Queensland 4870, Australia (fax 61-70-351-401)

Stony Corals:
　　Dr. J. E. N. Veron, Australian Institute of Marine Science, PMB No. 3 MC, Townsville, Queensland 4810, Australia (fax 61-77-725-852)

Polychaetes/Annelids:
　　Dr. Julie Bailey-Brock, Department of Zoology, University of Hawaii, Honolulu, HI 96822 (808-956-9812)

Sponges:
　　Dr. Michelle Kelly-Borges, Porifera Section, Zoology Department, The Natural History Museum, Cromwell Road, London SW7 5BD United Kingdom (fax 44-71-938-8754)

Mollusks:

Dr. E. Alison Kay, Department of Zoology, University of Hawaii, Honolulu, HI 96822 (808-956-9812)

Crustaceans:

Dr. Lucius Eldredge, Bishop Museum, P. O. Box 19000A, Honolulu, HI 96817 (fax 808-847-8252)

Echinoderms:

Dr. David Pawson, Department of Invertebrate Zoology (NHB W323), National Museum of Natural History, Smithsonian Institution, Washington, D.C. 20560 (fax 202-357-3043)

Fishes:

Dr. David Greenfield, Dean, Graduate Division, University of Hawaii at Manoa, 2540 Maile Way, Spalding Hall, Room 360, Honolulu, HI 96822 (phone 808-956-7541, fax 808-956-4261)

Dr. William Eschmeyer, California Academy of Sciences, Golden Gate Park, San Francisco, CA 94118 (fax 415-750-7346)

Marine Ecosystem Classification:

Paul Holthus, The Nature Conservancy, Pacific Region, 1116 Smith Street, Honolulu, HI 96817 (fax 808-545-2019)

November 3

Database management and geographic information systems for marine biodiversity (J. E. Maragos, Convener, fax 808-944-7298)

International Initiatives

UNEP-IOC-ASPEI Global Task Team on the Implications of Climate Change on Coral Reefs:

Dr. Charles Birkeland, Marine Laboratory, University of Guam, UOG Station, Mangilao, Guam 96923 (fax 671-734-6767)

GIS Equal-Area Grids:

Dr. Don E. McAllister, Canadian Centre of Biodiversity, Canadian Museum of Nature, c/o Ocean Voice International, 2883 Otterson Drive, Ottawa, Ontario KIV 7B2 Canada (phone 613-990-8819, fax 613-990-8818)

ReefBase:

Dr. John W. McManus, International Center for Living Aquatic Resources Management (ICLARM), MCPO Box 2631, Makati, Metro Manila 0718, Philippines (fax 63-2-816-3183)

CoralBase:

Kim F. Navin, Australian Institute of Marine Science, PMB No. 3 MC, Townsville, Queensland 4810, Australia (fax 61-77-78-9412)

FishBase:

> Dr. Rainer Froese, ICLARM, MCPO Box 2631, Makati, Metro Manila 0718, Philippines (fax 63-2-816-3183)

World Conservation Monitoring Centre—Marine and Coastal Information Resources:

> Mark Spalding, WCMC, 219 Huntingdon Road, Cambridge CB3 ODL, United Kingdom (fax 44-223-277-136)

ROSTSEA:

> Robin Harger, UNESCO/ROSTSEA Jalan M.H. Thamrin 14 Tromal Pos 1273/ Jkt. Jakarta 10012, Indonesia (fax 62-21-314-1308)

Regional/Local Initiatives

U.S. Coral Reef Initiative:

> Dr. James E. Maragos, Program on Environment, East-West Center, 1777 East-West Road, Honolulu, HI 96848 USA (fax 808-944-7298)

Johnston Atoll Biodiversity Database:

> Dr. Phillip Lobel, Woods Hole Oceanographic Institution, Woods Hole, MA 02543 USA (fax 508-457-2176)

Hawaii Biological Survey:

> Dr. Allen Allison and Scott Miller, Bishop Museum, P. O. Box 19000A, Honolulu, HI 96817 USA (fax 808-841-8968)

Marshall Atoll Marine Resources Information System (MARIS):

> Karin Meier, CORIAL, Suite 254, Manoa Innovation Center, 2800 Woodlawn Drive, Honolulu, HI 96822 USA (fax 808-539-3625)

AeGis and the Hawaii Environmental Risk-Ranking Project:

> Jim Laurel, Aspect Consulting Inc., Suite 250, Manoa Innovation Center, Honolulu, HI 96822 USA (808-539-3625)

Hawaii Heritage Program:

> Dr. Samuel Gon, The Nature Conservancy of Hawaii, 1116 Smith Street, Honolulu, HI 96817 USA (fax 808-545-2019)

Hawaii Coastal Zone Management Program GIS:

> Craig Tasaka, Hawaii Coastal Zone Management Program, Office of State Planning, P. O. Box 3540, Honolulu, HI 96811 USA (fax 808-587-2899)

The NOAA Fisheries Perspective on Marine/Coastal Biodiversity Database Management:

> Dr. George Boehlert, Pacific Fisheries Environmental Group, National Marine Fisheries Service, P. O. Box 831, Monterey, CA 93942 USA (fax 408-656-3319)

Coral Database for the Caribbean:

> Dr. Kathleen Sullivan, Marine Ecology Program of The Nature Conservancy, Department of Biology, University of Miami, P. O. Box 249118, Coral Gables, FL 33124 USA (fax 305-284-3039)

The SPREP Biodiversity Conservation Database:

> Gary Spiller, South Pacific Regional Environment Programme, P. O. Box 240, Apia, Western Samoa (fax 685-20-231)

November 4

Recommendations on an information management system for the Island Pacific (M. N. A. Peterson, Convener, fax 808-599-8690 or 619-755-2110)

Discussion groups and decision making on:

1. Assembling a compendium of nearshore marine species biodiversity for use in database development and management

2. Designing the elements of a database—species, ecosystem, biogeography, location, research subjects, author/reference sources, and uncertainty/quality of data sources

3. Designing the geographic components of a database—aerial photos and satellite imagery, map sources, map scales, map details, and analytical capabilities and interface with database component

4. Linking to other databases and GISs

5. Optimizing the design for an Oceania marine/coastal biodiversity database management GIS

OUTLINES AND INSTRUCTIONS FOR CONTRIBUTED PAPERS

Papers on the state of systematics and species richness for major marine organism groups:

- Describe the systematics and taxonomy of the group

- Discuss the level of certainty

- Include a hierarchy from phylum through genus

- Include a list of recognized species, if possible

- Include as junior synonyms other nominal species

- Recommend inclusion of species-level data in an information management system

Papers on the marine/coastal information management systems:

- Describe the scope of the system including geography, species, or ecosystem types, numbers, list of all inputs

- Describe mapping details, scale and software sources

- Describe strengths and weaknesses of the system

- Discuss how the system could be improved

- Describe potential for compatibility with or linkages to other systems

- Recommend specifics on designing a marine/coastal database system for Oceania

Applicable to all papers:

■ Maximum paper length should be 15–20 pages, including a "literature cited" of important (but not all) sources, but excluding species listings.

■ The review should be a synthesis or summary of the available or extant information on the status of the category; technical details should be tabulated to keep the papers short.

Notes from the Convenors and Summary of Working Group Sessions

Workshop contributors and participants met to develop an action plan for species systematics and information management priorities for marine biodiversity conservation in the tropical island Pacific region on 4 November 1994, following the first two days of prepared paper presentations. The purpose of the first plenary session was to identify the concerns, objectives, and goals to help focus the two working group discussions. After the first plenary session, the workshop contributors/participants, also on 4 November 1994, divided into two working groups, one of species systematics and the other on information management. The principal points raised during the first plenary session are described below.

The Importance of Species Systematics and Taxonomy

The central component of any biodiversity information management system is species-level taxonomic and systematic information. The species level is the lowest in the conservation management regime and is directly linked to other ecological, biological, and management information. Core information for species-level biodiversity includes the following quadruplet: species name, date, locality, and source of data.

Cultural Dimensions

Although the workshop had a science-driven agenda and budgetary constraints, all participants realized the importance of including Pacific islanders and cultural information in both taxonomic and database initiatives. In particular, there is the urgent need to preserve and document local taxonomy and nomenclature data among a diversity of island cultures before knowledgeable elders pass on, and to train islanders on scientific approaches to taxonomy. Any database management system needs to be designed with Pacific islanders being a principal user of the system(s) commensurate with their skills, capabilities, and knowledge. A goal of any database management system must include preservation of cultural heritage. Local vernacular names for species and ecosystems must be linked to universal scientific names and classifications for species and ecosystems.

Ecosystem Components

Information on ecosystems should be a major part of any marine biodiversity database for the tropical insular Pacific. Information on individual species should be linked to the type of ecosystems in which they are found. Ecosystems are real features but dynamic in nature and difficult to define because the concept is used at several spatial scales. The SPREP-based ecosystem classification should be used to clarify and build consistency in the use of ecosystem concepts and terminology. Use of the classification system will also allow ecosystems to be mapped at different scales, an important requirement for any biodiversity database management system.

Boundaries and Dimensions of Biodiversity Conservation Units

Any biodiversity database management system needs to have geographical units and boundaries based upon natural features such as watersheds, islands, and individual reefs. These units and boundaries should also be tied to units of conservation and management. Policy makers need to keep in mind the species, ecosystem, and geographical dimensions of biodiversity conservation. Those should be linked in a database system. Biogeographic units should also be considered, either as an alternative or integrated within a species- and ecosystem-based system. A grid system could also be considered.

Taxonomic Research

Scientific research on species taxonomy and systematics has declined relative to other more "glamorous" fields of inquiry. Taxonomists need to spend more time on taxonomy, including completing treatment of important groups of species in order to assist in database management and conservation actions for marine biodiversity. Many organism groups (e.g., fish, mangroves, seagrasses) can be finished without much additional effort. We also need to synthesize what we know and what we don't know.

Setting Priorities for Species-Level Biodiversity

Taxonomic information and a database management system must be useful for helping decision makers set priorities at the species level. However, it is difficult to set priorities because there is not much information on many species groups, and any system will be biased toward species for which we have more information. Nevertheless rare species, endangered and threatened species, and endemic species need to be given preferential attention in any biodiversity database management system to be used to set conservation priorities.

Setting Priorities for Ecosystems and Geographical Areas

Although the workshop focused on nearshore and shallow marine ecosystems, deep-water and offshore ecosystems are also important, but we have much less biodiversity information about them. Likewise we have much more data for some island groups (e.g., Hawaii, Palau) than others (e.g., Solomon Islands, New Caledonia). Therefore, database management systems need to flag areas where there is little available information. In areas where there is good information at the geographic, ecosystem, and species levels, then both "hot spots" (high priority areas) and "cold spots" (low priority areas) could also be flagged.

Unpublished and Anecdotal Information

Database management needs to address how to incorporate informal information such as unpublished and anecdotal data. Often such information is the best available for biodiversity conservation; yet, there is also the need to filter or corroborate the validity of such data. To a similar extent information from "grey" literature and early (eighteenth–nineteenth century) literature needs to be incorporated.

Coordination Among Databases

One of the goals of the workshop was to compare notes among the variety of biodiversity database managers that are now developing systems applicable to the tropical Pacific region. Continuous coordination and integration will be needed for the successful development of a marine biodiversity database for the region.

Relationship to Environmental Monitoring and Change

A marine biodiversity database system must be useful for supporting networks of global and regional environmental monitoring protocols and programs now being developed for coral reefs and other nearshore tropical ecosystems. This is especially important for linking time, species, location, climate, and ecosystem data.

Evolution and Species Taxonomy

A biodiversity database system must also be flexible enough to accommodate changes in the way that scientists view species and taxonomy. Emerging species concepts such as genetic conductivity, environmental influences, as well as natural selection (the basis for alpha-taxonomy), should be accommodated.

Role of Museums

Museums serve as the repository of most "type" or voucher specimens for the multitude of plant and animal species described in the scientific literature since Linnaeus initiated the system of binomial nomenclature in mid-eighteenth century. The importance of museums to biodiversity conservation and taxonomy needs to be better appreciated by the public. In turn museum data need to be more accessible to the public as well as to conservation and database management groups and the media.

Use of Biodiversity Data for Management

Biodiversity database management systems should address both immediate and long-range needs including protection of endangered species and their critical habitats, protection of important ecosystems, preservation of cultural-based information and management systems, and career opportunities for Pacific islanders in conservation, restoration of ecosystems, and education. Biodiversity database systems must also include or be linked to data fields that describe the importance and value of specific species, ecosystems, and biogeographic units. Biodiversity data must also be accessible to the media and a variety of nonscience users. A system developed for the Pacific must emphasize the unique qualities of the region including its vast diversity and size.

Analytical Capabilities for Databases

Provided that monitoring data are accessible to the biodiversity database management system(s), there is an opportunity to conduct analyses on changes in status including threats to species and ecosystems over time. Databases should also be able to identify holes in data for species ecosystems, regions, trends, and so forth. There are also the opportunities to assess the degree of restoration, recovery, and protection of species and ecosystems. Modern management must also transcend obsolete conservation strategies (i.e., conservation with a fence around it). Access to a variety of data types and sources will help to educate policy makers and formulate more effective conservation and management strategies. There should not be any technical problems in extending the database management system to accommodate these additional functions. Future advances in computer technology will make virtually anything possible. Compact disk (CD) technology already allows large databases to be efficiently stored and retrieved in a small and durable format.

Geographic Information Systems (GISs)

Database management systems need to include attribute data on species, ecosystems, environmental conditions, monitoring, distribution, etc., that are tied to mapped data. Maps of all reefs and islands and at various scales should be digitized to accommodate the need to identify the specific sites where attribute data were collected. Maps also allow resources to be mapped in one-, two-, and three-dimensional scales. There are several different GIS systems being developed now with strong applicability to natural resources management and biodiversity conservation. These systems need to be linked, accessible, and "friendly" to a variety of users. A customized system is desired for the marine and coastal biodiversity of the tropical Pacific that is fully compatible with more global systems. Customized systems are also needed for individual island groups/cultures including the use of local languages to facilitate use and create a sense of ownership.

Institutional Level Collaboration

Collaboration is strongly needed among the scientific, cultural, and educational systems within the region and outside the region to facilitate development and use of taxonomic and database management systems for marine and coastal biodiversity conservation.

SUMMARY OF WORKING GROUP SESSION ON SPECIES SYSTEMATICS

This group decided not to limit discussions to species systematics and covered a variety of biodiversity issues, which are summarized below.

Minimum Database Needs

Species systematics data must include what, where, when, and why data were recovered. Standards and formats need to be developed, and there needs to be compatibility among software for various database management systems as well as compatibility within the taxonomic system(s) used.

Public Support and Education

Public support is needed for taxonomy and biodiversity conservation. Despite differences in management philosophy and culture, Pacific islanders must support the need for conservation in a modern changing world. Education is one means; people need to understand the social and economic consequences, including visual aids. The justification on the importance of taxonomy needs to be publicized. Education can also lead to jobs supporting taxonomic research and teaching taxonomy, including more jobs for women. People need to be trained to be taxonomists and recognize places important to biodiversity conservation. In turn, taxonomists need to explain taxonomy in a way that the general public can understand and value it. Education should involve computers and begin at a young age.

Will Data Acquisition Slow Down Biodiversity Conservation?

There is a parallel need to get critical information out to the public and to policy makers in order to advance conservation programs and to solicit financial support for taxonomic research. Important goals for taxonomic research include achieving stability in nomenclature and revisiting the rules under which taxonomic decisions are based. Taxonomists also need to develop simplified field identification guides to accelerate the pace of assessment, conservation actions, and educational programs. Parataxonomists can also be trained to identify species in the field. However, the reliability of all taxonomic data needs to be appraised to ensure accuracy.

Systematics and Computer Technology

Taxonomists and systematists need to work closely with computer specialists in order to develop data storage and analytical capabilities for taxonomic data. Computers, if used effectively, provide unsurpassed opportunities for efficiencies in completing systematic research and disseminating the results to users. Computers will help in bridging the disciplines of taxonomy/systematics, information technology, education/training management (including conservation), and follow-up research.

Global Cooperation

There is a strong and obvious need for compatible and user-friendly worldwide databases for biodiversity. Computer experts need to understand all the dimensions that need to be included, the complexity of ecosystems, and the abundance and distribution of organisms. Focusing on key significant species and communities may help to ensure cooperation.

Taxonomy and Biodiversity

One participant remarked, "Biodiversity is the variability of living things as individuals, species, and other assemblages. It occurs as local variants in space and time. Taxonomy is the interface between the variants of living things and information technology that makes it possible to code diversity into a form amenable to information management."

Taxonomic Research and Capabilities

There is an obvious shortage in the number of taxonomists now active; how do we get more taxonomists? New fields are emerging including molecular taxonomy, which can provide support to traditional "morphological" taxonomy. More people need to be trained. The U.S. National Biological Survey is also engaged in cataloging species, and similar efforts could be initiated in other countries. International cooperation could include exchanges of organ information. Taxonomic reference books could be loaded into electronic databases and more widely disseminated. Electronic formats can also be more readily corrected and updated as time goes on. Taxonomy occurs in both scientific and nonscientific regimes. Depending on goals and purposes, nonscientists can be trained to assist in monitoring surveys and other conservation-related activities. Not all those involved need a Ph.D. to do taxonomy.

Elements of a Taxonomic Database Management System

Taxonomists should work with and rely on computer experts/information management specialists to design a system. One important capability for a system is quality control. The quality and validity of data sources entering the system need evaluation and coding. The system should include a catalog of resources including access to taxonomic experts. Illustrations, maps, and diagrams should also be included. The system also needs to be designed to be "aware" of outside sources of data and how to make them available. The system could concentrate on taxonomic data serving as "core" data, but the system should also be capable of handling more than systematic and taxonomic information. Because the costs in setting up and maintaining data in database systems may be high, systems are often undersold. User charges or fees may need to be established for specific products delivered by the system. The products need to be categorized and generated quickly. The system capability and availability may also need to be advertised.

Resources Requiring Priority Attention

Coastal areas and species groups need priority attention due to the proximity of humans and their impacts to nearshore and coastal biodiversity. In addition, priority attention is needed for certain organism groups including bacteria, sponges, and crustaceans. Priority attention is also needed for deep sea benthos where 25–50 percent of all biodiversity may be located.

SUMMARY OF WORKING GROUP SESSIONS ON INFORMATION MANAGEMENT

The consensus of the discussions of this working group can be summarized as follows:

1. The working group recognized that there are several possible information systems focusing on fields such as taxonomy/biodiversity, biology, ecology, and management, and that this sequence represents a kind of hierarchy (i.e., you cannot do sound management without knowledge about the species, their biology and ecology; see Figure B–1).

2. The working group agreed that it would focus on information systems at least at the taxonomy/biodiversity level.

3. The working group agreed that the minimum information required for a single record to be acceptable in a biodiversity database were species (i.e., genus or family is not good enough); locality (preferably in three-dimensional coordinates or depth, longitude, and latitude); date (and time of collection, if available); and scientific source (i.e., evidence for the correctness of the information such as references, name of identifier, museum samples, photos [see Figure B–2]).

4. The working group agreed that a relational database structure should be applied with major tables including (1) a taxonomic dictionary, which would also provide a link to biological information, photos, common names, and indigenous knowledge; (2) a table recording the occurrence of specimens in space and time, similar to a curatorial database; (3) a gazetteer, providing standards for locality names; and (4) a link to ecosystem information (see Figure B–3 for a draft layout of a database design).

5. The working group recognized that there are millions of existing records that fulfill the criteria in point 3 above, mainly in museum collections but also in the scientific literature, live specimen tagging databases, survey databases, reports, etc. This kind of information continues to be collected by numerous projects, programs, and monitoring efforts. However, these data are widely scattered, normally not available outside the specific projects, and often in danger of being lost.

6. The working group agreed that it would be of enormous value for biodiversity studies to have access to these data, preferably in one coherent database. CD-ROM technology was regarded as the proper medium for storing and distributing such a database. IBM-compatible PCs were suggested as the widest available computer system, which are powerful enough to deal with the foreseen amount of data.

7. The working group supported an implementation strategy where a relatively small project-based core group would create the aforementioned database, contact institutions with relevant data sets, negotiate agreements for the distribution of the data, write a translation routine to import all data that fulfill the criteria in point 3 above, plus additional quality criteria (e.g., likelihood of a species occurring at the indicated locality), put the data and analytical routines on the CD-ROM, and distribute it to all participating parties for free and to others on request for a reasonable fee (<$100).

8. The working group envisaged that institutions participating in this exercise would form a network, with some institutions taking on coordinating tasks for certain taxonomic groups or regions (see Figure B–4).

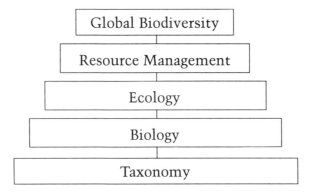

Figure B–1 The biodiversity pyramid

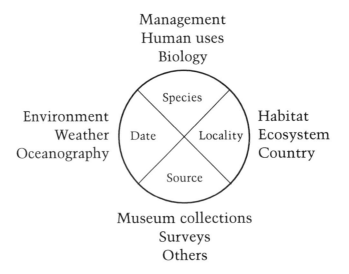

Figure B–2 The biodiversity quadruple

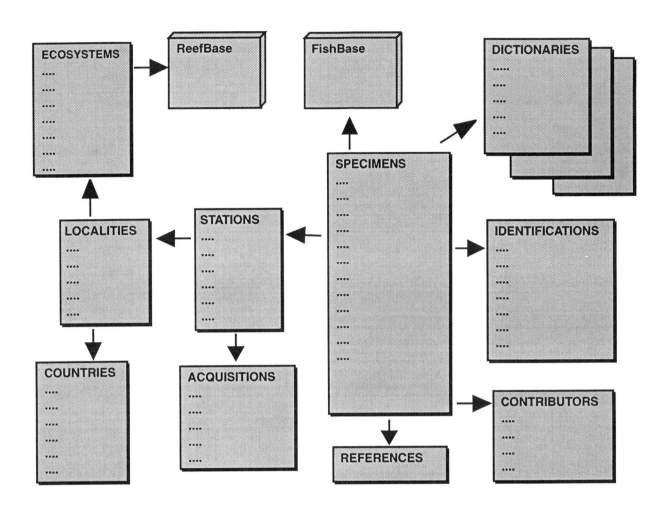

Figure B–3 Layout of a database design (Source: R. Froese and D. Pauly)

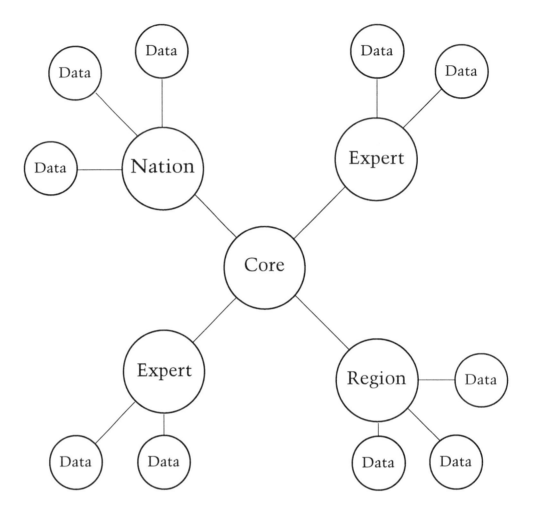

Figure B–4 Coordinating tasks proposed for network of institutions

Glossary of Technical Terms and Acronyms

AIMS	Australian Institute of Marine Science
algae	a plantlike organism of any of several divisions, including algae in green, yellow-green, brown, red, and blue-green
angiosperm	a flowering plant
arthropod	any of a phylum of invertebrate animals including insects and crustaceans
biosystematics	experimental taxonomy
biota	the flora and fauna of a region
bivalve	an animal, as a clam, protected by paired shells; a class of mollusks consisting of animals with paired shells
CCB	Canadian Centre for Biodiversity
CoralBase	a taxonomic and biogeographic information system for Scleractinian corals being developed by AIMS
CRI	Coral Reef Initiative
crustaceans	any of a large phylum of mostly aquatic arthropods like animals encased in a shell, including shrimps and crabs
database	a large collection of data organized for rapid search and retrieval as by a computer
echinoderms	members of this phylum are encased in a skeleton of armorlike plates with body parts symmetrically positioned around a central point like spokes in a wheel (includes starfishes, sea urchins, and related forms)
ecosystem	a community of organisms and its environment functioning as an ecological unit in nature
endemic	restricted or peculiar to a locality or region

ENGO	environmental non-governmental organization
FAO	Food and Agriculture Organization of the United Nations
FishBase	an electronic encyclopedia on fish being developed at ICLARM with FAO support
gastropods	any of a large class of mollusks, as snails and slugs, usually with a single shell or none and a head with sensory organs
genus	a class, kind, or group marked by common characteristics or by one common characteristic; a group of closely related species
GIS	geographic information system
HINHP	Hawaii Natural Heritage Program
ICLARM	International Center for Living Aquatic Resources Management
Indo-West Pacific	the tropical and subtropical portions of the Indian Ocean and the western and central part of the Pacific
invertebrates	animals without backbones
IUCN	International Union for Conservation of Nature and Natural Resources
mollusks	any of a large phylum of invertebrate animals as squids, clams, snails, octopi, tooth shells, and chitons
NGO	non-governmental organization
NOAA	National Oceanic and Atmospheric Administration
Oceania	includes the Pacific islands of Melanesia, Micronesia, Polynesia, and at times New Zealand, Australia, and the Malay Archipelago
polychaete	any of a class of chiefly elongated and segmented marine invertebrate worms, belonging to the phylum Annelida
ReefBase	an international database on the coral reefs of the world being developed by ICLARM
seagrasses	flowering plants that can live immersed in seawater
seamounts	submerged undersea mountains
SPREP	South Pacific Regional Environment Programme
systematics	the classification and study of organisms regarding their natural relationships, including taxonomy
taxonomy	study of principles of scientific classification
tropical island Pacific	the tropical islands of Melanesia, Micronesia, and Polynesia

UNDP	United Nations Development Programme
UNEP	United Nations Environment Programme
vertebrates	animals with backbones
WCMC	World Conservation Monitoring Centre

Workshop Participants and Contributors

MARINE/COASTAL BIODIVERSITY
IN THE TROPICAL ISLAND PACIFIC REGION
I. SPECIES SYSTEMATICS AND INFORMATION MANAGEMENT PRIORITIES

Hawaii Imin International Conference Center at Jefferson Hall
East-West Center, Honolulu, Hawaii, 2–4 November 1994

Isabella ABBOTT
G. P. Wilder Professor of Botany
University of Hawaii
3190 Maile Way
Honolulu, HI 96822 USA
Phone 808-956-8073
Fax 808-956-3923
E-mail: as,botany@cvaxdm

Allen ALLISON
Assistant Director, Research
Bishop Museum
P.O. Box 19000A
Honolulu, HI 96817-0916 USA
Phone 808-848-4145
Fax 808-847-8252
E-mail: allison@uhunix.uhcc.hawaii.edu

Julie BAILEY-BROCK
Professor
Department of Zoology
University of Hawaii
2538 The Mall
Honolulu, HI 96822 USA
Phone 808-956-8678
Fax 808-956-9812

John E. BARDACH
Senior Fellow Emeritus
East-West Center
1777 East-West Road
Honolulu, HI 96848 USA
Phone 808-944-7285
Fax 808-944-7298
E-mail: bardachj@ewc.bitnet

Charles BIRKELAND
Professor of Marine Biology
Marine Laboratory
University of Guam
Mangilao, Guam 96923 USA
Phone 671-734-2421
Fax 671-734-6767
E-mail: birkelan@uog9.uog.edu

Robert G. COLES
Principal Fisheries Scientist
Northern Fisheries Centre
Department of Primary Industries
P.O. Box 5396
Cairns, Queensland 4870, Australia
Phone 61-70-529830
Fax 61-70-351401

Grace U. CORONADO
ICLARM
MCPO Box 2631
Makati, Metro Manila 0718, Philippines
Phone 632-818-0466, 632-818-9283
Fax 632-816-3183
E-mail: iclarm@cgnet.com

Michael P. CROSBY
National Research Coordinator
Office of Ocean and Coastal Resource
    Management
NOAA, SSMC–4, Room 11536
1305 East West Highway
Silver Spring, MD 20910 USA
Phone 301-713-3155
Fax 301-713-4012

Sylvia A. EARLE
President
Deep Ocean Research and Exploration
12812 Skyline Blvd.
Oakland, CA 94619 USA
Phone 510-562-9300
Fax 510-430-8249

Lucius ELDREDGE
Executive Secretary
Pacific Science Association
P.O. Box 17801
Honolulu, HI 96817 USA
Phone 808-848-4139
Fax 808-847-8252
E-mail: psa@bishop.bishop.hawaii.org

Joanna C. ELLISON
Research Fellow
Australian Institute of Marine Science
PMB No. 3 MC
Townsville, Queensland 4810, Australia
Phone 61-77-534211
Fax 61-77-725852

William N. ESCHMEYER
Senior Curator
California Academy of Sciences
Golden Gate Park
San Francisco, CA 94118-4599 USA
Phone 415-221-5100
Fax 415-750-7346

Rainer FROESE
Research Scientist, ICLARM
MCPO Box 2631
Makati, Metro Manila 0718, Philippines
Phone 632-818-0466, 632-817-5255
Fax 632-816-3183

Samuel M. GON III
Heritage Coordinator/Ecologist
Hawaii Natural Heritage Program
1116 Smith Street, Suite 201
Honolulu, HI 96817 USA
Phone 808-537-4508
Fax 808-545-2019

Paul F. HOLTHUS
The Nature Conservancy
1116 Smith Street, Suite 201
Honolulu, HI 96817 USA
Phone 808-537-4508
Fax 808-545-2019

Roy KAM
Database Coordinator
Hawaii Natural Heritage Program
1116 Smith Street, Suite 201
Honolulu, HI 96817 USA
Phone 808-537-4508
Fax 808-545-2019

E. Alison KAY
Professor
Department of Zoology
University of Hawaii
2538 The Mall
Edmondson Hall 351
Honolulu, HI 96822 USA
Phone 808-956-8620
Fax 808-956-9812

Michelle KELLY-BORGES
Zoology Department
Natural History Museum
Cromwell Road
London SW7 5BD
United Kingdom
Phone 44-171-938-9524
Fax 44-171-938-8754
E-mail: mkb@nhm.ac.uk
and
Coral Reef Research Foundation
P.O. Box 70, Chuuk
Federated States of Micronesia, 96942

John KUO
Associate Professor
University of Western Australia
Centre for Electron Microscopy and
    Microanalysis
The University of Western Australia
Nedlands WA 6009
Australia
Phone 61-9-3802765
Fax 61-9-3801087

Jim LAUREL
Software Engineering, Inc.
Manoa Innovation Center, Suite 250
2800 Woodlawn Drive
Honolulu, HI 96822 USA
Phone 808-539-3783
Fax 808-539-3625

James MARAGOS
Senior Fellow
Program on Environment
East-West Center
1777 East-West Road
Honolulu, HI 96848 USA
Phone 808-944-7271
Fax 808-944-7298
E-mail: maragosj@ewc.bitnet

Don E. MCALLISTER
President
Ocean Voice International
Box 37026
3332 McCarthy Road
Ottawa, ON K1V 0W0, Canada
Phone 613-990-8819
Fax 613-521-4205
E-mail: ahl94@freenet.carleton.ca

John W. MCMANUS
ReefBase Project Leader
Coastal & Coral Reef Resource Systems
    Program, ICLARM
MCPO Box 2631
Makati, Metro Manila 0718, Philippines
Phone 632-818-0466, 632-818-9283
Fax 632-816-3183
E-mail: J.McManus@cgnet.com

Lambert A. B. MEÑEZ
ICLARM
MCPO Box 2631
Makati, Metro Manila 0718, Philippines
Phone 632-818-0466, 632-818-9283
Fax 632-816-3183
E-mail: iclarm@cgnet.com

Scott E. MILLER
Bishop Museum
P.O. Box 19000A
Honolulu, HI 96817-0916 USA
Phone 808-847-3511
Fax 808-841-8968

Kim F. NAVIN
Senior Experimental Scientist (CoralBase)
Australian Institute of Marine Science
PMB No. 3 MC
Townsville, Queensland 4810, Australia
Phone 61-77-534-412
Fax 61-77-725852
E-mail: k.navin@aims.gov.au

Gordon M. NISHIDA
Bishop Museum
P.O. Box 19000A
Honolulu, HI 96817-0916 USA
Phone 808-847-3511
Fax 808-841-8968

Maria Lourdes D. PALOMARES
c/o ICLARM
MCPO Box 2631
Makati, Metro Manila 0718, Philippines

David L. PAWSON
Senior Research Scientist
National Museum of Natural History
Smithsonian Institution
Washington, DC 20560 USA
Phone 202-786-2127
Fax 202-357-3043
E-mail: mnhiv070@sivm.si.edu

M. N. A. PETERSON
Ocean Policy Institute
1001 Bishop Street/Pauahi Tower, Suite 1150
Honolulu, HI 96813 USA
Phone 808-521-6745, 619-755-7071
Fax 808-599-8690, 619-755-2110

Richard L. PYLE
Ichthyology/Bishop Museum
P.O. Box 19000A
Honolulu, HI 96817-0916 USA
Phone 808-847-3511
Fax 808-841-8968

John E. RANDALL
Bishop Museum
P.O. Box 19000A
Honolulu, HI 96817-0916 USA
Phone 808-848-4130
Fax 808-841-8968

Michael L. SMITH
Center for Marine Conservation
1725 DeSales Street, NW, Suite 500
Washington, DC 20036
Phone 202-429-5609
Fax 202-872-0619

Mark D. SPALDING
Research Officer
World Conservation Monitoring Centre
219 Huntingdon Road
Cambridge CB3 ODL
United Kingdom
Phone 44-1223-277314
Fax 44-1223-277136
E-mail: mark.spalding@wcmc.org.uk
and
Department of Geography
University of Cambridge
Downing Place
Cambridge, CB2 3EN
United Kingdom

Benjamin M. VALLEJO Jr.
ICLARM
MCPO Box 2631
Makati, Metro Manila 0718, Philippines
Phone 632-818-0466, 632-818-9283
Fax 632-816-3183
E-mail: iclarm@cgnet.com

J. E. N. VERON
Senior Principal Research Scientist
Australian Institute of Marine Science
PMB No. 3 MC
Townsville, Queensland 4810, Australia
Phone 61-77-534274
Fax 61-77-725852
E-mail: j.veron@aims.gov.au

Observers

Timothy ADAMS
Fisheries Resource Adviser
South Pacific Commission (SPC)
Boite Postale D5
Noumea Cedex 98848
New Caledonia
Phone 687-26-20-00
Fax 687-26-38-18
E-mail: tbap@bixom

Alyssa MILLER
c/o Department of Geography
University of Hawaii at Manoa
Porteus Hall
Honolulu, HI 96822 USA
Phone 808-988-5490

Bruce L. MUNDY
Fishing Biologist
National Marine Fisheries Service
Southwest Fisheries Center
Honolulu Laboratory
2570 Dole Street
Honolulu, HI 96822-2396 USA
Phone 808-943-1212
Fax 808-943-1290

John NAUGHTON
National Marine Fisheries Service
Southwest Fisheries Center
Honolulu Laboratory
2570 Dole Street
Honolulu, HI 96822-2396 USA
Phone 808-973-2939
Fax 808-973-2941

Ying-Qian QIAN
Institute of Botany
Chinese Academy of Science
Beijing, China

Bertrand RICHER de FORGES
Researcher
ORSTOM
B. P. A5, Noumea Cedex
New Caledonia
Phone 687-26-10-00
Fax 687-26-43-26
E-mail: richer@noumea.orstom.nc

Sandra ROMANO
Department of Zoology
University of Hawaii at Manoa
2538 The Mall
Edmondson Hall
Honolulu, HI 96822 USA
Phone 808-539-7311

Susana SIAR
EWC Degree Fellow
Mindanao, Philippines

Randy THAMAN
Professor
Pacific Island Geography
University of the South Pacific
P.O. Box 1168
Suva, Fiji
Phone 679-313900
Fax 679-301487
E-mail: thaman_r@usp.ac.fj

Graduate Students

Jesse AMATORE
Barbara BANNISTER
Norma BUSTOS
Bob CAVILLA
Molly HURST
Lance JEFFERY
Melissa RAWLINSON
Steve YOUNG

Marine Option Program
School of Ocean and Earth Sciences and
    Technology
University of Hawaii at Manoa
1000 Pope Road, Room 229
Honolulu, HI 96822 USA
Phone 808-956-8433
Fax 808-956-2417